Solutions Manual
for
Quantitative Chemical Analysis

Tenth Edition

Daniel C. Harris
Michelson Laboratory, China Lake, California

Charles A. Lucy
University of Alberta, Edmonton, Alberta

Macmillan Learning
Austin | Boston | New York | Plymouth

ISBN-13: 978-1-319-33024-8
ISBN-10: 1-319-33024-X

© 2020, 2016, 2010, 2007 by W. H. Freeman and Company

All rights reserved.

Printed in the United States of America.

Macmillan Learning
One New York Plaza
Suite 4600
New York, NY 10004-1562
www.macmillanlearning.com

Contents

Chapter 0	The Analytical Process	1
Chapter 1	Chemical Measurements	3
Chapter 2	Tools of the Trade	12
Chapter 3	Experimental Error	19
Chapter 4	Statistics	27
Chapter 5	Quality Assurance and Calibration Methods	41
Chapter 6	Chemical Equilibrium	59
Chapter 7	Let the Titrations Begin	68
Chapter 8	Activity and the Systematic Treatment of Equilibrium	79
Chapter 9	Monoprotic Acid-Base Equilibria	93
Chapter 10	Polyprotic Acid-Base Equilibria	105
Chapter 11	Acid-Base Titrations	120
Chapter 12	EDTA Titrations	150
Chapter 13	Advanced Topics in Equilibrium	166
Chapter 14	Fundamentals of Electrochemistry	183
Chapter 15	Electrodes and Potentiometry	197
Chapter 16	Redox Titrations	213
Chapter 17	Electroanalytical Techniques	229
Chapter 18	Fundamentals of Spectrophotometry	247
Chapter 19	Applications of Spectrophotometry	264
Chapter 20	Spectrophotometers	284
Chapter 21	Atomic Spectroscopy	300
Chapter 22	Mass Spectrometry	316
Chapter 23	Introduction to Analytical Separations	336
Chapter 24	Gas Chromatography	358
Chapter 25	High-Performance Liquid Chromatography	379
Chapter 26	Chromatographic Methods and Capillary Electrophoresis	402
Chapter 27	Gravimetric and Combustion Analysis	423
Chapter 28	Sample Preparation	434

CHAPTER 0
THE ANALYTICAL PROCESS

0-1. Qualitative analysis finds out what is in a sample. Quantitative analysis measures how much is in a sample.

0-2. Steps in a chemical analysis:
   (1) Formulate the question: Convert a general question into a specific one that can be answered by a chemical measurement.
   (2) Select the appropriate analytical procedure.
   (3) Obtain a representative sample.
   (4) Sample preparation: Convert the representative sample into a sample suitable for analysis. If necessary, concentrate the analyte and remove or mask interfering species.
   (5) Analysis: Measure the unknown concentration in replicate analyses.
   (6) Produce a clear report of results, including estimates of uncertainty.
   (7) Draw conclusions: Based on the analytical results, decide what actions to take.

0-3. Masking converts an interfering species to a noninterfering species.

0-4. A calibration curve shows the response of an analytical method as a function of the known concentration of analyte in standard solutions. Once the calibration curve is known, then the concentration of an unknown can be deduced from a measured response.

0-5. (a) A homogeneous material has the same composition everywhere. In a heterogeneous material, the composition is not the same everywhere.

   (b) In a segregated heterogeneous material, the composition varies on a large scale. There could be large patches with one composition and large patches with another composition. The differences are segregated into different regions. In a random heterogeneous material, the differences occur on a fine scale. If we collect a "reasonable-size" portion, we will capture each of the different compositions that are present.

(c) To sample a *segregated heterogeneous material*, we take representative amounts from each of the obviously different regions. In panel b in Box 0-1, 66% of the area has composition A, 14% is B, and 20% is C. To construct a representative bulk sample, we could take 66 randomly selected samples from region A, 14 from region B, and 20 from region C. To sample a *random heterogeneous material*, we divide the material into imaginary segments and collect random segments with the help of a table of random numbers.

0-6. We are apparently observing *interference* by Mn^{2+} in the I^- analysis by method A. The result of the I^- analysis is affected by the presence of Mn^{2+}. The greater the concentration of Mn^{2+} in the mineral water, the greater is the apparent concentration of I^- found by method A. Method B is not subject to the same interference, so the concentration of I^- is low and independent of addition of Mn^{2+}. There could be some Mn^{2+} in the original mineral water, which causes method A to give a higher result than method B even when no Mn^{2+} is deliberately added. The result from Method B (0.009 mg/L) is more reliable.

CHAPTER 1
CHEMICAL MEASUREMENTS

A note from Dan and Chuck: Don't worry if your numerical answers are slightly different from ours. You or we may have rounded intermediate results. In general, retain many extra digits for intermediate answers and save your roundoff until the end. We'll study this process in Chapter 3.

1-1. (a) meter (m), kilogram (kg), second (s), ampere (A), kelvin (K), mole (mol)
(b) hertz (Hz), newton (N), pascal (Pa), joule (J), watt (W)

1-2. See Table 1-3. Abbreviations above kilo are capitalized: M (mega, 10^6), G (giga, 10^9), T (tera, 10^{12}), P (peta, 10^{15}), E (exa, 10^{18}), Z (zetta, 10^{21}), and Y (yotta, 10^{24}).

1-3.
(a) mW = milliwatt = 10^{-3} watt
(b) pm = picometer = 10^{-12} meter
(c) kΩ = kiloohm = 10^3 ohm
(d) μF = microfarad = 10^{-6} farad
(e) TJ = terajoule = 10^{12} joule
(f) ns = nanosecond = 10^{-9} second
(g) fg = femtogram = 10^{-15} gram
(h) dPa = decipascal = 10^{-1} pascal

1-4.
(a) 100 fJ or 0.1 pJ   (d) 0.1 nm or 100 pm
(b) 43.172 8 nF   (e) 21 TW
(c) 299.79 THz or 0.299 79 PHz   (f) 0.483 amol or 483 zmol

1-5. (a) $8 \text{ Pg} = 8 \times 10^{15}$ g. $8 \times 10^{15} \text{ g} \times \dfrac{1 \text{ kg}}{1\,000 \text{ g}} = 8 \times 10^{12}$ kg of C

(b) The formula mass of CO_2 is $12.011 + 2(15.999) = 44.009$

$$8 \times 10^{12} \text{ kg C} \times \dfrac{44.009 \text{ kg } CO_2}{12.011 \text{ kg C}} = 2.9 \times 10^{13} \text{ kg } CO_2$$

(c) 2.9×10^{13} kg $CO_2 \times \dfrac{1 \text{ ton}}{1\,000 \text{ kg}} = 2.9 \times 10^{10}$ tons of CO_2

$$\dfrac{2.9 \times 10^{10} \text{ tons } CO_2}{7 \times 10^9 \text{ people}} = 4 \text{ tons } CO_2 \text{ per person}$$

1-6. We will convert ounces of tuna to grams of tuna and find out how many µg of mercury are in one can (6 oz) of tuna. For a body mass of 68 kg, we will compute how many days are allowed between eating tuna so that the average dose does not exceed 0.1 µg Hg/kg body weight per day.

Table 1-4 tells us that 1 lb = 0.453 6 kg

6 oz = (6/16) lb = (6/16)(0.453 6 kg) = 0.170 kg = 170 g

One part per million means 1 µg Hg per gram of tuna

There is 0.6 ppm of mercury in chunk *white* tuna = 0.6 µg Hg/g tuna.

A 6-oz can contains $(170 \text{ g tuna})\left(\dfrac{0.6 \text{ µg Hg}}{\text{g tuna}}\right) = 102$ µg Hg

A dose of 0.1 µg Hg/kg body weight per day for a 68 kg person is

$$\left(\dfrac{0.1 \text{ µg Hg}}{\text{kg} \cdot \text{day}}\right)(68 \text{ kg}) = 6.8 \left(\dfrac{\text{µg Hg}}{\text{day}}\right)$$

If I eat 102 µg Hg in one day from one can of tuna, I have eaten the amount of mercury allowed in $(102 \text{ µg Hg})\left(\dfrac{1 \text{ day}}{6.8 \text{ µg Hg}}\right) = 15$ days

I should wait 15 days before consuming my next can of tuna so that my average intake does not exceed 6.8 µg Hg/day.

Chunk *light* tuna contains 0.14 ppm Hg = 0.14 µg Hg/g tuna. Substituting this number for 0.6 µg Hg/g tuna in the sequence of calculations gives a period of 3.5 days. I could eat 2 cans of chunk light tuna per week.

1-7. Table 1-4 tells us that 1 horsepower = 745.700 W = 745.700 J/s.

100.0 horsepower = $(100.0 \text{ horsepower})\left(\dfrac{745.700 \text{ J/s}}{\text{horsepower}}\right) = 7.457 \times 10^4$ J/s.

$$\dfrac{7.457 \times 10^4 \, \frac{\text{J}}{\text{s}}}{4.184 \, \frac{\text{J}}{\text{cal}}} \times 3600 \, \frac{\text{s}}{\text{h}} = 6.416 \times 10^7 \, \frac{\text{cal}}{\text{h}}.$$

1-8. (a) $\dfrac{\left(2.2 \times 10^6 \, \frac{\text{cal}}{\text{day}}\right)\left(4.184 \, \frac{\text{J}}{\text{cal}}\right)\left(\dfrac{1 \text{ day}}{24 \text{ h}}\right)\left(\dfrac{1 \text{ h}}{3600 \text{ s}}\right)}{(120 \text{ pound})\left(0.453 6 \, \dfrac{\text{kg}}{\text{pound}}\right)} = 2.0$ J/(s·kg)

Chemical Measurements 5

$$= 2.0 \text{ W/kg}$$

Similarly, $3.4 \times 10^3 \dfrac{\text{kcal}}{\text{day}} \Rightarrow 3.0 \text{ J/(s·kg)} = 3.0 \text{ W/kg}$.

(b) The office worker's power output is

$$\left(2.2 \times 10^6 \dfrac{\text{cal}}{\text{day}}\right)\left(4.184 \dfrac{\text{J}}{\text{cal}}\right)\left(\dfrac{1 \text{ day}}{24 \text{ h}}\right)\left(\dfrac{1 \text{ h}}{3\,600 \text{ s}}\right) = 1.1 \times 10^2 \dfrac{\text{J}}{\text{s}} = 1.1 \times 10^2 \text{ W}$$

The person's power output is greater than that of the 100 W light bulb.

1-9. (a) True

(b) $\left(1\,000 \dfrac{\text{m}}{\text{km}}\right)\left(\dfrac{1 \text{ inch}}{0.025\,4 \text{ m}}\right)\left(\dfrac{1 \text{ foot}}{12 \text{ inch}}\right)\left(\dfrac{1 \text{ mile}}{5\,280 \text{ foot}}\right) = 0.621\,37 \dfrac{\text{mile}}{\text{km}}$

(c) $\left(\dfrac{1 \text{ mg } CO_2}{\text{mile} \cdot \text{pound}}\right)\left(\dfrac{0.621\,37 \text{ mile}}{\text{km}}\right)\left(\dfrac{\text{pound}}{0.453\,592 \text{ kg}}\right) = \dfrac{1.37 \text{ mg } CO_2}{\text{km} \cdot \text{kg}}$

(d) 1.5 metric tons = 1 500 kg

$$\left(\dfrac{1.37 \text{ mg } CO_2}{\text{km} \cdot \text{kg}}\right)(150\,000 \text{ km})(1\,500 \text{ kg}) = 3.08 \times 10^8 \text{ mg } CO_2$$

$$(109)(3.08 \times 10^8 \text{ mg } CO_2)\left(\dfrac{1 \text{ g}}{1\,000 \text{ mg}}\right)\left(\dfrac{1 \text{ kg}}{1\,000 \text{ g}}\right)\left(\dfrac{1 \text{ metric ton}}{1\,000 \text{ kg}}\right)$$

$$= 33.6 \text{ metric tons } CO_2 \text{ for gasoline car}$$

$$(20)(3.08 \times 10^8 \text{ mg } CO_2)\left(\dfrac{1 \text{ g}}{1\,000 \text{ mg}}\right)\left(\dfrac{1 \text{ kg}}{1\,000 \text{ g}}\right)\left(\dfrac{1 \text{ metric ton}}{1\,000 \text{ kg}}\right)$$

$$= 6.2 \text{ metric tons } CO_2 \text{ for electric car in California}$$

1-10. (a) A microliter is 10^{-6} L and a picogram is 10^{-12} g. First find grams in 60 µL:

$$\left(60 \times 10^{-6} \text{ L}\right)\left(\dfrac{1\,000 \text{ mL}}{\text{L}}\right)\left(\dfrac{200 \times 10^{-12} \text{ g}}{\text{mL}}\right) = 1.2 \times 10^{-11} \text{ g}$$

Then convert grams to moles:

$$\dfrac{1.2 \times 10^{-11} \text{ g}}{150\,000 \text{ g/mol}} = 8.0 \times 10^{-17} \text{ mol} = 80 \times 10^{-18} \text{ mol} = 80 \text{ amol (attomol)}$$

(b) A nanogram is 10^{-9} g.

$$\left(\frac{100\times 10^{-9}\text{ g}}{150\,000\text{ g/mol}}\right)(6.022\times 10^{23}/\text{mol}) = 4\times 10^{11} \text{ molecules}$$

1-11. $\left(0.03\dfrac{\text{mg}}{\text{m}^2\cdot\text{day}}\right)\left(1\,000\dfrac{\text{m}}{\text{km}}\right)^2(535\text{ km}^2)\left(\dfrac{1\text{ g}}{1\,000\text{ mg}}\right)\times$

$\left(\dfrac{1\text{ kg}}{1\,000\text{ g}}\right)\left(\dfrac{1\text{ ton}}{1\,000\text{ kg}}\right)\left(365\dfrac{\text{day}}{\text{year}}\right) = 6\dfrac{\text{ton}}{\text{year}}$

1-12. (a) molarity = moles of solute / liter of solution

(b) molality = moles of solute / kilogram of solvent

(c) density = grams of substance / milliliter of substance

(d) weight percent = 100 × (mass of substance/mass of solution or mixture)

(e) volume percent = 100 × (volume of substance/volume of solution or mixture)

(f) parts per million = 10^6 × (grams of substance/grams of sample)

(g) parts per billion = 10^9 × (grams of substance/grams of sample)

(h) formal concentration = moles of formula/liter of solution

1-13. Acetic acid (CH_3CO_2H) is a weak electrolyte that is partially dissociated. When we dissolve 0.01 mol in a liter, the concentrations of CH_3CO_2H plus $CH_3CO_2^-$ add to 0.01 M. The concentration of CH_3CO_2H alone is less than 0.01 M.

1-14. 32.0 g / [(22.990 + 35.45) g/mol] = 0.548 mol NaCl

0.548 mol / 0.500 L = 1.10 M

1-15. $\left(1.71\dfrac{\text{mol CH}_3\text{OH}}{\text{L solution}}\right)(0.100\text{ L solution}) = 0.171$ mol CH_3OH

$(0.171\text{ mol CH}_3\text{OH})\left(\dfrac{32.04\text{ g}}{\text{mol CH}_3\text{OH}}\right) = 5.48$ g

1-16. (a) If atmospheric pressure is 1 bar, then a concentration of 1 ppb is 10^{-9} bar. A concentration of 39 ppb is 39×10^{-9} bar. There are exactly 10^5 Pa in a bar. So, first convert bar to Pa:

Chemical Measurements 7

$$(39 \times 10^{-9} \text{ bar}) \left(\frac{10^5 \text{ Pa}}{1 \text{ bar}} \right) = 39 \times 10^{-4} \text{ Pa}$$

Then convert Pa to mPa:

$$(39 \times 10^{-4} \text{ Pa}) \left(\frac{10^3 \text{ mPa}}{1 \text{ Pa}} \right) = 3.9 \text{ mPa}$$

The ground-level O_3 concentration of 3.9 mPa is about 20% of the stratospheric concentration of 19 mPa.

(b) The pressure of the atmosphere at 16 km altitude is 9.6 kPa. The pressure of O_3 is 19 mPa. From the definition of ppb, the O_3 pressure in ppb is

$$O_3 \text{ (ppb)} = \frac{O_3 \text{ pressure (Pa)}}{\text{atmospheric pressure (Pa)}} \times 10^9 = \frac{19 \times 10^{-3} \text{ Pa}}{9.6 \times 10^3 \text{ Pa}} \times 10^9$$
$$= 2.0 \times 10^3 \text{ ppb}$$

1-17. (a) 19 mPa = 19×10^{-3} Pa. $19 \times 10^{-3} \text{ Pa} \times \frac{1 \text{ bar}}{10^5 \text{ Pa}} = 1.9 \times 10^{-7}$ bar

(b) T (K) = 273.15 + °C = 273.15 − 70 = 203 K

$$\frac{n}{V} = \frac{P}{RT} = \frac{1.9 \times 10^{-7} \text{ bar}}{0.083\,14 \frac{L \cdot \text{bar}}{\text{mol} \cdot K} \times 203 \text{ K}} = 1.1 \times 10^{-8} \text{ M} = 11 \text{ nM}$$

1-18. (a) $PV = nRT$

$$(1.000 \text{ bar})(5.24 \times 10^{-6} \text{ L}) = n \left(0.083\,14 \frac{L \cdot \text{bar}}{\text{mol} \cdot K} \right)(298.15 \text{ K})$$

$\Rightarrow n = 2.11 \times 10^{-7}$ mol $\Rightarrow 2.11 \times 10^{-7}$ M

(b) Ar: 0.934% means 0.009 34 L of Ar per L of air

$PV = nRT$: $(1.000 \text{ bar})(0.009\,34 \text{ L}) = n \left(0.083\,14 \frac{L \cdot \text{bar}}{\text{mol} \cdot K} \right)(298.15 \text{ K})$

$\Rightarrow n = 3.77 \times 10^{-4}$ mol $\Rightarrow 3.77 \times 10^{-4}$ M

Kr: 1.14 ppm $\Rightarrow$ 1.14 μL Kr per L of air

$PV = nRT$: $(1.000 \text{ bar})(1.14 \times 10^{-6} \text{ L}) = n \left(0.083\,14 \frac{L \cdot \text{bar}}{\text{mol} \cdot K} \right)(298.15 \text{ K})$

$\Rightarrow n = 4.60 \times 10^{-8}$ M

Xe: 87 ppb $\Rightarrow$ 87 nL Xe per L of air

$PV = nRT$: $(1.000 \text{ bar})(87 \times 10^{-9} \text{ L}) = n\left(0.083\,14 \dfrac{\text{L} \cdot \text{bar}}{\text{mol} \cdot \text{K}}\right)(298.15 \text{ K})$

$\Rightarrow n = 3.5 \times 10^{-9}$ M

1-19. 1 ppm = $\dfrac{1 \text{ g solute}}{10^6 \text{ g solution}}$. Since 1 L of dilute solution $\approx 10^3$ g,

1 ppm = 10^{-3} g solute/L (= 10^{-3} g solute / 10^3 g solution).
Since 10^{-3} g = 10^3 µg, 1 ppm = 10^3 µg/L or 1 µg/mL.
Since 10^{-3} g = 1 mg, 1 ppm = 1 mg/L.

1-20. 0.2 ppb means 0.2×10^{-9} g of $C_{20}H_{42}$ per g of rainwater

$= 0.2 \times 10^{-6} \dfrac{\text{g } C_{20}H_{42}}{1\,000 \text{ g rainwater}} \approx \dfrac{0.2 \times 10^{-6} \text{ g } C_{20}H_{42}}{\text{L rainwater}}$.

$\dfrac{0.2 \times 10^{-6} \text{ g/L}}{282.56 \text{ g/mol}} = 7 \times 10^{-10} \dfrac{\text{mol}}{\text{L}} = 7 \times 10^{-10}$ M

1-21. $\left(0.705 \dfrac{\text{g HClO}_4}{\text{g solution}}\right)(37.6 \text{ g solution}) = 26.5 \text{ g HClO}_4$

37.6 g solution – 26.5 g HClO$_4$ = 11.1 g H$_2$O

1-22. (a) $\left(1.67 \dfrac{\text{g solution}}{\text{mL}}\right)\left(1\,000 \dfrac{\text{mL}}{\text{L}}\right) = 1.67 \times 10^3$ g solution in 1.000 L

(b) $\left(0.705 \dfrac{\text{g HClO}_4}{\text{g solution}}\right)(1.67 \times 10^3 \text{ g solution}) = 1.18 \times 10^3 \text{ g HClO}_4$

(c) $(1.18 \times 10^3 \text{ g}) / (100.45 \text{ g/mol}) = 11.7$ mol

1-23. molality = $\dfrac{\text{mol KI}}{\text{kg solvent}}$

20.0 wt% KI = $\dfrac{200 \text{ g KI}}{1\,000 \text{ g solution}} = \dfrac{200 \text{ g KI}}{800 \text{ g H}_2\text{O}}$

To find the grams of KI in 1 kg of H$_2$O, we set up a proportion:

$\dfrac{200 \text{ g KI}}{800 \text{ g H}_2\text{O}} = \dfrac{x \text{ g KI}}{1\,000 \text{ g H}_2\text{O}} \Rightarrow x = 250$ g KI

250 g KI / 166.00 g/mol = 1.51 mol KI, so the molality is 1.51 m.

Chemical Measurements

1-24. (a) $\dfrac{150 \times 10^{-15} \text{ mol/cell}}{2.5 \times 10^4 \text{ vesicles/cell}} = 6.0 \times 10^{-18} \dfrac{\text{mol}}{\text{vessicle}} = 6.0 \dfrac{\text{amol}}{\text{vessicle}}$

(b) $(6.0 \times 10^{-18} \text{ mol})\left(6.022 \times 10^{23} \dfrac{\text{molecules}}{\text{mol}}\right) = 3.6 \times 10^6 \text{ molecules}$

(c) Volume = $\dfrac{4}{3}\pi(200 \times 10^{-9} \text{ m})^3 = 3.35 \times 10^{-20} \text{ m}^3$;

$\dfrac{3.35 \times 10^{-20} \text{ m}^3}{10^{-3} \text{ m}^3/\text{L}} = 3.35 \times 10^{-17} \text{ L}$

(d) $\dfrac{10 \times 10^{-18} \text{ mol}}{3.35 \times 10^{-17} \text{ L}} = 0.30 \text{ M}$

1-25. $\dfrac{80 \times 10^{-3} \text{ g}}{180.2 \text{ g/mol}} = 4.4 \times 10^{-4} \text{ mol}$; $\dfrac{4.4 \times 10^{-4} \text{ mol}}{0.1 \text{ L}} = 4.4 \times 10^{-3} \text{ M}$;

Similarly, 120 mg/100 L = 6.7×10^{-3} M.

1-26. (a) Mass of 1.000 L = $1.046 \dfrac{\text{g}}{\text{mL}} \times 1000 \dfrac{\text{mL}}{\text{L}} \times 1.000 \text{ L} = 1046 \text{ g}$

Grams of $C_2H_6O_2$ per liter = $6.067 \dfrac{\text{mol}}{\text{L}} \times 62.07 \dfrac{\text{g}}{\text{mol}} = 376.6 \dfrac{\text{g}}{\text{L}}$

(b) 1.000 L contains 376.6 g of $C_2H_6O_2$ and 1 046 – 376.6 = 669 g of H_2O
= 0.669 kg

Molality = $\dfrac{6.067 \text{ mol } C_2H_6O_2}{0.669 \text{ kg } H_2O} = 9.07 \dfrac{\text{mol } C_2H_6O_2}{\text{kg } H_2O} = 9.07 \, m$

1-27. Shredded wheat: 1.000 g contains 0.099 g protein + 0.799 g carbohydrate

$0.099 \text{ g} \times 4.0 \dfrac{\text{Cal}}{\text{g}} + 0.799 \text{ g} \times 4.0 \dfrac{\text{Cal}}{\text{g}} = 3.6 \text{ Cal}$

Doughnut: 1.000 g contains 0.046 g protein + 0.514 g carbohydrate + 0.186 g fat

$0.046 \text{ g} \times 4.0 \dfrac{\text{Cal}}{\text{g}} + 0.514 \text{ g} \times 4.0 \dfrac{\text{Cal}}{\text{g}} + 0.186 \text{ g} \times 9.0 \dfrac{\text{Cal}}{\text{g}} = 3.9 \text{ Cal}$

In a similar manner, we find $2.8 \dfrac{\text{Cal}}{\text{g}}$ for hamburger and $0.48 \dfrac{\text{Cal}}{\text{g}}$ for apple.

There are 16 ounces in 1 pound, which Table 1-4 says is equal to 453.592 37 g
$\Rightarrow$ 453.592 37 g / 16 ounce = $28.35 \dfrac{\text{g}}{\text{ounce}}$.

To convert Cal/g to Cal/ounce, multiply (Cal/g) times (28.35 g/ounce):

	Shredded Wheat	Doughnut	Hamburger	Apple
Cal/g	3.6	3.9	2.8	0.48
Cal/ounce	102	111	79	14

1-28. $2.00 \cancel{L} \times 0.0500 \dfrac{\cancel{mol}}{\cancel{L}} \times 61.83 \dfrac{g}{\cancel{mol}} = 6.18$ g in a 2 L volumetric flask

1-29. Weigh out 2×0.0500 mol $= 0.100$ mol $= 6.18$ g B(OH)$_3$ and dissolve in 2.00 kg H$_2$O.

1-30. $M_{con} \cdot V_{con} = M_{dil} \cdot V_{dil}$

$\left(0.80 \dfrac{mol}{\cancel{L}}\right)(1.00 \cancel{L}) = \left(0.25 \dfrac{mol}{L}\right) V_{dil} \Rightarrow V_{dil} = 3.2$ L

1-31. We need $1.00 \cancel{L} \times 0.10 \dfrac{mol}{\cancel{L}} = 0.10$ mol NaOH $= 4.0$ g NaOH

$\dfrac{4.0 \text{ g NaOH}}{0.50 \dfrac{\text{g NaOH}}{\text{g solution}}} = 8.0$ g solution

1-32. (a) $V_{con} = V_{dil} \dfrac{M_{dil}}{M_{con}} = 1\,000$ mL $\left(\dfrac{1.00 \text{ M}}{18.0 \text{ M}}\right) = 55.6$ mL

(b) One liter of 98.0% H$_2$SO$_4$ contains $(18.0 \cancel{mol})(98.07 \text{ g/}\cancel{mol}) = 1.77 \times 10^3$ g of H$_2$SO$_4$. Since the solution contains 98.0 wt% H$_2$SO$_4$, and the mass of H$_2$SO$_4$ per mL is 1.77 g, the mass of solution per milliliter (the density) is

$\dfrac{1.77 \text{ g H}_2\text{SO}_4/\text{mL}}{0.980 \text{ g H}_2\text{SO}_4/\text{g solution}} = 1.80$ g solution/mL

1-33. 2.00 L of 0.169 M NaOH $= 0.338$ mol NaOH $= 13.5$ g NaOH

density $= \dfrac{\text{g solution}}{\text{mL solution}}$

$= \dfrac{13.5 \text{ g NaOH}}{(16.7 \text{ mL solution})\left(0.534 \dfrac{\text{g NaOH}}{\text{g solution}}\right)} = 1.52 \dfrac{g}{mL}$

Chemical Measurements 11

1-34. FM of Ba(NO$_3$)$_2$ = 261.34   4.35 g of solid with 23.2 wt% Ba(NO$_3$)$_2$ contains
[0.232 g Ba(NO$_3$)$_2$ / g solid](4.35 g solid) = 1.01 g Ba(NO$_3$)$_2$

$$\text{mol Ba}^{2+} = \frac{(1.01 \text{ g Ba(NO}_3)_2)}{(261.34 \text{ g Ba(NO}_3)_2/\text{mol})} = 3.86 \times 10^{-3} \text{ mol}$$

mol H$_2$SO$_4$ = mol Ba^{2+} = 3.86 × 10^{-3} mol

$$\text{volume of H}_2\text{SO}_4 = \frac{(3.86 \times 10^{-3} \text{ mol})}{(3.00 \text{ mol/L})} = 1.29 \text{ mL}$$

1-35. 25.0 mL of 0.023 6 M La^{3+} contains
(0.025 0 L)(0.023 6 M) = 5.90 × 10^{-4} mol La^{3+}
mol HF required for stoichiometric reaction = 3 × mol La^{3+} = 1.77 × 10^{-3} mol
50% excess = 1.50(1.77 × 10^{-3} mol) = 2.66 × 10^{-3} mol HF
Required mass of pure HF = (2.66 × 10^{-3} mol)(20.01 g/mol) = 0.053 1 g

$$\text{Mass of 0.491 wt\% HF solution} = \frac{(0.053\,1 \text{ g HF})}{(0.004\,91 \text{ g HF /g solution})} = 10.8 \text{ g}$$

1-36. (a) Acetic acid C$_2$H$_4$O$_2$ = 60.05 g/mol   NaHCO$_3$ = 84.01 g/mol

(b) (5 g NaHCO$_3$)/(84.01 g/mol) = 0.0595 mol
1 mol NaHCO$_3$ reacts with 1 mol CH$_3$CO$_2$H, so we need
(0.0595 mol CH$_3$CO$_2$H)(60.05 g/mol) = 3.57 g CH$_3$CO$_2$H

(c) But vinegar is ~5 wt% acetic acid, so we need

$$\frac{(3.57 \text{ g CH}_3\text{CO}_2\text{H})}{(0.05 \text{ g CH}_3\text{CO}_2\text{H/g vinegar})} = 71.5 \text{ g acetic acid}$$

If the density is 1.0 g/mL, 71.5 g vinegar = 71.5 mL vinegar

(d) Vinegar is the limiting reagent because 71.5 mL are required, but only 50 mL are available.

(e) Vinegar is the limiting reagent. The amount of CH$_3$CO$_2$H in 50 mL (≈50 g) of vinegar is (0.05 g CH$_3$CO$_2$H/g vinegar)(50 g vinegar) = 2.5 g CH$_3$CO$_2$H. Mol CH$_3$CO$_2$H = 2.5 g (CH$_3$CO$_2$H)/(60.05 g/mol) = 0.041 6 mol which will create 0.041 6 mol of CO$_2$ whose volume is

$$V = \frac{nRT}{P} = \frac{(0.041\,6 \text{ mol CO}_2)(0.083\,14 \text{ L}\cdot\text{bar/(mol}\cdot\text{K)})(300\text{K})}{1 \text{ bar}} = 1.04 \text{ L}$$

If the air space in the bottle is 0.5 L, and it contains 0.5 L of air plus 1.04 L of CO$_2$, we have 1.54 L of gas in 0.5 L of volume, giving a pressure of 1.54/0.5 ≈ 3 bar. The cork pops before this much pressure builds up.

CHAPTER 2
TOOLS OF THE TRADE

2-1. (a) The primary rule is to familiarize yourself with the hazards of what you are about to do and do not carry out a dangerous procedure without adequate precautions.

2-2. Nonpolar organic liquids might penetrate through rubber gloves. Concentrated hydrochloric acid is a polar aqueous solution that is not likely to penetrate through rubber gloves.

2-3. Dichromate ($Cr_2O_7^{2-}$) is soluble in water and contains carcinogenic Cr(VI). Reducing Cr(VI) to Cr(III) decreases the toxicity of the metal. Converting aqueous Cr(III) to solid $Cr(OH)_3$ decreases the solubility of the metal and therefore decreases its ability to be spread by water. Evaporation produces the minimum volume of waste.

2-4. Green chemistry is a set of principles intended to change our behavior in a manner that will help sustain the habitability of Earth. Green chemistry seeks to design chemical products and processes to reduce the use of resources and energy and the generation of hazardous waste.

2-5. The lab notebook must: (1) state what was done; (2) state what was observed; and (3) be understandable to a stranger.

2-6. An object placed on the electronic balance pushes the pan down with a force $m \times g$, where m is the mass of the object and g is the acceleration of gravity. The pan pushes down on a load receptor attached to parallel guides. The force of the sample pushes one side of the force-transmitting lever down and moves the other side of the lever up. A null position sensor detects movement of the lever arm away from its equilibrium (null) position. When the null sensor detects displacement of the lever arm, a servo amplifier sends electric current through the force compensation wire coil in the field of a permanent magnet. Electric current in the coil interacts with the magnetic field to produce a downward force that exactly compensates for the upward force on the lever arm to maintain a null position. Current flowing through the coil is measured and ultimately converted to a readout in grams. The conversion between current and mass is accomplished by measuring the current required to balance an internal calibration mass.

2-7. The buoyancy correction is 1 when the substance being weighed has the same density as the weights used to calibrate the balance.

2-8. $m = \dfrac{(14.82 \text{ g})\left(1 - \dfrac{0.001\,2 \text{ g/mL}}{8.0 \text{ g/mL}}\right)}{\left(1 - \dfrac{0.001\,2 \text{ g/mL}}{0.626 \text{ g/mL}}\right)} = 14.85 \text{ g}$

2-9. The smallest correction will be for PbO_2, whose density is closest to 8.0 g/mL. The largest correction will be for the least dense substance, lithium.

2-10. $m = \dfrac{4.236\,6 \text{ g}\left(1 - \dfrac{0.001\,2 \text{ g/mL}}{8.0 \text{ g/mL}}\right)}{\left(1 - \dfrac{0.001\,2 \text{ g/mL}}{1.636 \text{ g/mL}}\right)} = 4.239\,1 \text{ g}$

Without correcting for buoyancy, we would think the mass of primary standard is less than the actual mass and we would think the molarity of base reacting with the standard is also less than the actual molarity. The percentage error would be

$\dfrac{\text{true mass} - \text{measured mass}}{\text{true mass}} \times 100 = \dfrac{4.239\,1 - 4.236\,6}{4.239\,1} \times 100 = 0.06\%$.

2-11. $m' = \dfrac{1.267 \text{ g}\left(1 - \dfrac{0.001\,2 \text{ g/mL}}{3.988 \text{ g/mL}}\right)}{\left(1 - \dfrac{0.001\,2 \text{ g/mL}}{8.0 \text{ g/mL}}\right)} = 1.266_8 \text{ g}$

The buoyancy correction is negligible because the density of CsCl is high.

2-12. (a) One mol of He (= 4.003 g) occupies a volume of

$V = \dfrac{nRT}{P} = \dfrac{(1 \text{ mol})\left(0.083\,14 \dfrac{\text{L} \cdot \text{bar}}{\text{mol} \cdot \text{K}}\right)(293.15 \text{ K})}{1 \text{ bar}} = 24.37 \text{ L}$

Density = 4.003 g / 24.37 L = 0.164 g/L = 0.000 164 g/mL

(b) $m = \dfrac{(0.823 \text{ g})\left(1 - \dfrac{0.000\,164 \text{ g/mL}}{8.0 \text{ g/mL}}\right)}{\left(1 - \dfrac{0.000\,164 \text{ g/mL}}{0.97 \text{ g/mL}}\right)} = 0.823 \text{ g}$

(c) The buoyancy correction in the graph is ~1.0011 for a density of 0.97 g/mL.
$m = (1.0011)(0.823 \text{ g}) = 0.824 \text{ g}$.

2-13. (a) $(0.42)(2\,330 \text{ Pa}) = 979 \text{ Pa}$

(b) Air density (g/L) = $(0.003\,485\,B - 0.001\,318\,v)/T =$
$$\frac{(0.003\,485)(94\,000) - (0.001\,318)(979)}{293.15} = 1.11 \text{ g/L} = 0.001\,1 \text{ g/mL}$$

(c) Mass = $1.000\,0 \text{ g} \left(\dfrac{1 - \dfrac{0.001\,1 \text{ g/mL}}{8.0 \text{ g/mL}}}{1 - \dfrac{0.001\,1 \text{ g/mL}}{1.00 \text{ g/mL}}} \right) = 1.001\,0 \text{ g}$

2-14. (a) $m_b = m_a \dfrac{r_a^2}{r_b^2} = (100.000\,0 \text{ g}) \dfrac{(6\,370\,000 \text{ m})^2}{(6\,370\,030 \text{ m})^2} = 99.999\,1 \text{ g}$

(b) The calibrate function measures a known internal calibration mass and applies a factor to the electronic response to indicate the known mass. This calibration factor accounts for the change in gravitational force on going from one elevation to another.

2-15. (a) Area of gold electrode = $(3.3 \text{ mm}^2)(0.1 \text{ cm/mm})^2 = 0.033 \text{ cm}^2$

Moles DNA bound = $(1.2 \text{ pmol/cm}^2)(0.033 \text{ cm}^2)$
$= (1.2 \times 10^{-12} \text{ mol/cm}^2)(0.033 \text{ cm}^2) = 3.96 \times 10^{-14} \text{ mol}$

Mass of added cytosine = $(3.96 \times 10^{-14} \text{ mol})(287.2 \text{ g/mol})$
$= 1.14 \times 10^{-11} \text{ g} = 11 \text{ pg}$

(b) Binding of 1 ng/cm² gives a shift of –10 Hz. The observed shift of –4.4 Hz corresponds to
$$\frac{-4.4 \text{ Hz}}{-10 \text{ Hz/(ng/cm}^2)} = 0.44 \text{ ng/cm}^2 \text{ of bound cytosine}$$

$(0.44 \text{ ng/cm}^2 \text{ of bound cytosine})(0.033 \text{ cm}^2) = 14 \text{ pg}$, which is approximately equal to the expected binding of 11 pg of cytosine. The measured frequency shift corresponds to $(14/11) = 1.3$ cytosine nucleotides.

2-16. The volume of the boat is 85 L and the mass is 5 kg. The boat will float until its density is 1 kg/L. An 85-L boat will float until its total mass is 85 kg. If the empty mass is 5 kg, the payload it can carry is 85 – 5 = 80 kg. The boat could carry Dan, who weighs ~70 kg.

2-17. TD means "to deliver" and TC means "to contain."

2-18. Dissolve $(0.2500 \text{ L})(0.1500 \text{ mol/L}) = 0.03750$ mol of K_2SO_4 (= 6.534 g, FM 174.25 g/mol) in less than 250 mL of water in a 250-mL volumetric flask. Add more water and mix. Dilute to the 250.0 mL mark and invert the flask many times for complete mixing.

2-19. The plastic flask is needed for trace analysis of ionic analytes (especially cations) at ppb levels that might be lost by adsorption on the glass surface or contaminated by leaching of ions from the glass.

2-20. (a) With a suction device, suck liquid up past the 5.00 mL mark. Discard one or two pipet volumes of liquid to rinse the pipet. Take up a third volume past the calibration mark and quickly replace the bulb with your index finger. (Alternatively, use an automatic suction device that remains attached to the pipet.) Wipe excess liquid off the outside of the pipet with a clean tissue. Touch the tip of the pipet to the side of a beaker and drain liquid until the bottom of the meniscus reaches the center of the mark. Transfer the pipet to a receiving vessel and drain it by gravity while holding the tip against the wall. After draining stops, hold the pipet to the wall for a few more seconds to complete draining. Do not blow out the last drop. The pipet should be nearly vertical at the end of delivery.

(b) Transfer pipet.

2-21. (a) Adjust the knob for 50.0 µL. Place a fresh tip tightly on the barrel. Depress the plunger to the first stop, corresponding to 50.0 µL. Hold the pipet vertically, dip it 3–5 mm into reagent solution, and slowly release the plunger to suck up liquid. Leave the tip in the liquid for a few more seconds. Withdraw the pipet vertically. Take up and discard three squirts of reagent to clean and wet the tip and fill it with vapor. To dispense liquid, hold the pipet nearly vertical, touch the tip to the wall of the receiver, and gently depress the plunger to the first stop. After a few seconds, depress the plunger further to squirt out the last liquid.

(b) The procedure in (a) is called *forward mode*. For a foaming or viscous liquid, use *reverse* mode. Depress the plunger beyond the 50.0 μL stop and take in more than 50.0 μL. To deliver 50.0 μL, depress the plunger to the first stop and not beyond.

2-22. The trap prevents liquid filtrate from being sucked into the vacuum system. The watchglass keeps dust out of the sample.

2-23. Phosphorus pentoxide

2-24. (a) A 100.0 mM solution is the same as 0.100 0 M. To make 250.0 mL of 0.100 0 M solution requires (0.100 0 M)(0.250 0 L) = 0.025 00 mol benzoic acid. Required mass = (0.025 00 mol)(122.12 g/mol) = 3.053_0 g.

(b) $m' = \dfrac{3.053_0 \text{ g}\left(1 - \dfrac{0.001\ 2 \text{ g/mL}}{1.27 \text{ g/mL}}\right)}{\left(1 - \dfrac{0.001\ 2 \text{ g/mL}}{8.0 \text{ g/mL}}\right)} = 3.050_6 \text{ g}$

(c) We want to make a 50.0 μM solution from a 100.0 mM solution, which represents a dilution by a factor of (100.0 mM)/(50.0 μM) = $(100.0 \times 10^{-3}$ M)/$(50.0 \times 10^{-6}$ M) = 2 000. You could make a 2 000-fold dilution with a 20-fold dilution followed by a 100-fold dilution (20 × 100 = 2 000). You could do this by first diluting 5.00 mL of 100.0 mM solution up to 100.0 mL to get (5/100)(100.0 mM) = 5.00 mM. Then dilute 10.00 mL of the resulting solution up to 1000.0 mL to get (10/1 000)(5.00 mM) = 50.0 μM.

2-25. 20.214 4 g − 10.263 4 g = 9.951 0 g. Table 2-7 tells us that the true volume is (9.951 0 g)(1.002 9 mL/g) = 9.979 9 mL.

2-26. Expansion = $\dfrac{\text{density at } 15°}{\text{density at } 25°}$ = $\dfrac{0.999\ 102}{0.997\ 047}$ = 1.002 061 ≈ 0.2%. Densities were taken from Table 2-7. The 0.500 0 M solution at 25° would be (0.500 0 M)/(1.002) = 0.499 0 M.

2-27. In Table 2-7 density is for true mass measured in vacuum. Therefore, we can write: mass in vacuum = $(50.037 \text{ mL})(0.998\,207 \text{ g/mL}) = 49.947$ g.

In Table 2-7 the volume of 1 g of water is for mass measured in air. Therefore, we can write: mass in air = $\dfrac{(50.037 \text{ mL})}{(1.002\,9 \text{ mL/g})} = 49.892$ g.

2-28. When the solution is cooled to 20°C, the concentration will be higher than the concentration at 24°C by a factor of $\dfrac{\text{density at 20°C}}{\text{density at 24°C}}$. Therefore, the concentration needed at 24° will be lower than the concentration at 20°C.

Desired concentration at 24°C = $(1.000 \text{ M})\left(\dfrac{0.997\,299\,5 \text{ g/mL}}{0.998\,207\,1 \text{ g/mL}}\right) = 0.999\,1$ M

(using the quotient of densities from Table 2-7). The true mass of KNO_3 needed is $(0.500\,0 \text{ L})\left(0.999\,1 \dfrac{\text{mol}}{\text{L}}\right)\left(101.102 \dfrac{\text{g}}{\text{mol}}\right) = 50.506$ g.

$$m' = \dfrac{50.506 \text{ g}\left(1 - \dfrac{0.001\,2 \text{ g/mL}}{2.109 \text{ g/mL}}\right)}{\left(1 - \dfrac{0.001\,2 \text{ g/mL}}{8.0 \text{ g/mL}}\right)} = 50.484 \text{ g}$$

2-29. Procedure (ii) with the 10-mL pipet and 1-L flask is more accurate than procedure (i) with the 1-mL pipet and 100-mL flask. Relative uncertainties in the larger pipet and larger flask are less than the relative uncertainties in the smaller pipet and smaller flask. You can improve the accuracy of either procedure by calibrating the specific pipet and specific volumetric flask that you use so that you know how much volume they actually contain. The calibration uncertainty should be smaller than the manufacturer's tolerance for the glassware.

2-30. (a) Fraction within specifications = $e^{-t(\ln 2)/t_m}$. If $t_m = 2$ yr and $t = 2$ yr, then fraction within specifications = $e^{-2(\ln 2)/2} = e^{-\ln 2} = \frac{1}{2}$.

(b) Fraction within specifications = $0.95 = e^{-t(\ln 2)/2 \text{ yr}}$

To solve for t, take the natural logarithm of both sides:

$\ln(0.95) = -t(\ln 2)/2 \Rightarrow t = -2 \ln(0.95)/\ln 2 = 0.148$ yr = 54 days ≈ 8 weeks

2-31. Al extracted from glass = $(0.200 \text{ L})(5.2 \times 10^{-6} \text{ M}) = 1.04 \times 10^{-6}$ mol

mass of Al = $(1.04 \times 10^{-6} \text{ mol})(26.98 \text{ g/mol}) = 28.1$ μg

This much Al was extracted from 0.50 g of glass, so

$$\text{wt\% Al extracted} = 100 \times \frac{28.1 \times 10^{-6} \text{ g}}{0.50 \text{ g}} = 0.005\,6_2 \text{ wt\%}$$

$$\text{Fraction of Al extracted by EDTA} = \frac{0.005\,6_2 \text{ wt\%}}{0.80 \text{ wt\%}} = 0.007\,0 \text{ (or 0.70\% of Al)}$$

2-32.

	A	B	C	D	E	F	G	H	I	J
1	Buoyancy Correction with Equation 2-1									
2	(from the delightful book by Dan and Chuck)									
3	Constants:									
4	d_a =	0.0012	g/mL							
5	d_w =	8.0	g/mL							
6	Substance weighed:			Li	acetic acid	tris	CCl_4	S	$AgNO_3$	PbO_2
7	Density (g/mL) of substance:			0.53	1.05	1.33	1.59	2.07	4.45	9.4
8	Observed mass (m', g) in air:			100.00	100.00	100.00	100.00	100.00	100.00	100.00
9	True mass (m, g) in vacuum:			100.212	100.099	100.075	100.061	100.043	100.012	99.998
10	m = m'*(numerator/denominator)									
11	numerator = $(1-d_a/d_w)$ =			0.99985	0.99985	0.99985	0.99985	0.99985	0.99985	0.99985
12	denominator = $(1-d_a/d)$ =			0.997736	0.998857	0.999098	0.999245	0.99942	0.99973	0.999872
13	Formulas:	D11 = 1-$B$4/$B$5								
14		D12 = 1-$B$4/D7								
15		D9 = D8*D11/D12								
16										
17	Density	correction								
18	0.53	1.002119	Li							
19	1.05	1.000994	acetic acid							
20	1.33	1.000753	tris							
21	1.59	1.000605	CCl_4							
22	2.07	1.000430	S							
23	4.45	1.000120	$AgNO_3$							
24	9.4	0.999978	PbO2							
25	11.4	0.999955	Pb							
26	13.5	0.999939	Hg							
27	22.5	0.999903	Ir							

CHAPTER 3
EXPERIMENTAL ERROR

3-1. (a) 5   (b) 4   (c) 3

3-2. (a) 1.237   (b) 1.238   (c) 0.135   (d) 2.1   (e) 2.00

3-3. (a) 0.217   (b) 0.216   (c) 0.217

3-4. (a) 3.71   (b) 10.7   (c) 4.0×10^1 (d) 2.85×10^{-6}
(e) 12.625 1   (f) 6.0×10^{-4} (g) 242

3-5. (a) SrF_2 = 87.62 + 2(18.998 403 163) = 125.62 because the atomic mass of Sr has only 2 decimal places.
(b) Na_2CO_3 = 2(22.989 769 28) + 1(12.011) + 4(15.999) = 105.988

3-6. (a) 12.3   (b) 75.5   (c) 5.520×10^3 (d) 3.04
(e) 3.04×10^{-10} (f) 11.9   (g) 4.600   (h) 4.9×10^{-7}

3-7. All measurements have some uncertainty, so there is no way to know true value.

3-8. Systematic error is always above or always below the "true value" if you make replicate measurements. In principle, you can find the source of this error and eliminate it in a better experiment so the measured mean equals the true mean. Random error is equally likely to be positive or negative. Random error can be reduced by better technique but cannot be completely eliminated. A blunder is an accidental error caused by a significant departure from the procedure. The error may be so significant that data must be rejected or the experiment redone.

3-9. (a) The apparent mass of product is systematically low because the initial mass of the (crucible plus moisture) is higher than the true mass of the crucible.

(b) The error is systematic. There is also always some random error superimposed on the systematic error.

(c) Correction of the procedure to say that the filter crucible should be thoroughly dried before weighing would eliminate this systematic error.

3-10. (a) The pipet has a systematic error. It always delivers more than it is rated for.

(b) The volume fluctuates by the random error of ±0.009 mL. Calibration eliminated the systematic error leaving only the random error.

(c) The numbers 1.98 and 2.03 mL are systematic errors. The buret delivers too little between 0 and 2 mL and too much between 2 and 4 mL. The observed variations ±0.01 and ±0.02 are random errors.

(d) The difference between 1.9839 and 1.9900 g is random error. The mass will probably be different the next time I try the same procedure.

(e) This is a blunder. Any attempt to estimate the volume beyond the 50.00 mL mark based on the distance would have an unacceptably large error. Starting titrations near 0.00 mL minimizes the chance of such blunders.

(f) Differences in peak area are random error based on inconsistent injection volume, inconsistent detector response, and probably other small variations in the condition of the instrument from run to run.

3-11 (a) Pentane has a low density. Its apparent mass would be less than its true mass unless a buoyancy correction was applied.

(b) The relative uncertainty due to the uncalibrated pipet is (0.006/2.000) = 0.3%. Systematic error could be removed by calibration. Alternately, the procedure could be redesigned to use a 20-mL pipet whose nominal volume is 20.00 ± 0.03 mL (± 0.15%).

(c) Micropipets are calibrated at sea level and deliver consistently lower values at altitude. Calibrate the pipet using weight or move your lab to the coast.

(d) Matrix can cause systematic errors in sample. Preparing calibration standards in shellfish matrix factors out systematic error. This paper used extracts from mussels homogenized in a blender to prepare their standards.

3-12. (a) Katniss   (b) Gale   (c) Peeta   (d) Haymitch

3-13 (a) Haymitch has large random error. Random error is too large to be able to detect any systematic error. Gale has moderate random error and no apparent systematic error. Katniss has small random error and no systematic error. Peeta has small random error with systematic bias.

(b)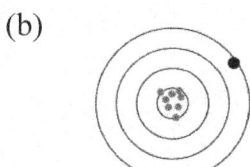

3-14. 3.124 (±0.005), 3.124 (±0.2%). It would also be reasonable to keep an additional digit: 3.123$_6$ (±0.005$_2$), 3.123$_6$ (±0.1$_7$%)

3-15. (a) 6.2 (±0.2) M − 4.1 (±0.1) M = 2.1 (± e)

$$e = \sqrt{(0.2 \text{ M})^2 + (0.1 \text{ M})^2} = 0.2_{24} \text{ M} \qquad \%e = \left(\frac{0.2_{24} \text{ M}}{2.1 \text{ M}}\right) \times 100 = 10._7\%$$

Answer: 2.1 ± 0.2 M (or 2.1 M ± 11%)

(b) 9.43 (±0.05) M × 0.016 (±0.001) L = 0.150 88 mol (± %e)

$$\%e = \sqrt{0.5_3\%^2 + 6._{25}\%^2} = 6._{272}\% = \text{Relative uncertainty}$$

Absolute uncertainty = 0.150 88 mol × 0.06$_2$$_{72}$ = 0.009$_{46}$ mol

Answer: 0.151 ± 0.009 mol   (or 0.151 mol ± 6%)

(**Real rule**: The first uncertain figure is the last significant figure.)

(c) The first term in brackets is the same as part (a), so we can rewrite the problem as 2.1 (±0.2$_{24}$) mmol ÷ 9.43 (±0.05) mL

= 2.1 mmol (±10.$_7$%) ÷ 9.43 mL (±0.5$_3$%) = 0.223 M (± %e)

$$\%e = \sqrt{10._7^2 + 0.5_3^2} = 10._7\% \quad (0.5_3\% \text{ is} < ¼ \text{ of } 10._7\%, \text{ and so is negligible})$$

Absolute uncertainty = 0.10$_7$ × 0.223 M = 0.02$_3$$_9$ M

Answer: 0.22$_3$ ± 0.02$_4$ M (±11%)

(d) The term in brackets is

6.2 (±0.2) × 10^{-3} L $e = \sqrt{(0.2 \text{ mL})^2 + (0.1 \text{ mL})^2} = 0.2_{24}$ mL

+ 4.1 (±0.1) × 10^{-3} L

10.3 (±0.2$_{24}$) × 10^{-3} L = 10.3 × 10^{-3} L (±2.$_{17}$%)

9.43 M (±0.5$_3$%) × 0.010 3 L (±2.$_{17}$%) = 0.097 13 mol ± 2.$_{23}$%

= 0.097 13 mol ± 0.002 $_{17}$ mol

Answer: 0.097$_1$ ± 0.002$_2$ mol (± 2.$_2$%)

3-16. (a) Uncertainty = $\sqrt{(0.03 \text{ mL})^2 + (0.02 \text{ mL})^2 + (0.06 \text{ mL})^2}$ = 0.07_0 mL

Answer: 10.18 (±0.07) mL (±0.7%)

(b) 91.3 mM (±1.1_0%) × 40.3 mL (±0.5_0%) / 21.1 mL (±0.9_{48}%) = 174.4 mM ± e%

%e = $\sqrt{1.1_0{}^2 + 0.5_0{}^2 + 0.9_{48}{}^2}$ = 1.5_4%; Absolute e = 174.4 × 0.015_4 = 2.7 mM

Answer: 174 (±3) mM (±2%) or $174._4$ (±$2._7$) mL (±1.5%)

(c) [4.97 (±0.05) mmol − 1.86 (±0.01) mmol] / [21.1 (±0.2) mL] = ?

= [3.11 (±0.05_{10} mmol)] / 21.1 (±0.2) mL = [3.11 (±1.6_4%)] / 21.1 (±0.9_{48}%)

= 0.147 M (±1.9_4%) = 0.147 (±0.003) M (±2%) or 0.147_4 (±0.002_8) M (±1.9%)

(d) 2.0164 (±0.0008) g
     1.233   (±0.002) g
  + 4.61   (±0.01) g
  ─────────────────────
     7.85_{94} $e = \sqrt{(0.0008 \text{ g})^2 + (0.002 \text{ g})^2 + (0.01 \text{ g})^2}$ = 0.01_{02} g

Answer: 7.86 (±0.01) g (±0.1%)

(e) 2 016.4 (±0.8) g
  + 123.3   (±0.2) g
  + 46.1   (±0.1) g
  ─────────────────────
     2 185.8 $e = \sqrt{(0.8 \text{ g})^2 + (0.2 \text{ g})^2 + (0.1 \text{ g})^2}$ = 0.8_3 g = 0.8 g

Answer: 2 185.8 (±0.8) g (±0.04%)

(f) $[3.14 (\pm 0.05)]^{1/3}$ = 1.464_3 ± ?

For $y = x^a$, %e_y = a(%e_x)

$x = 3.14 \pm 0.05 \Rightarrow$ %e_x = (0.05 / 3.14) × 100 = 1.5_{92}%

%e_y = $\frac{1}{3}(1.5_{92}\%)$ = 0.5_{31}% ; e_y = 0.005_{31} × 1.464_3 = 0.007_8

Answer: 1.464 ± 0.008 (±0.5%)

(g) log[3.14 (±0.05)] = 0.496_9 ± ?

For $y = \log x$, $e_y = 0.43429 \dfrac{e_x}{x}$

$x = 3.14 \pm 0.05 \Rightarrow e_y = 0.43429 \left(\dfrac{0.05}{3.14}\right)$ = 0.006_{915}

Answer: 0.497 ± 0.007 (± 1.4%)

3-17. (a) $y = x^{1/2} \Rightarrow \%e_y = a(\%e_x) = \frac{1}{2}\left(100 \times \frac{0.0011}{3.1415}\right) = 0.01751\%$

$e_y = (1.751 \times 10^{-4}) \times \sqrt{3.1415} = 3.1 \times 10^{-4}$ Answer: 1.7724 ± 0.0003

(b) $y = \log x \Rightarrow e_y = \frac{1}{\ln 10} \frac{e_x}{x} = 0.43429 \left(\frac{0.0011}{3.1415}\right) = 1.52 \times 10^{-4}$

Answer: 0.49714 ± 0.00015

(c) $y = \text{antilog } x = 10^x \Rightarrow e_y = y \times 2.3026\, e_x$

$e_y = (10^{3.1415})(2.3026)(0.0011) = 3._{51}$ Answer: $1385 \pm 0.004 \times 10^3$

(d) $y = \ln x \Rightarrow e_y = \frac{e_x}{x} = \frac{0.0011}{3.1415} = 3.5_{02} \times 10^{-4}$ Answer: 1.1447 ± 0.0004

(e) Numerator of log term: $y = x^{1/2} \Rightarrow e_y = a(\%e_x) = \frac{1}{2}\left(\frac{0.006}{0.104} \times 100\right) = 2._{88}\%$

$\frac{0.3225 \pm 2._{88}\%}{0.0511 \pm 0.0009} = \frac{0.3225 \pm 2._{88}\%}{0.0511 \pm 1._{76}\%} = 6.311 \pm 3._{38}\% = 6.3_{11} \pm 0.2_{13}$

For $y = \log x$, $e_y = \frac{1}{\ln 10} \frac{e_x}{x} = 0.43429 \frac{e_x}{x} = 0.43429 \left(\frac{0.2_{13}}{6.3_{11}}\right) = 0.01_{47}$

Answer: $0.80_0 \pm 0.01_5$

3-18. molarity $= \frac{\text{mol}}{L} = \frac{[2.634(\pm 0.002)\text{ g}]/[58.44\text{ g/mol}]}{0.10000\,(\pm 0.00008)\text{ L}}$

$= \frac{[2.634\text{ g }(\pm 0.07_6\%)]/58.44\text{ g/mol}}{0.10000\text{ L }(\pm 0.08\%)}$

relative error $= \sqrt{(0.07_6\%)^2 + (0.08\%)^2} = 0.11\%$

molarity $= 0.4507\,(\pm 0.0005)\text{ M}$

3-19. $m = \frac{m'\left(1 - \dfrac{d_a}{d_w}\right)}{1 - \dfrac{d_a}{d}} = \frac{[1.0346\,(\pm 0.0002)\text{ g}]\left(1 - \dfrac{0.0012\,(\pm 0.0001)\text{ g/mL}}{8.0\,(\pm 0.5)\text{ g/mL}}\right)}{1 - \dfrac{0.0012\,(\pm 0.0001)\text{ g/mL}}{0.9972995\text{ g/mL}}}$

$m = \frac{[1.0346\text{ g }(\pm 0.01_{93}\%)]\left(1 - \dfrac{0.0012\text{ g/mL }(\pm 8._{33}\%)}{8.0\text{ g/mL }(\pm 6._{25}\%)}\right)}{1 - \dfrac{0.0012\text{ g/mL }(\pm 8._{33}\%)}{0.997299\text{ g/mL }(\pm 0\%)}}$

$m = \frac{[1.0346\text{ g }(\pm 0.01_{93}\%)][1 - 0.000150\,(\pm 10._4\%)]}{[1 - 0.001203\,(\pm 8._{33}\%)]}$

$$m = \frac{[1.034\,6\text{ g }(\pm 0.01_9\,_3\%)][1-0.000\,150\,(\pm 0.000\,015\,_6)]}{[1-0.001\,203\,(\pm 0.000\,10_0)]}$$

$$m = \frac{[1.034\,6\text{ g }(\pm 0.01_9\,_3\%)][0.999\,850\,0\,(\pm 0.000\,015\,_6)]}{[0.998\,797\,(\pm 0.000\,10_0)]}$$

$$m = \frac{[1.034\,6\text{ g }(\pm 0.01_9\,_3\%)][0.999\,850\,0\,(\pm 0.001\,5_6\%)]}{[0.998\,797\,(\pm 0.010\,_0\%)]}$$

$$\text{relative error} = \sqrt{(0.01_9\,_3\%)^2 + (0.001\,5_6\%)^2 + (0.010\,_0\%)^2} = 0.02_1\,_8\%$$

$$m = 1.035\,7\text{ g }(\pm 0.02\%) = 1.035\,7\,(\pm 0.000\,2)\text{ g}$$

3-20. (a) mol H^+ = 2 × mol Na_2CO_3

$$\text{mol Na}_2\text{CO}_3 = \frac{0.967\,4\,(\pm 0.000\,9)\text{ g}}{105.988\text{ g/mol}} = \frac{0.967\,4\text{ g }(\pm 0.09_3\%)}{105.988\text{ g/mol }(\pm 0.0\%)}$$

$$= 0.009\,127\,4\,(\pm 0.09_3\%)\text{ mol}$$

mol H^+ = 2[0.009 127 4 (±0.09$_3$%)] = 0.018 255 (±0.09$_3$%) mol

(Relative error is not affected by the multiplication by 2 because mol H^+ and uncertainty in mol H^+ are both multiplied by 2.)

$$[\text{HCl}] = \frac{0.018\,255\text{ mol }(\pm 0.09_3\%)}{0.027\,35\,(\pm 0.000\,04)\text{ L}} = \frac{0.018\,255\text{ mol }(\pm 0.09_3\%)}{0.027\,35\text{ L }(\pm 0.14_6\%)}$$

$$= 0.667\,46\text{ M }(\pm 0.17_3\%) = 0.667\,_{46}\,(\pm 0.001\,_{12})\text{ M} = 0.667 \pm 0.001\text{ M}$$

(b) We can account for the uncertainty in purity of Na_2CO_3 by introducing an additional uncertainty of ± 0.05% into the mol of Na_2CO_3:

mol H^+ = 2 × mol Na_2CO_3 (1.000 0 ± 0.05%)

$$[\text{HCl}] = \frac{[0.018\,255\text{ mol }(\pm 0.09_3\%)] \times (1.000\,0 \pm 0.05\%)}{0.027\,35\text{ L }(\pm 0.14_6\%)}$$

$$\text{relative error} = \sqrt{(0.09_3\%)^2 + (0.05\%)^2 + (0.14_6\%)^2} = 0.18_0\%$$

[HCl] = 0.667 46 M (±0.18$_0$%) = 0.667$_{46}$ (±0.001$_{20}$) M = 0.667 ± 0.001 M

(c) The volume of HCl has the largest relative uncertainty (0.15%). If this error could be eliminated the overall relative error would be reduced to:

$$\text{relative error} = \sqrt{(0.09_3\%)^2 + (0.05\%)^2} = 0.11\%$$

whereas eliminating the error in the mass would leave an overall relative error of 0.15% and eliminating the error due to Na_2CO_3 purity has no significant effect on the overall error (compare answers for part *a* and *b*).

3-21. (a) Glassware tolerance:

1-mL pipet: 1.000 ± 0.006 mL   10-mL pipet: 10.00 ± 0.02 mL
100-mL volumetric flask: 100.00 ± 0.08 mL

We write the dilution on one line to simplify the propagation of uncertainty:

$$\text{Final concentration} = \text{initial concentration} \times \text{first dilution} \times \text{second dilution}$$

$$= \left(150.0 \pm 0.3 \frac{\mu g}{mL}\right)\left(\frac{10.00 \pm 0.02 \text{ mL}}{100.00 \pm 0.08 \text{ mL}}\right)\left(\frac{10.00 \pm 0.02 \text{ mL}}{100.00 \pm 0.08 \text{ mL}}\right)$$

$$\text{Final concentration} = \left(150.0 \pm 0.3 \frac{\mu g}{mL}\right)\left(\frac{1}{10}\right)\left(\frac{1}{10}\right) = 1.500 \frac{\mu g}{mL}$$

$$\text{Relative uncertainty} = \sqrt{\left(\frac{0.3}{150}\right)^2 + \left(\frac{0.02}{10}\right)^2 + \left(\frac{0.08}{100}\right)^2 + \left(\frac{0.02}{10}\right)^2 + \left(\frac{0.08}{100}\right)^2} = 0.003_{64}$$

Absolute uncertainty = $(0.003_{64})(1.500 \;\mu g/mL) = 0.005_{47} \;\mu g/mL$

Answer: $1.500_0 \pm 0.005_5 \;\mu g/mL$ or $1.500 \pm 0.005 \;\mu g/mL$

To compute relative uncertainty, you could have written each fraction in the square root as a percentage $\left(\text{such as } 100 \times \frac{0.03}{150}\right)$, and then convert percent uncertainty back to relative uncertainty by dividing by 100 at the end. Instead, we just used relative uncertainty without converting to and from percentage.

(b) $$\text{Final concentration} = \left(150.0 \pm 0.3 \frac{\mu g}{mL}\right)\left(\frac{1.000 \pm 0.006 \text{ mL}}{100.00 \pm 0.08 \text{ mL}}\right) = 1.500 \frac{\mu g}{mL}$$

$$\text{Relative uncertainty} = \sqrt{\left(\frac{0.3}{150}\right)^2 + \left(\frac{0.006}{1}\right)^2 + \left(\frac{0.08}{100}\right)^2} = 0.006_{37}$$

Absolute uncertainty = $(0.006_{37})(1.500 \;\mu g/mL) = 0.009_{56} \;\mu g/mL$

Answer: $1.500_0 \pm 0.009_6 \;\mu g/mL$ or $1.50 \pm 0.01 \;\mu g/mL$

The second method is less precise because the uncertainty in the 1-mL pipet is 0.6%, which is 3 times larger than any other uncertainty in either step.

3-22. (a) To find the uncertainty in l^3, we use the function $y = x^a$, where $x = l$ and $a = 3$. The uncertainty in l^3 is

$$\%e_y = a\,\%e_x = 3 \times 1.8_4 \times 10^{-7}\% = 5.5 \times 10^{-7}\%$$

The relative uncertainty in the mass of the silicon sphere is found from the relative uncertainties in the bracketed term, A_{Si}, V, and l^3. (There is no uncertainty in the number 8 atoms/unit cell.)

percent uncertainty in brackets = $4.7 \times 10^{-8}\%$

percent uncertainty in A_{Si} = $5.4 \times 10^{-7}\%$

percent uncertainty in V = $2.0_2 \times 10^{-6}\%$

percent uncertainty in l^3 = $5.5 \times 10^{-7}\%$ (calculated before)

$$\%\text{ uncertainty in } m_{\text{sphere}} = \sqrt{(\%e_{\text{brackets}})^2 + (\%e_{A_{Si}})^2 + (\%e_V)^2 + (\%e_{l^3})^2}$$

$$= \sqrt{(4.7\times10^{-8}\%)^2 + (5.4\times10^{-7}\%)^2 + (2.0_2\times10^{-6}\%)^2 + (5.5\times10^{-7}\%)^2}$$

$$= 2.1_6 \times 10^{-6}\% = 2.2 \times 10^{-6}\%$$

(If we assume volume uncertainty is dominant (> 4 times other uncertainties) we would estimate an error of $2.0 \times 10^{-6}\%$ with less effort.)

(b) $m'_{\text{sphere}} = 999.698\,336\,5\text{ g} - 3.8 \times 10^{-6}\text{ g} + 1.206 \times 10^{-4}\text{ g} = 999.698\,453\,3\text{ g}$

Abs. uncertainty of $m_{\text{sphere}} = 999.698\,336\,5\text{ g} \times \dfrac{2.1_6 \times 10^{-6}\%}{100} = 21._6\text{ μg}$

Abs. uncertainty of $m'_{\text{sphere}} = \sqrt{e_{\text{sphere}}^2 + e_{\text{defects}}^2 + e_{\text{oxide}}^2}$

$$= \sqrt{(21._6\text{ μg})^2 + (3.8\text{ μg})^2 + (8.9\text{ μg})^2} = 23._6\text{ μg}$$

Relative error = $\dfrac{23._6 \times 10^{-6}\text{ g}}{999.698\,336\,5\text{ g}} \times 100 = 2.4 \times 10^{-6}\%$

$m'_{\text{sphere}} = 999.698\,453\,(24)\text{ g}$ $(2.4 \times 10^{-6}\%)$

(c) Yes, the method is fit for purpose, as the relative error achieved ($2.4 \times 10^{-6}\%$) is below the target error ($2 \times 10^{-5}\%$).

CHAPTER 4
STATISTICS

4-1. Standard deviation is an inverse measure of precision. The larger the standard deviation of a measurement, the poorer the precision. There is no relationship between standard deviation and accuracy.

4-2. (a) $\mu \pm \sigma$ corresponds to $z = -1$ to $z = +1$. The area from $z = 0$ to $z = +1$ is 0.341 3. The area from $z = 0$ to $z = -1$ is also 0.341 3.
Total area (= fraction of population) from $z = -1$ to $z = +1 = 0.682\,6$.

(b) $z = -2$ to $z = +2 \Rightarrow$ area $= 2 \times 0.477\,3 = 0.954\,6$

(c) $z = 0$ to $z = +1 \Rightarrow$ area $= 0.341\,3$

(d) $z = 0$ to $z = 0.5 \Rightarrow$ area $= 0.191\,5$

(e) Area from $z = -1$ to $z = 0$ is 0.341 3. Area from $z = -0.5$ to $z = 0$ is 0.191 5. Area from $z = -1$ to $z = -0.5$ is $0.341\,3 - 0.191\,5 = 0.149\,8$.

4-3. (a) Mean $= \frac{1}{8}(1.526\,60 + 1.529\,74 + 1.525\,92 + 1.527\,31 + 1.528\,94 + 1.528\,04 + 1.526\,85 + 1.527\,93) = 1.527\,67$

(b) Standard deviation =
$$\sqrt{\frac{(1.526\,60 - 1.527\,67)^2 + \ldots + (1.527\,93 - 1.527\,67)^2}{8-1}} = 0.001\,26$$

If your answer is 0.001 18, you used the population standard deviation function $\sqrt{\left[\sum(x_i - \bar{x})^2\right]/n}$ on your calculator, rather than sample standard deviation. Try again.

(c) Variance $= (0.001\,26)^2 = 1.59 \times 10^{-6}$

(d) Standard deviation of the mean $= (0.001\,26)/\sqrt{8} = 0.000\,45$

(e) Significant figures: $\bar{x} \pm s = 1.527_7 \pm 0.001_3$ or 1.528 ± 0.001.

4-4. (a) Hare after 1 hour: Standard deviation of the mean $= (1.5\%)/\sqrt{10} = 0.47\%$
Tortoise after 1 hour: Standard deviation of the mean $= (0.5\%)/\sqrt{2} = 0.35\%$

(b) Hare after 2 hours: Standard deviation of the mean $= (1.5\%)/\sqrt{20} = 0.34\%$
Tortoise after 2 hours: Standard deviation of the mean $= (0.5\%)/\sqrt{4} = 0.25\%$

(c) Moral: Care and exact technique yield more precise results than speed, if speed comes at cost of precision. But precision improves with replicates, so be efficient so that you can get a good number of analyses done in a period.

4-5. (a) Mean of 16 means at left side of table = 0.890 2_0 g.
Mean of 16 means at right side of table = 0.896 4_9 g.

(b) Standard deviation of 16 means at left side of table = 0.027 8_5 g.
Standard deviation of 16 means at right side of table = 0.011 9_5 g.

(c) The standard deviation of the mean for sets of 4 candies is theoretically $\sigma/\sqrt{4}$, where σ is the population standard deviation for all candies. The standard deviation of the mean for sets of 16 candies is theoretically $\sigma/\sqrt{16}$. The quotient is theoretically $(\sigma/\sqrt{16})/(\sigma/\sqrt{4}) = \sqrt{4}/\sqrt{16} = 0.5$. The observed quotient of standard deviations in (b) is (0.011 9_5 g)/(0.027 8_5) = 0.429. If we measured many more sets of 4 and sets of 16 candies, we expect the quotient to approach 0.5.

4-6. (a) The area under the curve from $-\infty$ to 31 heads in Excel is NORM.DIST(31,25,3.54,TRUE) = 0.955. The area above 31 heads is $1 - 0.955 = 0.045$.

(b) The area under the curve from $-\infty$ to 18 heads in Excel is NORM.DIST(18,25,3.54,TRUE) = 0.024 0. The area under the curve from $-\infty$ to 22 heads in Excel is NORM.DIST(22,25,3.54,TRUE) = 0.198 4. The area from 18 to 22 heads is 0.198 4 – 0.024 0 = 0.174.

4-7.

	A	B	C	D	E	F	G
1	Gaussian curve for coin tosses (Fig 4-1b)						
2							
3	mean =	x (heads)	Freq of result	Formula for cell C4 = ($A$8/($A$6*$A$10))			
4	24.98	10	0.01	*EXP(-((B4-$A$4)^2/(2*$A$6^2)))			
5	std dev =	11	0.04				
6	3.71	12	0.09				
7	total tosses =	13	0.23				
8	400	14	0.54				
9	sqrt(2*pi()) =	15	1.15				
10	2.50662827	16	2.30				
11		17	4.26				
12		18	7.33				
13		19	11.73				
14		20	17.47				
15		21	24.19				
16		22	31.15				
33		39	0.03				
34		40	0.01				

4-8. Use the same spreadsheet as in the previous problem but vary the standard deviation. Here are the results:

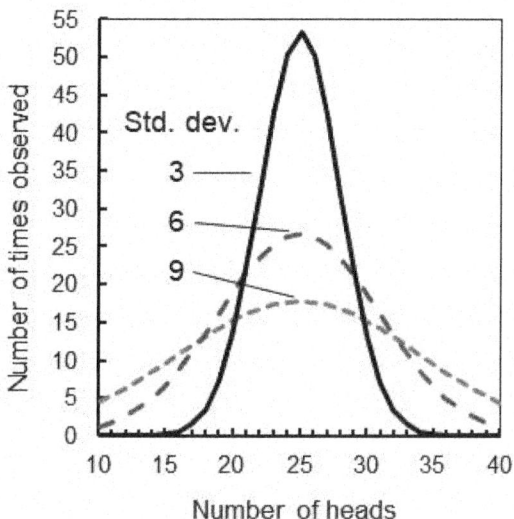

4-9. A confidence interval is a region around the measured mean in which the true mean is likely to lie: If we were to repeat a set of *n* measurements many times and compute the mean and standard deviation for each set, the 95% confidence interval would include the true population mean (whose value we do not know) in 95% of the sets of *n* measurements.

4-10. Bars are drawn at a 50% confidence level, so 50% of them ought to include the mean value if many experiments are performed. 90% of the 90% confidence bars should reach the mean value if we do enough experiments. The 90% bars must be longer than the 50% bars because more of the 90% bars must reach the mean.

4-11. Case 1: Comparing a measured result to a "known" value. See if the known value is included within the 95% confidence interval computed as in Equation 4-8.

Case 2: Comparing replicate measurements. Use the *F* test (Equation 4-6) to decide if the two standard deviations are significantly different.
Case 2a: If the two standard deviations are not significantly different, find the pooled standard deviation with Equation 4-10a and compute *t* with Equation 4-9a.
Case 2b: If two standard deviations are significantly different, find degrees of freedom with Equation 4-10b and compute *t* with Equation 4-9b.

Case 3: Comparing individual differences with Equations 4-11 and 4-12.

4-12. $\bar{x} = 0.14_8$, $s = 0.03_4$

90% confidence interval = $0.14_8 \pm \dfrac{(2.015)(0.03_4)}{\sqrt{6}} = 0.14_8 \pm 0.02_8$

99% confidence interval = $0.14_8 \pm \dfrac{(4.032)(0.03_4)}{\sqrt{6}} = 0.14_8 \pm 0.05_6$

4-13. 99% interval = $\bar{x} \pm \dfrac{(3.707)(0.000\,07)}{\sqrt{7}} = \bar{x} \pm 0.000\,10$ (1.527 83 to 1.528 03)

4-14. (a) dL = deciliter = 0.1 L = 100 mL

(b) $F_{\text{calculated}} = (0.05_3/0.04_2)^2 = 1.5_9 < F_{\text{table}} = 9.36$ (for $6-1 = 5$ degrees of freedom in numerator and $5-1 = 4$ degrees of freedom in denominator). Since $F_{\text{calculated}} < F_{\text{table}}$, we can use the following equations:

$s_{\text{pooled}} = \sqrt{\dfrac{0.53^2(5) + 0.42^2(4)}{6+5-2}} = 0.48_4$

$t = \dfrac{|14.5_7 - 13.9_5|}{0.48_4}\sqrt{\dfrac{6 \cdot 5}{6+5}} = 2.12 < 2.262$ (listed for 95% confidence and 9 degrees of freedom). The results agree and the trainee should be released.

4-15. (a) $F_{\text{calculated}} = 0.32^2/0.13^2 = 6.0_6 < F_{\text{table}} = 39.0$ (for 2 degrees of freedom in both numerator and denominator). Standard deviations are not significantly different at 95% confidence level.

(b) Because $F_{\text{calculated}} < F_{\text{table}}$, we use Equations 4-10a and 4-9a.

$s_{\text{pooled}} = \sqrt{\dfrac{0.32^2(3-1) + 0.13^2(3-1)}{3+3-2}} = 0.24_4$

$t_{\text{calculated}} = \dfrac{|4.70 - 4.84|}{0.24_4}\sqrt{\dfrac{3 \times 3}{3+3}} = 0.70_2 < 2.776$ (for $n_1 + n_2 - 2 = 4$ degrees of freedom). Difference is <u>not</u> significant at 95% confidence level.

(c) $F_{\text{calculated}} = 0.32^2/0.17^2 = 3.5_4 < F_{\text{table}} = 39.0$ (for 2 degrees of freedom in both numerator and denominator). Standard deviations are not significantly different at 95% confidence level.

Because $F_{\text{calculated}} < F_{\text{table}}$, we use Equations 4-10a and 4-9a.

$s_{\text{pooled}} = \sqrt{\dfrac{0.32^2(3-1) + 0.17^2(3-1)}{3+3-2}} = 0.25_6$

$t_{\text{calculated}} = \dfrac{|4.70 - 3.81|}{0.25_6}\sqrt{\dfrac{3 \times 3}{3+3}} = 4.2_5 > 2.776$ (for $n_1 + n_2 - 2 = 4$ degrees of freedom). Difference <u>is</u> significant at the 95% confidence level.

Statistics

4-16.

	A	B	C	D	E	F
1	Paired t test of two methods					
2						
3	Sample	Old Method	New Method	d_i		
4	Cat 1	84.9	86.2	1.3	= C4-B4	
5	Cat 2	73.5	81.8	8.3		
6	Cat 3	173.0	186.0	13.0		
7	Cat 4	62.7	73.4	10.7		
8	Cat 5	154.0	138.0	-16.0		
9	Dog 1	80.1	72.5	-7.6		
10	Dog 2	185.0	203.0	18.0		
11			mean =	3.96	=AVERAGE(D4:D10)	
12			stdev =	12.13	=STDEV.S(D4:D10)	
13			$t_{calculated}$ =	0.863	=D11/D12*SQRT(7)	
14			t_{table} =	2.447	=T.INV.2T(0.05,6)	

$t_{calculated} = 0.863 < 2.447$ (Student's t for 95% confidence and $7 - 1 = 6$ degrees of freedom). The difference is <u>not</u> significant.

4-17. In the following spreadsheet, we find $|t_{calculated}| = 0.86$ (labeled *t Stat* in cell F10) is less than $t_{table} = 2.45$ (*t Critical two-tail* in cell F14). Therefore, the difference between the methods is <u>not</u> significant.

The probability *P(T<=t) two-tail* in cell F13 is 0.42. There is a 42% chance of finding the observed difference between equivalent methods by random variations in results. The probability would have to be ≤0.05 for us to conclude that the methods differ.

	A	B	C	D	E	F	G
1	Paired t test ot two methods				t-Test: Paired Two Sample for Means		
2							
3	Sample	Old Method	New Method			Variable 1	Variable 2
4	Cat 1	84.9	86.2		Mean	116.17143	120.12857
5	Cat 2	73.5	81.8		Variance	2726.159	3099.729
6	Cat 3	173.0	186.0		Observations	7	7
7	Cat 4	62.7	73.4		Pearson Correlation	0.9767717	
8	Cat 5	154.0	138.0		Hypothesized Mean Difference	0	
9	Dog 1	80.1	72.5		df	6	
10	Dog 2	185.0	203.0		t Stat	-0.863413	
11					P(T<=t) one-tail	0.2105394	
12	Absolute value of calculated t Statistic				t Critical one-tail	1.9431803	
13	in cell F10 is less than critical t in cell				P(T<=t) two-tail	0.4210787	
14	F14. Therefore the difference between				t Critical two-tail	2.4469119	
15	methods is not significant.						

4-18. $F_{\text{calculated}} = s_B^2/s_A^2 = (0.039)^2/(0.025)^2 = 2.43$. Larger s always in numerator.

$F_{\text{table}} = 15.4$ for $4 - 1 = 3$ degrees of freedom in the numerator and denominator
Since $F_{\text{calculated}} < F_{\text{table}}$, the difference in standard deviation is not significant.

$$s_{\text{pooled}} = \sqrt{\frac{s_1^2(n_1-1) + s_2^2(n_2-1)}{n_1 + n_2 - 2}} = \sqrt{\frac{0.025^2(4-1) + 0.039^2(4-1)}{4 + 4 - 2}} = 0.032\,8$$

$$t_{\text{calculated}} = \frac{|\bar{x}_1 - \bar{x}_2|}{s_{\text{pooled}}}\sqrt{\frac{n_1 n_2}{n_1 + n_2}} = \frac{|1.382 - 1.346|}{0.032\,8}\sqrt{\frac{4 \cdot 4}{4 + 4}} = 1.55$$

t_{table} ($4 + 4 - 2 = 6$ degrees of freedom) $= 2.447$
Since $t_{\text{calculated}} < t_{\text{table}}$, the difference is <u>not</u> significant.

4-19. For Method A, $\bar{x}_A = 0.082\,605_2$, $s_A = 0.000\,013_4$.
For Method B, $\bar{x}_B = 0.082\,00_5$, $s_B = 0.000\,12_9$.

The two standard deviations differ by approximately a factor of 10. We should use the F test to compare the two standard deviations:

$F_{\text{calculated}} = s_B^2/s_A^2 = (0.000\,12_9)^2/(0.000\,013_4)^2 = 92.7$

F_{table} (5, 4) $= 9.36$. Since $F_{\text{calculated}} > F_{\text{table}}$, we use Equations for Case 2b in comparison of means. The following spreadsheet shows $t_{\text{calculated}} = 11.3$ (cell E9) and $t_{\text{table}} = 2.57$ (cell E13).

$t_{\text{calculated}} > t_{\text{table}}$, so the difference <u>is</u> significant at the 95% confidence level.

	A	B	C	D	E	F
1	Two sample t-test with			t-Test: Two-Sample Assuming Unequal Variances		
2	unequal variances					
3					Variable 1	Variable 2
4	Method A	Method B		Mean	0.082605	0.082005
5	0.082601	0.08183		Variance	1.8E-10	1.67E-08
6	0.082621	0.08186		Observations	5	6
7	0.082589	0.08205		Hypothesized Mean Difference	0	
8	0.082617	0.08206		df	5	
9	0.082598	0.08215		t Stat	11.31371	
10		0.08208		P(T<=t) one-tail	4.72E-05	
11				t Critical one-tail	2.015048	
12				P(T<=t) two-tail	9.43E-05	
13				t Critical two-tail	2.570582	

4-20. 90% confidence interval $= \bar{x} \pm \dfrac{(2.353)(1\%)}{\sqrt{4}} = \bar{x} \pm 1.1_8\% < 1.2\%$.

The answer is yes.

4-21. For indicators A and B: $F_{calculated} = (0.00225/0.00098)^2 = 5.2_7 > F_{table}$ (which is in the range 2.62 to 2.50 for 27 degrees of freedom in the numerator and 17 degrees of freedom in the denominator). Since $F_{calculated} > F_{table}$, we use the following equations:

$$\text{Degrees of freedom} = \frac{(s_1^2/n_1 + s_2^2/n_2)^2}{\frac{(s_1^2/n_1)^2}{n_1-1} + \frac{(s_2^2/n_2)^2}{n_2-1}}$$

$$= \frac{(0.00225^2/28 + 0.00098^2/18)^2}{\frac{(0.00225^2/28)^2}{28-1} + \frac{(0.00098^2/18)^2}{18-1}} = 39.8 \text{ (round to 40)}$$

$$t_{calculated} = \frac{|\bar{x}_1 - \bar{x}_2|}{\sqrt{s_1^2/n_1 + s_2^2/n_2}} = \frac{|0.09565 - 0.08686|}{\sqrt{0.00225^2/28 + 0.00098^2/18}} = 18.2$$

$t_{calculated}$ is much greater than t_{table} for ~40 degrees of freedom and 95% confidence, which is 2.02. The difference is significant.

For indicators B and C: $F_{calculated} = (0.00113/0.00098)^2 = 1.3_3 < F_{table}$ (which is in the range 2.50 to 2.62 for 28 degrees of freedom in the numerator and 17 degrees of freedom in the denominator). Since $F_{calculated} < F_{table}$, we use the following equations:

$$s_{pooled} = \sqrt{\frac{s_1^2(n_1-1) + s_2^2(n_2-1)}{n_1+n_2-2}} = 0.0010758$$

$$t_{calculated} = \frac{|\bar{x}_1 - \bar{x}_2|}{s_{pooled}}\sqrt{\frac{n_1 n_2}{n_1+n_2}} = 1.39 < 2.02 \text{ (~43 degrees of freedom)}$$

$\Rightarrow$ difference is not significant.

4-22. The standard deviations are almost identical, so we use the equations

$$s_{pooled} = \sqrt{\frac{30.0^2(31) + 29.8^2(31)}{32+32-2}} = 29.9$$

$$t = \frac{52.9-31.4}{29.9}\sqrt{\frac{32 \cdot 32}{32+32}} = 2.88. \text{ The table gives } t = 2.00 \text{ (95\% confidence) and}$$

$t = 2.66$ (99% confidence) for 60 degrees of freedom, which is close to 62. The difference is significant at the 95 and 99% levels.

4-23. $\bar{x} = 97.0_0$, $s = 1.6_6$

95% confidence interval $= \bar{x} \pm \dfrac{ts}{\sqrt{n}} = 97.0_0 \pm \dfrac{(2.776)(1.6_6)}{\sqrt{5}} = 97.0_0 \pm 2.0_6$

Range = 94.9_4 to 99.0_6

The 95% confidence interval does not include the certified value of 94.6 ppm, so the difference <u>is</u> significant at the 95% confidence level.

If we make one more measurement, the results are $\bar{x} = 96.5_8$, $s = 1.8_0$

95% confidence interval $= 96.5_8 \pm \dfrac{(2.571)(1.8_0)}{\sqrt{6}} = 96.5_8 \pm 1.8_9$

Range = 94.6_9 to 98.4_7

The 95% confidence interval still does not include the certified value of 94.6 ppm, so the difference <u>is</u> still significant at the 95% confidence level.

4-24. $\bar{x} = 201.8$; $s = 9.34$

$G_{calculated} = |216 - 201.8| / 9.34 = 1.52$

$G_{table} = 1.672$ for five measurements

Because $G_{calculated} < G_{table}$, we should retain 216.

4-25. $\bar{x} = 32.58_8$; $s = 0.10_2$

$G_{calculated} = |32.77 - 32.58_8|/0.10_2 = 1.7_8$

$G_{table} = 1.672$ for five measurements

Because $G_{calculated} > G_{table}$, we should reject 32.77 and recalculate $\bar{x}$ and s.

For the remaining 4 readings, $\bar{x} = 32.54_2$; $s = 0.01_3$

4-26. Statement (i) is true. The null hypothesis for the F test is that the two sets of data are drawn from populations with the same population standard deviation ($\sigma_1 = \sigma_2$). There must be strong evidence to *reject* the null hypothesis, or we do not reject it. We retain the null hypothesis if there is more than a 5% chance that $\sigma_1 = \sigma_2$. The statement that "there is at least a 95% probability that the two sets of data are drawn from populations with $\sigma_1 = \sigma_2$" is not being tested and is a much stronger statement than the one that is being tested.

4-27. Slope $= -1.298\,72 \times 10^4\ (\pm 0.001\,319\,0 \times 10^4)$
$= -1.299\ (\pm 0.001) \times 10^4$ or $-1.298_7\ (\pm 0.001_3) \times 10^4$

Intercept $= 256.695\ (\pm 323.57) = 3\ (\pm 3) \times 10^2$

4-28.

x_i	y_i	$x_i y_i$	x_i^2	d_i	d_i^2
0	1	0	0	0.07143	0.00510
2	2	4	4	−0.21429	0.04592
3	3	9	9	0.14286	0.02041
sums: 5	6	13	13	0	0.07143

$$m = \frac{n\Sigma(x_i y_i) - \Sigma x_i \Sigma y_i}{n\Sigma(x_i^2) - (\Sigma x_i)^2} = \frac{3 \times 13 - 5 \times 6}{3 \times 13 - 5^2} = \frac{9}{14} = 0.64286$$

$$b = \frac{\Sigma(x_i^2)\Sigma y_i - \Sigma(x_i y_i)\Sigma x_i}{n\Sigma(x_i^2) - (\Sigma x_i)^2} = \frac{13 \times 6 - 13 \times 5}{3 \times 13 - 5^2} = \frac{13}{14} = 0.92857$$

$$s_y = \sqrt{\frac{\Sigma(d_i^2)}{n-2}} = \sqrt{\frac{0.07143}{3-2}} = 0.26726$$

$$u_m = s_y \sqrt{\frac{n}{D}} = (0.26726)\sqrt{\frac{3}{14}} = 0.12371$$

$$u_b = s_y \sqrt{\frac{\Sigma(x_i^2)}{D}} = (0.26726)\sqrt{\frac{13}{14}}$$

$$= 0.25754$$

slope = $0.6_4 \pm 0.1_2$ intercept = $0.9_3 \pm 0.2_6$

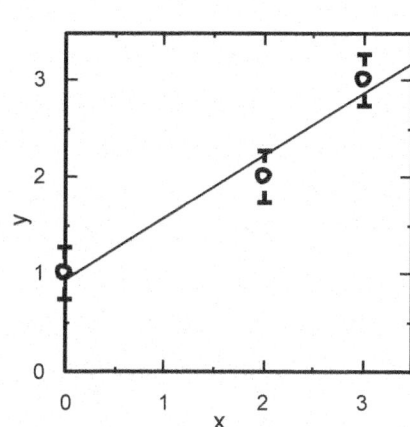

4-29.

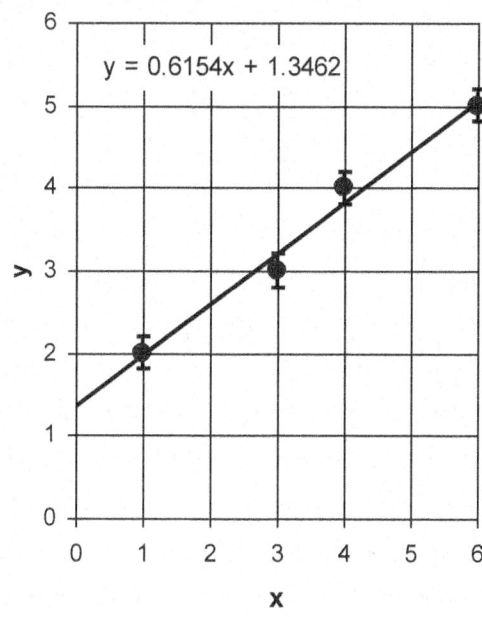

y = 0.6154x + 1.3462

4-30.

	A	B	C	D
1	x	y		
2	3.0	-0.074		
3	10.0	-1.411		
4	20.0	-2.584		
5	30.0	-3.750		
6	40.0	-5.407		
7				
8		LINEST output		
9	m	-0.13789	0.195343	b
10	u_m	0.006635	0.162763	u_b
11	R^2	0.993102	0.197625	s_y
12	Highlight cell B9:C11			
13	Type			
14	=LINEST(B2:B6,A2:A6,TRUE,TRUE)"			
15	Press CTRL+SHIFT+ENTER on PC			
16	Press CTRL+SHIFT+RETURN on Mac			

4-31. We must measure how an analytical procedure responds to a known quantity of analyte (or a known quantity of a related compound) before the procedure can be used for an unknown. Therefore, we must be able to measure out the analyte (or a related compound) in pure form to use as a calibration standard.

4-32. Hopefully, the negative value is within experimental error of 0. If so, no detectable analyte is present. If the negative concentration is beyond experimental error, there is something wrong with your analysis. The same is true for a value above 100% of the theoretical maximum concentration of an analyte. Another possible way to get values below 0 or above 100% is if you extrapolated the calibration curve past the range covered by standards, and the curve is not linear.

4-33. For 8 degrees of freedom, $t_{90\%}$ = 1.860 and $t_{99\%}$ = 3.355.

90% confidence interval: 15.2_2 ($\pm 1.860 \times 0.4_6$) = $15.2_2 \pm 0.8_6$ µg

99% confidence interval: 15.2_2 ($\pm 3.355 \times 0.4_6$) = $15._2 \pm 1._5$ µg

4-34. (a) $x = \dfrac{y-b}{m} = \dfrac{2.58 - 1.3_5}{0.61_5} = 2.00$

$\bar{y} = (2+3+4+5)/4 = 3.5 \qquad \bar{x} = (1+3+4+6)/4 = 3.5$

$\Sigma(x_i - \bar{x})^2 = (1-3.5)^2 + (3-3.5)^2 + (4-3.5)^2 + (6-3.5)^2 = 13.0$

$u_x = \dfrac{s_y}{|m|}\sqrt{\dfrac{1}{k} + \dfrac{1}{n} + \dfrac{(y-\bar{y})^2}{m^2 \Sigma(x_i-\bar{x})^2}} = \dfrac{0.196\,12}{|0.615\,38|}\sqrt{\dfrac{1}{1} + \dfrac{1}{4} + \dfrac{(2.58-3.5)^2}{(0.615\,38)^2(13.0)}} = 0.38$

Answer: $2.0_0 \pm 0.3_8$

(b) For $k = 4$ replicate measurements,

$$u_x = \frac{0.196\,12}{|0.615\,38|}\sqrt{\frac{1}{4}+\frac{1}{4}+\frac{(2.58-3.5)^2}{(0.615\,38)^2(13.0)}} = 0.26$$

Answer: $2.0_0 \pm 0.2_6$

(c) There are only 4 calibration points, so there are $4 - 2 = 2$ degrees of freedom. Student's t for 95% confidence and 2 degrees of freedom is 4.303. For (a), 95% confidence interval = $\pm(4.303)(0.38) = \pm 1.6$. For (b), 95% confidence interval = $\pm(4.303)(0.26) = \pm 1.1$. You really need more than 4 points in a calibration curve to cut down the breadth of the confidence interval.

4-35. (a) Corrected absorbance = $0.264 - 0.095 = 0.169$

Equation of line: $0.169 = 0.016\,30x + 0.004\,7 \Rightarrow x = 10.1$ µg

(b) In the spreadsheet below, $x = 10.082$ µg in cell B26 and $u_x = 0.204\,5$ µg in cell B27. For $14 - 2 = 12$ degrees of freedom and 95% confidence, Student's $t = 2.179$ in cell B28. The 95% confidence interval is ± 0.4455 µg in cell B29. A reasonable answer is 10.1 ± 0.4 µg or $10.0_8 \pm 0.4_5$ µg.

	A	B	C	D
1	Least-squares spreadsheet			
2		x	y	
3	Highlight cells B18:C20	0	-0.0003	
4	Type "= LINEST(C3:C16,	0	-0.0003	
5	B3:B16,TRUE,TRUE)	0	0.0007	
6	For PC, press	5	0.0857	
7	CTRL+SHIFT+ENTER	5	0.0877	
14	For Mac, press	20	0.3257	
15	CTRL+SHIFT+RETURN	20	0.3257	
16		20	0.3307	
17	LINEST output:			
18	m	0.0163	0.0047	b
19	u_m	0.0002	0.0026	u_b
20	R^2	0.9978	0.0059	s_y
21	$n =$	14	B21 = COUNT(B3:B16)	
22	Mean $y =$	0.1618	B22 = AVERAGE(C3:C16)	
23	$\Sigma(x_i - \text{mean } x)^2 =$	723.21	B23 = DEVSQ(B3:B16)	
24	Measured $y =$	0.169	Input	
25	$k =$ Number of replicate measurements of $y =$	4	Input	
26	Derived $x =$	10.082	B26 = (B24-C18)/B18	
27	$u_x =$	0.2045	B27=(C20/ABS(B18))*SQRT((1/B25)+(1/B21)+((B24-B22)^2)/(B18^2*B23))	
28	$t =$	2.1788	B28 = T.INV.2T(0.05,12) (t for 95% confidence, 12 degrees of freedom)	
29	Confidence interval = $\pm$	0.4455	B29 = B28*B27 = $t*u_x$	

Graph: $y = 0.0163x + 0.0047$, $R^2 = 0.9978$

4-36. (a)

	A	B	C	D	E	F	G	H
1	Least-Squares Spreadsheet for Methane							
2								
3		x	$y_{corrected}$	y				
4		0	0	9.1				
5		0.062	38.4	47.5				
6		0.122	86.5	95.6				
7		0.245	184.7	193.8				
8		0.486	378.4	387.5				
9		0.971	803.4	812.5				
10		1.921	1662.8	1671.9				
11	LINEST output:							
12	m	869.13	-22.0852	b				
13	u_m	10.6422	8.9474	u_b				
14	R^2	0.9993	18.0527	s_y				
15								
16	$n =$	7	B16 = COUNT(B4:B10)					
17	Mean $y =$	450.6	B17 = AVERAGE(C4:C10)					
18	$\Sigma(x_i - \text{mean } x)^2 =$	2.87757	B18 = DEVSQ(B4:B10)					
19								
20	Measured $y =$	145.0	Input					
21	No. of replicates of $y = k =$	4	Input					
22	Derived x	0.19224	B22 = (B20-C12)/B12					
23	$u_x =$	0.0137	B23 = (C14/B12)*SQRT((1/B21)+(1/B16)+((B20-B17)^2)/(B12^2*B18))					
24	$t =$	2.57058	B24=T.INV.2T(0.05, 5) (t for 95% confidence, 5 degrees of freedom)					
25	Conf. interval =	0.03525	B25 = B24*B23 = tu_x					

(b) Corrected signal = 154.0 – 9.0 = 145.0 mV

(c) Cell B22 gives [CH$_4$] = 0.0192$_2$ vol%

Cell B23 gives u_x = 0.013$_7$ vol%

Cell B25 gives confidence interval = $t^* u_x$ = 0.035$_2$ vol%

Rounded answers: $x \pm u_x$ = 0.19$_2$ (±0.01$_4$) vol%

95% confidence interval: 0.19$_2$ (±0.03$_5$) vol%

4-37. $0.350 = -1.17 \times 10^{-4} x^2 + 0.018\,58\,x - 0.000\,7$

$1.17 \times 10^{-4} x^2 - 0.018\,58\,x + 0.350\,7 = 0$

$$x = \frac{+0.018\,58 \pm \sqrt{0.018\,58^2 - 4(1.17 \times 10^{-4})(0.350\,7)}}{2(1.17 \times 10^{-4})} = 137 \text{ or } 21.9 \text{ μg}$$

Correct answer is 21.9 μg.

4-38. (a) The logarithmic graph spreads out the data and is linear over the entire range.

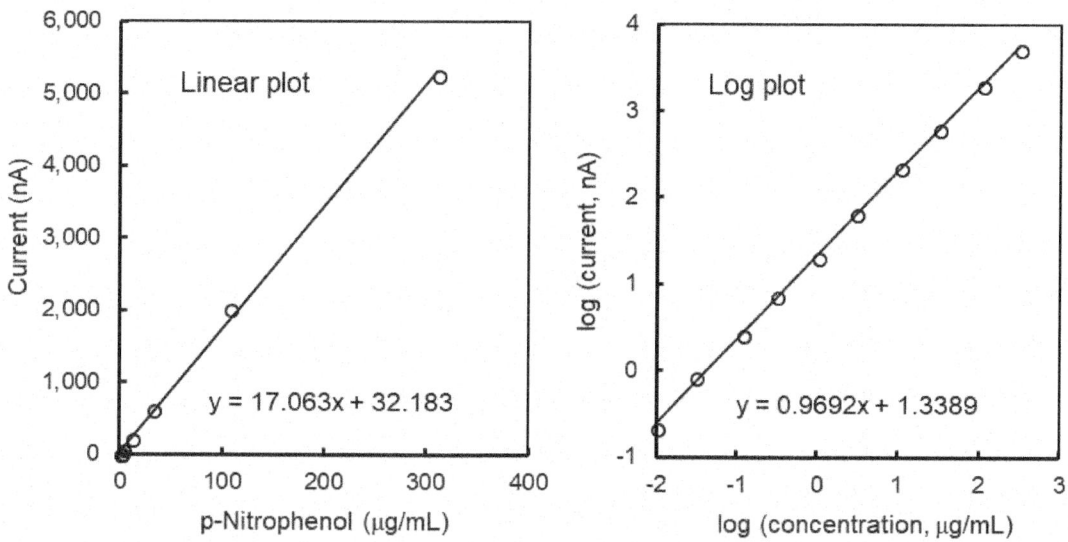

(b) log (current, nA) = 0.969 2 log (concentration, µg/mL) + 1.338 9

(c) log (99.9) = 0.969 2 log [X] + 1.338 9
$\Rightarrow$ log [X] = 0.681 6 $\Rightarrow$ [X] = 4.80 µg/mL

(d) concentration = y = $10^{0.681\,6 \pm 0.044\,0}$

The function is of the form $y = 10^x$, for which Table 3-1 states
$e_y/y = (\ln 10)e_x = (2.302\,6)(0.044\,0) = 0.101\,3$

Substituting $y = 10^{0.681\,6} = 4.804$ for y gives
$e_y/4.804 = 0.101\,3 \Rightarrow e_y = 0.487$

Answer: concentration = 4.80 (±0.49) µg/mL

The uncertainty is an estimate of the standard deviation of the mean. If you need to find a confidence interval, multiply (±0.49) µg/mL by Student's t.

4-39. (a) and (b) In the spreadsheet, unsorted molar mass data is in column B and sorted molar mass data is in column C. Mean, standard deviation, and number of data are calculated in cells B102:C104 and should be unchanged by sorting.

The suspect data are 60.4 and 1 232 g/mol. The 1 232 g/mol is left off the graph to allow closer inspection of the rest of the data. The scatter plot shows results are clustered around molar masses of about 105, 135, and 150 g/mol.

	A	B	C	D	E	F	G	H	I
1	Molar mass data for diprotic acids from University of Sydney								
2									
3		Original molar mass data	Sorted molar mass data						
4		127.9	60.4						
5		149.4	100.0						
6		134.0	102.8						
7		150.3	102.8						
8		134.7	103.2						
9		102.8	103.5						
10		151.5	103.9						
11		133.8	104.1						
12		104.3	104.1						
100		134.6	160.7						
101		104.9	1232						
102	Average =	140.8591837	140.8591837	B102 = AVERAGE(B4:B101)					
103	Std dev =	113.1126515	113.1126515	B103 = STDEV.S(B4:B101)					
104	n =	98	98	B104 = COUNT(B4:B101)					

(c) Data for the low molar mass diprotic acid are the 32 analyses from 60.4 to 113.5 g/mol. The 60.4 looks like an outlier. Let's check using the Grubbs test. The mean and standard deviation of these data are 103.9_8 and 8.3_3. So

$$G_{calculated} = \frac{|\text{questionable value} - \bar{x}|}{s} = \frac{|60.4 - 103.9_8|}{8.3_3} = 5.2_3$$

$G_{calculated}$ is larger than the critical G values of 2.745 for 30 observations in Table 4-6, and even the 2.956 for 50 observations. Therefore 60.4 is a clearly an outlier and is rejected. The mean and standard deviation for the remaining 31 points are 105.3_8 and 2.5_2. The next furthest datum from the mean is 113.5. However, the Grubbs test cannot be used to reject more than one outlier. (Statistics are no substitute for careful technique and record keeping.) For $n = 31$, the 95% confidence interval is

$$\text{95\% confidence interval} = \bar{x} \pm \frac{ts}{\sqrt{n}} = 105.3_8 \pm \frac{(2.042)(2.5_2)}{\sqrt{31}}$$
$$= 105.3_8 \pm 0.9_2$$

(d) The 95% confidence interval of 104.4_6 to 106.3_0 g/mol does not include the molecular mass of malonic acid. There is a systematic bias in the analysis.

CHAPTER 5
QUALITY ASSURANCE AND CALIBRATION METHODS

5-1. Get the right data: Measure what is relevant to the question at hand.
Get the data right: Sampling and analytical procedures must be satisfactory to measure what we intend to measure.
Keep the data right: Record keeping should document that samples were collected properly and data has demonstrated reliability.

5-2. The three parts of quality assurance are defining use objectives, setting specifications, and assessing results.
Use objectives:
Question: Why do I want the data and results and how will I use them?
Actions: Write use objectives.
Specifications:
Question: How good do the numbers have to be?
Actions: Write specifications and pick an analytical method to meet the specifications. Consider requirements for sampling, precision, accuracy, selectivity, sensitivity, detection limit, robustness, and allowed rate of false results. Plan to employ blanks, fortification, calibration checks, quality control samples, and control charts. Write and follow standard operating procedures.
Assessment:
Question: Did I meet the specifications?
Actions: Compare data and results with specifications. Document procedures and keep records suitable for meeting use objectives. Verify that the use objectives were met and that the method is fit for purpose.

5-3. *Precision* is demonstrated by the repeatability of analyses of replicate samples and replicate portions of the same sample. Accuracy is demonstrated by analyzing certified reference materials, by comparing results from different analytical methods, by fortification (spike) recovery, by standard additions, by calibration checks, blanks, and quality control samples (blind samples).

5-4. *Raw data* are directly measured quantities, such as peak area in a chromatogram or volume from a buret. *Treated data* are concentrations or amounts found by applying a calibration method to the raw data. *Results*, such as the mean and standard deviation, are what we ultimately report after applying statistics to treated data.

5-5. Methods: (a) Plot calibration data with least-squares line and inspect how data deviate from line. (b) Graph residual $[y_i - (mx_i + b)]$ versus concentration x_i of standards. A pattern in the residual plot indicates curvature. (c) R^2 should be very close to 1. (d) y-Intercept should be close to 0. (e) Is spike recovery acceptable, particularly at low concentrations?
Rank: Spike recovery > residual plot > inspecting graph > non-zero intercept > R^2

5-6. A *calibration check* is an analysis of a solution *formulated by the analyst* to contain a known concentration of analyte. It is the analyst's own check that procedures and instruments are functioning correctly. A *performance test sample* is an analysis of a solution *formulated by someone other than the analyst* to contain a known concentration of analyte. It is a test to see if the analyst gets the right answer when the analyst does not know the right answer.

5-7. A *blank* is a sample intended to contain no analyte. Its purpose is to find the response of a method when no analyte is deliberately present. Positive response to the blank arises from analyte impurities in reagents and equipment and from interference by other species. A *method blank* is taken through all steps in a chemical analysis. A *reagent blank* is the same as a method blank, but it has not been subjected to all sample preparation procedures. A *field blank* is similar to a method blank, but it has been taken into the field and exposed to the same environment as samples collected in the field and transported to the lab.

5-8. *Linear range* is the analyte concentration interval over which the analytical signal is proportional to analyte concentration. *Dynamic range* is the concentration range over which there is a useable response to analyte, even if it is not linear. *Range* is the analyte concentration interval over which an analytical method has specified linearity, accuracy, and precision.

5-9. A *false positive* is a conclusion that analyte exceeds a certain limit when, in fact, it is below the limit. A *false negative* is a conclusion that analyte is below a certain limit when, in fact, it is above the limit.

5-10. ~1% of the area under the curve for blanks lies to the right of the detection limit. Therefore, ~1% of samples containing no analyte will give a signal above the detection limit. 50% of the area under the curve for samples containing analyte at the detection limit lies below the detection limit. Therefore, 50% of samples

Quality Assurance and Calibration Methods 43

containing analyte at the detection limit will be reported as not containing analyte at a level above the detection limit.

5-11. A control chart tracks the performance of a process to see if it remains within expected bounds. Indications that a process might be out of control are (1) a reading outside the action lines; (2) 2 out of 3 consecutive readings between the warning and action lines; (3) 8 consecutive measurements all above or all below the center line; (4) 6 consecutive measurements, all steadily increasing or all steadily decreasing, wherever they are located; (5) 14 consecutive points alternating up and down, regardless of where they are located; and (6) an obvious nonrandom pattern.

5-12. Question (iii) is correct. The purpose of the analysis is to see if concentrations of haloacetates are in compliance with (i.e., do not exceed) levels set by a certain rule. The purpose is not just to achieve a specified precision and accuracy (question i) or to just see if detectable levels of haloacetates are present (question ii).

5-13. (a) Metals and trace metals in water and waste (sometimes called biosolids).

(b) E2866. D7597 is not correct. D7697 is for analysis of same analyte using the same equipment, but in water samples.

(c) 996.06 fat (Total, saturated, and unsaturated) in foods

(d) Water (moisture) in pharmaceuticals by Karl Fischer titration.

5-14. The *instrument detection limit* is obtained by replicate measurements of aliquots from one sample. The *method detection limit* is obtained by preparing and analyzing many independent samples. There is more variability in the latter procedure, so the method detection limit should be higher than the instrument detection limit.

Robustness is the ability of an analytical method to be unaffected by small, deliberate changes in operating parameters. *Intermediate precision* is the variation observed when an assay is performed by different people on different instruments on different days in the same lab. Each analysis might incorporate independently prepared reagents and different lots of the same chromatography column from one manufacturer. When demonstrating intermediate precision, experimental conditions are intended to be the same in each analysis. When measuring robustness, conditions are intentionally varied by small amounts.

5-15. *Repeatability* describes the spread in results when one person uses one procedure to analyze the same sample by the same method multiple times. *Reproducibility* describes the spread in results when different people in different labs using different instruments each try to follow the same procedure.

Instrument precision is the reproducibility observed when the same quantity of one sample is repeatedly introduced into an instrument.

Intra-assay precision is evaluated by analyzing aliquots of a homogeneous material several times by one person on one day with the same equipment.

Intermediate precision is the variation observed when an assay is performed by different people on different instruments on different days in the same lab.

Interlaboratory precision is the reproducibility observed when aliquots of the same sample are analyzed by different people in different laboratories at different times using equipment and reagents belonging to each lab.

5-16. Criteria:
- Observations outside action lines — no
- 2 out of 3 consecutive measurements between warning and action lines — no
- 8 consecutive measurements all above or all below the center line — YES: days 19–29 are all above the center line
- 6 consecutive measurements steadily increasing or steadily decreasing — no
- 14 consecutive points alternating up and down — no
- Obvious nonrandom pattern — no

5-17. (a) pg/µL = ng/mL = 10^{-9} g/g = parts per billion

(b) $y = 109.55x + 616.56$ with $R^2 = 0.9968$

(c) Least-squares trendline passes near all data, but there is a pattern to the residuals. The y-intercept is > 2% of highest concentration standard. Both characteristics suggest slight curvature to the calibration data.

(d) $y = mx + b \Rightarrow x = (y - b)/m = (540 - 616.56)/109.55 = -0.70$ pg/µL (?!) This impossible concentration is significantly different from the 4 pg/µL standard that yielded a signal of 540. Curvature in a calibration curve causes greatest relative error at low concentrations.

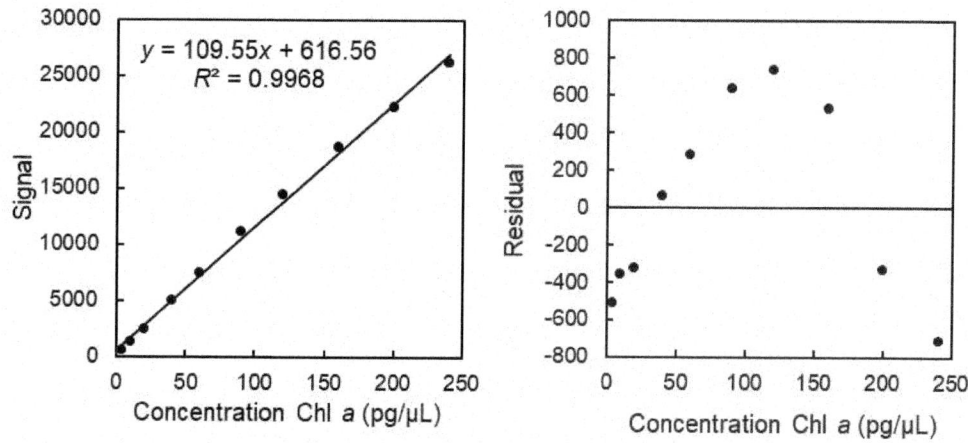

(e) The residuals plot shows a distinct upside-down U pattern, indicating that there is curvature in the calibration data.

(f)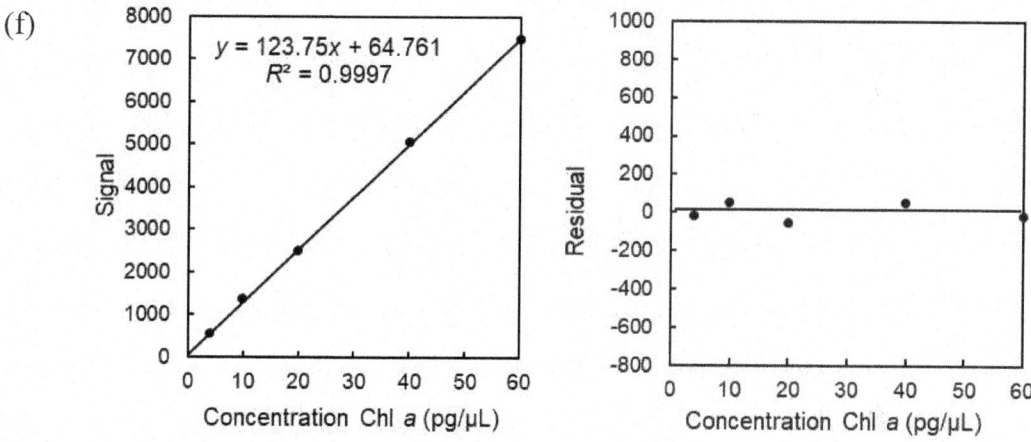

$y = 123.75x + 64.761$ with $R^2 = 0.9997$. The y-intercept is <1% of the highest concentration standard. Both the calibration curve and residuals plot show residuals that are small and randomly distributed.

$y = mx + b \Rightarrow x = (y - b)/m = (540 - 64.76)/123.75 = 3.84$ pg/µL

This value is in good agreement with the 4 pg/µL standard that yielded a signal of 540. The narrower concentration range provides a linear calibration.

(g) $540 = -0.095x^2 + 131.5x - 14.3 \Rightarrow 0 = -0.095x^2 + 131.5x - 554.3$

$$x = \frac{-b \pm \sqrt{b^2 - 4ac}}{2a} = 4.23, 1380$$

The calibration plot tells us that the 4.23 pg/µL, and is in good agreement with the signal observed for the 4 pg/µL calibration standard.

5-18. LINEST gives m, b, u_m, u_b, R^2, and s_y in cells C16:D18. TRENDLINE produces the same value of m, b, and R^2, which are printed inside the graph. The 95% confidence interval for y is computed in cell C21.

	A	B	C	D	E	F	G	H	I	J
1	Graphing y =26.4x+1.37 with random Gaussian noise (std dev = 80)									
2										
3	x	y								
4	0	14								
5	10	350								
6	20	566								
7	30	957								
8	40	1067								
9	50	1354								
10	60	1573								
11	70	1732								
12	80	2180								
13	90	2330								
14	100	2508								
15			LINEST output			To add error bars using Excel 2016:				
16		m	24.80727	89.72727	b	Double click anywhere in graph				
17		u_m	0.683532	40.43827	u_b	In Chart Tools ribbon, click Add Chart Element				
18		R^2	0.993214	71.6894	s_y	Select Error Bars, and More Error Bars Options				
19						For Error Amount, choose Custom, and Specify Value				
20		Student's t =	2.262157	=T.INV.2T(0.05,9)		For both Postive Value and Negative value enter C21				
21		t^*s_y =	162.1727	=C20*D18		Click on any x error bar and delete to remove these.				

Graph shows $y = 24.807x + 89.727$, $R^2 = 0.9932$.

(d) The 95% confidence range of signals for a spike of 20 is $529 \pm ts_y$
= $529 \pm 162.17 = 366._8$ to $691._2$.

Using the equation for the least-squares line determined in the spreadsheet, signals of $366._8$ and $691._2$ correspond to

$$x = \frac{y - 89.727}{24.807} = \frac{366._8 - 89.727}{24.807} = 11.1_7 \qquad x = \frac{691._2 - 89.727}{24.807} = 24.2_5$$

The % recovery for the low end of the 95% confidence interval is

$$\% \text{ recovery} = \frac{C_{\text{spiked blank}} - C_{\text{unspiked blank}}}{C_{\text{added}}} \times 100 = \frac{11.1_7 - 0}{20} = 55.8\%$$

The % recovery for the high end of the 95% confidence interval is 121.2%. Spike recoveries do not meet the use objective of recoveries of $100 \pm 10\%$. The method is not fit for purpose, even though the R^2 is quite high. The noise in the data is too high.

Quality Assurance and Calibration Methods 47

5-19. (a) For the fortification level of 22.2 ng/mL, the mean of the 5 values is 23.6_6 ng/mL and the standard deviation is 5.6_3 ng/mL.

$$\text{Precision} = 100 \times \frac{5.63}{23.66} = 23.8\%.$$

$$\text{Error} = 100 \times \frac{23.66 - 22.2}{22.2} = 6.6\%$$

For the fortification level of 88.2 ng/mL, the mean of the 5 values is 82.4_8 ng/mL and the standard deviation is 11.4_9 ng/mL.

$$\text{Precision} = 100 \times \frac{11.49}{82.48} = 13.9\%.$$

$$\text{Error} = 100 \times \frac{82.48 - 88.2}{88.2} = -6.5\%$$

For the fortification level of 314 ng/mL, the mean of the 5 values is $302._8$ ng/mL and the standard deviation is 23.5_1 ng/mL.

$$\text{Precision} = 100 \times \frac{23.51}{302.8} = 7.8\%.$$

$$\text{Error} = 100 \times \frac{302.8 - 314}{314} = -3.6\%$$

(b) Standard deviation of 10 samples: $s = 28._2$; mean blank: $y_{\text{blank}} = 45._0$
Signal detection limit $= y_{\text{blank}} + 3s = 45._0 + (3)(28._2) = 129._6$

$$\text{Concentration detection limit} = \frac{3s}{m} = \frac{(3)(28._2)}{1.75 \times 10^9 \text{ M}^{-1}} = 4.8 \times 10^{-8} \text{ M}$$

$$\text{Lower limit of quantitation} = \frac{10s}{m} = \frac{(10)(28._2)}{1.75 \times 10^9 \text{ M}^{-1}} = 1.6 \times 10^{-7} \text{ M}$$

5-20. (a) 1 wt% $\Rightarrow C = 0.01$: RSD(%) $\approx 2^{(1-0.5\log 0.01)} = 2^2 = 4\%$
If $C = 10^{-6}$, RSD(%) $\approx 2^{(1-0.5\log 0.000001)} \approx 2^4 = 16\%$
For $C < 10^{-7}$, RSD(%) $\approx 22\%$. Similarly if $C = 10^{-11}$, RSD(%) $\approx 22\%$.

(b) If class RSD(%) is 50% of the value given by the Horwitz curve, it would be
$0.5 \times 2^{(1-0.5\log 0.1)} = 1.4\%$.

5-21. Mean = 0.38_3 µg/L and standard deviation = 0.021_4 µg/L

$$\% \text{ recovery} = \frac{0.38_3 \text{ µg/L}}{0.40 \text{ µg/L}} \times 100 = 96\%$$

The concentration detection limit is 3 times the standard deviation = 3(0.0214 µg/L) = 0.064 µg/L which is rounded to 0.06 µg/L.

5-22. For a concentration of 0.2 µg/L, the relative standard deviation of 14.4% corresponds to (0.144)(0.2 µg/L) = 0.028 8 µg/L. The detection limit is 3(0.028 8 µg/L) = 0.086 µg/L. Here are the results for the other concentrations:

Concentration (µg/L)	Relative standard deviation (%)	Concentration standard deviation (µg/L)	Detection limit (µg/L)
0.2	14.4	0.028 8	0.086
0.5	6.8	0.034 0	0.102
1.0	3.2	0.032 0	0.096
2.0	1.9	0.038 0	0.114
		mean detection limit:	0.10

5-23. If an athlete tests positive for drugs, the test should be repeated with a second sample that was drawn at the same time as the first sample and preserved in an appropriate manner. If there is a 1% chance of a false positive in each test, the chances of observing a false positive twice in a row are 1% of 1% or 0.01%. Instead of falsely accusing 1% of innocent athletes, we would be falsely accusing 0.01% of innocent athletes.

5-24. *Comparison of Lab C with Lab A:*

First, use the F test to see if the standard deviations are significantly different:

$F_{calculated} = s_C^2 / s_A^2 = 0.78^2/0.14^2 = 31.0 > F_{table} = 5.10$ (with 2 degrees of freedom for s_C and 12 degrees of freedom for s_A)

Standard deviations are not equivalent, so use the following t test:

$$\text{Degrees of freedom} = \frac{(s_1^2/n_1 + s_2^2/n_2)^2}{\frac{(s_1^2/n_1)^2}{n_1-1} + \frac{(s_2^2/n_2)^2}{n_2-1}} = \frac{(0.14^2/13 + 0.78^2/3)^2}{\frac{(0.14^2/13)^2}{13-1} + \frac{(0.78^2/3)^2}{3-1}} = 2.03 \approx 2$$

$$t_{calculated} = \frac{|\bar{x}_1 - \bar{x}_2|}{\sqrt{s_1^2/n_1 + s_2^2/n_2}} = \frac{|1.59 - 2.68|}{\sqrt{0.14^2/13 + 0.78^2/3}} = 2.4_1$$

For 2 degrees of freedom, $t_{table} = 4.303$ for 95% confidence. Since $t_{calculated} < t_{table}$, we conclude that the difference between Lab C and Lab A is not significant.

Quality Assurance and Calibration Methods

Comparison of Lab C with Lab B:

$F_{\text{calculated}} = s_C^2 / s_B^2 = 0.78^2/0.56^2 = 1.9_4 < F_{\text{table}} = 6.54$ (with 2 degrees of freedom for s_C and 7 degrees of freedom for s_A). The standard deviations are not significantly different, so we use the following t test:

$$s_{\text{pooled}} = \sqrt{\frac{0.56^2(8-1) + 0.78^2(3-1)}{8+3-2}} = 0.61_6$$

$$t_{\text{calculated}} = \frac{|1.65 - 2.68|}{0.61_6}\sqrt{\frac{8 \cdot 3}{8+3}} = 2.4_7$$

$t_{\text{table}} = 2.262$ for 95% confidence and $8 + 3 - 2 = 9$ degrees of freedom. $t_{\text{calculated}} > t_{\text{table}}$, so the difference is significant at the 95% confidence level.

It makes no sense to conclude that Lab C [2.68 ± 0.78 (3)] > Lab B [1.65 ± 0.56 (8)], but Lab C = Lab A [1.59 ± 0.14 (13)]. The problem with the comparison of Labs C and A is that the standard deviation of C is much greater than the standard deviation of A and the number of replicates for C is much less than the number of replicates for A. The conclusion is biased by a large standard deviation and a small number of degrees of freedom. I would tentatively conclude that results from Lab C are greater than results from Labs B and A. I would also ask for more replicate results from Lab C. With just 3 replications, it is hard to reach any statistically significant conclusions.

5-25. A small volume of standard will not change the sample matrix very much, so matrix effects remain nearly constant. If large, variable volumes of standard are used, the matrix is different in every mixture and the matrix effects will be different in every sample.

5-26. (a) $[Cu^{2+}]_f = [Cu^{2+}]_i \dfrac{V_i}{V_f} = 0.950\,[Cu^{2+}]_i$

(b) $[S]_f = [S]_i \dfrac{V_i}{V_f} = (100.0 \text{ ppm})\left(\dfrac{1.00 \text{ mL}}{100.0 \text{ mL}}\right) = 1.00 \text{ ppm}$

(c) $\dfrac{[Cu^{2+}]_i}{1.00 \text{ ppm} + 0.950[Cu^{2+}]_i} = \dfrac{0.262}{0.500} \Rightarrow [Cu^{2+}]_i = 1.04 \text{ ppm}$

5-27. (a) All solutions were made up to the same final volume. So, we prepare a graph of signal versus concentration of added standard. The line in the graph was drawn by the method of least squares with the spreadsheet.

	A	B	C	D	E	F
1	Standard addition constant volume spreadsheet					
2						
3	$[Sr]_f$ (ng/mL)	Signal				
4	0	28				
5	2.5	34.3	Highlight cells B10:C12			
6	5	42.8	Type "=LINEST(B4:B8,A4:A8,TRUE,TRUE)"			
7	7.5	51.5	PC: CONTROL + SHIFT + ENTER			
8	10	58.6	Mac: CONTROL + SHIFT + RETURN			
9						
10	slope	3.136	27.36	y-intercept		
11	u_m	0.0945445	0.578965	u_b		
12	R^2	0.9972807	0.74744	s_y		
13	x-intercept= -b/m=	-8.724	B13=-C10/B10			
14						
15	n=	5	B15=COUNT(B4:B8)			
16	Mean y =	43.04	B16=AVERAGE(B4:B8)			
17	$\Sigma(x_i-\text{mean } x)^2=$	62.5	B17=DEVSQ(A4:A8)			
18	u_x =	0.427	B18=C12/ABS(B10)*			
19		SQRT(1/B15+B16^2/(B10^2*B17))				

R^2 in cell B12 is 0.997. Data points are randomly scattered about the least-squares line. This is a standard addition calibration, so we cannot use y-intercept to check linearity. The standard addition plot is linear.

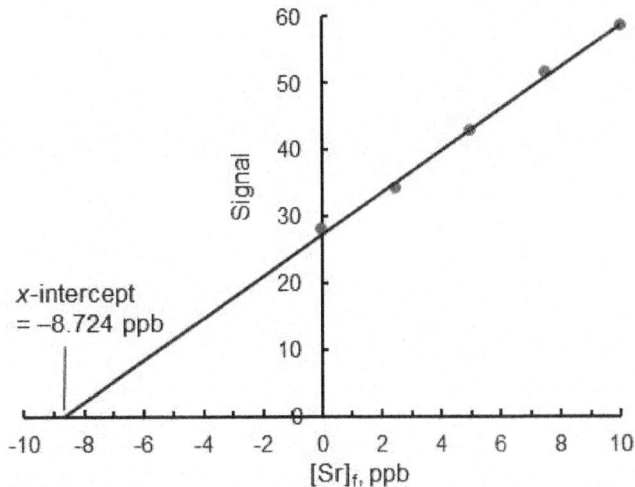

(b) The x-intercept, 8.72 ppb, is the concentration of unknown in the 10-mL solution. In cell B18 of the spreadsheet, the standard uncertainty of the x-intercept is 0.427 ppm. A reasonable answer is $8.7_2 \pm 0.4_3$ ppb.

(c) Unknown solution volume = 10.0 mL with Sr = 8.7_2 ppb = 8.7_2 ng/mL. In 10.0 mL, there are (10 mL)(8.7_2 ng/mL) = $87._2$ ng. Solution was made from 0.750 mg of tooth enamel. Sr (ppm) in tooth enamel is

$$\text{Concentration} = \frac{\text{mass of Sr}}{\text{mass of enamel}} \times 10^6 = \frac{87._2 \times 10^{-9} \text{ g}}{0.750 \times 10^{-3} \text{ g}} \times 10^6 = 116 \text{ ppm}$$

(d) In cell B18 we calculate the standard uncertainty in the intercept $u_x = 0.43$. Relative uncertainty = $100 \times 0.43/8.72 = 4.9\%$, which leads to a 4.9% uncertainty in the concentration of Sr in the tooth enamel. 0.049×116 ppm = 5.7 ppm. Final answer: 116 ± 6 ppm.

(e) Student's t for $n - 2 = 5 - 2 = 3$ degrees of freedom and 95% confidence is 3.182. We found standard deviation = 5.7 ppm. 95% confidence interval is $\pm tu_x = (3.182)(5.7 \text{ ppm}) = 18.1$ ppm. Answer: 116 ± 18 ppm.

5-28. (a) Visual inspection shows that data have only minor deviations from the least-squares lines. Deviations appear random. Standard addition always has a positive intercept due to the analyte present in the sample, so it is not possible to check linearity based on intercept. Both plots are linear.

(b) The intercept for tap water is –6.0 mL, corresponding to an addition of (6.0 mL)(0.152 ng/mL) = 0.91_2 ng Eu(III). This much Eu(III) is in 10.00 mL of tap water, so the concentration is 0.91_2 ng/10.00 mL = 0.091 ng/mL. For pond water, the intercept of –14.6 mL corresponds to an addition of (14.6 mL)(15.2 ng/mL) = 2.22×10^2 ng/10.00 mL pond water = 22.2 ng/mL.

(c) Added standard Eu(III) gives a response of 3.03 units/ng for tap water and 0.0822 units/ng for pond water. The relative response is 3.03/0.0822 = 36.9 times greater in tap water than in pond water. There is probably a *matrix effect* in which something in pond water decreases the Eu(III) emission. By using standard addition, we measure the response in the actual sample matrix. Even though Eu(III) in pond water and tap water do not give equal signals, we measure the actual signal in each matrix and can therefore carry out accurate analyses.

5-29. (a) and (b) In spreadsheet below the standard addition graph is a plot of peak area versus mg methamphetamine added using data in cells A5:B13. We use LINEST in cells B16:C18 to compute slope, intercept, and uncertainties.

	A	B	C	D	E	F	G
1	Taylor standard addition of evidence bag GCS-47B-PT130106-1						
2							
3	mg meth	Peak					
4	added	area					
5	0	138.29					
6	20	203.59					
7	40	264.88					
8	60	320.59					
9	80	373.68					
10	100	425.06					
11	120	467.89					
12	140	510.05					
13	160	539.66					
14							
15		LINEST(B5:B13,A5:A13,TRUE,TRUE)					
16	slope	2.529458	158.0533	intercept			
17	u_m	0.094992	9.045066	u_b			
18	R^2	0.990224	14.71613	s_y			
19	x-intercept=	−62.48505	B19=−C16/B16				
20	n=	9	B20=COUNT(B5:B13)				
21	Mean y =	360.41	B21=AVERAGE(B5:B13)				
22	$\Sigma(x_i - \text{mean } x)^2 =$	24000	B22=DEVSQ(A5:A13)				
23	$u_x =$	5.692	B23=C18/ABS(B16)*				
24		SQRT(1/B20+B21^2/(B16^2*B22))					

Graph: $y = 2.5295x + 158.05$, $R^2 = 0.9902$, x-intercept = −62.49 mg

(c) *x*-Intercept, 62.49 mg, is mass of methamphetamine in 100 mL of solution. In cell B23, standard uncertainty of *x*-intercept is 5.69 mg. Each flask contains 1/100 of original sample. Mass of methamphetamine in sample is *x*-intercept × 100 = 62.49 mg × 100 = 6.249 g. Standard uncertainty is 5.69 mg × 100 = 0.569 g. A reasonable answer would be 6.2 ± 0.6 g.

(d) Data in A5:B8 yield the graph below. LINEST slope = 3.041 ± 0.076 mg^{-1} and intercept = 140.6 ± 2.8 mg. *x*-Intercept = 46.24 ± 1.99 mg. A reasonable answer is 4.6 ± 0.2 g methamphetamine in the evidence bag.

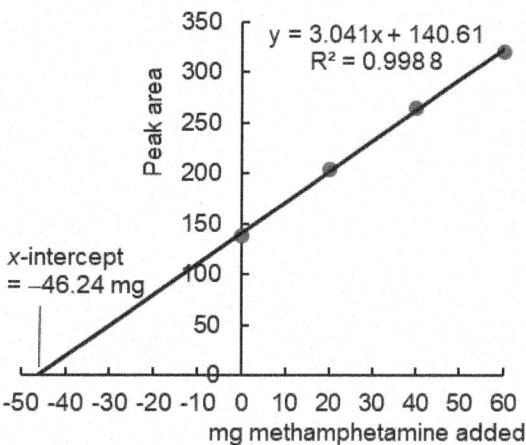

(e) Blaine used only smaller additions of methamphetamine to ensure standard addition plot was linear. Curvature is evident in Taylor's plot in part (a). Blaine's calibration is more linear, but linearity alone does not validate a method. Blaine demonstrated her procedure was fit for purpose by validating it using a certified reference material. I leave it to you to debate with your classmates and instructor whether Taylor made an honest procedural error or deliberately used a flawed procedure to bias results high.

(f) Standard operating procedures need to state the range of standard additions, and be validated using certified reference materials. Analysts need to show they can analyze a control sample with acceptable accuracy and precision. Control samples should be in each batch of evidence bags, and results monitored with control charts.

5-30. (a) In the spreadsheet, data appear in columns B and C. The abscissa and ordinate functions are in columns D and E. R^2 is 0.999 2. Data in the graph are randomly scattered closely about the line. As this is a standard addition plot, we cannot use the intercept to judge linearity. The data are linear.

(b) The negative x-intercept (0.025 4 ppm) is the concentration of Pb(II) in the initial sample volume V_i = 4.60 mL. The concentration of Pb(II) in the 1.00-mL extract is found from the dilution formula:
(x ppm)(1.00 mL) = (0.025 4 ppm)(4.60 mL) $\Rightarrow$ x = 0.116 9 ppm

(c) The standard uncertainty in the intercept in cell B18 is 0.000 98 ppm. With 4 points and 2 degrees of freedom, Student's t for 95% confidence is 4.303, so the 95% confidence interval is (4.303)(0.000 98 ppm) = 0.004 2 ppm. The relative uncertainty in the intercept is $100 \times (0.004\ 2\ \text{ppm}/0.025\ 4\ \text{ppm})$ = 16.6%. If other sources of uncertainty are negligible, the relative uncertainty in concentration in the 1.00-mL extract is also 16.6%, giving an absolute uncertainty of (0.166)(0.116 9 ppm) = 0.019 ppm. A reasonable expression of the answer is 0.117 ± 0.019 ppm or 0.12 ± 0.02 ppm.

	A	B	C	D	E
1	Lead standard addition: Add 2.50 ppm standard Pb to river sediment extract				
2					
3	V_i (mL) =	V_s = mL Pb	I_{x+s} =	x-axis function	y-axis function
4	4.60	std added	signal	S_i*V_s/V_i (ppm)	$I_{x+s}*(V_i+V_s)/V_i$
5	$[S]_i$ (ppm) =	0.000	1.10	0.000	1.100
6	2.50	0.025	1.66	0.014	1.669
7		0.050	2.20	0.027	2.224
8		0.075	2.81	0.041	2.856
9			D5=$A$6*B5/$A$4		E5=C5*($A$4+B5)/$A$4
10		Linest output			
11	slope	42.8524	1.088836957	y-intercept	
12	u_m	0.8720355	0.022166155	u_b	
13	R^2	0.99917246	0.026493623	s_y	
14	x-intercept =	-0.025409	B14=-C11/B11		
15	n =	4	B15=COUNT(B5:B8)		
16	Mean y =	1.9621875	B16=AVERAGE(E5:E8)		
17	$\Sigma(x_i-\text{mean } x)^2$=	0.001	B17=DEVSQ(D5:D8)		
18	u_x =	0.00098174	B18=C13/ABS(B11)*		
19			SQRT(1/B15+B16^2/(B11^2*B17))		

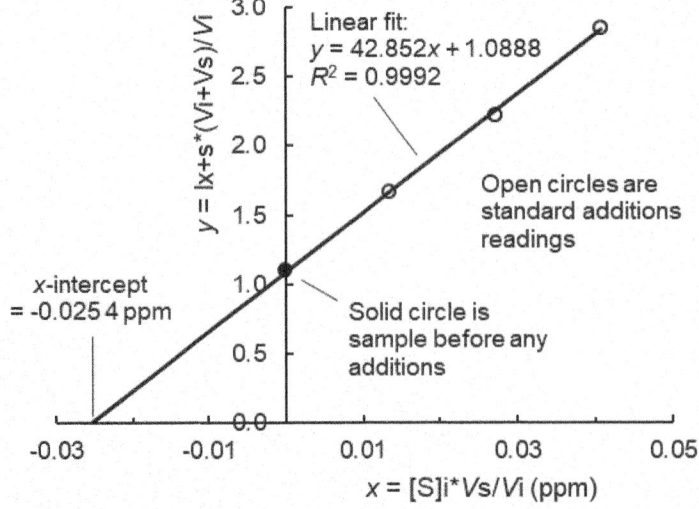

5-31. (a) Volume is not constant, so we follow procedure of Figure 5-6 and 5-7 to find μg Tl$^+$/L in the 5.00-mL sample. Data in the spreadsheet appear in columns B and C. Abscissa and ordinate functions are in columns D and E. $R^2 = 0.997$ and data are randomly scattered about line. As this is a standard addition plot, we cannot use the intercept to judge linearity. The calibration is linear.

(b) In cell B15 the negative x-intercept (0.279 μg/L) is the concentration of Tl$^+$ in the initial sample volume $V_i = 5.00$ mL.

(c) The standard uncertainty (0.016 9 μg/L) is in cell B19. The 95% confidence interval is $tu_x = (3.182)(0.016\ 9\ \mu g/L) = 0.054$ μg/L thallium, where Student's t is taken for 95% confidence and $5 - 2 = 3$ degrees of freedom. A reasonable answer is 0.28 ± 0.05 μg/L Tl$^+$.

	A	B	C	D	E
1	Thallium standard addition: Add 100 μg/L Tl(I) to 5.00 mL sample				
2					
3	V_i (mL) =	V_s = mL Tl	I_{x+s} =	x-axis function	y-axis function
4	5.00	added	signal (μA)	$S_i * V_s / V_i$	$I_{x+s}*(V_i+V_s)/V_i$
5	$[S]_i$ (μg/L) =	0.000	0.0300	0.000	0.0300
6	100	0.005	0.0418	0.100	0.0418
7		0.010	0.0539	0.200	0.0540
8		0.015	0.0621	0.300	0.0623
9		0.020	0.0744	0.400	0.0747
10			D5=$A$6*B5/$A$4		E5=C5*($A$4+B5)/$A$4
11		Linest output			
12	slope	0.1098397	0.03059876	y-intercept	
13	u_m	0.00370926	0.00090858	u_b	
14	R^2	0.99659047	0.001172972	s_y	
15	x-intercept =	-0.2785765	B15=-C12/B12		
16	n =	5	B16=COUNT(B5:B9)		
17	Mean y =	0.0525667	B17=AVERAGE(E5:E9)		
18	$\Sigma(x_i$-mean x$)^2$ =	0.100	B18=DEVSQ(D5:D9)		
19	u_x =	0.01685228	B19=C14/ABS(B12)*		
20			SQRT(1/B16+B17^2/(B12^2*B18))		
21	95% interval =	0.05363149	B21=B19*T.INV.2T(0.05,B16-2)		
22			3.182446305		

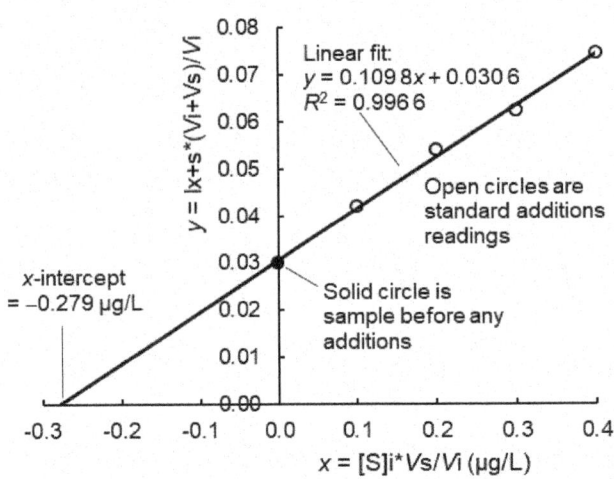

5-32. Standard addition is appropriate when the sample matrix is unknown or complex and hard to duplicate, and unknown matrix effects are anticipated. An internal standard can be added to an unknown at the start of a procedure in which uncontrolled losses of sample will occur or when instrument conditions vary from run to run. Variations affect the analyte and standard equally, so their relative signals remain constant. An internal standard that is chemically similar to the analyte will also correct for unknown matrix effects.

5-33. (a) $\dfrac{A_X}{[X]_f} = F\left(\dfrac{A_S}{[S]_f}\right) \Rightarrow \dfrac{3\,473}{[3.47 \text{ mM}]} = F\left(\dfrac{10\,222}{[1.72 \text{ mM}]}\right) \Rightarrow F = 0.168_4$

(b) $[S]_f = (8.47 \text{ mM})\left(\dfrac{1.00 \text{ mL}}{10.0 \text{ mL}}\right) = 0.847 \text{ mM}$

(c) $\dfrac{A_X}{[X]_f} = F\left(\dfrac{A_S}{[S]_f}\right) \Rightarrow \dfrac{5\,428}{[X]_f} = 0.168_4\left(\dfrac{4\,431}{[0.847 \text{ mM}]}\right) \Rightarrow [X]_f = 6.16 \text{ mM}$

(d) The original concentration of [X] was twice as great as the diluted concentration, so [X] = 12.3 mM.

5-34. For the standard mixture:

$\dfrac{A_X}{[X]_f} = F\left(\dfrac{A_S}{[S]_f}\right) \Rightarrow \dfrac{10.1 \text{ μA}}{[0.800 \text{ mM}]} = F\left(\dfrac{15.3 \text{ μA}}{[0.500 \text{ mM}]}\right) \Rightarrow F = 0.412_6$

Chloroform added to unknown = $(10.2 \times 10^{-6} \text{ L})(1\,484 \text{ g/L}) = 0.015\,1_4 \text{ g} = 0.126_8$ mmol in 0.100 L = 1.26_8 mM

For the unknown mixture:

$$\frac{A_X}{[X]_f} = F\left(\frac{A_S}{[S]_f}\right) \Rightarrow \frac{8.7\ \mu A}{[X]_f} = 0.412_6 \left(\frac{29.4\ \mu A}{[1.26_8\ mM]}\right) \Rightarrow [X]_f = 0.909\ mM$$

$$[DDT]\ \text{in unknown} = (0.909\ mM)\left(\frac{100\ mL}{10.0\ mL}\right) = 9.09\ mM$$

5-35. (a) and (b) In the spreadsheet at the end of this problem, the internal standard graph is a plot of A_X/A_S versus $[X]_f/[S]_f$ using data in cells A5:B10. We use LINEST in cells B12:C14 to compute slope, intercept, and uncertainties.

(c) Data appear randomly scattered about the line, and the intercept is near 0. We will check if the intercept is statistically equivalent to 0 in part (e). For now the data appears linear, so let's proceed.

(d) For $A_X/A_S = 1.98$, cell B20 computes $[X]_f/[S]_f = (1.98 - b)/m = 0.560$. Standard uncertainty in the quotient $[X]_f/[S]_f$ is computed in cell B21:

$$u_x = \frac{s_y}{|m|}\sqrt{\frac{1}{k} + \frac{1}{n} + \frac{(y-\bar{y})^2}{m^2\Sigma(x_i - \bar{x})^2}} = 0.024\ 6$$

where $s_y = 0.072$, $m = 3.47$, $k =$ number of replicate measurements of unknown $= 1$, $n =$ number of points in calibration line $= 6$, $y = 1.98$, $\bar{y}$ is the mean value of y for the points on the calibration line $= 1.179$ in cell B16, and $\Sigma(x_i - \bar{x})^2 = 0.233$ in cell B17. For 4 degrees of freedom and 95% confidence, Student's t is 2.776, which you can find in the table of Student t values or you can calculate in Excel with the formula "=T.INV.2T(0.05,4)". The 95% confidence interval for $[X]_f/[S]_f$ is $\pm tu_x = \pm(2.776)(0.024\ 6) = \pm 0.068$. A reasonable expression of the mole ratio is $[X]_f/[S]_f = 0.56 \pm 0.07$.

(e) The intercept with its 95% confidence interval is $b \pm tu_b = 0.038 \pm (2.776)(0.057\ 3) = 0.038 \pm 0.159$. The 95% interval for the intercept is -0.12 to $+0.20$, which includes 0.

(f) $\text{Error} = 100 \times \dfrac{\text{value found} - \text{known value}}{\text{known value}} = 100 \times \dfrac{0.560 - 0.582}{0.582} = -3.8\%$

$\text{Precision} = 100 \times \dfrac{0.024\ 6}{0.560} = 4.4\%$

The error and precision are within the criteria specified in the use objectives. The method is fit for purpose even though R^2 was low.

	A	B	C	D	E	F	G
1	Poly(ethylene-co-vinyl acetate) internal standard calibration curve						
2	$[X]_f/[S]_f =$	$A_x/A_s =$					
3	(mol% vinyl acetate	A_{1020}					
4	/ (mol% ethylene)	$/ A_{720}$					
5	0.099	0.291					
6	0.163	0.656					
7	0.220	0.800					
8	0.333	1.235					
9	0.493	1.808					
10	0.667	2.284					
11	{B12:C14} = LINEST(B5:B10,C5:C10,TRUE,TRUE)						
12	m	3.465798	0.038175	b			
13	u_m	0.149438	0.057347	u_b			
14	R^2	0.992618	0.07221	s_y			
15	n =		6	B15 = COUNT(B5:B10)			
16	Mean y =		1.179	B16 = AVERAGE(B5:B10)			
17	$\Sigma(x - \text{mean } x)^2 =$		0.233493	B17 = DEVSQ(A5:A10)			
18	Measured y =		1.98	Input			
19	k = number of replicates of y =		1	Input			
20	Derived x=	0.560282	B20 = (B18-C12)/B12				
21	$u_x =$	0.024612	B21 = (C14/ABS(B12))				
22			*SQRT((1/B19)+(1/B15)+((B18-B16)^2)/(B12^2*B17))				

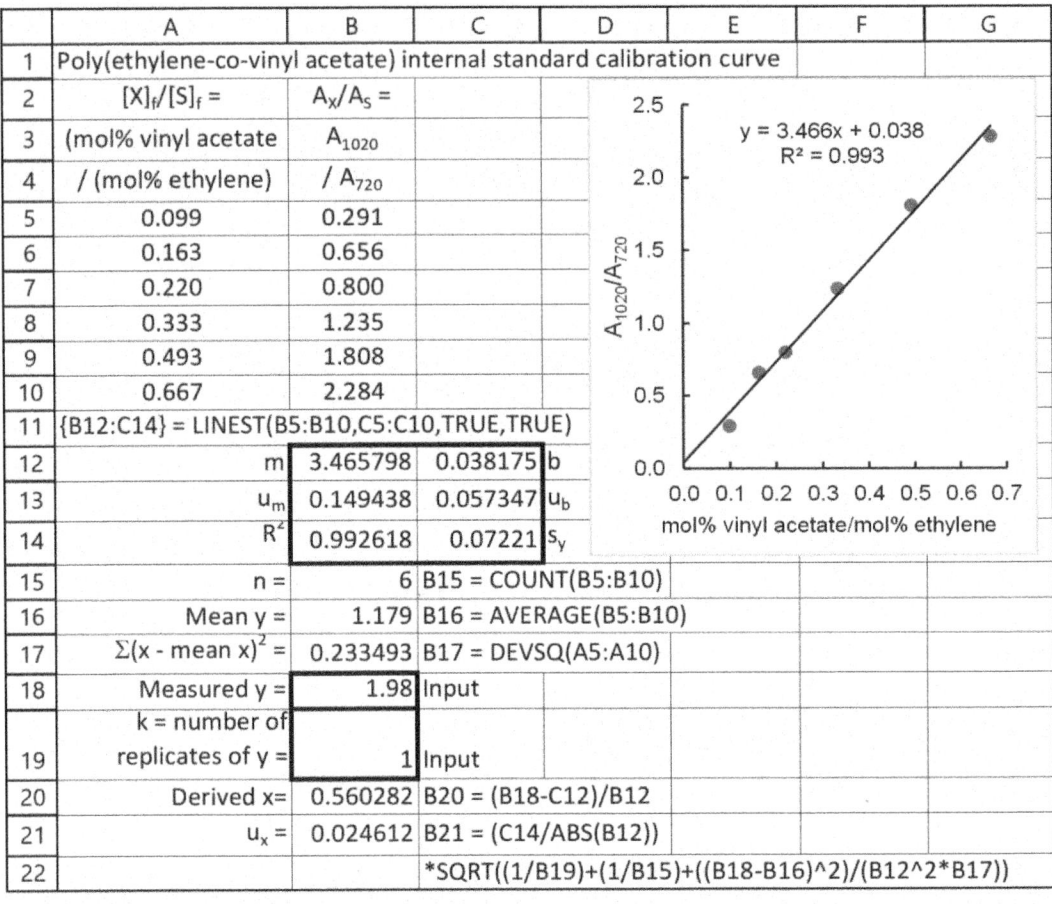

5-36. To use the internal standard, a known quantity of internal standard is added to a sample of sewage. Analyte signal is measured relative to the signal from internal standard. Since the internal standard is so similar to analyte (differing only by the substitution of D for H), matrix effects are likely to be essentially identical for both compounds. If matrix enhances the response to analyte, it enhances the response to internal standard to the same extent. By measuring unknown relative to the known quantity of internal standard, an accurate measurement of unknown can be made. The key is to use an internal standard that is almost identical to the analyte in chemical properties.

CHAPTER 6
CHEMICAL EQUILIBRIUM

6-1. Concentrations in an equilibrium constant are really dimensionless <u>ratios</u> of actual concentrations divided by standard state concentrations. Since standard states are 1 M for solutes, 1 bar for gases, and pure substances for solids and liquids, these are the units we must use. A solvent is approximated as a pure liquid.

6-2. All concentrations in equilibrium constants are expressed as dimensionless ratios of actual concentrations divided by standard-state concentrations.

6-3. Predictions based on free energy or Le Châtelier's principle tell us which way a reaction will go (thermodynamics), but not how long it will take (kinetics). A reaction could be over instantly or it could take forever.

6-4. (a) $K = 1/[Ag^+]^3 [PO_4^{3-}]$ (b) $K = P_{CO_2}^6 / P_{O_2}^{15/2}$

6-5. $K = \dfrac{P_E^3}{P_A^2 [B]} = \dfrac{\left(\dfrac{3.6 \times 10^4 \text{ Torr}}{760 \text{ Torr/atm}} \times 1.013 \dfrac{\text{bar}}{\text{atm}}\right)^3}{\left(\dfrac{2.8 \times 10^3 \text{ Pa}}{10^5 \text{ Pa/bar}}\right)^2 (1.2 \times 10^{-2} \text{M})} = 1.2 \times 10^{10}$

6-6.
$$\begin{array}{rll}
\text{HOBr} + \text{OCl}^- & \rightleftharpoons \text{HOCl} + \text{OBr}^- & K_1 = 1/15 \\
\text{HOCl} & \rightleftharpoons \text{H}^+ + \text{OCl}^- & K_2 = 3.0 \times 10^{-8} \\
\hline
\text{HOBr} & \rightleftharpoons \text{H}^+ + \text{OBr}^- & K = K_1 K_2 = 2.0 \times 10^{-9}
\end{array}$$

6-7. (a) Decrease (b) give off
(c) $\Delta G = \Delta H - T\Delta S$; ΔG is negative for a spontaneous change

6-8. $K = e^{-(59.0 \times 10^3 \text{ J/mol})/(8.314\,472 \text{ J/(K·mol)})(298.15\text{K})} = 5 \times 10^{-11}$

6-9. (a) Right because reactant is added

(b) Right because product is removed

(c) Neither because solid graphite does not appear in the reaction quotient

(d) Right. The pressure of reactant and product both increase by a factor of 8. However, reactant appears to the second power in the reaction quotient and product only appears to the first power. Increasing each pressure by the same factor decreases the reaction quotient.

(e) Smaller. An exothermic reaction liberates heat. Adding heat is like adding a product.

6-10. (a) $K = P_{H_2O} = e^{-\Delta G°/RT} = e^{-(\Delta H° - T\Delta S°)/RT}$

$= e^{-\{[(63.11 \times 10^3 \text{ J/mol}) - (298.15 \text{K})(148 \text{ J K}^{-1} \text{ mol}^{-1})]/(8.314 \text{ J K}^{-1} \text{ mol}^{-1})(298.15 \text{ K})\}}$

$= 4.7 \times 10^{-4}$ bar

(b) $P_{H_2O} = 1 = e^{-(\Delta H° - T\Delta S°)/RT} \Rightarrow \Delta H° - T\Delta S°$ must be zero.

$\Delta H° - T\Delta S° = 0 \Rightarrow T = \dfrac{\Delta H°}{\Delta S°} = 426 \text{ K} = 153°C$

6-11. (a) Let's designate the equilibrium constant at temperature T_1 as K_1 and the equilibrium constant at temperature T_2 as K_2.

$K_1 = e^{-\Delta G°/RT_1} = e^{-(\Delta H° - T\Delta S°)/RT_1} = e^{-\Delta H°/RT_1} \cdot e^{\Delta S°/R}$

Similarly, $K_2 = e^{-\Delta H°/RT_2} \cdot e^{\Delta S°/R}$

Dividing K_1 by K_2 gives $\dfrac{K_1}{K_2} = e^{-(\Delta H°/R)(1/T_1 - 1/T_2)}$

$\Rightarrow \Delta H° = \left(\dfrac{1}{T_2} - \dfrac{1}{T_1}\right)^{-1} R \ln \dfrac{K_1}{K_2}$

Putting in $K_1 = 1.479 \times 10^{-5}$ at $T_1 = 278.15$ K and $K_2 = 1.570 \times 10^{-5}$ at $T_2 = 283.15$ K gives $\Delta H° = +7.82$ kJ / mol.

(b) $K = e^{-\Delta H°/RT} \cdot e^{\Delta S°/R}$

$\underbrace{\ln K}_{y} = \underbrace{-\dfrac{\Delta H°}{R}}_{m} \underbrace{\left(\dfrac{1}{T}\right)}_{x} + \underbrace{\dfrac{\Delta S°}{R}}_{b}$

$\boxed{\text{A graph of } \ln K \text{ vs. } 1/T \text{ will have a slope of } -\Delta H°/R}$

6-12. (a) $Q = \left(\dfrac{48.0 \text{ Pa}}{10^5 \text{ Pa/bar}}\right)^2 \Big/ \left(\dfrac{1\,370 \text{ Pa}}{10^5 \text{ Pa/bar}}\right)\left(\dfrac{3\,310 \text{ Pa}}{10^5 \text{ Pa/bar}}\right)$

$= 5.08 \times 10^{-4} < K$ The reaction will go to the right.

It was not really necessary to convert Pa to bar, since the units cancel.

Chemical Equilibrium

(b)
$$H_2 + Br_2 \rightleftharpoons 2HBr$$

Initial pressure: 1 370   3 310   48.0
Final pressure: 1 370 – x 3 310 – x 48.0 + 2x

Note that 2x Pa of HBr are formed when x Pa of H_2 are consumed.

$$\frac{(48.0+2x)^2}{(1\,370-x)(3\,310-x)} = 7.2 \times 10^{-4} \Rightarrow x = 4.50 \text{ Pa}$$

$P_{H_2} = 1\,366$ Pa, $P_{Br_2} = 3\,306$ Pa, $P_{HBr} = 57.0$ Pa

(c) Neither, since Q is unchanged.

(d) HBr will be formed, since $\Delta H°$ is positive.

6-13. The concentration of MTBE in solution is 100 µg/mL = 100 mg/L. The molarity is
$[MTBE] = \dfrac{0.100 \text{ g/L}}{88.15 \text{ g/mol}} = 1.13_4 \times 10^{-3}$ M. The pressure in the gas phase is $P =$
$[MTBE]/K_h = (1.13_4 \times 10^{-3} \text{ M})/(1.71 \text{ M/bar}) = 0.663$ mbar.

6-14. $[Cu^+][Br^-] = K_{sp}$
$[Cu^+][0.10] = 5 \times 10^{-9} \Rightarrow [Cu^+] = 5 \times 10^{-8}$ M

6-15. $[Ag^+]^4[Fe(CN)_6^{4-}] = K_{sp}$
$[1.0 \times 10^{-6}]^4[Fe(CN)_6^{4-}] = 8.5 \times 10^{-45} \Rightarrow [Fe(CN)_6^{4-}] = 8.5 \times 10^{-21}$ M
$= 8.5$ zM

6-16. If we let $x = [Cu^{2+}]$, then $[SO_4^{2-}] = \frac{1}{4}x$.

$K = [Cu^{2+}]^4 [OH^-]^6 [SO_4^{2-}] = (x)^4 (1.0 \times 10^{-6})^6 \left(\dfrac{1}{4}x\right) = 2.3 \times 10^{-69}$

$\Rightarrow x = [Cu^{2+}] = \left(\dfrac{(4)(2.3 \times 10^{-69})}{(1.0 \times 10^{-6})^6}\right)^{1/5} = 3.9 \times 10^{-7}$ M

6-17. (a) $[Zn^{2+}]^2[Fe(CN)_6^{4-}] = (0.000\,10)^2[Fe(CN)_6^{4-}] = 2.1 \times 10^{-16}$
$\Rightarrow [Fe(CN)_6^{4-}] = 2.1 \times 10^{-8}$ M

(b) $[Zn^{2+}]^2[Fe(CN)_6^{4-}] = (5.0 \times 10^{-7})^2[Fe(CN)_6^{4-}] = 2.1 \times 10^{-16}$
$\Rightarrow [Fe(CN)_6^{4-}] = 8.4 \times 10^{-4}$ M

6-18. BX_2 coprecipitates with AX_3. This means that some BX_2 is trapped in or on AX_3 during precipitation of AX_3.

6-19. For $CaSO_4$, $K_{sp} = 2.4 \times 10^{-5}$. For Ag_2SO_4, $K_{sp} = 1.5 \times 10^{-5}$.

Removing 99% of the Ca^{2+} reduces $[Ca^{2+}]$ to 0.000 500 M. The concentration of SO_4^{2-} in equilibrium with 0.000 500 M Ca^{2+} is $2.4 \times 10^{-5}/0.000\,500 = 0.048$ M. This much SO_4^{2-} <u>will</u> precipitate Ag_2SO_4, because $Q = [Ag^+]^2[SO_4^{2-}] = (0.030\,0)^2(0.048) = 4.3 \times 10^{-5} > K_{sp}$. The separation is not feasible.

When Ag^+ first precipitates, $[SO_4^{2-}] = 1.5 \times 10^{-5}/(0.030\,0)^2 = 1.67 \times 10^{-2}$ M. $[Ca^{2+}] = 2.4 \times 10^{-5}/1.67 \times 10^{-2} = 0.001\,4$ M.   97% of the Ca^{2+} has precipitated.

6-20. $BaCrO_4(s) \rightleftharpoons Ba^{2+} + CrO_4^{2-}$ $K_{sp} = 2.1 \times 10^{-10}$

$Ag_2CrO_4(s) \rightleftharpoons 2Ag^+ + CrO_4^{2-}$ $K_{sp} = 1.2 \times 10^{-12}$

The stoichiometries are not identical, so it is not clear that the salt with lower K_{sp} will precipitate first. Let's try each possibility. Suppose that $BaCrO_4$ precipitates first. The concentration of CrO_4^{2-} that will reduce Ba^{2+} to 0.10% of its initial concentration is

$[Ba^{2+}][CrO_4^{2-}] = [1.0 \times 10^{-5}][CrO_4^{2-}] = 2.1 \times 10^{-10} \Rightarrow [CrO_4^{2-}] = 2.1 \times 10^{-5}$ M.

Will this much chromate precipitate 0.010 M Ag^+? We test by evaluating the reaction quotient for Ag_2CrO_4:

$Q = [Ag^+]^2[CrO_4^{2-}] = (0.010)^2(2.1 \times 10^{-5}) = 2.1 \times 10^{-9} > K_{sp}$ for Ag_2CrO_4

Since $Q > K_{sp}$ for Ag_2CrO_4, Ag^+ will precipitate.

Let's try the reverse calculation. If Ag_2CrO_4 precipitates first, the concentration of CrO_4^{2-} that will reduce Ag^+ to 1.0×10^{-5} M is

$[Ag^+]^2[CrO_4^{2-}] = [1.0 \times 10^{-5}]^2[CrO_4^{2-}] = 1.2 \times 10^{-12} \Rightarrow [CrO_4^{2-}] = 0.012$ M.

This concentration of CrO_4^{2-} exceeds the concentration required to precipitate 99.90% of Ba^{2+}. Neither Ag^+ nor Ba^{2+} can be 99.90% precipitated without precipitating the other ion.

6-21.

Salt	K_{sp}	[Ag$^+$] (M, in equilibrium with 0.1 M anion)	
AgCl	1.8×10^{-10}	$K_{sp}/0.10$	$= 1.8 \times 10^{-9}$
AgBr	5.0×10^{-13}	$K_{sp}/0.10$	$= 5.0 \times 10^{-12}$
AgI	8.3×10^{-17}	$K_{sp}/0.10$	$= 8.3 \times 10^{-16}$
Ag_2CrO_4	1.2×10^{-12}	$\sqrt{K_{sp}/0.10}$	$= 3.5 \times 10^{-6}$

Chemical Equilibrium 63

I⁻ requires the lowest concentration of Ag⁺ to begin precipitating, so I⁻ precipitates first. The order of precipitation is: I⁻ before Br⁻ before Cl⁻ before CrO_4^{2-}.

6-22. Measure Pb in the precipitate. If there is no coprecipitation, there would be no Pb in the precipitate. A small amount of Pb observed in the precipitate tells us that there was a small amount of coprecipitation. The mother liquor contains ~0.01 M $Pb(NO_3)_2$. A small amount of coprecipitation might not change the concentration of Pb^{2+} in the liquid enough to measure the difference in Pb^{2+} before and after addition of I⁻.

6-23. At low I⁻ concentration, $[Pb^{2+}]$ decreases with increasing $[I^-]$ because of the reaction $Pb^{2+} + 2I^- \rightarrow PbI_2(s)$. Concentrations of other Pb^{2+}–I⁻ species are negligible. At high I⁻ concentration, complex ions form by reactions such as $PbI_2(s) + I^- \rightarrow PbI_3^-$.

6-24. (a) BF_3 (b) AsF_5

6-25. $\dfrac{[SnCl_2(aq)]}{[Sn^{2+}][Cl^-]^2} = \beta_2 \Rightarrow [SnCl_2(aq)] = \beta_2[Sn^{2+}][Cl^-]^2 = (12)(0.20)(0.20)^2 = 0.096$ M

6-26. $[Zn^{2+}] = K_{sp}/[OH^-]^2 = (3 \times 10^{-16})/(3.2 \times 10^{-7})^2 = 2.93 \times 10^{-3}$ M

$[ZnOH^+] = \beta_1[Zn^{2+}][OH^-] = \beta_1 K_{sp}/[OH^-] = 9 \times 10^{-6}$ M

$[Zn(OH_2(aq)] = \beta_2[Zn^{2+}][OH^-]^2 = \beta_3 K_{sp}[OH^-] = 6 \times 10^{-6}$ M

$[Zn(OH)_3^-] = \beta_3[Zn^{2+}][OH^-]^3 = \beta_3 K_{sp}[OH^-] = 8 \times 10^{-9}$ M

$[Zn(OH)_4^{2-}] = \beta_4[Zn^{2+}][OH^-]^4 = \beta_4 K_{sp}[OH^-]^2 = 9 \times 10^{-14}$ M

6-27.

	Na^+ +	OH^- ⇌	$NaOH(aq)$
Initial concentration:	1	1	0
Final concentration:	$1-x$	$1-x$	x

$\dfrac{x}{(1-x)^2} = 0.2 \Rightarrow x = 0.15$ M

$x = 0.2 - 0.4x + 0.2x^2$

$\underset{a}{0.2x^2} - \underset{b}{1.4x} + \underset{c}{0.2} = 0$

64   Chapter 6

$$x = \frac{-b \pm \sqrt{b^2 - 4ac}}{2a} = \frac{1.4 \pm \sqrt{1.4^2 - 4(0.2)(0.2)}}{2(0.2)} = 6.85 \text{ or } 0.15$$

x cannot be greater than 1 (the initial or formal concentration of NaOH), so the correct answer must be 0.15. That is, 15% of the sodium is in the form NaOH(aq).

6-28.
$$PbI_2(s) \underset{}{\overset{K_{sp}}{\rightleftharpoons}} Pb^{2+} + 2I^- \qquad K_{sp} = [Pb^{2+}][I^-]^2 = 7.9 \times 10^{-9}$$

$$+ \; Pb^{2+} + 2I^- \underset{}{\overset{\beta_2}{\rightleftharpoons}} PbI_2(aq) \qquad \beta_2 = [PbI_2(aq)] / [Pb^{2+}][I^-]^2 = 1.4 \times 10^3$$

$$PbI_2(s) \rightleftharpoons PbI_2(aq) \qquad K = K_{sp}\beta_2 = 1.1 \times 10^{-5} = [PbI_2(aq)]$$

6-29. Lewis acids and bases are electron pair acceptors and donors, respectively:

$$F_3B + :\ddot{O}(CH_3)_2 \rightarrow F_3\overset{-}{B}-\overset{+}{\ddot{O}}(CH_3)_2$$

Lewis   Lewis   Addict
acid   base

Brønsted-Lowry acids and bases are proton donors and acceptors, respectively:

H$_2$S + ⟨C$_6$H$_5$⟩N: → ⟨C$_6$H$_5$⟩NH$^+$ + HS$^-$

Brønsted   Brønsted
acid   base

6-30. (a) An adduct   (b) dative or coordinate covalent

(c) conjugate   (d) [H$^+$] > [OH$^-$]; [OH$^-$] > [H$^+$]

6-31. Dissolved CO$_2$ from the atmosphere lowers the pH by reacting with water to form carbonic acid. Water can be distilled under an inert atmosphere to exclude CO$_2$, or most CO$_2$ can be removed by boiling the distilled water.

6-32.

```
           H
           |
         H-C-H
     H     |     H
     |     |     |
   H-C —— N⁺ —— C-H
     |     |     |
     H     |     H
         H-C-H
           |
           H
```

:Ö—H (with negative charge)

There is no place for OH$^-$ to bond to (CH$_3$)$_4$N$^+$.

6-33. (a) HI   (b) H$_2$O

Chemical Equilibrium 65

6-34. $2H_2SO_4 \rightleftharpoons HSO_4^- + H_3SO_4^+$

6-35.

	acid	base
(a)	H_3O^+	H_2O
(a)	$H_3\overset{+}{N}CH_2CH_2\overset{+}{N}H_3$	$H_3\overset{+}{N}CH_2CH_2NH_2$
(b)	$C_6H_5CO_2H$	$C_6H_5CO_2^-$
(b)	$C_5H_5NH^+$	C_5H_5N

6-36. (a) $[H^+] = 0.010$ M $\Rightarrow$ pH $= -\log [H^+] = 2.00$

(b) $[OH^-] = 0.035$ M $\Rightarrow [H^+] = K_w/[OH^-] = 2.8_6 \times 10^{-13}$ M $\Rightarrow$ pH $= 12.54$

(c) $[H^+] = 0.030$ M $\Rightarrow$ pH $= 1.52$

(d) $[H^+] = 3.0$ M $\Rightarrow$ pH $= -0.48$

(e) $[OH^-] = 0.010$ M $\Rightarrow [H^+] = 1.0 \times 10^{-12}$ M $\Rightarrow$ pH $= 12.00$

6-37. (a) $K_w = [H^+][OH^-] = 1.01 \times 10^{-14}$ at 25°C

$$\underset{x}{} \quad \underset{x}{}$$

$x^2 = 1.01 \times 10^{-14} \Rightarrow x = [H^+] = 1.00_5 \times 10^{-7}$ M $\Rightarrow$ pH $= -\log [H^+] = 6.998$

(b) At 100°C, $K_w = 5.43 \times 10^{-13} \Rightarrow$ pH $= 6.132$

6-38. Since $[H^+][OH^-] = 1.0 \times 10^{-14}$, $K = [H^+]^4 [OH^-]^4 = 1.0 \times 10^{-56}$

6-39. $[La^{3+}][OH^-]^3 = K_{sp} = 2 \times 10^{-21}$

$[OH^-]^3 = K_{sp} / (0.010) \Rightarrow [OH^-] = 5.8 \times 10^{-7}$ M $\Rightarrow$ pH $= 7.8$

6-40. (a) At 25°C, K_w increases as temperature increases $\Rightarrow$ endothermic

(b) At 100°C, K_w increases as temperature increases $\Rightarrow$ endothermic

(c) At 300°C, K_w decreases as temperature increases $\Rightarrow$ exothermic

6-41. See Table 6-2.

6-42. Weak acids: RCO_2H (Carboxylic acids), R_3NH^+X (Ammonium ions), M^{n+} (Metal ions)

Weak bases: R_3N: (Amines), $RCO_2^-M^+$ (Carboxylate ions)

6-43. In the reaction $(H_2O)_6Fe^{3+} \rightleftharpoons (OH)(H_2O)_5Fe^{2+} + H^+$, positive charge concentrated in $(H_2O)_6Fe^{3+}$ is reduced when H^+ dissociates. This process is favored by Coulombic repulsion of like charges. In the reaction $(H_2O)_6Cl^- \rightleftharpoons (OH)(H_2O)_5Cl^{2-} + H^+$, negative charge is concentrated in the product when H^+ dissociates. This process is disfavored by Coulombic repulsion of like charges.

6-44. $Cl_3CCO_2H \rightleftharpoons Cl_3CCO_2^- + H^+$

$C_6H_5-NH_3^+ \rightleftharpoons C_6H_5-NH_2 + H^+$

$La^{3+} + H_2O \rightleftharpoons LaOH^{2+} + H^+$

6-45. $C_5H_5N + H_2O \rightleftharpoons C_5H_5NH^+ + OH^-$

$HOCH_2CH_2S^- + H_2O \rightleftharpoons HOCH_2CH_2SH + OH^-$

6-46. K_a: $HCO_3^- \rightleftharpoons H^+ + CO_3^{2-}$ K_b: $HCO_3^- + H_2O \rightleftharpoons H_2CO_3 + OH^-$

6-47. (a) $H_3\overset{+}{N}CH_2CH_2\overset{+}{N}H_3 \overset{K_{a1}}{\rightleftharpoons} H_2NCH_2CH_2\overset{+}{N}H_3 + H^+$

$H_2NCH_2CH_2\overset{+}{N}H_3 \overset{K_{a2}}{\rightleftharpoons} H_2NCH_2CH_2NH_2 + H^+$

(b) $^-O_2CCH_2CO_2^- + H_2O \overset{K_{b1}}{\rightleftharpoons} HO_2CCH_2CO_2^- + OH^-$

$HO_2CCH_2CO_2^- + H_2O \overset{K_{b2}}{\rightleftharpoons} HO_2CCH_2CO_2H + OH^-$

6-48. (a), (c) (each has a greater equilibrium constant than the other choice)

6-49. $CN^- + H_2O \rightleftharpoons HCN + OH^-$ $K_b = \dfrac{K_w}{K_a} = 1.6 \times 10^{-5}$

6-50. $H_2PO_4^- \overset{K_{a2}}{\rightleftharpoons} HPO_4^{2-} + H^+$ $HC_2O_4^- + H_2O \overset{K_{b2}}{\rightleftharpoons} H_2C_2O_4 + OH^-$

6-51. $K_{a1} = \dfrac{K_w}{K_{b3}} = 7.04 \times 10^{-3}$ $K_{a2} = \dfrac{K_w}{K_{b2}} = 6.25 \times 10^{-8}$

$K_{a3} = \dfrac{K_w}{K_{b1}} = 4.3 \times 10^{-13}$

6-52. Add the two reactions and multiply their equilibrium constants to get $K = 3.0 \times 10^{-6}$.

Chemical Equilibrium

6-53. (a) $Ca(OH)_2 (s) \rightleftharpoons Ca^{2+} + 2OH^-$
  $ x \phantom{^{2+} + } 2x$

  $x(2x)^2 = K_{sp} = 10^{-5.19} \Rightarrow x = 1.2 \times 10^{-2}$ M

(b) Since some Ca^{2+} reacts with OH^- to form $CaOH^+$, the K_{sp} reaction will be drawn to the right, and the solubility of $Ca(OH)_2$ will be greater than we would expect just on the basis of K_{sp}.

6-54. Reversing the first reaction and then adding the four reactions gives

$Ca^{2+} + CO_2(g) + H_2O(l) \rightleftharpoons CaCO_3(s) + 2H^+ \qquad K = K_{CO_2}K_1K_2/K_{sp}$

$K = (3.4 \times 10^{-2})(4.5 \times 10^{-7})(4.7 \times 10^{-11})/(6.0 \times 10^{-9}) = 1.2 \times 10^{-10}$

$\dfrac{[H^+]^2}{[Ca^{2+}]P_{CO_2}} = \dfrac{(1.8 \times 10^{-7})^2}{[Ca^{2+}][0.10]} = K = 1.2 \times 10^{-10}$

$[Ca^{2+}] = 2.7 \times 10^{-3}$ M $= 0.22$ g/2.00 L

CHAPTER 7
LET THE TITRATIONS BEGIN

7-1. Concentrations of reagents used in an analysis are determined either by weighing out supposedly pure primary standards or by reaction with such standards. If the standards are not pure, none of the concentrations will be correct.

7-2. The equivalence point occurs when the exact stoichiometric quantities of reagents have been mixed. The end point, which comes near the equivalence point, is marked by a sudden change in a physical property brought about by the disappearance of a reactant or appearance of a product. We try to select a property to measure for the end point that is as close as possible to the equivalence point.

7-3. In a blank titration, the quantity of titrant required to reach the end point in the absence of analyte is measured. By subtracting this quantity from the amount of titrant needed in the presence of analyte, we reduce the systematic error.

7-4. In a direct titration, titrant reacts directly with analyte. In a back titration, a known excess of reagent that reacts with analyte is used. The excess is then measured with a second titrant.

7-5. Primary standards are purer than reagent-grade chemicals. The assay of a primary standard must be very close to the nominal value (such as 99.95–100.05%), whereas the assay on a reagent chemical might be only 99%. Primary standards must have very long shelf lives.

7-6. Since a relatively large amount of acid might be required to dissolve a small amount of sample, we cannot tolerate even modest amounts of impurities in the acid for trace analysis. Otherwise, the quantity of impurity could be greater than quantity of analyte in the sample.

7-7. 40.0 mL of 0.0400 M $Hg_2(NO_3)_2$ = 1.60 mmol of Hg_2^{2+}, which will require 3.20 mmol of KI. This is contained in volume = $\dfrac{3.20 \text{ mmol}}{0.100 \text{ mmol/mL}}$ = 32.0 mL.

7-8. 108.0 mL of 0.1650 M oxalic acid = 17.82 mmol, which requires
$\left(\dfrac{2 \text{ mol } MnO_4^-}{5 \text{ mol } H_2C_2O_4}\right)(17.82 \text{ mol } H_2C_2O_4)$ = 7.128 mmol of MnO_4^-.

7.128 mmol / (0.165 0 mmol/mL) = 43.20 mL of KMnO$_4$.

Another way to see this is to note that the reagents are both 0.165 0 M. Therefore, volume of MnO$_4^-$ = $\frac{2}{5}$ (volume of oxalic acid).

For the second part of the question,

volume of oxalic acid = $\frac{5}{2}$ (volume of MnO$_4^-$) = 270.0 mL.

7-9. 1.69 mg NH$_3$ (FM 17.03) = 0.099 2 mmol NH$_3$. This will react with $\frac{3}{2}$(0.099 2 mmol) = 0.149 mmol of OBr$^-$. The molarity of OBr$^-$ is 0.149 mmol/1.00 mL = 0.149 M.

7-10. mol sulfamic acid = $\dfrac{0.333\ 7\ g}{97.088\ g/mol}$ = 3.437$_1$ mmol

molarity of NaOH = $\dfrac{3.437_1\ mmol}{34.26\ mL}$ = 0.100 3 M

7-11. HCl added to powder = (10.00 mL)(1.396 M) = 13.96 mmol
NaOH required = (39.96 mL)(0.100 4 M) = 4.012 mmol
HCl consumed by carbonate = 13.96 – 4.012 = 9.94$_8$ mmol
mol CaCO$_3$ = $\frac{1}{2}$ mol HCl consumed = 4.97$_4$ mmol = 0.497$_8$ g CaCO$_3$

wt% CaCO$_3$ = $\dfrac{0.497_8\ g\ CaCO_3}{0.541\ 3\ g\ limestone}$ × 100 = 92.0 wt%

7-12. (a) Theoretical molarity = 3.214 g/158.032 g/mol = 0.020 33$_8$ M.

(b) 25.00 mL of 0.020 333$_8$ M KMnO$_4$ = 0.508 4$_4$ mmol. But 2 mol of MnO$_4^-$ react with 5 mol of H$_3$AsO$_3$, which comes from $\frac{5}{2}$ moles of As$_2$O$_3$.

The moles of As$_2$O$_3$ needed to react with 0.508 4$_4$ mmol of MnO$_4^-$ = ($^1/_2$)($^5/_2$)(0.508 5 mmol) = 0.635 5$_5$ mmol = 0.125 7$_4$ g of As$_2$O$_3$.

(c) $\dfrac{0.508\ 4_4\ mmol\ KMnO_4}{0.125\ 7_4\ g\ As_2O_3}$ = $\dfrac{x\ mmol\ KMnO_4}{0.146\ 8\ g\ As_2O_3}$

⇒ x = 0.593 6$_0$ mmol KMnO$_4$ in (29.98 – 0.03) = 29.95 mL

⇒ [KMnO$_4$] = $\dfrac{0.593\ 6_0\ mmol}{29.95\ g\ mL}$ = 0.019 82 M.

7-13. FM of NaCl = 58.440. FM of KBr = 119.002. 48.40 mL of 0.048 37 M Ag^+ = 2.341 1 mmol. This must equal the mmol of (Cl^-+ Br^-). Let x = mass of NaCl and y = mass of KBr. $x + y = 0.2386$ g.

$$\underbrace{\frac{x}{58.440}}_{\text{mol Cl}^-} + \underbrace{\frac{y}{119.002}}_{\text{mol Br}^-} = 2.3411 \times 10^{-3} \text{ mol}$$

Substituting $x = 0.2386 - y$ gives $y = 0.2000$ g of KBr = 1.681 mmol of KBr = 1.681 mmol of Br = 0.1343 g of Br = 56.28% of the sample.

7-14. Let x = mg of $FeSO_4 \cdot (NH_4)_2SO_4 \cdot 6H_2O$ and $(54.85 - x)$ = mg of $FeCl_2 \cdot 6H_2O$.
mmol of Ce^{4+} = mmol $FeSO_4 \cdot (NH_4)_2SO_4 \cdot 6H_2O$ + mmol $FeCl_2 \cdot 6H_2O$.

$$(13.39 \text{ mL})(0.01234 \text{ M}) = 0.16523 \text{ mmol} = \frac{x \text{ mg}}{392.12 \text{ mg/mmol}} + \frac{(54.85-x)}{234.84 \text{ mg/mmol}}$$

$\Rightarrow x = 40.01$ mg $FeSO_4 \cdot (NH_4)_2SO_4 \cdot 6H_2O$.

mass of $FeCl_2 \cdot 6H_2O$ = 14.84 mg = 0.06319 mmol = 4.48 mg Cl.

wt% Cl = $\frac{4.48 \text{ mg}}{54.85 \text{ mg}} \times 100 = 8.17\%$.

7-15. 30.10 mL of Ni^{2+} reacted with 39.35 mL of 0.01307 M EDTA. Therefore, the Ni^{2+} molarity is

$$[Ni^{2+}] = \frac{(39.35 \text{ mL})(0.01307 \text{ mol/L})}{30.10 \text{ mL}} = 0.01709 \text{ M}.$$

25.00 mL of Ni^{2+} contains 0.4272 mmol of Ni^{2+}. 10.15 mL of EDTA = 0.1327 mmol of EDTA. The amount of Ni^{2+} which must have reacted with CN^- was 0.4272 − 0.1327 = 0.2945 mmol. The cyanide which reacted with Ni^{2+} must have been (4)(0.2945) = 1.178 mmol. $[CN^-]$ = 1.178 mmol/12.73 mL = 0.09254 M.

7-16. (a) mol O_2 = $(2.9 \times 10^6 \text{ L})(2.2 \times 10^{-4} \text{ M}) = 638$ mol

Reaction 1 requires 2 mol CH_3OH for 3 mol O_2, so the required CH_3OH is (2/3)(638 mol) = 425 mol CH_3OH = 13.6 kg CH_3OH.

volume of CH_3OH = $\frac{13.6 \text{ kg}}{0.791 \text{ kg/L}}$ = 17.2 L

(b) $6NO_3^- + 2CH_3OH \rightarrow 6NO_2^- + 2CO_2 + 4H_2O$

$6NO_2^- + 3CH_3OH \rightarrow 3N_2 + 3CO_2 + 3H_2O + 6OH^-$

net: $6NO_3^- + 5CH_3OH \rightarrow 3N_2 + 5CO_2 + 7H_2O + 6OH^-$

$$\text{mol NO}_3^- = (2.9 \times 10^6 \text{ L})(8.1 \times 10^{-3} \text{ M}) = 2.35 \times 10^4 \text{ mol}$$

Net reaction requires 5 mol CH_3OH for 6 mol NO_3^-, so the required CH_3OH is

$(5/6)(2.35 \times 10^4 \text{ mol}) = 1.96 \times 10^4 \text{ mol } CH_3OH = 627 \text{ kg } CH_3OH$.

$$\text{volume of } CH_3OH = \frac{627 \text{ kg}}{0.791 \text{ kg/L}} = 7.9_3 \times 10^2 \text{ L}$$

(c) total volume required = $1.30 (17._2 \text{ L} + 7.9_3 \times 10^2 \text{ L}) = 1.05 \times 10^3 \text{ L}$

7-17. A known solution of I^- is titrated with known volumes of a known concentration of Ag^+. Before the equivalence point, the reaction is $I^- + Ag^+ \rightarrow AgI(s)$. There is a known excess of I^- left after each addition of Ag^+, so $[Ag^+] = K_{sp}/[I^-]$. At the equivalence point, just enough Ag^+ has been added to react with all of I^- to precipitate $AgI(s)$. The concentrations of dissolved Ag^+ and I^- are equal and are the amounts in equilibrium with $AgI(s)$: $[Ag^+][I^-] = [Ag^+]^2 = K_{sp}$. After the equivalence point, there is a known excess of Ag^+, so $[I^-] = K_{sp}/[Ag^+]$.

7-18. At V_e, moles of Ag^+ = moles of I^-

$(V_e \text{ mL})(0.051\,1 \text{ M}) = (25.0 \text{ mL})(0.082\,3 \text{ M}) \Rightarrow V_e = 40.26 \text{ mL}$

(a) When $V_{Ag^+} = 39.00$ mL,

$$[I^-] = \left(\frac{40.26 - 39.00}{40.26}\right)(0.082\,30 \text{ M})\left(\frac{25.00 \text{ mL}}{(25.00+39.00) \text{ mL}}\right) = 1.006 \times 10^{-3} \text{ M}.$$

$[Ag^+] = K_{sp}/[I^-] = 8.3 \times 10^{-14} \text{ M} \Rightarrow pAg^+ = 13.08$.

(b) When $V_{Ag^+} = V_e$, $[Ag^+][I^-] = x^2 = K_{sp} \Rightarrow x = [Ag^+] = 9.1 \times 10^{-9} \text{ M}$
$\Rightarrow pAg^+ = 8.04$.

(c) When $V_{Ag^+} = 44.30$ mL, there is an excess of $(44.30 - 40.26) = 4.04$ mL Ag^+.

$$[Ag^+] = \left(\frac{4.04 \text{ mL}}{(25.00+44.30) \text{ mL}}\right)(0.051\,10) = 2.98 \times 10^{-3} \text{ M} \Rightarrow pAg^+ = 2.53.$$

7-19. Moles of Ca^{2+} = moles of $C_2O_4^{2-}$

$(V_e)(0.025\,7 \text{ M}) = (25.00 \text{ mL})(0.031\,1 \text{ M}) \Rightarrow V_e = 30.25 \text{ mL}$

(a) The fraction of $C_2O_4^{2-}$ remaining when 10.00 mL of Ca^{2+} have been added is

$(30.25 - 10.00)/(30.25) = 0.669\,4$.

$$[C_2O_4^{2-}] = (0.669\,4)(0.031\,10 \text{ M})\left(\frac{25.00 \text{ mL}}{35.00 \text{ mL}}\right) = 0.014\,87 \text{ M}$$

$[Ca^{2+}] = K_{sp}/[C_2O_4^{2-}] = (1.3 \times 10^{-8})/(0.014\ 87) = 8.7 \times 10^{-7}$

$\Rightarrow pCa^{2+} = -\log(8.7 \times 10^{-7}) = 6.06$

(b) At the equivalence point, there are equal numbers of moles of Ca^{2+} and $C_2O_4^{2-}$ dissolved. Call each concentration x:

$[Ca^{2+}][C_2O_4^{2-}] = (x)(x) = K_{sp} \Rightarrow x = \sqrt{K_{sp}} = 1.1_4 \times 10^{-4}$ M

$pCa^{2+} = -\log(1.1_4 \times 10^{-4}) = 3.94$

(c) $[Ca^{2+}] = (0.025\ 70\ M)\left(\dfrac{(35.00-30.25)\ \text{mL}}{60.00\ \text{mL}}\right) = 0.002\ 03\ M \Rightarrow pCa^{2+} = 2.69$

7-20. Equilibrium constants for ion pair formation:

$$\dfrac{[AgX(aq)]}{[Ag^+][X^-]} = \begin{cases} K_f = 10^{3.31} & (X = Cl) \\ K_f = 10^{4.6} & (X = Br) \\ K_f = 10^{6.6} & (X = I) \end{cases}$$

Calling the ion pair formation constant K_f, we can write $[AgX(aq)] = K_f[Ag^+][X^-]$. But the product $[Ag^+][X^-]$ is just K_{sp}. So, $[AgX(aq)] = K_f K_{sp}$. Putting in the values $K_{sp} = 10^{-9.74}$ for AgCl, $10^{-12.30}$ for AgBr, and $10^{-16.08}$ for AgI gives

$[AgCl(aq)] = 10^{3.31} \times 10^{-9.74} = 10^{-6.43}$ M = 370 nM

$[AgBr(aq)] = 10^{4.6} \times 10^{-12.30} = 10^{-7.7}$ M = 20 nM

$[AgI(aq)] = 10^{6.6} \times 10^{-16.08} = 10^{-9.5}$ M = 0.32 nM

7-21. (a) Titration reaction: SO_4^{2-} [from soil] + Ba^{2+} [from $BaCl_2(s)$] $\rightarrow BaSO_4(s)$

(b) Prior to adding $BaCl_2$, the 25-mL solution contained 0.000 19 M Cl^-.
At the end point, the solution contained 0.009 6 M Cl^- from addition of $BaCl_2$.
$[Cl^-]$ added from $BaCl_2$ = 0.009 6 − 0.000 19 M = 0.009 4_1 M
mmol Cl^- = (0.009 4_1 M)(25 mL) = 0.23$_5$ mmol
$BaCl_2$ contains 1 mol Ba^{2+} for every 2 mol Cl^-, so
mmol Ba^{2+} added = $\tfrac{1}{2}$(0.23$_5$ mmol) = 0.11$_8$ mmol

(c) One mole of Ba^{2+} consumes one mole of SO_4^{2-} in the titration. Therefore, 0.11$_8$ mmol SO_4^{2-} must have been present in the 25-mL aqueous extract.

(d) Mass of SO_4^{2-} in extract = $(0.11_8 \text{ mmol } SO_4^{2-})(96.06 \text{ mg/mmol}) = 11._3$ mg

wt% SO_4^{2-} in soil = $100 \times (11._3 \text{ mg } SO_4^{2-})/(1\,000 \text{ mg soil}) = 1.1$ wt%

7-22. (i) $I^-(\text{excess}) + Ag^+ \rightarrow AgI(s)$ $[Ag^+] = K_{sp} \text{ (for AgI)} / [I^-]$

(ii) A stoichiometric quantity of Ag^+ has been added that would be just equivalent to I^-, if no Cl^- were present. Instead, a tiny amount of AgCl precipitates and a slight amount of I^- remains in solution.

(iii) $Cl^-(\text{excess}) + Ag^+ \rightarrow AgCl(s)$ $[Ag^+] = K_{sp} \text{ (for AgCl)} / [Cl^-]$

(iv) Virtually all I^- and Cl^- have precipitated.
$[Ag^+] \approx [Cl^-] \Rightarrow [Ag^+] = \sqrt{K_{sp} \text{ (for AgCl)}}$

(v) There is excess Ag^+ delivered from the buret.

$$[Ag^+] = [Ag^+]_{titrant} \cdot \left(\frac{\text{volume added past 2nd equivalence point}}{\text{total volume}} \right)$$

7-23. At the equivalence point, $[Ag^+][I^-] = K_{sp} \Rightarrow (x)(x) = 8.3 \times 10^{-17}$

$\Rightarrow [Ag^+] = 9.1 \times 10^{-9}$ M. The concentration of Cl^- in the titration solution is the initial concentration (0.050 0 M) corrected for dilution from an initial volume of 40.00 mL up to ~63.85 mL at the equivalence point:

$$[Cl^-] = (0.050\,0 \text{ M}) \left(\frac{40.00 \text{ mL}}{63.85 \text{ mL}} \right) = 0.031\,3 \text{ M}$$

Is the solubility of AgCl exceeded? The reaction quotient is $Q = [Ag^+][Cl^-]$ = $(9.1 \times 10^{-9})(0.031\,3) = 2.8 \times 10^{-10}$, which is greater than K_{sp} for AgCl ($= 1.8 \times 10^{-10}$). Therefore, AgCl begins to precipitate before AgI finishes precipitating. If the concentration of Cl^- were about two times lower, AgCl would not precipitate prematurely.

7-24. mmol of $BrCH_2CH_2CH_2CH_2Cl$ = $\dfrac{82.67 \text{ mg}}{171.46 \text{ mg/mmol}} = 0.482\,2$ mmol

There will be 0.482 2 mmol of Cl^- and 0.482 2 mmol of Br^- liberated by reaction with $CH_3O^-Na^+$.

$$Ag^+ \text{ required for } Br^- = \frac{0.482\,2 \text{ mmol}}{0.025\,70 \text{ mmol/mL}} = 18.76 \text{ mL}$$

The same amount of Ag^+ is required to react with Cl^-, so the second equivalence point is at $18.76 + 18.76 = 37.52$ mL.

7-25. Titration of 40.00 mL of 0.050 2 M KI + 0.050 0 M KCl with 0.084 5 M $AgNO_3$

$I^- + Ag^+ \rightarrow AgI(s)$ $V_{e1} = (40.00 \text{ mL}) \left(\dfrac{0.050\ 2\ M}{0.084\ 5\ M} \right) = 23.76$ mL

$Cl^- + Ag^+ \rightarrow AgCl(s)$ $V_{e2} = (40.00 \text{ mL}) \left(\dfrac{(0.050\ 2 + 0.050\ 0)\ M}{0.084\ 5\ M} \right) = 47.43$ mL

The figure gives $V_{e2} = 47.41$ mL, which we will use as a more accurate value.

(a) 10.00 mL: A fraction of the I^- has reacted.

$$[I^-] = \underbrace{\left(\dfrac{23.76 - 10.00}{23.76} \right)}_{\text{Fraction remaining}} \underbrace{(0.050\ 2\ M)}_{\text{Initial concentration}} \underbrace{\left(\dfrac{40.00\ \text{mL}}{50.00\ \text{mL}} \right)}_{\text{Dilution factor}} = 0.023\ 3\ M$$

$[Ag^+] = \dfrac{K_{sp}(AgI)}{[I^-]} = \dfrac{8.3 \times 10^{-17}}{0.023\ 3} = 3.57 \times 10^{-15}$ M

$\Rightarrow pAg^+ = -\log [Ag^+] = 14.45$

(b) 20.00 mL: A fraction of the I^- has reacted.

$$[I^-] = \left(\dfrac{23.76 - 20.00}{23.76} \right) (0.050\ 2\ M) \left(\dfrac{40.00\ \text{mL}}{60.00\ \text{mL}} \right) = 0.005\ 30\ M$$

$[Ag^+] = \dfrac{8.3 \times 10^{-17}}{0.005\ 30} = 1.57 \times 10^{-14}$ M $\Rightarrow pAg^+ = 13.80$

(c) 30.00 mL: I^- has been consumed and a fraction of Cl^- has reacted.

$$[Cl^-] = \underbrace{\left(\dfrac{47.41 - 30.00}{47.41 - 23.76} \right)}_{\text{Fraction remaining}} \underbrace{(0.050\ 0\ M)}_{\text{Initial concentration}} \underbrace{\left(\dfrac{40.00\ \text{mL}}{70.00\ \text{mL}} \right)}_{\text{Dilution factor}} = 0.021\ 0\ M$$

$[Ag^+] = \dfrac{K_{sp}(AgCl)}{[Cl^-]} = \dfrac{1.8 \times 10^{-10}}{0.021\ 0} = 8.56 \times 10^{-9}$ M $\Rightarrow pAg^+ = 8.07$

(d) $[Ag^+][Cl^-] = x^2 = 1.8 \times 10^{-10} \Rightarrow [Ag^+] = 1.3 \times 10^{-5} \Rightarrow pAg^+ = 4.87$

(e) 50.00 mL: There is excess Ag^+.

$$[Ag^+] = \underbrace{(0.084\ 5\ M)}_{\text{Initial concentration}} \underbrace{\left(\dfrac{(50.00 - 47.41)\ \text{mL}}{90.00} \right)}_{\text{Dilution factor}} = 0.002\ 43\ M \Rightarrow pAg^+ = 2.61$$

7-26. (a) $Hg_2^{2+} + 2CN^- \rightarrow Hg_2(CN)_2(s)$ $K_{sp} = 5 \times 10^{-40}$

$Ag^+ + CN^- \rightarrow AgCN(s)$ $K_{sp} = 2.2 \times 10^{-16}$

K_{sp} for $Hg_2(CN)_2$ is much smaller than K_{sp} for AgCN, so it looks like $Hg_2(CN)_2$ will precipitate first. We must check that assumption. The equivalence point occurs at 20.00 mL. The second equivalence point is at 30.00 mL. At 5.00, 10.00, 15.00, and 19.90 mL, there is excess Hg_2^{2+}.

At 5.00 mL, $[Hg_2^{2+}] = \left(\dfrac{20.00-5.00}{20.00}\right)(0.1000 \text{ M})\left(\dfrac{10.00 \text{ mL}}{(10.00+5.00) \text{ mL}}\right)$

$= 0.05000 \text{ M}$

$[CN^-] = \sqrt{K_{sp}(\text{for } Hg_2(CN)_2/([Hg_2^{2+}])} = 1.0 \times 10^{-19} \Rightarrow pCN^- = 19.00$

Now let's check to be sure that this much cyanide does not precipitate AgCN:

Q for $AgCN(s) = [Ag^+][CN^-] = \dfrac{10.00 \text{ mL}}{15.00 \text{ mL}}(0.100 \text{ M})(1.0 \times 10^{-19} \text{ M})$

$= 7 \times 10^{-21} < K_{sp}$. Since $Q < K_{sp}$, AgCN does not precipitate.

By similar calculations we find 10.00 mL: $pCN^- = 18.85$
15.00 mL: $pCN^- = 18.65$
19.90 mL: $pCN^- = 17.76$

At 20.10 mL, AgCN has begun to precipitate. The Ag^+ remaining is

$[Ag^+] = \left(\dfrac{30.00-20.10}{10.00}\right)(0.1000 \text{ M})\left(\dfrac{10.00 \text{ mL}}{(10.00+20.10) \text{ mL}}\right) = 0.0329 \text{ M}$

$[CN^-] = K_{sp}(\text{for AgCN})/[Ag^+] = 6.7 \times 10^{-15} \text{ M} \Rightarrow pCN^- = 14.17$

By similar reasoning we find $pCN^- = 13.81$ at 25.00 mL.

At the second equivalence point (30.00 mL), $[Ag^+] = [CN^-] = x$

$\Rightarrow x^2 = K_{sp}(\text{for AgCN}) \Rightarrow [CN^-] = 1.5 \times 10^{-8} \text{ M} \Rightarrow pCN^- = 7.83$.

At 35.00 mL, there are 5.00 mL of excess CN^-.

$[CN^-] = \left(\dfrac{5.00 \text{ mL}}{(10.00+35.00) \text{ mL}}\right)(0.1000 \text{ M}) = 0.0111 \text{ M} \Rightarrow pCN^- = 1.95$

(b) Will Ag$^+$ precipitate when 19.90 mL of CN$^-$ have been added? We calculated above that pCN$^-$ = 17.76 ($\Rightarrow$ [CN$^-$] = 1.7 × 10^{-18} M) at 19.90 mL if no Ag$^+$ had precipitated. We can check to see whether the solubility product of AgCN is exceeded if [CN$^-$] = 1.7 × 10^{-18} M and

$$[Ag^+] = \left(\frac{10.00 \text{ mL}}{(10.00+19.90) \text{ mL}}\right)(0.1000 \text{ M}) = 0.0334 \text{ M}.$$

[Ag$^+$][CN$^-$] = 5.7 × 10^{-20} < K_{sp} (for AgCN).

The Ag$^+$ will not precipitate at 19.90 mL.

7-27. $\text{M}^+ + \text{X}^- \rightleftharpoons \text{MX}(s)$
Analyte   Titrant
C_M^o, V_M^o C_X^o, V_X

Mass balance for M: $C_M^o V_M^o = [\text{M}^+](V_M^o + V_X) + \text{mol MX}(s)$

Mass balance for X: $C_X^o V_X = [\text{X}^-](V_M^o + V_X) + \text{mol MX}(s)$

Equating mol MX(s) from both mass balances gives

$$C_M^o V_M^o - [\text{M}^+](V_M^o + V_X) = C_X^o V_X - [\text{X}^-](V_M^o + V_X)$$

which can be rearranged to $V_X = V_M^o \left(\dfrac{C_M^o - [\text{M}^+] + [\text{X}^-]}{C_X^o + [\text{M}^+] - [\text{X}^-]}\right)$

7-28. Your graph should look like the figure in the text.

7-29. Mass balance for M: $C_M^o V_M = [\text{M}^{m+}](V_M + V_X^o) + x\{\text{mol M}_x\text{X}_m(s)\}$

Mass balance for X: $C_X^o V_X^o = [\text{X}^{x-}](V_M + V_X^o) + m\{\text{mol M}_x\text{X}_m(s)\}$

Equating mol M$_x$X$_m$ from the two equations gives

$$\frac{1}{x}\{C_M^o V_M - [\text{M}^{m+}](V_M + V_X^o)\} = \frac{1}{m}\{C_X^o V_X^o - [\text{X}^{x-}](V_M + V_X^o)\}$$

which can be rearranged to the required form.

7-30. Titration of chromate with Ag^+:

	A	B	C	D	E	F	G	H	I
1	K_{sp} =	pM	[M]	[X]	V_M				
2	1.2E-12	5.46	3.47E-06	9.98E-02	0.013		C2 = 10^-B2		
3	C_M =	5.4	3.98E-06	7.57E-02	1.932		D2 = $A$2/C2^2		
4	0.1	5.3	5.01E-06	4.78E-02	5.342		E2 = $A$8*(2*$A$6+C2-2*D2)		
5	C_X =	5.2	6.31E-06	3.01E-02	8.717			/($A$4-C2+2*D2)	
6	0.1	5.1	7.94E-06	1.90E-02	11.734				
7	V_X =	5	1.00E-05	1.20E-02	14.195				
8	10	4.9	1.26E-05	7.57E-03	16.057				
9		4.8	1.58E-05	4.78E-03	17.388				
10		4.6	2.51E-05	1.90E-03	18.908				
11		4.4	3.98E-05	7.57E-04	19.564				
12		4	1.00E-04	1.20E-04	19.958				
13		3.6	2.51E-04	1.90E-05	20.064				
14		3.2	6.31E-04	3.01E-06	20.189				
15		2.8	1.58E-03	4.78E-07	20.483				
16		2.4	3.98E-03	7.57E-08	21.244				
17		2.3	5.01E-03	4.78E-08	21.583				
18		2.2	6.31E-03	3.01E-08	22.020				
19		2.1	7.94E-03	1.90E-08	22.589				
20		2	1.00E-02	1.20E-08	23.333				
21		1.9	1.26E-02	7.57E-09	24.321				
22		1.8	1.58E-02	4.78E-09	25.650				

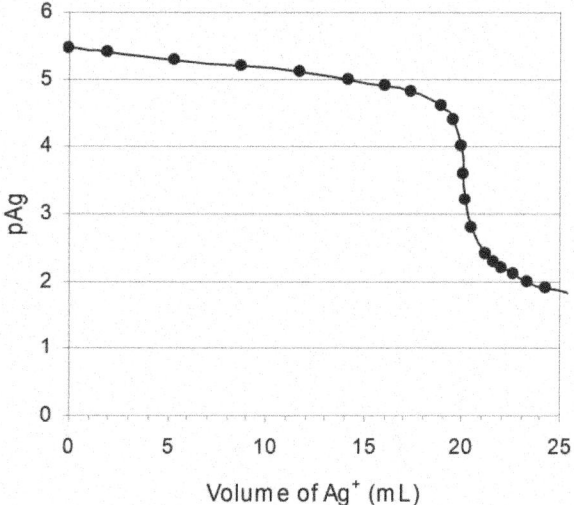

7-31. Consider the titration of C^+ (in a flask) by A^- (from a buret). Before the equivalence point, there is excess C^+ in solution. Selective adsorption of C^+ on the CA crystal surface gives the crystal a positive charge. After the equivalence point, there is excess A^- in solution. Selective adsorption of A^- on the CA crystal surface gives it a negative charge.

7-32. Beyond the equivalence point, there is excess $Fe(CN)_6^{4-}$ in solution. Selective adsorption of this ion by the precipitate will give the particles a negative charge.

7-33. Add a known excess of standard $AgNO_3$ solution to vigorously stirred unknown I^- in ~0.5 M HNO_3 to precipitate $AgI(s)$. Add some Fe^{3+} and titrate the excess Ag^+ with standard KSCN to precipitate $AgSCN(s)$. When Ag^+ is consumed, the next drop of SCN^- reacts with Fe^{3+} to form the red complex, $FeSCN^{2+}$.

7-34. 50.00 mL of 0.3650 M $AgNO_3$ = 18.25 mmol of Ag^+
3.60 mL of 0.2870 M KSCN = 1.03 mmol of SCN^-
Difference = 18.25 – 1.03 = 17.22 mmol of Br^-
HBr concentration = 17.22 mmol/30.00 mL = 0.5740 M
mg Br^- = (17.22 mmol)(79.904 mg/mmol) = 1 376 mg

7-35. Carbonate is a weak base that reacts with 2 mol HNO_3 to form carbonic acid (H_2CO_3), which dissociates to CO_2 and H_2O. $CO_2(g)$ bubbles out of the solution when HNO_3 is added. Even if all H_2CO_3 remained in solution, the concentration of free CO_3^{2-} would be so low that the solubility product for Ag_2CO_3 would not be exceeded. It would be necessary to precipitate Ag_2CO_3 in a neutral solution, separate and wash the precipitate, and then acidify the filtrate prior to titration with KSCN.

CHAPTER 8
ACTIVITY AND THE SYSTEMATIC TREATMENT OF EQUILIBRIUM

8-1. As ionic strength increases, the charges of the ionic atmospheres increase and the net ionic attractions decrease. There is less tendency for ions to bind to each other.

8-2. (a) true (b) true (c) true

8-3. $HBG^- \rightleftharpoons BG^{2-} + H^+$
yellow   blue

The reaction $HBG^- \rightleftharpoons BG^{2-} + H^+$ is an ion dissociation. Increasing ionic strength promotes ion dissociation by forming an ionic atmosphere around each ion, thus decreasing the attraction of H^+ for BG^{2-}. That is, adding NaCl drives the reaction to the right. The reactant is yellow and the product is blue. The initial pale green solution is an equilibrium mixture of yellow HBG^- and blue BG^{2-} (yellow + blue makes green). As NaCl is added, the equilibrium is displaced to the right, causing blue color to increase and yellow color to decrease.

8-4. (a) $\frac{1}{2}[0.008\,7 \cdot 1^2 + 0.008\,7 \cdot (-1)^2] = 0.008\,7$ M

(b) $\frac{1}{2}[0.000\,2 \cdot 3^2 + 0.000\,6 \cdot (-1)^2] = 0.001\,2$ M

8-5. (a) 0.660 (b) 0.54 (c) 0.18 (Eu^{3+} is a lanthanide ion) (d) 0.83

8-6. The ionic strength 0.030 M is halfway between the values 0.01 and 0.05 M. Therefore, the activity coefficient will be halfway between the tabulated values: $\gamma = \frac{1}{2}(0.914 + 0.86) = 0.88_7$.

8-7. (a) $\log \gamma = \dfrac{-0.51 \cdot 2^2 \cdot \sqrt{0.083}}{1 + (600\sqrt{0.083}/305)} = -0.375 \Rightarrow \gamma = 10^{-0.375} = 0.42_2$

(b) $\gamma = \left(\dfrac{0.083 - 0.05}{0.1 - 0.05}\right)(0.405 - 0.485) + 0.485 = 0.43_2$

8-8. $\gamma = \left(\dfrac{0.083 - 0.05}{0.1 - 0.05}\right)(0.18 - 0.245) + 0.245 = 0.20_2$

8-9. If $K = [\text{ether}(aq)]\gamma_{\text{ether}}$ is a constant and $[\text{ether}(aq)]$ becomes smaller as salt is added, then γ_{ether} must <u>increase</u> as salt is added so that K remains constant.

8-10. The solubility of Hg_2Br_2 is small with $K_{sp} = 5.6 \times 10^{-23}$, so Hg_2Br_2 contributes negligible Br^- to 0.001 00 M KBr.

$\mu = 0.001\ 00$ M, $[Br^-] = 0.001\ 00$ M, $\gamma_{Hg_2^{2+}} = 0.867$, $\gamma_{Br^-} = 0.964$

$K_{sp} = 5.6 \times 10^{-23} = [Hg_2^{2+}]\gamma_{Hg_2^{2+}}[Br^-]^2\gamma_{Br^-}^2$

$= [Hg_2^{2+}](0.867)(0.001\ 00)^2(0.964)^2 \Rightarrow [Hg_2^{2+}] = 7.0 \times 10^{-17}$ M

Check our assumption: Yes, Br^- from Hg_2Br_2 is negligible.

8-11. The solubility of $Ba(IO_3)_2$ is small, so we assume that $Ba(IO_3)_2$ contributes negligible IO_3^- to 0.100 M $(CH_3)_4NIO_3$.

$\mu = 0.100$ M, $[IO_3^-] = 0.100$ M, $\gamma_{Ba^{2+}} = 0.38$, $\gamma_{IO_3^-} = 0.775$

$K_{sp} = 1.5 \times 10^{-9} = [Ba^{2+}]\gamma_{Ba^{2+}}[IO_3^-]^2\gamma_{IO_3^-}^2$

$= [Ba^{2+}](0.38)(0.100)^2(0.775)^2 \Rightarrow [Ba^{2+}] = 6.6 \times 10^{-7}$ M

8-12. Ionic strength = 0.010 M (from HCl) + 0.040 M (from $KClO_4$ that gives $K^+ + ClO_4^-$) = 0.050 M. Using Table 8-1, $\gamma_{H^+} = 0.86$.

pH = $-\log([H^+]\gamma_{H^+}) = -\log[(0.010)(0.86)] = 2.07$.

8-13. Ionic strength = 0.010 M from NaOH + 0.012 M from $LiNO_3$ = 0.022 M. Interpolating in Table 8-1 gives $\gamma_{OH^-} = 0.873$.

$[H^+]\gamma_{H^+} = \dfrac{K_w}{[OH^-]\gamma_{OH^-}} = \dfrac{1.0 \times 10^{-14}}{(0.010)(0.873)} = 1.15 \times 10^{-12}$

pH = $-\log(1.15 \times 10^{-12}) = 11.94$

If we had neglected activities, pH $\approx -\log[H^+] = -\log\dfrac{K_w}{[OH^-]} = 12.00$

8-14. $\varepsilon = 79.755\ e^{-4.6 \times 10^{-3}(323.15 - 293.15)} = 69.474$

$\log \gamma = \dfrac{(-1.825 \times 10^6)[(69.474)(323.15)]^{-3/2}(-2)^2\sqrt{0.100}}{1 + \dfrac{400\sqrt{0.100}}{2.00\sqrt{(69.474)(323.15)}}}$

$= -0.4826 \Rightarrow \gamma = 0.329$ (In the table, $\gamma = 0.355$ at 25°C.)

8-15. The equilibrium constant is

$$HA \rightleftharpoons H^+ + A^- \qquad K_a = \frac{[H^+]\gamma_{H^+}[A^-]\gamma_{A^-}}{[HA]\gamma_{HA}}$$

From the table of activity coefficients, $\gamma_{H^+} = 0.83$ and $\gamma_{A^-} = 0.80$ at $\mu = 0.1$ M.

For HA, we estimate

$$\log \gamma_{HA} = k\mu = (0.2)(0.1) = 0.02, \text{ or } \gamma_{HA} = 10^{0.02} = 1.05.$$

Putting activity coefficients for $\mu = 0.1$ M into the equilibrium expression gives

$$K_a = \frac{[H^+]\gamma_{H^+}[A^-]\gamma_{A^-}}{[HA]\gamma_{HA}} = \frac{[H^+](0.83)[A^-](0.80)}{[HA](1.05)} = 0.63 \frac{[H^+][A^-]}{[HA]}$$

or $\dfrac{[H^+][A^-]}{[HA]} (\mu = 0.1 \text{ M}) = K_a/0.63$.

The activity coefficients at $\mu = 0$ are all 1, so the equilibrium expression is

$$K_a = \frac{[H^+]\gamma_{H^+}[A^-]\gamma_{A^-}}{[HA]\gamma_{HA}} = \frac{[H^+](1)[A^-](1)}{[HA](1)} = \frac{[H^+][A^-]}{[HA]} \ (\mu = 0).$$

The concentration quotient at the two different ionic strengths is

$$\text{Concentration quotient} = \frac{\dfrac{[H^+][A^-]}{[HA]}(\mu=0)}{\dfrac{[H^+][A^-]}{[HA]}(\mu=0.1M)} = \frac{K_a}{K_a/0.63} = 0.63$$

in agreement with the observed value of 0.63 ± 0.03.

8-16. The charge balance states the magnitude of positive charge equals the magnitude of negative charge in a solution. That is, the solution must be neutral. The mass balance states that atoms are conserved. If we deliver a certain number of atom A into a solution, then the sum of atom A in all species must equal the atoms of A delivered to the solution. If we deliver a certain ratio of atoms A and B into the solution, then the sum of atoms of A and B in all species must be in that same ratio.

8-17. Charge and the number of atoms are proportional to molarity, not to activity.

8-18. $[H^+] + 2[Ca^{2+}] + [Ca(HCO_3)^+] + [Ca(OH)^+] + [K^+]$
$= [OH^-] + [HCO_3^-] + 2[CO_3^{2-}] + [ClO_4^-]$

8-19. Charge: $[H^+] = [OH^-] + [HSO_4^-] + 2[SO_4^{2-}]$

8-20. $[H^+] = [OH^-] + [H_2AsO_4^-] + 2[HAsO_4^{2-}] + 3[AsO_4^{3-}]$

[Lewis structures of arsenate species shown]

8-21. (a) Charge balance: $2[Mg^{2+}] + [H^+] + [MgBr^+] + [MgOH^+] = [Br^-] + [OH^-]$

Mass balance: total Br = 2(total Mg)

$[MgBr^+] + [Br^-] = 2\{[Mg^{2+}] + [MgBr^+] + [MgOH^+]\}$

(b) $[Mg^{2+}] + [MgBr^+] + [MgOH^+] = 0.2$ M

$[MgBr^+] + [Br^-] = 0.4$ M

8-22. 250 mL of 1.0×10^{-6} M charge = 0.25×10^{-6} moles of charge.

$(0.25 \times 10^{-6} \text{ moles of charge})\left(9.648 \times 10^4 \dfrac{\text{coulombs}}{\text{mole of charge}}\right) = 0.024\,12$ C.

The dielectric constant of air is $\varepsilon = 1$ and the separation is 1.5 m.

Force = $-(8.988 \times 10^9)\dfrac{(0.024\,12)(-0.024\,12)}{(1)(1.5^2)} = 2.3 \times 10^6$ N.

$(2.3 \times 10^6 \text{ N})(0.224\,8 \text{ pounds/N}) = 5.2 \times 10^5$ pounds.

Two elephants do not weigh enough to keep the beakers apart.

8-23. $[CH_3CO_2^-] + [CH_3CO_2H] = 0.1$ M

8-24. Mass balance: $Y_{total} = \dfrac{3}{2} X_{total}$

$2[X_2Y_2^{2+}] + [X_2Y^{4+}] + 3[X_2Y_3] + [Y^{2-}] = \dfrac{3}{2}\{2[X_2Y_2^{2+}] + 2[X_2Y^{4+}] + 2[X_2Y_3]\}$

$[Y^{2-}] = [X_2Y_2^{2+}] + 2[X_2Y^{4+}]$

8-25. 3 (total Fe) = 2 (total sulfur)

$3\{[Fe^{3+}] + [Fe(OH)^{2+}] + [Fe(OH)_2^+] + 2[Fe_2(OH)_2^{4+}] + [FeSO_4^+]\}$
$= 2\{[FeSO_4^+] + [SO_4^{2-}] + [HSO_4^-]\}$

We write 2 in front of $[Fe_2(OH)_2^{4+}]$ because $Fe_2(OH)_2^{4+}$ contains 2 Fe.

8-26. Here is the spreadsheet after executing Goal Seek:

	A	B	C	D	E
1	Using Goal Seek for Ammonia Equilibrium				
2					
3	$pK_b =$	4.755	$K_b = 10\wedge(-pK_b) =$	1.76E-05	$= 10\wedge$-B3
4	$pK_w =$	14.00	$K_w = 10\wedge(-pK_w) =$	1.00E-14	$= 10\wedge$-B4
5	F =	0.05			
6					
7	pH =	10.96791342	Initial value is estimate		
8	$[H^+] =$	1.08E-11	$= 10\wedge$-B7		
9	$[NH_4^+] = K_w/[H^+] - [H^+] =$	9.29E-04	= D4/B8-B8		
10	$[OH^-] = K_w/[H^+] =$	9.29E-04	= D4/B8		
11	$[NH_3] = F - K_w/[H^+] + [H^+] =$	4.91E-02	= B5-D4/B8+B8		
12	$Q = [NH_4^+][OH^-]/[NH_3] =$	1.76E-05	= B9*B10/B11		
13	$K_b - [NH_4^+][OH^-]/[NH_3] =$	3.82E-14	= D3-B12		

Concentrations in cells B8:B11 are $[H^+] = 1.08 \times 10^{-11}$ M, $[NH_4^+] = 9.29 \times 10^{-4}$ M, $[OH^-] = 9.29 \times 10^{-4}$ M, and $[NH_3] = 4.91 \times 10^{-2}$ M. The pH is 10.97 and the fraction of hydrolysis is $[NH_4^+]/([NH_4^+] + [NH_3]) = 9.29 \times 10^{-4}$ M /0.05 M = 1.86%. Increasing the formal concentration of ammonia increased the pH and decreases the fraction of hydrolysis.

	F =	0.01	0.05
	pH =	10.6134	10.96791
	$[H^+] =$	2.44E-11	1.08E-11
$[NH_4^+] = K_w/[H^+] - [H^+] =$		0.000411	0.000929
$[OH^-] = K_w/[H^+] =$		0.000411	0.000929
$[NH_3] = F - K_w/[H^+] + [H^+] =$		0.009589	0.049071
$K_b = [NH_4^+][OH^-]/[NH_3] =$		1.76E-05	1.76E-05
% of hydrolysis = $100*[NH_4^+]/F$		4.11	1.86

8-27. Here is the spreadsheet after executing Solver:

	A	B	C	D	E	F
1	Ammonia equilibrium					F =
2	1. *Estimate* values of pC = -log[C] for NH_4^+ and OH^- in cells B6 and B7					0.01
3	2. Use Solver to adjust the values of pC to minimize the sum in cell F8					
4						
5	Species	pC			Mass and charge balances	$10^6 * b_i$
6	NH_4^+	3.38660381	4.1058E-04	C6 = 10^-B6	$b_1 = F - [NH_4^+] - [NH_3]$	5.29E-07
7	OH^-	3.38660378	4.1058E-04	C7 = 10^-B7	$b_2 = [NH_4^+] + [H^+] - [OH^-]$	-1.07E-07
8	NH_3		9.5894E-03	C8 = C6*C7/D12	$\Sigma(10^6 * b_i)^2 =$	2.91E-13
9	H^+		2.4356E-11	C9 = D13/C7	F6 = 1e6*(F2-C6-C8)	
10					F7 = 1e6*(C6+C9-C7)	
11					F8 = F6^2+F7^2	
12	$pK_b =$	4.755		$K_b =$ 1.76E-05	=10^-B12	
13	$pK_w =$	14.00		$K_w =$ 1.00E-14	=10^-B13	

Concentrations in cells C6:C9 are $[NH_4^+] = 4.11 \times 10^{-4}$ M, $[OH^-]$ = 4.11×10^{-4} M, $[NH_3] = 9.59 \times 10^{-3}$ M and $[H^+] = 2.44 \times 10^{-11}$ M.

The fraction of hydrolysis is $[NH_4^+]/([NH_4^+] + [NH_3]) = 4.11 \times 10^{-4}/0.01 = 4.11\%$.

8-28. (a) Pertinent reactions:

$$A^- + H_2O \underset{}{\overset{K_b}{\rightleftharpoons}} HA + OH^- \qquad K_b = \frac{[HA]\gamma_{HA}[OH^-]\gamma_{OH^-}}{[A^-]\gamma_{A^-}} = 10^{-9.244} \qquad (A)$$

$$H_2O \underset{}{\overset{K_w}{\rightleftharpoons}} H^+ + OH^- \qquad K_w = [H^+]\gamma_{H^+}[OH^-]\gamma_{OH^-} = 10^{-14.00} \qquad (B)$$

Charge balance: $[H^+] + [Na^+] = [OH^-] + [A^-]$ (C)

Mass balance: $[Na^+] = 0.01$ M $\equiv$ F   (D)

Mass balance: $[HA] + [A^-] = 0.01$ M $\equiv$ F   (E)

(b) For initial estimates in the spreadsheet, I chose pA = 2 and pOH = 5. Final concentrations appear in cells C8:C11. The pH is in cell B16 and μ is in cell B5. The fraction of hydrolysis is $[HA]/F = 2.4 \times 10^{-6}/0.010 = 0.024\%$.

	A	B	C	D	E	F	G	H
1	Sodium acetate hydrolysis						F = [Na$^+$] =	
2	1. *Estimate* values of pC = -log[C] for A$^-$ and OH$^-$ in cells B8 and B9						0.01	
3	2. Use Solver to adjust the values of pC to minimize the sum in cell H16							
4	Ionic strength							
5	μ	1.000E-02				Extended	Activity	
6				Size		Debye-Hückel	coefficient	
7	Species	pC	C (M)	α (pm)	Charge	log γ	γ	
8	A$^-$	2.00010375	9.998E-03	450	-1	-4.444E-02	9.027E-01	G8 = 10^F8
9	OH$^-$	5.62093677	2.394E-06	350	-1	-4.575E-02	9.000E-01	
10	HA		2.389E-06		0	0.000E+00	1.000E+00	
11	H$^+$		5.082E-09	900	1	-3.938E-02	9.133E-01	
12								
13	pK$_b$ =	9.244		K$_b$ =	5.70E-10	Mass and charge balances:		10^6*b$_i$
14	pK$_w$ =	14.00		K$_w$ =	1.00E-14	b$_1$ = F - [HA] - [A$^-$]		2.69E-08
15						b$_2$ = [Na$^+$] + [H$^+$] - [A$^-$] - [OH$^-$]		2.573E-07
16	pH = -log([H+]γ$_{H+}$)=		8.33				Σ(10^6*b$_i$)^2 =	6.69E-14
17							H14 = 1e6*(G2-C10-C8)	
18							H15 = 1e6*(G2+C11-C8-C9)	
19	Initial values:						H16 = H14^2 + H15^2	
20	pA$^-$ =	2					C8 = 10^-B8	
21	pOH =	5					C9 = 10^-B9	
22							C10 = D13*C8*G8/(C9*G9*G10)	
23							C11 = D14/(C9*G9*G11)	
24						F8 = -0.51*E8^2*SQRT($B$5)/(1+D8*SQRT($B$5)/305)		
25						B5 = 0.5*(G2+E8^2*C8+E9^2*C9+E10^2*C10+E11^2*C11)		

8-29. (a) Pertinent reactions:

$$Ca(OH)_2(s) \xrightleftharpoons{K_{sp}} Ca^{2+} + 2OH^- \quad K_{sp} = [Ca^{2+}]\gamma_{Ca^{2+}}[OH^-]^2\gamma_{OH^-}^2 = 10^{-5.19}$$

$$Ca^{2+} + OH^- \xrightleftharpoons{K_1} CaOH^+ \quad K_1 = \frac{[CaOH^+]\gamma_{CaOH^+}}{[Ca^{2+}]\gamma_{Ca^{2+}}[OH^-]\gamma_{OH^-}} = 10^{1.30}$$

$$H_2O \xrightleftharpoons{K_w} H^+ + OH^- \quad K_w = [H^+]\gamma_{H^+}[OH^-]\gamma_{OH^-} = 10^{-14.00}$$

Charge balance: $2[Ca^{2+}] + [CaOH^+] + [H^+] = [OH^-]$

$[OH^-] + [CaOH^+] = 2\{[Ca^{2+}] + [CaOH^+]\}$

Mass balance: $\underbrace{[OH^-] + [CaOH^+]}_{\text{species containing OH}^-} \underbrace{\{[Ca^{2+}] + [CaOH^+]\}}_{\text{species containing Ca}^{2+}} + [H^+]$

(Mass balance gives the same result as charge balance.)

There are 4 equations (3 equilibria and charge balance) and 4 unknowns: $[Ca^{2+}]$, $[CaOH^+]$, $[H^+]$, and $[OH^-]$.

(b) There are 4 unknowns and 3 equilibria, so we need to estimate 4 − 3 = 1 concentration. I choose $pCa^{2+} \approx 3$ as an initial estimate. The spreadsheet after executing Solver is shown below.

	A	B	C	D	E	F	G	H
1	Calcium hydroxide equilibria							
2	1. *Estimate* pCa^{2+} in cell B8							
3	2. Use Solver to adjust B8 to minimize sum in cell H15							
4	Ionic strength							
5	μ	5.201E-02				Extended	Activity	
6				Size		Debye-Hückel	coefficient	
7	Species	pC	C (M)	α (pm)	Charge	log γ	γ	
8	Ca^{2+}	1.8077757	1.557E-02	600	2	-3.212E-01	4.773E-01	G8 = 10^F8
9	OH^-		3.645E-02	350	-1	-9.219E-02	8.087E-01	
10	$CaOH^+$		5.311E-03	500	1	-8.466E-02	8.229E-01	
11	H^+		3.982E-13	900	1	-6.952E-02	8.521E-01	
12								
13	pK_{sp} =	5.19		K_{sp} =	6.46E-06	Mass and charge balances:		$10^6 * b_i$
14	pK_1 =	-1.30		K_1 =	2.00E+01	$b_1 = 2[Ca^{2+}]+[CaOH^+]+[H^+]-[OH^-]$		2.78E-11
15	pK_w =	14.00		K_w =	1.00E-14	$\Sigma(10^6 * b_{ij})^2$ =		7.70E-22
16						H14 = 1e6*(2*C8+C10+C11-C9)		
17	Ion size estimate:					H15 = H14^2		
18	$CaOH^+$ size ≈ 500 pm					C8 = 10^-B8		
19						C9 = SQRT(D13/(C8*G8))/G9		
20	Initial value:					C10 = D14*C8*G8*C9*G9/G10		
21	pCa^{2+} =	3				C11 = D15/(C9*G9*G11)		
22						F8 = -0.51*E8^2*SQRT($B$5)/(1+D8*SQRT($B$5)/305)		
23						B5 = 0.5*(E8^2*C8+E9^2*C9+E10^2*C10+E11^2*C11)		

Results: $[Ca^{2+}]$ = 0.015 6 M $[CaOH^+]$ = 0.005 3 M

$[OH^-]$ = 0.036 4 M $[H^+] = K_w/[OH^-]\gamma_{H^+}\gamma_{OH^-}$ = 3.98 × 10^{-13} M

Fraction of hydrolysis = $[CaOH^+]/\{[Ca^{2+}] + [CaOH^+]\}$ = 25%

Total dissolved Ca = 0.015 6 + 0.005 3 = 0.020 9 M

Measured total dissolved calcium = 0.019 8 M at 25°C (excellent agreement)

The formula mass of $Ca(OH)_2$ is 74.09 g/mol, so 0.020 9 M is 1.5$_5$ g/L. The calculated solubility makes sense because it is between the reported values of 1.85 g/L at 0°C and 0.77 g/L at 100°C.

8-30. The spreadsheet shows $[Na^+] = [Cl^-] = 0.02486$ M and $[NaCl(aq)] = 0.000143$ M in cells C8:C10. Ionic strength = 0.02486 M in cell B5. Fraction of ion pairing = $[NaCl(aq)]/F = 0.57\%$ in cell D15.

	A	B	C	D	E	F	G	H
1	Sodium Chloride ion pairing with activities							
2	1. *Estimate* pNa^+ and pCl^- in cells B8 and B9				Formal conc = F =		0.025	M
3	2. Use Solver to adjust B8 and B9 to minimize sum in cell H16							
4	Ionic strength							
5	μ	2.486E-02				Extended	Activity	
6					Size	Debye-Hückel	coefficient	
7	Species	pC	C (M)	α (pm)	Charge	log γ	γ	
8	Na^+	1.604555369	2.486E-02	450	1	-6.523E-02	0.861	G8 = 10^F8
9	Cl^-	1.604555369	2.486E-02	300	-1	-6.961E-02	0.852	
10	NaCl(aq)		1.432E-04		0	0.000E+00	1.000	
11								
12								
13	pK_{ip} =	0.50		K_{ip} =	3.16E-01	Mass and charge balances:		$10^6 \cdot b_i$
14						$b_1 = F - [Na^+] - [NaCl]$		5.55E-07
15	Ion pair fraction = $[NaCl(aq)]/F$ =			0.0057		$b_2 = [Na^+] - [Cl^-]$		-7.717E-08
16				D15 = C10/G2		$\Sigma(10^6 \cdot b_i)^2 =$		3.14E-13
17	Ion size estimate:							H14 = 1e6*(G2-C8-C10)
18								H15 = 1e6*(C8-C9)
19	Initial value:							H16 = H14^2+H15^2
20	pNa^+ =	2						C8 = 10^-B8
21	pCl^- =	2						C9 = 10^-B9
22	check:					C10 = D13*C8*G8*C9*G9/G10		
23	$[Na^+] + [NaCl(aq)]$ =		0.025000		F8 = -0.51*E8^2*SQRT($B$5)/(1+D8*SQRT($B$5)/305)			
24	$[Cl^-] + [NaCl(aq)]$ =		0.025000		B5 = 0.5*(E8^2*C8+E9^2*C9+E10^2*C10)			

8-31. The spreadsheet shows [Na$^+$] = 0.047 75 M, [SO$_4^{2-}$] 0.022 75M, and [NaSO$_4^-$(aq)] = 0.002 246 M in cells C8:C10. Ionic strength = 0.070 51 M in cell B5. Ion pair fraction = [NaSO$_4^-$(aq)]/F = 8.98% in cell D15.

	A	B	C	D	E	F	G	H
1	Sodium Sulfate ion pairing with activities							
2	1. *Estimate* pNa and PSO$_4$ in cells B8 and B9					Formal conc = F =	0.025	M
3	2. Use Solver to adjust B8 and B9 to minimize cell H16					[Na+] + [NaSO$_4^-$(aq)] = 2F		
4	Ionic strength					[SO4^{2-}] + [NaSO$_4^-$(aq)] = F		
5	μ	7.051E-02				Extended	Activity	
6				Size		Debye-Hückel	coefficient	
7	Species	pC	C (M)	α (pm)	Charge	log γ	γ	
8	Na$^+$	1.320987425	4.775E-02	450	1	-9.730E-02	0.799	G8 = 10^F8
9	SO$_4^{2-}$	1.642936328	2.275E-02	400	-2	-4.018E-01	0.396	
10	NaSO$_4^-$		2.246E-03	500	-1	-9.435E-02	0.805	
11								
12								
13	pK$_{ip}$ =	-0.72		K$_{ip}$ =	5.25E+00	Mass and charge balances:		10^6*b$_i$
14						b$_1$ = 2F - [Na$^+$] - [NaSO$_4^-$]		4.36E-07
15		Ion pair fraction = [NaSO$_4^-$(aq)]/F =		0.0898		b$_2$ = F - [SO$_4^{2-}$] - [NaSO$_4^-$]		-4.22E-07
16				D15 = C10/G2		Σ(10^6*b$_i$)^2 =		3.68E-13
17	Ion size estimate:					H14 = 1e6*(2*G2-C8-C10)		
18	NaSO$_4^-$ =		500			H15 = 1e6*(G2-C9-C10)		
19	Initial value:					H16 = H14^2 + H15^2		
20	pNa$^+$ =		1.4			C8 = 10^-B8		
21	pSO$_4^{2-}$ =		1.5			C9 = 10^-B9		
22	Check:					C10 = D13*C8*G8*C9*G9/G10		
23		[Na$^+$] + [NaSO$_4^-$] =		0.050000		F8 = -0.51*E8^2*SQRT($B$5)/(1+D8*SQRT($B$5)/305)		
24		[SO$_4^{2-}$] + [NaSO$_4^-$] =		0.025000		B5 = 0.5*(E8^2*C8+E9^2*C9 +E10^2*C10)		

8-32. (a) The spreadsheet shows [Mg^{2+}] = [SO$_4^{2-}$] = 0.016 16M and [MgSO$_4$(aq)] = 0.008 844 M in cells C8:C10. Ionic strength = 0.064 63 M in cell B5. Fraction of ion pairing = 35.4% in cell D15.

(b) $Mg^{2+} + OH^- \rightleftharpoons MgOH^+$ $K_1 = 10^{2.6}$ in Appendix I

$SO_4^{2-} + H_2O \rightleftharpoons HSO_4^- + OH^-$ $pK_b = 12.01$

For hydrolysis of Mg^{2+} with $[Mg^{2+}] = 0.016$ M and $[OH^-] = 10^{-7}$ M, $[MgOH^+] \approx K_1[Mg^{2+}][OH^-] = 10^{2.6}[0.016]10^{-7} = 6 \times 10^{-7}$ M, which is negligible in comparison with $[Mg^{2+}] = 0.016$ M.

For hydrolysis of SO_4^{2-}, $K_b = K_w/K_a$, with $pK_a = 1.99$ in Appendix G. $K_b = K_w/K_a = 10^{-14.00}/10^{-1.99} = 10^{-12.01}$. Inserting $[SO_4^{2-}] = 0.016$ M and $[OH^-] = 10^{-7}$ M, we estimate $[HSO_4^-] \approx K_b[SO_4^{2-}]/[OH^-] = 10^{-12.01}[0.016]/10^{-7} = 2 \times 10^{-7}$ M, which is negligible in comparison with $[SO_4^{2-}] = 0.016$ M.

	A	B	C	D	E	F	G	H
1	Magnesium Sulfate ion pairing with activities							
2	1. *Estimate* pMg^{2+} and pSO_4^{2-} in cells B8 and B9				Formal conc = F =		0.025	M
3	2. Use Solver to adjust B8 and B9 to minimize sum in cell H16							
4	Ionic strength							
5	μ	6.463E-02				Extended	Activity	
6				Size		Debye-Hückel	coefficient	
7	Species	pC	C (M)	α (pm)	Charge	log γ	γ	
8	Mg^{2+}	1.79165324	1.616E-02	800	2	-3.111E-01	4.885E-01	G8 = 10^F8
9	SO_4^{2-}	1.79165324	1.616E-02	400	-2	-3.889E-01	4.084E-01	
10	$MgSO_4(aq)$		8.844E-03			0.000E+00	1.000E+00	
11								
12								
13	pK_{ip} =	-2.23		K_{ip} =	1.70E+02	Mass and charge balances:		$10^6 * b_i$
14						$b_1 = F - [Mg^{2+}] - [MgSO_4]$		2.45E-07
15	Ion pair fraction = $[MgSO_4(aq)]/F =$			0.3537		$b_2 = [Mg^{2+}] - [SO4^{2-}]$		-2.94E-07
16				D15 = C10/G2			$\Sigma b_i^2 =$	1.47E-13
17	Ion size estimate:						H14 = 1e6*(G2-C8-C10)	
18							H15 = 1e6*(C8-C9)	
19	Initial value:						H16 = H14^2+H15^2	
20	pMg^{2+} =	1.9					C8 = 10^-B8	
21	pSO_4^{2-} =	1.9					C9 = 10^-B9	
22	Check:						C10 = D13*C8*G8*C9*G9/G10	
23	$[Mg^{2+}] + [MgSO_4(aq)] =$		0.025000		F8 = -0.51*E8^2*SQRT($B$5)/(1+D8*SQRT($B$5)/305)			
24						B5 = 0.5*(E8^2*C8+E9^2*C9)		

8-33. (a) Equilibria:

$$\text{LiF}(s) \rightleftharpoons \text{Li}^+ + \text{F}^- \qquad K_{sp} = [\text{Li}^+]\gamma_{\text{Li}^+}[\text{F}^-]\gamma_{\text{F}^-} \qquad pK_{sp} = 2.77$$

$$\text{LiF}(s) \rightleftharpoons \text{LiF}(aq) \qquad K_{\text{ion pair}} = [\text{LiF}(aq)]\gamma_{\text{LiF}(aq)} \qquad pK_{\text{ion pair}} = 2.54$$

$$\text{F}^- + \text{H}_2\text{O} \rightleftharpoons \text{HF} + \text{OH}^- \qquad K_b = \frac{K_w}{K_a} = \frac{[\text{HF}]\gamma_{\text{HF}}[\text{OH}^-]\gamma_{\text{OH}^-}}{[\text{F}^-]\gamma_{\text{F}^-}} \qquad pK_b = 10.83$$

$$\text{H}_2\text{O} \overset{K_w}{\rightleftharpoons} \text{H}^+ + \text{OH}^- \qquad K_w = [\text{H}^+]\gamma_{\text{H}^+}[\text{OH}^-]\gamma_{\text{OH}^-} \qquad pK_w = 14.00$$

Derivation of $pK_{\text{ion pair}}$:

$$\text{LiF}(s) \rightleftharpoons \text{Li}^+ + \text{F}^- \qquad pK_{sp} = 2.77 \text{ from Appendix F}$$

$$\underline{\text{Li}^+ + \text{F}^- \rightleftharpoons \text{LiF}(aq) \qquad pK_{\text{formation}} = -0.23 \text{ from Appendix J}}$$

$$\text{LiF}(s) \rightleftharpoons \text{LiF}(aq) \qquad K_{\text{ion pair}} = K_{sp}K_{\text{formation}}$$

$$pK_{\text{ion pair}} = pK_{sp} + pK_{\text{formation}} = 2.54$$

Charge balance: $[\text{Li}^+] + [\text{H}^+] = [\text{F}^-] + [\text{OH}^-]$

Mass balance: $[\text{Li}^+] + [\text{LiF}(aq)] = [\text{F}^-] + [\text{LiF}(aq)] + [\text{HF}]$

(b) There are six equations and six unknowns: $[\text{Li}^+]$, $[\text{F}^-]$, $[\text{LiF}(aq)]$, $[\text{HF}]$, $[\text{H}^+]$, and $[\text{OH}^-]$. For the spreadsheet, we need to estimate values for (6 unknowns) − (4 equilibria) = 2 unknowns. I choose to estimate $p\text{F}^-$ and $p\text{OH}^-$ because F^- and OH^- are involved in multiple equilibria. It does not work to chose F^- and Li^+ because either concentration fixes that of the other through the relation $K_{sp} = [\text{Li}^+]\gamma_{\text{Li}^+}[\text{F}^-]\gamma_{\text{F}^-}$. The equations in the spreadsheet are:

$$[\text{Li}^+] = K_{sp}/([\text{F}^-]\gamma_{\text{F}^-}\gamma_{\text{Li}^+}) \qquad [\text{HF}] = K_b[\text{F}^-]\gamma_{\text{F}^-}/([\text{OH}^-]\gamma_{\text{OH}^-}\gamma_{\text{F}^-})$$

$$[\text{LiF}(aq)] = K_{\text{ion pair}}/\gamma_{\text{LiF}(aq)} \qquad [\text{H}^+] = K_w/[\text{OH}^-]\gamma_{\text{OH}^-}\gamma_{\text{H}^+}$$

	A	B	C	D	E	F	G	H	I
1	Lithium fluoride equilibria								
2	1. *Estimate* values in cells B8 and B9								
3	2. Use Solver to adjust B8 and B9 to minimize sum in cell I19								
4	Ionic strength					Extended			
5	$\mu =$	0.050126				Debye-			
6				Size		Hückel	Activity		
7	Species	pC	C (M)	α (pm)	Charge	log γ	coefficient, γ		
8	F^-	1.299947	5.012E-02	350	-1	-9.084E-02	0.811		
9	OH^-	6.060800	8.694E-07	350	-1	-9.084E-02	0.811		
10	Li^+		5.013E-02	600	1	-7.927E-02	0.833		
11	LiF(aq)		2.884E-03		0	0.000E+00	1.000		
12	HF		8.528E-07		0	0.000E+00	1.000		
13	H^+		1.661E-08	900	1	-6.876E-02	0.854		
14									
15									
16							Mass and charge balances:		$10^6 * b_i$
17	$pK_{sp} =$	2.77	$K_{sp} =$	1.70E-03		$b_1 = [Li^+] - [F^-] - [HF]$			5.09E-04
18	$pK_{ion\ pair} =$	2.54	$K_{ion\ pair} =$	2.88E-03		$b_2 = [Li^+] + [H^+] - [F^-] - [OH^-]$			5.72E-04
19	$pK_{base} =$	10.83	$K_{base} =$	1.48E-11		$\Sigma(10^6 * b_i)^2 =$			5.87E-07
20	$pK_w =$	14.00	$K_w =$	1.00E-14					I17 = 1e6*(C10-C8-C12)
21									I18 = 1e6*(C10+C13-C8-C9)
22	Initial values:								I19 = I17^2 + I18^2
23	pF = 1.5	pOH = 6							C8 = 10^-B8
24									C9 = 10^-B9
25	Optimize both pF and pOH together for a few cycles								C10 = D17/(C8*G8*G10)
26	Then optimize just pF and just pOH alternately								C11 = D18
27	Continue optimization as long as cell I19 becomes smaller								C12 = D19*C8*G8/(G9*C9*G12)
28									C13 = D20/(C9*G9*G13)
29				B5 = 0.5*(E8^2*C8+E9^2*C9+E10^2*C10+E11^2*C11+E12^2*C12+E13^2*C13)					

8-34. (a)

$$CaCO_3(s) \rightleftharpoons Ca^{2+} + CO_3^{2-} \qquad K_{sp} = 4.5 \times 10^{-9}$$
$$CO_2(aq) + H_2O \rightleftharpoons HCO_3^- + H^+ \qquad K_1 = 4.46 \times 10^{-7}$$
$$CO_3^{2-} + H^+ \rightleftharpoons HCO_3^- \qquad 1/K_2 = 1/(4.69 \times 10^{-11})$$
$$\overline{CaCO_3(s) + CO_2(aq) + H_2O \rightleftharpoons Ca^{2+} + 2HCO_3^-} \qquad K = K_{sp} K_1 K_2$$
$$= 4.2_8 \times 10^{-5}$$

(b) The equilibrium constant for the net reaction is

$$\frac{[Ca^{2+}][HCO_3^-]^2}{[CO_2(aq)]} = K = 4.2_8 \times 10^{-5}$$

We can substitute into this equation $[HCO_3^-] = 2[Ca^{2+}]$ (from the mass balance) and $[CO_2(aq)] = K_{CO_2} P_{CO_2}$ (from Henry's law, where $K_{CO_2} = 0.032$ and $P_{CO_2} = 4.0 \times 10^{-4}$ bar):

$$\frac{[Ca^{2+}](2[Ca^{2+}])^2}{K_{CO_2} P_{CO_2}} = K = 4.2_8 \times 10^{-5} \Rightarrow [Ca^{2+}] = 5.1_5 \times 10^{-4} \text{ M}$$

$$= 21 \text{ mg/L}$$

(c) If $[Ca^{2+}] = 80$ mg/L $= 2.0 \times 10^{-3}$ M, then

$$P_{CO_2} = \frac{[Ca^{2+}](2[Ca^{2+}])^2}{K_{CO_2} K} = 0.023 \text{ bar}$$

The partial pressure of CO_2 in the river is about $(0.023 \text{ bar})/(4.0 \times 10^{-4} \text{ bar}) = 58$ times higher than the atmospheric pressure of CO_2. There must be a source of extra CO_2 such as respiration in the river or inflow of ground water that is very rich in CO_2 and not in equilibrium with the atmosphere.

CHAPTER 9
MONOPROTIC ACID-BASE EQUILIBRIA

9-1. HBr (or any other acid or base) drives the reaction $H_2O \rightleftharpoons H^+ + OH^-$ to the left, according to Le Châtelier's principle. If, for example, the solution contains 10^{-4} M HBr, the concentration of OH^- from H_2O is $K_w/[H^+] = 10^{-10}$ M. The concentration of H^+ from H_2O must also be 10^{-10} M, since H^+ and OH^- are created in equimolar quantities.

9-2. (a) $pH = -\log[H^+] = -\log(1.0 \times 10^{-3}) = 3.00$

(b) $[H^+] = K_w/[OH^-] = (1.0 \times 10^{-14})/(1.0 \times 10^{-2}) = 1.0 \times 10^{-12}$ M
$pH = -\log[H^+] = 12.00$

9-3. Charge balance: $[H^+] = [OH^-] + [ClO_4^-] \Rightarrow [OH^-] = [H^+] - 5.0 \times 10^{-8}$
Mass balance is the same as charge balance.
Equilibrium: $[H^+][OH^-] = K_w$
$[H^+]([H^+] - 5.0 \times 10^{-8}) = 1.0 \times 10^{-14} \Rightarrow [H^+] = 1.28 \times 10^{-7}$ M
$pH = -\log[H^+] = 6.89$
$[OH^-] = K_w/[H^+] = 7.8 \times 10^{-8}$ M $\Rightarrow [H^+]$ from $H_2O = 7.8 \times 10^{-8}$ M

Fraction of $[H^+]$ from $H_2O = \dfrac{7.8 \times 10^{-8} \text{ M}}{1.28 \times 10^{-7} \text{ M}} = 0.61$

9-4. (a) $pH = -\log[H^+]\gamma_{H^+}$
$1.092 = -\log(0.100)\gamma_{H^+} \Rightarrow \gamma_{H^+} = 0.809$
The tabulated activity coefficient is 0.83.

(b) $2.102 = -\log(0.0100)\gamma_{H^+} \Rightarrow \gamma_{H^+} = 0.791$

(c) The activity coefficient depends somewhat on the identity of the other ions.

9-5. (a) $\text{Ph–CO}_2\text{H} \rightleftharpoons \text{Ph–CO}_2^- + H^+$ $\quad K_a$

(b) $\text{Ph–CO}_2^- + H_2O \rightleftharpoons \text{Ph–CO}_2H + OH^-$ $\quad K_b$

(c) $\text{Ph–NH}_2 + H_2O \rightleftharpoons \text{Ph–NH}_3^+ + OH^-$ $\quad K_b$

(d) $\text{Ph–NH}_3^+ \rightleftharpoons \text{Ph–NH}_2 + H^+$ $\quad K_a$

93

9-6. Let $x = [H^+] = [A^-]$ and $0.100 - x = [HA]$.

$$\frac{x^2}{0.100-x} = 1.00 \times 10^{-5} \Rightarrow x = 9.95 \times 10^{-4} \text{ M} \Rightarrow \text{pH} = -\log x = 3.00$$

$$\alpha = \frac{[A^-]}{[A^-]+[HA]} = \frac{9.95 \times 10^{-4}}{0.100} = 9.95 \times 10^{-3}$$

9-7. $$BH^+ \underset{}{\overset{K_a}{\rightleftharpoons}} B + H^+ \quad K_a = K_w/K_b = 1.00 \times 10^{-10}$$
$$0.100-x \qquad x \quad x$$

$$\frac{x^2}{0.100-x} = 1.00 \times 10^{-10} \Rightarrow x = [B] = [H^+] = 3.16 \times 10^{-6} \text{ M} \Rightarrow \text{pH} = 5.50$$

9-8. The initial spreadsheet follows (on the left). Guess a value for x in cell A4. The formula in cell B4 is "=A4^2/(A6-A4)". Before using Goal Seek in Excel 2016, click the File menu and select Options. In the Options window, select Formulas. Check Enable iterative calculation and Set Maximum Change to 1e-15 to find an answer with high precision. Highlight cell B4 and select Goal Seek from the Data ribbon and What-If Analysis. Set cell B4 To value 1e-5 By changing cell A4. Click OK and Goal Seek finds the solution in the second spreadsheet (on the right). The value $x = 9.95 \times 10^{-5}$ makes the quotient $x^2/(F-x)$ equal to 1.00×10^{-5}.

	A	B
1	Using Excel GOAL SEEK	
2		
3	x =	x²/(F-x) =
4	0.01	1.1111E-03
5	F =	
6		0.1

	A	B
1	Using Excel GOAL SEEK	
2		
3	x =	x²/(F-x) =
4	0.00099501	1.0000E-05
5	F =	
6		0.1

Before executing Goal Seek   After executing Goal Seek

9-9. $(CH_3)_3NH^+ \rightleftharpoons (CH_3)_3N + H^+ \quad K_a = 1.59 \times 10^{-10}$
    $F - x \qquad\qquad x \quad\quad x$

$$\frac{x^2}{0.060-x} = K_a \Rightarrow x = 3.0_9 \times 10^{-6} \Rightarrow \text{pH} = 5.51$$

$[(CH_3)_3N] = x = 3.1 \times 10^{-6}$ M, $[(CH_3)_3NH^+] = F - x = 0.060$ M

9-10. C₆H₅—CO₂H ⇌ C₆H₅—CO₂⁻ + H⁺

$F - 10^{-2.78} \qquad\qquad 10^{-2.78} \qquad\quad 10^{-2.78}$

Monoprotic Acid-Base Equilibria

$$K_a = \frac{(10^{-2.78})^2}{0.0450 - 10^{-2.78}} = 6.35 \times 10^{-5} = pK_a = 4.20$$

9-11. $HA \rightleftharpoons H^+ + A^-$ $\alpha = 0.0060 = \frac{x}{F}$
  $F-x \quad\quad x \quad\quad x$

$F = 0.0450$ M and $x = (0.0060)(0.0450$ M$) = 2.7 \times 10^{-4}$ M

$$\Rightarrow K_a = \frac{x^2}{F-x} = \frac{(2.7 \times 10^{-4})^2}{F - 2.7 \times 10^{-4}} = 1.6 \times 10^{-6} \Rightarrow pK_a = 5.79$$

9-12. (a) $HA \rightleftharpoons H^+ + A^-$
  $F-x \quad\quad x \quad\quad x$

$$\frac{x^2}{F-x} = K_a \quad \frac{x^2}{0.010-x} = 9.8 \times 10^{-5} \Rightarrow x = 9.4 \times 10^{-4}$$

$\Rightarrow pH = 3.03$

$$\alpha = \frac{[A^-]}{[HA]+[A^-]} = \frac{x}{F} = 9.4\%$$

(b) pH = 7.00 because the acid is so dilute. From the K_a equilibrium we write

$$[A^-] = \frac{K_a}{[H^+]}[HA] = \frac{9.8 \times 10^{-5}}{1.0 \times 10^{-7}}[HA] = 980 [HA]$$

$$\alpha = \frac{[A^-]}{[HA]+[A^-]} = \frac{980[HA]}{[HA]+980[HA]} = \frac{980}{981} = 99.9\%$$

Notice that activity coefficients do not appear in the expression for α.

9-13. Phenol is a weak acid, so it will contribute negligible ionic strength. The ionic strength of the solution is 0.050 M.

  $HA \rightleftharpoons H^+ + A^- \quad\quad K_a = 1.01 \times 10^{-10}$
  $F-x \quad\quad x \quad\quad x$

$$\frac{[H^+]\gamma_{H^+}[A^-]\gamma_{A^-}}{[HA]\gamma_{HA}} = K_a \Rightarrow \frac{(x)(0.86)(x)(0.835)}{(0.0500-x)(1.00)} = 1.01 \times 10^{-10}$$

$\Rightarrow x = 2.65 \times 10^{-6}$

$pH = -\log[H^+]\gamma_{H^+} = -\log[2.65 \times 10^{-6}](0.86) = 5.64$

$$\alpha = \frac{[A^-]}{[HA]+[A^-]} = \frac{2.65 \times 10^{-6}}{0.0500} = 5.30 \times 10^{-5} = 0.0053\%$$

9-14.
$$Cr^{3+} + H_2O \underset{}{\overset{K_{a1}}{\rightleftharpoons}} Cr(OH)^{2+} + H^+$$
$$\underset{0.010-x}{\phantom{Cr^{3+}}} \qquad \underset{x}{\phantom{Cr(OH)^{2+}}} \quad \underset{x}{}$$

$$\frac{x^2}{0.010-x} = 10^{-3.66} \Rightarrow x = 1.3_7 \times 10^{-3} \text{ M}$$

$$pH = -\log x = 2.86 \qquad \alpha = \frac{x}{0.010} = 0.14$$

9-15.
$$HNO_3 \rightleftharpoons H^+ + NO_3^-$$
$$\underset{F-x}{} \quad \underset{x}{} \quad \underset{x}{}$$

$$\frac{x^2}{F-x} = 26.8 \Rightarrow x = 0.099\,6 \text{ M when F} = 0.100 \text{ M} \Rightarrow \alpha = \frac{x}{F} = 99.6\%$$

$$\Rightarrow x = 0.965 \text{ M when F} = 1.00 \text{ M} \Rightarrow \alpha = \frac{x}{F} = 96.5\%$$

9-16. The "fishy" smell comes from volatile amines (R_3N). Lemon juice protonates the amines, giving less volatile ammonium ions (R_3NH^+).

9-17. Let $x = [OH^-] = [BH^+]$ and $0.100 - x = [B]$. $\dfrac{x^2}{0.100-x} = 1.00 \times 10^{-5}$

$\Rightarrow x = 9.95 \times 10^{-4}$ M $\Rightarrow [H^+] = \dfrac{K_w}{x} = 1.005 \times 10^{-11} \Rightarrow$ pH = 11.00

$$\alpha = \frac{[BH^+]}{[B]+[BH^+]} = \frac{9.95 \times 10^{-4}}{0.100} = 9.95 \times 10^{-3}$$

9-18.
$$(CH_3)_3N + H_2O \rightleftharpoons (CH_3)_3NH^+ + OH^- \qquad K_b = K_w/K_a = 6.3 \times 10^{-5}$$
$$\underset{F-x}{} \quad \underset{x}{} \quad \underset{x}{}$$

$$\frac{x^2}{0.060-x} = K_b \Rightarrow x = 1.9_1 \times 10^{-3} \Rightarrow pH = -\log \frac{K_w}{x} = 11.28$$

$[(CH_3)_3NH^+] = x = 1.9_1 \times 10^{-3}$ M, $[(CH_3)_3N] = F - x = 0.058$ M

9-19.
$$CN^- + H_2O \rightleftharpoons HCN + OH^- \qquad K_b = K_w/K_a = 1.6 \times 10^{-5}$$
$$\underset{F-x}{} \quad \underset{x}{} \quad \underset{x}{}$$

$$\frac{x^2}{0.050-x} = K_b \Rightarrow x = 8.9 \times 10^{-4} \Rightarrow pH = -\log \frac{K_w}{x} = 10.95$$

Monoprotic Acid-Base Equilibria

9-20. $CH_3CO_2^- + H_2O \rightleftharpoons CH_3CO_2H + OH^-$ $K_b = K_w/K_a = 5.7 \times 10^{-10}$
$F - x$ $\qquad\qquad\qquad x \qquad\qquad x$

$$\frac{x^2}{(1.00 \times 10^{-1}) - x} = K_b \Rightarrow x = 7.6 \times 10^{-6} \Rightarrow \alpha = \frac{x}{F} = 0.0076\%$$

$$\frac{x^2}{(1.00 \times 10^{-2}) - x} = K_b \Rightarrow x = 2.4 \times 10^{-6} \Rightarrow \alpha = \frac{x}{F} = 0.024\%$$

For 1.00×10^{-12} M sodium acetate, pH = 7.00 and we can say

$$[HA] = \frac{K_b[A^-]}{[OH^-]} = \frac{(5.7 \times 10^{-10})[A^-]}{1.0 \times 10^{-7}} = (5.7 \times 10^{-3})[A^-]$$

$$\alpha = \frac{[HA]}{[HA] + [A^-]} = \frac{(5.7 \times 10^{-3})[A^-]}{(5.7 \times 10^{-3})[A^-] + [A^-]} = 0.57\%$$

The more dilute the solution, the greater is α.

9-21. $\qquad\qquad\quad B \quad + \quad H_2O \rightleftharpoons BH^+ \quad + \quad OH^-$
$\qquad\qquad F - (K_w/10^{-9.28}) \qquad\qquad\quad K_w/10^{-9.28} \quad K_w/10^{-9.28}$

$$K_b = \frac{(K_w/10^{-9.28})^2}{F - (K_w/10^{-9.28})} = \frac{(K_w/10^{-9.28})^2}{0.10 - (K_w/10^{-9.28})} = 3.6 \times 10^{-9}$$

9-22. $\qquad B + H_2O \rightleftharpoons BH^+ + OH^- \qquad \alpha = 0.020 = \frac{x}{F} \Rightarrow x = 2.0 \times 10^{-3}$ M
$\qquad\quad 0.10 - x \qquad\qquad\quad x \quad\quad x$

$$K_b = \frac{x^2}{0.10 - x} = \frac{(2.0 \times 10^{-3})^2}{0.10 - (2.0 \times 10^{-3})} = 4.1 \times 10^{-5}$$

9-23. I would weigh out 0.0200 mol of acetic acid (=1.201 g) and place it in a beaker with ~75 mL of water. While monitoring the pH with a pH electrode, I would add 3 M NaOH (~4 mL is required) until the pH is exactly 5.00. I would then pour the solution into a 100-mL volumetric flask and wash the beaker several times with a few milliliters of distilled water. Each washing would be added to the volumetric flask, to ensure quantitative transfer from the beaker to the flask. After swirling the volumetric flask to mix the solution, I would carefully add water up to the 100 mL mark, insert the cap, and invert 20 times to ensure complete mixing.

98   Chapter 9

9-24. The inside cover of the book states that 67.6 mL of 28 wt% ammonia are required to make 1 L of ~1.0 M solution. To make 250 mL of 1 M NH_3 will require 1/4 as much = 16.9 mL of 28 wt% ammonia. (Note that the composition of 28 wt% ammonia is only approximate and the resulting buffer molarity will only be approximate.) Add 16.9 mL of 28 wt% ammonia to ~160 mL of H_2O in a beaker containing a pH electrode and stir bar. We need somewhat more than half the moles of HCl to lower the pH of ammonia to 9, which is below pK_a = 9.24. 250 mL of 1 M NH_3 = 0.25 mol NH_3, which will require a little more than 0.125 mol HCl. The back of the book tells us that 82.4 mL of 37.2% HCl contains 1 mol HCl. We will need somewhat more than 1/8 of this volume of HCl ≈ 10 mL of HCl. While measuring the pH of the solution, add about 9 mL of HCl. Then add HCl dropwise until the pH is 9.00. Transfer the liquid to a 250-mL volumetric flask and wash the beaker and stir bar many times with small quantities of H_2O. Pour all of the washings into the volumetric flask to make a quantitative transfer. Then dilute to exactly 250 mL. The buffer will only be ~1.0 M because the concentration of 28 wt% NH_3 is not precisely known.

9-25. The solution contains 100 mL of ~0.050 0 M $B(OH)_3$ and ~0.050 0 M $B(OH)_4^-$, which amounts to ~5 mmol of $B(OH)_3$ and ~5 mmol of $B(OH)_4^-$. We do not want the acid generated by the chemical reaction to consume more than half of the $B(OH)_4^-$, which would be about 2.5 mmol of acid. The pH would be reduced from pH = pK_a = 9.24 down to pH = pK_a + log [$B(OH)_4^-$]/[$B(OH)_3$]
= 9.24 + log (2.5 mmol/7.5 mmol) = 8.76.

9-26. The pH of a buffer depends on the ratio of the concentrations of HA and A^- (pH = pK_a + log [A^-]/[HA]). When the volume of solution is changed, both concentrations are affected equally and their ratio does not change.

9-27. Buffer capacity measures the ability to maintain the original [A^-]/[HA] ratio when acid or base is added. A more concentrated buffer has more A^- and HA, so a smaller fraction of A^- or HA is consumed by a given addition of acid or base. The higher the buffer concentration, the smaller is the change in the ratio [A^-]/[HA].

9-28. At very low or very high pH, there is so much acid or base in the solution already that small additions of acid or base have little effect. At low pH, the buffer is H_3O^+/H_2O; and at high pH, the buffer is H_2O/OH^-.

Monoprotic Acid-Base Equilibria

9-29. When pH = pK_a, the ratio of concentrations [A$^-$]/[HA] is unity. A given increment of added acid or base has the least effect on the ratio [A$^-$]/[HA] when the concentrations of A$^-$ and HA are initially equal.

9-30. The Henderson-Hasselbalch is just a rearranged form of the K_a equilibrium expression, which is always true. When we make the approximation that [HA] and [A$^-$] are unchanged from what we added, we are neglecting acid dissociation and base hydrolysis, which can change the concentrations in dilute solutions of moderately strong acids or bases.

9-31.

acid	pK_a	
hydrogen peroxide	11.65	
propanoic acid	4.87	
cyanoacetic acid	2.47	
4-aminobenzenesulfonic acid	3.23	← most suitable because pK_a is closest to desired pH

H_2O_2 would never be good as a buffer because it is reactive and unstable. Propanoic acid is flammable and harmful in case of contact. Cyanoacetic acid is hygroscopic, so it would not be easy to weigh. It also burns skin and eyes on contact. 4-Amino-benzenesulfonic acid is hazardous in case of skin contact or inhalation.

9-32. $\text{pH} = \text{p}K_a + \log\dfrac{[\text{A}^-]}{[\text{HA}]} = 5.00 + \log\dfrac{0.050}{0.100} = 4.70$

9-33. $\text{pH} = 3.744 + \log\dfrac{[\text{HCO}_2^-]}{[\text{HCO}_2\text{H}]}$

pH:	3.000	3.744	4.000
[HCO$_2^-$]/[HCO$_2$H]:	0.180	1.00	1.80

9-34. $\text{pH} = 3.57 + \log\dfrac{[\text{HCO}_2^-]}{[\text{HCO}_2\text{H}]}$, where 3.57 is p$K_a$ at μ = 0.1 M

Substituting pH = 3.744 gives [HCO$_2^-$]/[HCO$_2$H] = 1.5

9-35. $\text{pH} = \text{p}K_a + \log\dfrac{[\text{NO}_2^-]}{[\text{HNO}_2]}$, where p$K_a$ = 14.00 − pK_b = 3.15

(a) If pH = 2.00, [HNO$_2$]/[NO$_2^-$] = 14

(b) If pH = 10.00, [HNO$_2$]/[NO$_2^-$] = 1.4 × 10^{-7}

9-36. (a) HEPES is an acid with pK_a = 7.56. When it is dissolved in water, the solution will be acidic and will require NaOH to raise the pH to 7.45.

(b) 1. Weigh out (0.250 L)(0.0500 M) = 0.0125 mol of HEPES and dissolve in ~200 mL.

2. With stirring, adjust pH to 7.45 with NaOH using a pH meter.

3. Dilute to 250 mL in a volumetric flask.

9-37. 2,2′-Bipyridine (B) is a weak base whose conjugate acid is BH$^+$ with acid dissociation pK_a (which is pK_{BH^+}) = 4.34.

We are asked to add x mol of strong acid to 213 mL of 0.00666 M 2,2′-bipyridine (= 1.41$_9$ mmol base) to make the mixture of B and BH$^+$ with a pH of 4.19.

	2,2′-bipyridine (B)	+	H$^+$	→	2,2′-bipyridineH$^+$ (BH$^+$)
Initial mmol:	1.41$_9$		x		—
Final mmol:	1.41$_9$ – x		—		x

From the table, the mixture will have (1.41$_9$ – x) mol B and x mol BH$^+$.

$$pH = pK_a + \log \frac{[B]}{[BH^+]} = pK_{BH^+} + \log \frac{[B]}{[BH^+]}$$

$$4.19 = 4.34 + \log \frac{1.41_9 - x}{x}$$

$$(4.19 - 4.34) = -0.15 = \log \frac{1.41_9 - x}{x}$$

Raising 10 to the number on each side of the equation gives

$$10^{-0.15} = 10^{\log[(1.41_9 - x)/x]}$$

$$0.707946 = \frac{1.41_9 - x}{x} \Rightarrow x = 0.831 \text{ mmol of HNO}_3 \text{ required}$$

The concentration of HNO$_3$ is 0.246 M, so volume = $\dfrac{0.831 \text{ mmol}}{0.246 \text{ mmol/mL}}$

= 3.38 mL

Monoprotic Acid-Base Equilibria   101

9.38. (a) imidazole + H$_2$O ⇌ imidazole-H$^+$ + OH$^-$ K_b

imidazole-H$^+$ ⇌ imidazole + H$^+$ K_a

(b) FM of imidazole = 68.08.   FM of imidazole hydrochloride = 104.54.

$$\text{pH} = 6.993 + \log \frac{(1.00\,\text{g})/(68.08\,\text{g/mol})}{(1.00\,\text{g})/(104.54\,\text{g/mol})} = 7.18$$

(c)

	B	+	H$^+$	→	BH$^+$
Initial mmol:	14.6$_9$		2.46		9.57
Final mmol:	12.2$_3$		—		12.0$_3$

$$\text{pH} = 6.993 + \log \frac{12.2_3}{12.0_3} = 7.00$$

(d) The imidazole must be half neutralized to obtain pH = pK_a = 6.993

14.6$_9$ mmol imidazole requires $\frac{1}{2}$(14.6$_9$) = 7.34 mmol of HClO$_4$ = 6.86 mL.

9-39. (a) $\text{pH} = 2.865 + \log \dfrac{0.040\,0}{0.080\,0} = 2.56$

(b) Using Equations (9-21) and (9-22), and neglecting [OH$^-$] because the solution is acidic (so [H$^+$] >> [OH$^-$]), we can write

$$K_a = 1.36 \times 10^{-3} = \frac{[\text{H}^+](0.040\,0 + [\text{H}^+])}{0.080\,0 - [\text{H}^+]} \Rightarrow [\text{H}^+] = 2.48 \times 10^{-3}\,\text{M}$$

$\Rightarrow$ pH = 2.61

(c) 0.080 mol of HNO$_3$ + 0.080 mol of Ca(OH)$_2$ react completely, leaving an excess of 0.080 mol of OH$^-$. This much OH$^-$ converts 0.080 mol of ClCH$_2$CO$_2$H into 0.080 mol of ClCH$_2$CO$_2^-$. The final concentrations are [ClCH$_2$CO$_2^-$] = 0.020 + 0.080 = 0.100 M and [ClCH$_2$CO$_2$H] = 0.180 − 0.080 = 0.100 M. So pH = pK_a = 2.86.

9-40.

	HA	+	OH$^-$	→	A$^-$	+	H$_2$O
Initial moles:	0.022 4		x		—		
Final moles:	0.022 4 − x		—		x		

$$\text{pH} = 7.40 = \text{p}K_a + \log\frac{[\text{A}^-]}{[\text{HA}]} = 7.48 + \log\frac{x}{0.022\,4 - x} \Rightarrow x = 0.010\,17 \text{ mol.}$$

$$\text{volume} = \frac{0.010\,17 \text{ mol}}{0.626 \text{ M}} = 16.2 \text{ mL}$$

9-41. (a) Since pK_a for acetic acid is 4.756, we expect the solution to be acidic and will ignore [OH$^-$] in comparison to [H$^+$].

[HA] = 0.002 0 − [H$^+$]   [A$^-$] = 0.004 00 + [H$^+$]

$$K_a = 1.75 \times 10^{-5} = \frac{[\text{H}^+](0.004\,00 + [\text{H}^+])}{0.002\,00 - [\text{H}^+]} \Rightarrow [\text{H}^+] = 8.69 \times 10^{-6} \text{ M}$$

⇒ pH = 5.06   [HA] = 0.001 99 M   [A$^-$] = 0.004 01 M

If you used $K_a = 10^{-4.756}$ instead of rounding to 1.75×10^{-5}, then [H$^+$] = 8.71×10^{-6} M.

(b) Use Goal Seek to vary cell B5 until cell D4 is equal to K_a.

	A	B	C	D	E
1	Ka = 10^-pKa =	1.75E-05		Reaction quotient	
2	Kw =	1.00E-14		for Ka =	
3	FHA =	0.002000		[H+][A-]/[HA] =	
4	FA =	0.004000		1.75E-05	
5	H =	8.693E-06		<-Goal Seek solution	
6	OH = Kw/H =	1.15E-09		D4 = H*(FA+H-OH)/(FHA-H+OH)	
7	pH = -logH =	5.0608262			
8	[HA] =	0.0019913		B8 = FHA-H+OH	
9	[A-] =	0.0040087		B9 = FA+H-OH	

9-42. (a) If we dissolve B and BH$^+$ Br$^-$ (where Br$^-$ is an inert anion), the mass balance is $F_{\text{BH}^+} + F_\text{B} = [\text{BH}^+] + [\text{B}]$ and the charge balance is [Br$^-$] + [OH$^-$] = [BH$^+$] + [H$^+$]. Noting that [Br$^-$] = F_{BH^+}, the charge balance can be rewritten as

$$[\text{BH}^+] = F_{\text{BH}^+} + [\text{OH}^-] - [\text{H}^+] \quad \text{(A)}$$

Substituting this expression into the mass balance gives

$$[\text{B}] = F_\text{B} - [\text{OH}^-] + [\text{H}^+] \quad \text{(B)}$$

If we assume that [B] = 0.010 0 M and [BH$^+$] = 0.020 0 M, we calculate

$$\text{pH} = \text{p}K_a + \log\frac{[\text{B}]}{[\text{BH}^+]} = 12.00 + \log\frac{0.010\,0}{0.020\,0} = 11.70$$

Monoprotic Acid-Base Equilibria

If we do not assume that [B] = 0.0100 M and [BH$^+$] = 0.0200 M, we use Equations A and B. Since the solution is basic, we neglect [H$^+$] relative to [OH$^-$] and write [B] = 0.0100 − x and [BH$^+$] = 0.0200 + x, where x = [OH$^-$].

Then we can say $K_b = 10^{-2.00} = \dfrac{[BH^+][OH^-]}{[B]} = \dfrac{(0.0200+x)(x)}{(0.0100-x)}$

$\Rightarrow x = 0.00303\ M \qquad pH = -\log \dfrac{K_w}{x} = 11.48.$

(b) We use Goal Seek to vary cell B5 until cell D4 is equal to K_b.

	A	B	C	D	E
1	Kb =	1.00E-02		Reaction quotient	
2	Kw =	1.00E-14		for Kb =	
3	FBH =	0.02		[OH-][BH+]/[B] =	
4	FB =	0.01		0.01	
5	OH =	3.028E-03		<-Goal Seek solution	
6	H = Kw/OH =	3.303E-12		D4 = OH*(FBH-H+OH)/(FB+H-OH)	
7	pH = -logH =	11.481121			
8	BH =	0.0230278		C8 = FBH-H+OH	
9	B =	0.0069722		C9 = FB+H-OH	

9-43. $K_a = \dfrac{[HPO_4^{2-}][H^+]\gamma_{HPO_4^{2-}}\gamma_{H^+}}{[H_2PO_4^-]\gamma_{H_2PO_4^-}} = 10^{-7.20}$

To find pH, rearrange the K_a expression to solve for the activity of H$^+$, which is $[H^+]\gamma_{H^+}$:

$[H^+]\gamma_{H^+} = \dfrac{K_a[H_2PO_4^-]\gamma_{H_2PO_4^-}}{[HPO_4^{2-}]\gamma_{HPO_4^{2-}}}$

At μ = 0.1 M, γ_{H^+} = 0.83, $\gamma_{H_2PO_4^-}$ = 0.775, and $\gamma_{HPO_4^{2-}}$ = 0.355, so

$[H^+]\gamma_{H^+} = \dfrac{10^{-7.20}[H_2PO_4^-](0.775)}{[HPO_4^{2-}](0.355)} = 1.38 \times 10^{-7}$ when $[H_2PO_4^-] = [HPO_4^{2-}]$

$pH = -\log \mathcal{A}_{H^+} = -\log[H^+]\gamma_{H^+} = -\log(1.38 \times 10^{-7}) = 6.86$

9-44. As the pH of the solution is adjusted by addition of acid or base, the reaction is

$$\text{Pyridine} \; (C_5H_5N:) + H^+ \rightleftharpoons \text{pyridinium ion} \; (C_5H_5NH^+)$$

Pyridine at high pH   pyridinium ion at low pH

The chemical shift of H_4 at low pH levels off near 8.67 ppm in the graph. This value would be the chemical shift of H_4 of pyridinium ion. The chemical shift of H_4 at high pH levels off near 7.89, which would be the chemical shift of H_4 of pyridine. The fraction of pyridine in each solution would be the fraction of the chemical shift change from 8.67 to 7.89. When pH = pKa, there will be equal amounts of pyridine and pyridinium ion. The chemical shift will be half way between 8.67 and 7.89, which is 8.28 ppm. Draw a horizontal line on the graph at a chemical shift of 8.28. This horizontal line intersects a smooth curve drawn between the data points at approximately pH = 5.2. We estimate pK_a ≈ 5.2 for the pyridinium ion, which agrees with the value in the appendix.

CHAPTER 10
POLYPROTIC ACID-BASE EQUILIBRIA

10-1. The K_a reaction, with a much greater equilibrium constant than K_b, releases H$^+$:

$$HA^- \rightleftharpoons H^+ + A^{2-} \qquad K_a$$

Each mole of H$^+$ reacts with one mole of OH$^-$ from the K_b reaction:

$$HA^- + H_2O \rightleftharpoons H_2A + OH^-.$$

The K_b reaction is driven almost as far toward completion as the K_a reaction.

10-2. $\overset{+}{H_3N}-\overset{R}{\underset{|}{C}}H-CO_2^-$ pK values apply to $-NH_3^+$, $-CO_2H$, and, in some cases, R.

10-3. (pyrrolidine-CO_2^-) + H_2O $\rightleftharpoons$ (pyrrolidinium-CO_2^-) + OH^- $K_{b1} = \dfrac{K_w}{K_2} = 4.37 \times 10^{-4}$

(pyrrolidinium-CO_2^-) + H_2O $\rightleftharpoons$ (pyrrolidinium-CO_2H) + OH^- $K_{b2} = \dfrac{K_w}{K_1} = 8.93 \times 10^{-13}$

10-4. (a) $\dfrac{x^2}{0.100-x} = K_1 \Rightarrow x = 3.11 \times 10^{-3} = [H^+] = [HA^-] \Rightarrow pH = 2.51$

$[H_2A] = 0.100 - x = 0.0969\,M$ $[A^{2-}] = \dfrac{K_2[HA^-]}{[H^+]} = 1.00 \times 10^{-8}\,M$

(b) $[H^+] \approx \sqrt{\dfrac{K_1 K_2 F + K_1 K_w}{K_1 + F}} = 1.00 \times 10^{-6} \Rightarrow pH = 6.00$

$[HA^-] \approx 0.100\,M$

$[H_2A] = \dfrac{[H^+][HA^-]}{K_1} = 1.00 \times 10^{-3}\,M;\quad [A^{2-}] = \dfrac{K_2[HA^-]}{[H^+]} = 1.00 \times 10^{-3}\,M$

(c) $\dfrac{x^2}{0.100-x} = \dfrac{K_w}{K_2} \Rightarrow x = [OH^-] = [HA^-] = 3.16 \times 10^{-4}\,M \Rightarrow pH = 10.50$

$[A^{2-}] = 0.100 - x = 9.97 \times 10^{-2}\,M$ $[H_2A] = \dfrac{[H^+][HA^-]}{K_1} = 1.00 \times 10^{-10}\,M$

	pH	[H$_2$A]	[HA$^-$]	[A^{2-}]
0.100 M H$_2$A	2.51	9.69×10^{-2}	3.11×10^{-3}	1.00×10^{-8}
0.100 M NaHA	6.00	1.00×10^{-3}	1.00×10^{-1}	1.00×10^{-3}
0.100 M Na$_2$A	10.50	1.00×10^{-10}	3.16×10^{-4}	9.97×10^{-2}

10-5. (a) $H_2M \rightleftharpoons H^+ + HM^-$ $K_1 = 1.42 \times 10^{-3}$
$F-x$ x x

$$\frac{x^2}{0.100-x} = K_1 \Rightarrow x = 1.12 \times 10^{-2} \Rightarrow pH = -\log x = 1.95$$

$[H_2M] = 0.100 - x = 0.089$ M

$[HM^-] = x = 1.12 \times 10^{-2}$ M; $[M^{2-}] = \dfrac{[HM^-]K_2}{[H^+]} = 2.01 \times 10^{-6}$ M

(b) $[H^+] = \sqrt{\dfrac{K_1K_2(0.100) + K_1K_w}{K_1 + 0.100}} = 5.30 \times 10^{-5} \Rightarrow pH = 4.28$

$[HM^-] \approx 0.100$ M; $[H_2M] = \dfrac{[HM^-][H^+]}{K_1} = 3.7 \times 10^{-3}$ M

$[M^{2-}] = \dfrac{K_2[HM^-]}{[H^+]} = 3.8 \times 10^{-3}$ M

The method of Box 10-2 gives more accurate answers, since $[HM^-]$ is not that much greater than $[H_2M]$ or $[M^{2-}]$ in this case. Successive approximations give pH = 4.28, $[H_2M] = 0.003\,44$ M, $[HM^-] = 0.093\,1$ M, $[M^{2-}] = 0.003\,53$ M.

(c) $M^{2-} + H_2O \rightleftharpoons HM^- + OH^-$ $K_{b1} = K_w/K_{a2} = 4.98 \times 10^{-9}$
$F-x$ x x

$$\dfrac{x^2}{0.100-x} = K_{b1} \Rightarrow x = 2.23 \times 10^{-5} \Rightarrow pH = -\log \dfrac{K_w}{x} = 9.35$$

$[M^{2-}] = 0.100 - x = 0.100$ M $[HM^-] = x = 2.23 \times 10^{-5}$ M

$[H_2M] = \dfrac{[H^+][HM^-]}{K_1} = 7.04 \times 10^{-12}$ M

10-6. $HNNH + H_2O \rightleftharpoons HN{}^+NH_2 + OH^-$ $K_{b1} = \dfrac{K_w}{K_2} = 5.38 \times 10^{-5}$
$F-x$ x x

$$\dfrac{x^2}{0.300-x} = K_{b1} \Rightarrow x = 3.99 \times 10^{-3} \text{ M} \Rightarrow pH = -\log K_w/x = 11.60$$

$[B] = 0.300 - x = 0.296$ M $[BH^+] = x = 3.99 \times 10^{-3}$ M

$[BH_2^{2+}] = \dfrac{[BH^+][H^+]}{K_1} = 2.15 \times 10^{-9}$ M

10-7. For H_2A, $K_1 = 5.62 \times 10^{-2}$ and $K_2 = 5.42 \times 10^{-5}$

First approximation ($[HA^-]_1 \approx 0.001\,00$ M):

$$[H^+]_1 = \sqrt{\frac{K_1 K_2 (0.001\,00) + K_1 K_w}{K_1 + 0.001\,00}} = 2.31 \times 10^{-4}\,M \Rightarrow pH_1 = 3.64$$

$$[H_2A]_1 = \frac{[H^+]_1 [HA^-]_1}{K_1} = 4.10 \times 10^{-6}\,M$$

$$[A^{2-}]_1 = \frac{K_2 [HA^-]_1}{[H^+]} = 2.35 \times 10^{-4}\,M$$

Second approximation:

$[HA^-]_2 \approx 0.001\,00 - [H_2A]_1 - [A^{2-}]_1 = 0.000\,761$ M

$$[H^+]_2 = \sqrt{\frac{K_1 K_2 (0.000\,761) + K_1 K_w}{K_1 + 0.000\,761}} = 2.02 \times 10^{-4}\,M \Rightarrow pH_2 = 3.70$$

$$[H_2A]_2 = \frac{[H^+]_2 [HA^-]_2}{K_1} = 2.73 \times 10^{-6}\,M$$

$$[A^{2-}]_2 = \frac{K_2 [HA^-]_2}{[H^+]_2} = 2.04 \times 10^{-4}\,M$$

Third approximation:

$[HA^-]_3 \approx 0.001\,00 - [H_2A]_2 - [A^{2-}]_2 = 0.000\,793$ M

$$[H^+]_3 = \sqrt{\frac{K_1 K_2 (0.000\,793) + K_1 K_w}{K_1 + 0.000\,793}} = 2.06 \times 10^{-4}\,M \Rightarrow pH_3 = 3.69$$

$$[H_2A]_3 = \frac{[H^+]_3 [HA^-]_3}{K_1} = 2.90 \times 10^{-6}\,M$$

$$[A^{2-}]_3 = \frac{K_2 [HA^-]_3}{[H^+]_3} = 2.09 \times 10^{-4}\,M$$

10-8. (a) Charge balance: $[K^+] + [H^+] = [OH^-] + [HP^-] + 2[P^{2-}]$ (1)

Mass balance: $[K^+] = [H_2P] + [HP^-] + [P^{2-}]$ (2)

Equilibria:
$$K_1 = \frac{[H^+]\gamma_{H^+}[HP^-]\gamma_{HP^-}}{[H_2P]\gamma_{H_2P}} \qquad (3)$$

$$K_2 = \frac{[H^+]\gamma_{H^+}[P^{2-}]\gamma_{P^{2-}}}{[HP^-]\gamma_{HP^-}} \qquad (4)$$

$$K_w = [H^+]\gamma_{H^+}[OH^-]\gamma_{OH^-} \qquad (5)$$

Solving for $[K^+]$ in Equations (1) and (2) and equating the results gives

$$[H_2P] + [H^+] - [P^{2-}] - [OH^-] = 0$$

Making substitutions from Equations (3), (4), and (5), we can write

$$\frac{[H^+]\gamma_{H^+}[HP^-]\gamma_{HP^-}}{K_1\,\gamma_{H_2P}} + [H^+] - \frac{K_2[HP^-]\gamma_{HP^-}}{[H^+]\gamma_{H^+}\,\gamma_{P^{2-}}} - \frac{K_w}{[H^+]\gamma_{H^+}\gamma_{OH^-}} = 0$$

which can be rearranged to

$$[H^+] = \sqrt{\dfrac{\dfrac{K_1 K_2[HP^-]\gamma_{HP^-}\gamma_{H_2P}}{\gamma_{H^+}\gamma_{P^{2-}}} + \dfrac{K_1 K_w \gamma_{H_2P}}{\gamma_{H^+}\gamma_{OH^-}}}{K_1\gamma_{H_2P} + [HP^-]\gamma_{H^+}\gamma_{HP^-}}} \qquad (6)$$

(b) The ionic strength of 0.050 M KHP is 0.050 M because the only major ions are K^+ and HP^-.

$[HP^-] \approx 0.050\ M$, $\gamma_{HP^-} = 0.835$, $\gamma_{P^{2-}} = 0.485$, $\gamma_{H_2P} \approx 1.00$,

$\gamma_{H^+} = 0.86$, $\gamma_{OH^-} = 0.81$. Using these values in Eq. (6) gives

$[H^+] = 1.09 \times 10^{-4} \Rightarrow pH = -\log[H^+]\gamma_{H^+} = 4.03$.

10-9. Case (a) is shown in column G and case (b) is in column H.

	A	B	C	D	E	F	G	H
1	Successive approximations by circular reference							
2	for intermediate form of diprotic acid							
3								
4	H_2A malic acid							
5	K_1 =	3.50E-04					1.00E-04	1.00E-04
6	K_2 =	7.90E-06					1.00E-08	1.00E-05
7	K_w =	1.00E-14					1.00E-14	1.00E-14
8	F =	1.000E-03					1.000E-02	1.000E-02
9	$[H^+]$ =	4.356E-05	=SQRT((K_1*K_2*[HA$^-$]+K_1*K_w)/(K_1+[HA$^-$]))				9.950E-07	3.137E-05
10	$[H_2A]$ =	9.532E-05	= $[H^+]$[HA$^-$]/K_1				9.755E-05	1.921E-03
11	$[HA^-]$ =	7.658E-04	=F-[H_2A]-[A^{2-}]				9.804E-03	6.126E-03
12	$[A^{2-}]$ =	1.389E-04	= K_2[HA$^-$]/$[H^+]$				9.853E-05	1.953E-03
13	pH =	4.360887	=-log[H^+]				6.002182	4.503516
14								
15	1. Set Excel Options Formulas to Enable Iterative Calculation							
16	with Maximum Change = 1E-12							
17	2. Begin by setting [HA$^-$] = F							
18	3. Write correct formulas in for other concentrations							
19	4. Then change [HA$^-$] to =F-[H_2A]-[A^{2-}]							
20	5. Correct answers now appear in all cells							

Polyprotic Acid-Base Equilibria 109

10-10. ["H_2CO_3"] = [$CO_2(aq)$] = $K_H P_{CO_2}$ = $10^{-1.5} \cdot 10^{-3.4}$ = $10^{-4.9}$ M

$$H_2CO_3 \rightleftharpoons HCO_3^- + H^+ \qquad K_{a1} = 4.46 \times 10^{-7}$$

$10^{-4.9} - x \qquad\qquad x \qquad\quad x$

$$\frac{x^2}{10^{-4.9} - x} = K_{a1} \Rightarrow x = 2.1_6 \times 10^{-6} \text{ M} \Rightarrow \text{pH} = 5.67$$

10-11. (a) From Equation C, $[CO_3^{2-}] = \dfrac{K_{a2}[HCO_3^-]}{[H^+]}$ \hfill (F)

From Equation B, $[HCO_3^-] = \dfrac{K_{a1}[CO_2(aq)]}{[H^+]}$ \hfill (G)

Substituting [HCO_3^-] from Equation G into Equation F gives

$$[CO_3^{2-}] = \frac{K_{a2} K_{a1}[CO_2(aq)]}{[H^+]^2} \tag{H}$$

Substituting for [$CO_2(aq)$] from Equation A into Equation H gives

$$[CO_3^{2-}] = \frac{K_{a2} K_{a1} K_H P_{CO_2}}{[H^+]^2} \tag{I}$$

(b) For P_{CO_2} = 800 μbar, pH = 7.8, 0°C, we find

$$[CO_3^{2-}] = \frac{K_{a2} K_{a1} K_H P_{CO_2}}{[H^+]^2} =$$

$$\frac{10^{-9.376\,2} \text{ mol kg}^{-1} \, 10^{-6.100\,4} \text{ mol kg}^{-1} \, 10^{-1.207\,3} \text{ mol kg}^{-1} \text{ bar}^{-1} (800 \times 10^{-6} \text{ bar})}{[10^{-7.8} \text{ mol kg}^{-1}]^2}$$

= 6.6×10^{-5} mol kg^{-1}

For P_{CO_2} = 800 μbar, pH = 7.8, 30°C, we find

$$[CO_3^{2-}] = \frac{K_{a2} K_{a1} K_H P_{CO_2}}{[H^+]^2} =$$

$$\frac{10^{-8.832\,4} \text{ mol kg}^{-1} \, 10^{-5.800\,8} \text{ mol kg}^{-1} \, 10^{-1.604\,8} \text{ mol kg}^{-1} \text{bar}^{-1} (800 \times 10^{-6} \text{bar})}{[10^{-7.8} \text{ mol kg}^{-1}]^2}$$

= 1.8×10^{-4} mol kg^{-1}

(c) The equilibrium expressions for aragonite and calcite are

$$CaCO_3(s, aragonite) \rightleftharpoons Ca^{2+} + CO_3^{2-} \tag{D}$$

$$K_{sp}^{arg} = [Ca^{2+}][CO_3^{2-}] = 10^{-6.1113} \text{ mol}^2 \text{ kg}^{-2} \text{ at } 0°C$$

$$= 10^{-6.1391} \text{ mol}^2 \text{ kg}^{-2} \text{ at } 30°C$$

$$CaCO_3(s, calcite) \rightleftharpoons Ca^{2+} + CO_3^{2-} \qquad (E)$$

$$K_{sp}^{cal} = [Ca^{2+}][CO_3^{2-}] = 10^{-6.3652} \text{ mol}^2 \text{ kg}^{-2} \text{ at } 0°C$$

$$= 10^{-6.3713} \text{ mol}^2 \text{ kg}^{-2} \text{ at } 30°C$$

The reaction quotient at 0°C is

$$[Ca^{2+}][CO_3^{2-}] = [0.010 \text{ mol kg}^{-1}][6.6 \times 10^{-5} \text{ mol kg}^{-1}]$$

$$= 6.6 \times 10^{-7} \text{ mol}^2 \text{ kg}^{-2} = 10^{-6.18} \text{ mol}^2 \text{ kg}^{-2} < 10^{-6.1113} \text{ mol}^2 \text{ kg}^{-2},$$

so aragonite will dissolve.

But $10^{-6.18}$ mol² kg⁻² > $10^{-6.3652}$ mol² kg⁻², so calcite will not dissolve.

The reaction quotient at 30°C is

$$[Ca^{2+}][CO_3^{2-}] = [0.010 \text{ mol kg}^{-1}][1.84 \times 10^{-4} \text{ mol kg}^{-1}]$$

$$= 1.84 \times 10^{-6} \text{ mol}^2 \text{ kg}^{-2} = 10^{-5.74} \text{ mol}^2 \text{ kg}^{-2} > 10^{-6.1113} \text{ mol}^2 \text{ kg}^{-2},$$

so neither aragonite nor calcite dissolves.

10-12. $pH = pK_a + \log \dfrac{[CO_3^{2-}]}{[HCO_3^-]}$

$10.00 = 10.329 + \log \dfrac{(x \text{ g})/(105.99 \text{ g/mol})}{(5.00 \text{ g})/(84.01 \text{ g/mol})} \Rightarrow x = 2.96 \text{ g}$

10-13. We begin with $(25.0 \text{ mL})(0.023\ 3 \text{ M}) = 0.582_5$ mmol salicylic acid (H_2A, $pK_1 = 2.972$, $pK_2 = 13.7$). At pH 3.50, there will be a mixture of H_2A and HA^-.

	H_2A	+	OH^-	→	HA^-	+	H_2O
Initial mmol:	0.582_5		x		—		
Final mmol:	$0.582_5 - x$		—		x		

$3.50 = 2.972 + \log \dfrac{x}{0.582_5 - x} \Rightarrow x = 0.449_3$ mmol

$(0.449_3 \text{ mmol})/(0.202 \text{ M}) = 2.223$ mL NaOH

10-14. Picolinic acid is HA, the intermediate form of a diprotic system with $pK_1 = 1.01$ and $pK_2 = 5.39$. To achieve pH 5.50, we need a mixture of $HA + A^-$.

	HA	+	OH^-	→	A^-
Initial mmol:	10.0		x		—
Final mmol:	$10.0 - x$		—		x

$5.50 = 5.39 + \log \dfrac{x}{10.0 - x} \Rightarrow x = 5.63$ mmol ≈ 5.63 mL NaOH

Procedure: Dissolve 10.0 mmol (1.23 g) picolinic acid in ~75 mL H$_2$O in a beaker. Add NaOH (~5.63 mL) until the measured pH is 5.50. Transfer to a 100 mL volumetric flask and use small portions of H$_2$O to rinse the contents of the beaker into the flask. Dilute to 100.0 mL and mix well.

10-15. At pH 2.80, we have a mixture of SO$_4^{2-}$ and HSO$_4^-$, since pK_a for HSO$_4^-$ is 1.987.

$$2.80 = 1.987 + \log \frac{[\text{SO}_4^{2-}]}{[\text{HSO}_4^-]} \Rightarrow \text{HSO}_4^- = 0.153_8 \,[\text{SO}_4^{2-}]$$

The reaction between H$_2$SO$_4$ and SO$_4^{2-}$ produces 2 moles of HSO$_4^-$:

$$\text{H}_2\text{SO}_4 + \text{SO}_4^{2-} \rightarrow 2\text{HSO}_4^-$$

Initial mmol: x y —
Final mmol: — $y-x$ $2x$

The Henderson-Hasselbalch equation told us that [HSO$_4^-$] = 0.153$_8$ [SO$_4^{2-}$] $\Rightarrow 2x = 0.153_8\,(y-x)$. Since the total sulfur is 0.200 M, $x + y = 0.200$ mol. Substituting $x = 0.200 - y$ into the equation $2x = 0.153_8\,(y-x)$ gives Na$_2$SO$_4$ = y = 0.186$_7$ mol = 26.5$_2$ g and H$_2$SO$_4$ = x = 0.013 3 mol = 1.30 g.

10-16. pK_2 for phosphoric acid is 7.20, so it has a high buffer capacity at pH 7.45 (from the buffer pair H$_2$PO$_4^-$/HPO$_4^{2-}$). At pH 8.5, the buffer capacity of phosphate would be low and it would not be very useful. The other two pK_a values for phosphoric acid are pK_1 = 2.15 and pK_3 = 12.38. Neither is near pH 8.5.

10-17.

$$\underset{\text{Glutamic acid}}{\overset{+\text{NH}_3}{\underset{\text{CO}_2\text{H}}{|}}\text{CHCH}_2\text{CH}_2\text{CO}_2\text{H} \;\overset{K_1}{\rightleftharpoons}\; \overset{+\text{NH}_3}{\underset{\text{CO}_2^-}{|}}\text{CHCH}_2\text{CH}_2\text{CO}_2\text{H} \;\overset{K_2}{\rightleftharpoons}\; \overset{+\text{NH}_3}{\underset{\text{CO}_2^-}{|}}\text{CHCH}_2\text{CH}_2\text{CO}_2^- \;\overset{K_3}{\rightleftharpoons}\; \overset{\text{NH}_2}{\underset{\text{CO}_2^-}{|}}\text{CHCH}_2\text{CH}_2\text{CO}_2^-}$$

Tyrosine:

+NH$_3$–HC(CO$_2$H)–CH$_2$–C$_6$H$_4$–OH $\overset{K_1}{\rightleftharpoons}$ +NH$_3$–HC(CO$_2^-$)–CH$_2$–C$_6$H$_4$–OH $\overset{K_2}{\rightleftharpoons}$

NH$_2$–HC(CO$_2^-$)–CH$_2$–C$_6$H$_4$–OH $\overset{K_3}{\rightleftharpoons}$ NH$_2$–HC(CO$_2^-$)–CH$_2$–C$_6$H$_4$–O$^-$

10-18. (a) For 0.050 0 M KH$_2$PO$_4$, $[H^+] = \sqrt{\dfrac{K_1K_2(0.050\,0) + K_1K_w}{K_1 + 0.050\,0}}$

$= 1.99 \times 10^{-5} \Rightarrow$ pH = 4.70

$4.70 = 2.148 + \log \dfrac{[H_2PO_4^-]}{[H_3PO_4]} \Rightarrow \dfrac{[H_3PO_4]}{[H_2PO_4^-]} = 2.8 \times 10^{-3}$

(b) For 0.050 0 M K$_2$HPO$_4$, $[H^+] = \sqrt{\dfrac{K_2K_3(0.050\,0) + K_2K_w}{K_2 + 0.050\,0}}$

$= 1.99 \times 10^{-10} \Rightarrow$ pH = 9.70

$9.70 = 2.148 + \log \dfrac{[H_2PO_4^-]}{[H_3PO_4]} \Rightarrow \dfrac{[H_3PO_4]}{[H_2PO_4^-]} = 2.8 \times 10^{-8}$

10-19. (a) $H_3PO_4 \xrightarrow{pK_1=2.148} H_2PO_4^- \xrightarrow{pK_2=7.198} HPO_4^{2-} \xrightarrow{pK_3=12.375} PO_4^{3-}$

$\qquad\qquad\qquad\uparrow\qquad\qquad\qquad\uparrow$
$\qquad\qquad$ pH ≈ (2.148+7.198)/2   pH ≈ (7.198+12.375)/2
$\qquad\qquad\qquad= 4.67 \qquad\qquad\qquad = 9.79$

pH 7.45 corresponds to a mixture of NaH$_2$PO$_4$ and Na$_2$HPO$_4$. (You could get the same result by mixing other combinations such as H$_3$PO$_4$ and Na$_3$PO$_4$ or H$_3$PO$_4$ and Na$_2$HPO$_4$.)

(b) pH = pK_2 + log $\dfrac{[HPO_4^{2-}]}{[H_2PO_4^-]}$

$7.45 = 7.198 + \log \dfrac{[HPO_4^{2-}]}{[H_2PO_4^-]} \Rightarrow \dfrac{[HPO_4^{2-}]}{[H_2PO_4^-]} = 1.78_6$

Combining this last result with [HPO$_4^{2-}$] + [H$_2$PO$_4^-$] = 0.050 0 M gives [HPO$_4^{2-}$] = 0.032 0$_5$ M and [H$_2$PO$_4^-$] = 0.017 9$_5$ M. Use 4.55 g of Na$_2$HPO$_4$ and 2.15 g of NaH$_2$PO$_4$.

(c) Here is one of several ways: Weigh out 0.050 0 mol Na$_2$HPO$_4$ and dissolve it in 900 mL of water. Add HCl while monitoring the pH with a pH electrode. When the pH is 7.45, stop adding HCl and dilute to exactly 1 L with H$_2$O.

10-20. Lysine hydrochloride (H$_2$L$^+$) is

$$\begin{array}{c} ^+NH_3 \quad Cl^- \\ | \\ HC-CH_2CH_2CH_2\overset{+}{N}H_3 \\ | \\ CO_2^- \end{array}$$

for which $[H^+] = \sqrt{\dfrac{K_1K_2(0.010\,0) + K_1K_w}{K_1 + 0.010\,0}} = 2.32 \times 10^{-6}$ M $\Rightarrow$ pH = 5.64

Polyprotic Acid-Base Equilibria   113

$$[H_2L^+] = 0.0100 \text{ M}$$

$$[H_3L^{2+}] = \frac{[H^+][H_2L^+]}{K_1} = 1.36 \times 10^{-6} \text{ M}$$

$$[HL] = \frac{K_2[H_2L^+]}{[H^+]} = 3.68 \times 10^{-6} \text{ M} \qquad [L^-] = \frac{K_3[HL]}{[H^+]} = 2.40 \times 10^{-11} \text{ M}$$

10-21. H_3His^{2+} $\xrightleftharpoons{pK_1 = 1.6}$ H_2His^+ $\xrightleftharpoons{pK_2 = 5.97}$ HHis (Histidine) $\xrightleftharpoons{pK_3 = 9.28}$ His$^-$

Histidine hydrochloride (FM 191.62) is $H_2His^+Cl^-$, the intermediate form between pK_1 and pK_2. A pH of 9.30 requires neutralizing all of the H_2His^+ to HHis and then adding more KOH to create a mixture of HHis and His$^-$. Therefore, we must add 1 mol KOH for each mol of His·HCl to get to HHis and then add some more KOH to obtain the mixture of HHis and His$^-$. Initial mol of His·HCl = 10.0 g/(191.62 g/mol) = 0.052 1$_9$ mol. We require 0.052 1$_9$ mol of KOH plus the amount x in the following table to obtain the correct mixture:

	HHis	+	OH$^-$	→	His$^-$
Initial mol:	0.052 1$_9$		x		—
Final mol:	0.052 1$_9$ – x		—		x

$$pH = pK_3 + \log \frac{[His^{2-}]}{[HHis^-]} \Rightarrow 9.30 = 9.28 + \log \frac{x}{0.052\ 1_9 - x}$$

$$\Rightarrow x = 0.026\ 7_0$$

Total mol KOH required = 0.052 1$_9$ + 0.026 7$_0$ = 0.078 9 mol

= 78.9 mL of 1.00 M KOH

10-22. (a) $pH = pK_3 \text{ (citric acid)} + \log \frac{[C^{3-}]\gamma_{C^{3-}}}{[HC^{2-}]\gamma_{HC^{2-}}}$

$$pH = 6.396 + \log \frac{(1.00)(0.405)}{(2.00)(0.665)} = 5.88$$

(b) If the ionic strength is raised to 0.10 M,
$$pH = 6.396 + \log \frac{(1.00)(0.115)}{(2.00)(0.37)} = 5.59$$

10-23. (a) HA   (b) A⁻

(c) $pH = pK_a + \dfrac{[A^-]}{[HA]}$

$7.00 = 7.00 + \log \dfrac{[A^-]}{[HA]} \Rightarrow [A^-]/[HA] = 1.0$

$6.00 = 7.00 + \log \dfrac{[A^-]}{[HA]} \Rightarrow [A^-]/[HA] = 0.10$

10-24. (a) 4.00   (b) 8.00   (c) H_2A (d) HA^- (e) A^{2-}

10-25. (a) 9.00   (b) 9.00   (c) BH^+

(d) $12.00 = 9.00 + \log \dfrac{[B]}{[BH^+]} \Rightarrow [B]/[BH^+] = 1.0 \times 10^3$

10-26.

[Structure of pyridoxal phosphate: phosphate group $^-O-P(=O)(O^-)-O-CH_2-$ attached to a pyridinium ring with CHO, O^-, CH_3 substituents and N⁺–H]

10-27. Fraction in form HA = $\alpha_{HA} = \dfrac{[H^+]}{[H^+]+K_a} = \dfrac{10^{-5}}{10^{-5}+10^{-4}} = 0.091$.

Fraction in form A⁻ = $\alpha_{A^-} = \dfrac{K_a}{[H^+]+K_a} = 0.909$.

$\dfrac{[A^-]}{[HA]} = \dfrac{\alpha_{A^-}}{\alpha_{HA}} = 10$, which makes sense.

10-28. $\alpha_{H_2A} = \dfrac{[H^+]^2}{[H^+]^2 + [H^+]K_1 + K_1K_2}$, where $[H^+] = 10^{-7.00}$, $K_1 = 10^{-8.00}$, and $K_2 = 10^{-10.00} \Rightarrow \alpha_{H_2A} = 0.91$

10-29. Use Equations for a diprotic acid with $pK_1 = 8.85$ and $pK_2 = 10.43$.

$$\alpha_{H_2A} = \frac{[H_2A]}{F} = \frac{[H^+]^2}{[H^+]^2 + [H^+]K_1 + K_1K_2}$$

$$\alpha_{HA^-} = \frac{[HA^-]}{F} = \frac{K_1[H^+]}{[H^+]^2 + [H^+]K_1 + K_1K_2}$$

$$\alpha_{A^{2-}} = \frac{[A^{2-}]}{F} = \frac{K_1K_2}{[H^+]^2 + [H^+]K_1 + K_1K_2}$$

	pH 8.00	pH 10.00
α_{H_2A}	0.876	0.0491
α_{HA^-}	0.124	0.693
$\alpha_{A^{2-}}$	4.60×10^{-4}	0.258

10-30.

pH:	1.00	1.92	6.00	6.27	10.00
α_{H_2A}	0.893	0.500	5.41×10^{-5}	2.23×10^{-5}	1.55×10^{-12}
α_{HA^-}	0.107	0.500	0.651	0.500	1.86×10^{-4}
$\alpha_{A^{2-}}$	5.76×10^{-7}	2.23×10^{-5}	0.349	0.500	0.9998

10-31. (a) The derivation follows the outline of Equations 10-19 through 10-21. The results are

$$\alpha_{H_3A} = \frac{[H_3A]}{F} = \frac{[H^+]^3}{[H^+]^3 + [H^+]^2 K_1 + [H^+]K_1K_2 + K_1K_2K_3}$$

$$\alpha_{H_2A^-} = \frac{[H_2A^-]}{F} = \frac{[H^+]^2 K_1}{[H^+]^3 + [H^+]^2 K_1 + [H^+]K_1K_2 + K_1K_2K_3}$$

$$\alpha_{HA^{2-}} = \frac{[HA^{2-}]}{F} = \frac{[H^+]K_1K_2}{[H^+]^3 + [H^+]^2 K_1 + [H^+]K_1K_2 + K_1K_2K_3}$$

$$\alpha_{A^{3-}} = \frac{[A^{3-}]}{F} = \frac{K_1K_2K_3}{[H^+]^3 + [H^+]^2 K_1 + [H^+]K_1K_2 + K_1K_2K_3}$$

(b) For phosphoric acid, $pK_1 = 2.148$, $pK_2 = 7.198$, and $pK_3 = 12.375$. At pH = 7.00, expressions above give $\alpha_{H_3A} = 8.6 \times 10^{-6}$, $\alpha_{H_2A^-} = 0.61$, $\alpha_{HA^{2-}} = 0.39$, and $\alpha_{A^{3-}} = 1.6 \times 10^{-6}$.

10-32. $\text{pH} = \text{p}K_{\text{NH}_4^+} + \log \dfrac{[\text{NH}_3]}{[\text{NH}_4^+]} \Rightarrow 9.00 = 9.245 + \log \dfrac{[\text{NH}_3]}{[\text{NH}_4^+]} \Rightarrow \dfrac{[\text{NH}_3]}{[\text{NH}_4^+]} = 0.56_9$

$$\text{Fraction unprotonated} = \dfrac{[\text{NH}_3]}{[\text{NH}_3] + [\text{NH}_4^+]} = \dfrac{0.56_9}{0.56_9 + 1} = 0.36$$

The presence of other acids and bases has no bearing on the fraction of ammonia that is unprotonated.

10-33. The quantity of morphine in the solution is negligible compared to the quantity of cacodylic acid. The pH is determined by the reaction of cacodylic acid (HA) with NaOH:

	HA	+	OH$^-$	→	A$^-$	+	H$_2$O
Initial mmol:	1.000		0.800		—		—
Final mmol:	0.200		—		0.800		—

$$\text{pH} = \text{p}K_a + \log \dfrac{[\text{A}^-]}{[\text{HA}]} = 6.19 + \log \dfrac{0.800}{0.200} = 6.79$$

For morphine (B), $K_a = K_w/K_b = 1.0 \times 10^{-14}/1.6 \times 10^{-6} = 6.2_5 \times 10^{-9}$

$\Rightarrow \text{p}K_a = 8.20$

At pH 6.79, we can write $\text{pH} = \text{p}K_{\text{BH}^+} + \log \dfrac{[\text{B}]}{[\text{BH}^+]} \Rightarrow$

$6.79 = 8.20 + \log \dfrac{[\text{B}]}{[\text{BH}^+]} \Rightarrow \dfrac{[\text{B}]}{[\text{BH}^+]} = 0.039 \Rightarrow [\text{B}] = 0.039\,[\text{BH}^+]$

$$\text{Fraction in form BH}^+ = \dfrac{[\text{BH}^+]}{[\text{B}] + [\text{BH}^+]} = \dfrac{[\text{BH}^+]}{0.039[\text{BH}^+] + [\text{BH}^+]} = 96\%$$

10-34.

	A	B	C	D	E	F
1	Fractional composition for diprotic acid					
2						
3	K1 =	pH	[H+]	α(H2A)	α(HA$^-$)	α(A^{2-})
4	9.55E-04	1	0.1	9.91E-01	0.00946	3.13E-06
5	K2 =	2	0.01	0.912562	0.087149	0.000289
6	3.31E-05	3	0.001	0.503369	0.480713	0.015918
7	pK1 =	4	0.0001	0.072928	0.696455	0.230618
8	3.02	5	0.00001	0.002423	0.231386	0.766191
9	pK2 =	6	0.000001	3.07E-05	0.029313	0.970656
10	4.48	7	1E-07	3.15E-07	0.003011	0.996989
11						
12	A4 = 10^-A8	D4 = $C4^2/($C4^2+$C4*$A$4+$A$4*$A$6)				
13	A6 = 10^-A10	E4 = $C4*$A$4/($C4^2+$C4*$A$4+$A$4*$A$6)				
14	C4 = 10^-B4	F4 = $A$4*$A$6/($C4^2+$C4*$A$4+$A$4*$A$6)				

10-35.

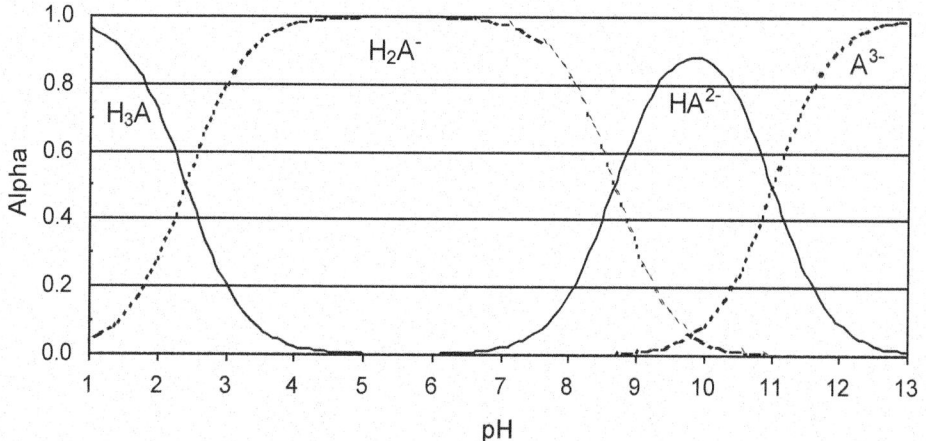

	A	B	C	D	E	F	G
1	Fractional composition for triprotic acid						
2							
3	K1 =	pH	[H+]	α(H3A)	α(H2A⁻)	α(HA²⁻)	α(A³⁻)
4	3.89E-03	1	1.00E-01	9.63E-01	3.74E-02	8.01E-10	7.82E-20
5	K2 =	2	1.00E-02	7.20E-01	2.80E-01	5.99E-08	5.85E-17
6	2.14E-09	3	1.00E-03	2.04E-01	7.96E-01	1.70E-06	1.66E-14
7	K3 =	4	1.00E-04	2.51E-02	9.75E-01	2.08E-05	2.04E-12
8	9.77E-12	5	1.00E-05	2.56E-03	9.97E-01	2.13E-04	2.08E-10
9	pK1 =	6	1.00E-06	2.56E-04	9.98E-01	2.13E-03	2.08E-08
10	2.41	7	1.00E-07	2.52E-05	9.79E-01	2.09E-02	2.05E-06
11	pK2 =	8	1.00E-08	2.12E-06	8.24E-01	1.76E-01	1.72E-04
12	8.67	9	1.00E-09	8.14E-08	3.17E-01	6.77E-01	6.61E-03
13	pK3 =	10	1.00E-10	1.05E-09	4.09E-02	8.74E-01	8.54E-02
14	11.01	11	1.00E-11	6.07E-12	2.36E-03	5.05E-01	4.93E-01
15		12	1.00E-12	1.12E-14	4.34E-05	9.28E-02	9.07E-01
16		13	1.00E-13	1.22E-17	4.74E-07	1.01E-02	9.90E-01
17	A4 = 10^-A10						
18	C4 = 10^-B4						
19	D4 = $C4^3/($C4^3+$C4^2*$A$4+$C4*$A$4*$A$6+$A$4*$A$6*$A$8)						
20	E4 = $C4^2*$A$4/($C4^3+$C4^2*$A$4+$C4*$A$4*$A$6+$A$4*$A$6*$A$8)						
21	F4 = $C4*$A$4*$A$6/($C4^3+$C4^2*$A$4+$C4*$A$4*$A$6+$A$4*$A$6*$A$8)						
22	G4 = $A$4*$A$6*$A$8/($C4^3+$C4^2*$A$4+$C4*$A$4*$A$6+$A$4*$A$6*$A$8)						

10-36. (a) Fractional composition in tetraprotic system

	A	B	C	D	E	F	G	H	I
1	Fractional composition in tetraprotic system								
2									
3	Ka1 =	pH	[H+]	Denom.	Alph(H4A)	Alph(H3A)	Alph(H2A)	Alph(HA)	Alph(A)
4	1.58E-04	1	1E-01	1.0E-04	1.0E+00	1.6E-03	6.3E-09	2.5E-14	9.9E-25
5	Ka2 =	2	1E-02	1.0E-08	9.8E-01	1.6E-02	6.2E-07	2.5E-11	9.8E-21
6	3.98E-07	3	1E-03	1.2E-12	8.6E-01	1.4E-01	5.4E-05	2.2E-08	8.6E-17
7	Ka3 =	4	1E-04	2.6E-16	3.9E-01	6.1E-01	2.4E-03	9.7E-06	3.9E-13
8	3.98E-07	5	1E-05	1.7E-19	5.7E-02	9.1E-01	3.6E-02	1.4E-03	5.7E-10
9	Ka4 =	6	1E-06	2.5E-22	4.1E-03	6.4E-01	2.5E-01	1.0E-01	4.0E-07
10	3.98E-12	7	1E-07	3.3E-24	3.0E-05	4.8E-02	1.9E-01	7.6E-01	3.0E-05
11		8	1E-08	2.6E-25	3.9E-08	6.2E-04	2.4E-02	9.7E-01	3.9E-04
12		9	1E-09	2.5E-26	4.0E-11	6.3E-06	2.5E-03	9.9E-01	4.0E-03
13		10	1E-10	2.6E-27	3.8E-14	6.1E-08	2.4E-04	9.6E-01	3.8E-02
14		11	1E-11	3.5E-28	2.9E-17	4.5E-10	1.8E-05	7.2E-01	2.8E-01
15		12	1E-12	1.2E-28	8.0E-21	1.3E-12	5.0E-07	2.0E-01	8.0E-01
16		13	1E-13	1.0E-28	9.8E-25	1.5E-15	6.2E-09	2.5E-02	9.8E-01
17									
18	C4 = 10^-B4								
19	D4 = C4^4+$A$4*C4^3+$A$4*$A$6*C4^2+$A$4*$A$6*$A$8*C4								
20		+$A$4*$A$6*$A$8*$A$10							
21	E4 = C4^4/D4								
22	F4 = $A$4*C4^3/D4				H4 = $A$4*$A$6*$A$8*C4/D4				
23	G4 = $A$4*$A$6*C4^2/D4				I4 = $A$4*$A$6*$A$8*$A$10/D4				

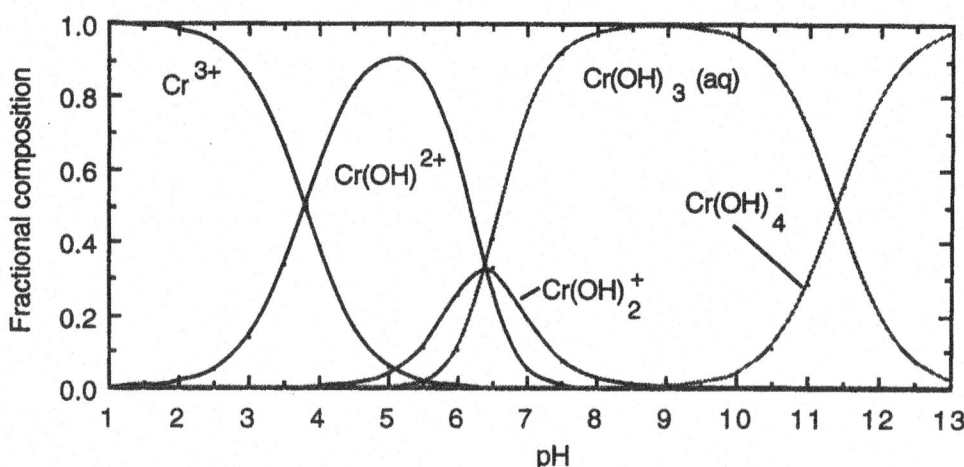

(b) $K = 10^{-6.84} = [Cr(OH)_3(aq)]$, so $[Cr(OH)_3(aq)] = 10^{-6.84}$ M $= 1.4_5 \times 10^{-7}$ M

(c) $K_{a3} = 10^{-6.40} = \dfrac{[Cr(OH)_3(aq)][H^+]}{[Cr(OH)_2^+]} = \dfrac{[10^{-6.84}][10^{-4.00}]}{[Cr(OH)_2^+]}$

$\Rightarrow [Cr(OH)_2^+] = \dfrac{[10^{-6.84}][10^{-4.00}]}{10^{-6.40}} = 10^{-4.44}$ M

Polyprotic Acid-Base Equilibria

$$K_{a2} = 10^{-6.40} = \frac{[Cr(OH)_2^+][H^+]}{[Cr(OH)^{2+}]} = \frac{[10^{-4.44}][10^{-4.00}]}{[Cr(OH)^{2+}]}$$

$$\Rightarrow [Cr(OH)^{2+}] = \frac{[10^{-4.44}][10^{-4.00}]}{10^{-6.40}} = 10^{-2.04} \text{ M}$$

10-37. Acidic substituents: aspartic acid, cysteine, glutamic acid, tyrosine

Basic substituents: arginine, histidine, lysine

10-38. The isoelectric pH is the pH at which the protein has no net charge, even though it has many positive and negative sites. The isoionic pH is the pH of a solution containing only protein, H^+, and OH^-.

10-39. The <u>average</u> charge is zero. There is no pH at which <u>all</u> molecules have zero charge.

10-40. Isoelectric pH $= \dfrac{pK_1 + pK_2}{2} = 5.59$

Isoionic $[H^+] = \sqrt{\dfrac{K_1 K_2 (0.010) + K_1 K_w}{K_1 + (0.010)}} \Rightarrow \text{pH} = 5.72$

10-41. A mixture of proteins is exposed to a strong electric field in a medium with a pH gradient. Positively charged molecules move toward the negative pole and negatively charged molecules move toward the positive pole. Each protein migrates until it reaches the point where the pH is the same as its isoelectric pH. At this point, the protein has no net charge and no longer moves. Each protein is therefore focused in one region at its isoelectric pH. If a protein diffuses out of its isoelectric zone, it becomes charged and migrates back into the zone.

CHAPTER 11
ACID-BASE TITRATIONS

11-1. The equivalence point occurs when the quantity of titrant is exactly the stoichiometric amount needed for complete reaction with analyte. The end point occurs when there is an abrupt change in a physical property, such as pH or indicator color. Ideally, the end point is chosen to occur at the equivalence point.

11-2.
V_a	0	1	5	9	9.9	10	10.1	12
pH	13.00	12.95	12.68	11.96	10.96	7.00	3.04	1.75

Representative calculations:
$(V_e \text{ (mL)})(1.000 \text{ M HBr}) = (100.0 \text{ mL})(0.100 \text{ M NaOH}) \Rightarrow V_e = 10.00 \text{ mL}$

0 mL: $\quad \text{pH} = -\log \dfrac{K_w}{[\text{OH}^-]} = -\log \dfrac{10^{-14}}{0.100} = 13.00$

1 mL: $\quad [\text{OH}^-] = \underbrace{\dfrac{9 \text{ mL}}{10 \text{ mL}}}_{\substack{\text{Fraction} \\ \text{remaining}}} \underbrace{(0.100 \text{ M})}_{\substack{\text{Initial} \\ \text{concentration}}} \underbrace{\dfrac{100 \text{ mL}}{101 \text{ mL}}}_{\substack{\text{Dilution} \\ \text{factor}}} = 0.0891 \text{ M} \Rightarrow \text{pH} = 12.95$

10 mL: $\quad [\text{OH}^-] = [\text{H}^+] = 10^{-7} \text{ M}$

10.1 mL: $\quad [\text{H}^+] = \underbrace{\left(\dfrac{0.1 \text{ mL}}{110.1 \text{ mL}}\right)}_{\substack{0.1 \text{ mL excess H}^+ \\ \text{diluted to } 110.1 \text{ mL}}} \underbrace{(1.00 \text{ M})}_{\substack{\text{Initial} \\ \text{concentration} \\ \text{of H}^+}} = 9.08 \times 10^{-4} \text{ M} \Rightarrow \text{pH} = 3.04$

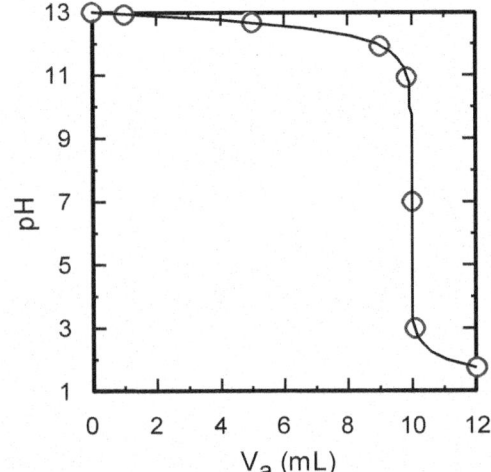

11-3. If we titrate strong acid with strong base, the concentration of H^+ is close to 1% of its initial value when we are 99% of the way to the equivalence point (i.e., when $V_b = 0.99 V_e$). (This statement would be exactly true if there were no dilution. We will neglect dilution.) If the initial acid concentration were, say, 0.1 M, then $[\text{H}^+] =$

Acid-Base Titrations   121

1% of 0.1 M = 0.001 M at $V_b = 0.99\ V_e$. The pH is $-\log(0.001) = 3$. At 99.9% completion, $[H^+] = 0.1\%$ of 0.1 M = 0.0001 M and the pH is 4. When the titration is 0.1% past the equivalence point, $[OH^-] = 0.0001$ M, pH = $-\log(K_w/0.0001) = 10$. The pH jumps from 4 to 10 in the interval from $V_b = 0.999\ V_e$ to $1.001\ V_e$. Even though the concentration of H^+ hardly changes, its logarithm changes rapidly around the equivalence point because $[H^+]$ decreases by orders of magnitude with tiny additions of OH^- when there is hardly any H^+ present.

11-4. The sketch should look like Figure 11-2. Before base is added, pH is determined by acid dissociation of HA. Between the initial point and the equivalence point, each mole of OH^- converts an equivalent quantity of HA into A^-. The resulting buffer containing HA and A^- determines the pH. At the equivalence point, all HA has been converted to A^-. The pH is controlled by hydrolysis of A^- with H_2O. After the equivalence point, excess OH^- is being added to the solution. To a good approximation, pH is determined just by the concentration of excess OH^-.

11-5. If analyte is too weak or too dilute, there is little change in pH at the equivalence point, making it difficult or impossible to identify an end point.

11-6.

V_b	0	1	5	9	9.9	10	10.1	12
pH	3.00	4.05	5.00	5.95	7.00	8.98	10.96	12.25

Representative calculations:

$(V_e\ (\text{mL}))(1.000\ \text{M KOH}) = (100.0\ \text{mL})(0.100\ \text{M HA}) \Rightarrow V_e = 10.0\ \text{mL}$

<u>0 mL</u>: $\underset{0.100-x}{\text{HA}} \rightleftharpoons \underset{x}{H^+} + \underset{x}{A^-}$ $\dfrac{x^2}{0.100-x} = 10^{-5.00} \Rightarrow x = 9.95 \times 10^{-4}$ M

$\Rightarrow$ pH = 3.00

<u>1 mL</u>: pH = $pK_a + \log\dfrac{[A^-]}{[HA]} = 5.00 + \log\dfrac{1}{9} = 4.05$

<u>10 mL</u>: $\underset{\left(\frac{100\ \text{mL}}{110\ \text{mL}}\right)(0.100)-x}{A^-} + H_2O \rightleftharpoons \underset{x}{HA} + \underset{x}{OH^-}$ $\dfrac{x^2}{0.0909-x} = \dfrac{K_w}{K_a}$

$\Rightarrow x = 9.53 \times 10^{-6}$ $\Rightarrow [H^+] = \dfrac{K_w}{x} \Rightarrow$ pH = 8.98

10.1 mL: $[OH^-] = \underbrace{\left(\dfrac{0.1 \text{ mL}}{110.1 \text{ mL}}\right)}_{\substack{0.1 \text{ mL excess } OH^- \\ \text{diluted to } 110.1 \text{ mL}}} \underbrace{(1.00 \text{ M})}_{\substack{\text{Initial} \\ \text{concentration} \\ \text{of } OH^-}} = 9.08 \times 10^{-4} \text{ M} \Rightarrow \text{pH} = 10.96$

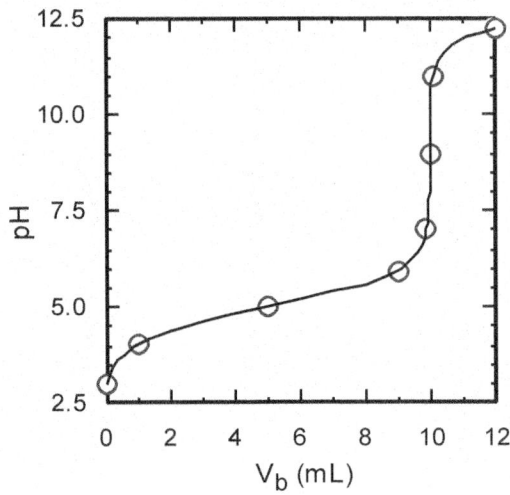

11-7. $\text{pH} = pK_a + \log\dfrac{[A^-]}{[HA]}$ $\qquad pK_a - 1 = pK_a + \log\dfrac{[A^-]}{[HA]} \Rightarrow \dfrac{[A^-]}{[HA]} = \dfrac{1}{10}$

If the ratio $\dfrac{[A^-]}{[HA]}$ is to be $\dfrac{1}{10}$, then $\dfrac{1}{11}$ of the initial HA must remain as HA.

At this point, $[A^-]/[HA] = (1/11)/(10/11) = 1/10$. So pH = $pK_a - 1$ when $V_b = V_e/11$.

In a similar manner, pH = $pK_a + 1$ when $V_b = 10V_e/11$.

For anilinium ion, $pK_a = 4.601$. For the titration of 100 mL of 0.100 M anilinium ion with 0.100 M OH⁻, the reaction is

C₆H₅–NH₃⁺ + OH⁻ → C₆H₅–NH₂ + H₂O and $V_e = 100$ mL.

0 mL: C₆H₅–NH₃⁺ $\underset{}{\overset{K_a}{\rightleftharpoons}}$ C₆H₅–NH₂ + H⁺
        0.100 – x   x

$\dfrac{x^2}{0.100 - x} = K_a = 10^{-4.60} \Rightarrow x = 1.57 \times 10^{-3} \Rightarrow \text{pH} = 2.80$

$V_e/11 = 9.09$ mL: pH = $pK_a - 1$ = 3.60

$V_e/2 = 50.0$ mL: pH = pK_a = 4.60

$10V_e/11 = 90.91$ mL: pH = $pK_a + 1$ = 5.60

$V_e = 100.0$ mL: The anilinium ion (BH⁺) has been converted to its basic form B, whose pH is governed by base hydrolysis:

Acid-Base Titrations

$$\underset{F-x}{B} + H_2O \underset{}{\overset{K_b}{\rightleftharpoons}} \underset{x}{BH^+} + \underset{x}{OH^-} \quad K_b = \frac{K_w}{K_a} = \frac{x^2}{\left(\frac{100}{200}\right)(0.100) - x}$$

$$\Rightarrow x = 4.46 \times 10^{-6} \text{ M} \quad pH = -\log \frac{K_w}{x} = 8.65$$

$\underline{1.2 V_e = 120.0 \text{ mL}}$: There are 20.0 mL of excess NaOH.

$$[OH^-] = \left(\frac{20 \text{ mL}}{220 \text{ mL}}\right)(0.100 \text{ M}) = 9.09 \times 10^{-3} \text{ M} \Rightarrow pH = 11.96$$

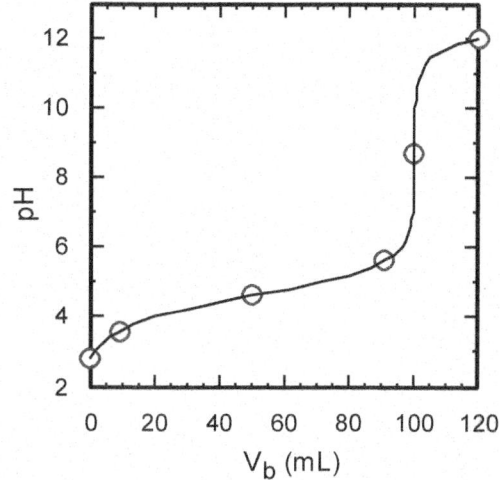

11-8. The titration reaction is $HA + OH^- \rightarrow A^- + H_2O$. A volume of V mL of HA will require $2V$ mL of KOH to reach the equivalence point, because [HA] = 0.100 M and [KOH] = 0.050 0 M. The formal concentration of A^- at the equivalence point will be $\left(\frac{V}{V+2V}\right)(0.100 \text{ M}) = 0.033\,3$ M. The pH is found by writing

$$\underset{0.033\,3-x}{A^-} + H_2O \rightleftharpoons \underset{x}{HA} + \underset{x}{OH^-} \quad \frac{x^2}{0.033\,3-x} = K_b = \frac{K_w}{K_a} = \frac{1.0 \times 10^{-14}}{1.48 \times 10^{-4}}$$

$$\Rightarrow x = 1.50 \times 10^{-6} \text{ M} \Rightarrow pH = 8.18$$

11-9.

$$\text{O}\bigcirc^{+}\text{NH}-\text{SO}_3^- \overset{K_a}{\rightleftharpoons} \text{O}\bigcirc\text{N}-\text{SO}_3^- + H^+ \quad K_a = 10^{-6.27}$$

$$H^+ + OH^- \rightleftharpoons H_2O \quad 1/K_w = 10^{14}$$

$$\text{O}\bigcirc^{+}\text{NH}-\text{SO}_3^- + OH^- \rightleftharpoons \text{O}\bigcirc\text{N}-\text{SO}_3^- + H_2O$$

MES $\qquad K = \frac{K_a}{K_w} = 5.4 \times 10^7$

11-10.

	HA	+	OH⁻	→	A⁻	+	H₂O
Initial mmol:	5.857		x		—		
Final mmol:	$5.857 - x$		—		x		

$$\text{pH} = 9.24 = pK_a + \log \frac{[A^-]}{[HA]} = 9.39 + \log \frac{x}{5.857 - x} \Rightarrow x = 2.4_{28} \text{ mmol}$$

$$[OH^-] = \frac{2.4_{28} \text{ mmol}}{22.63 \text{ mL}} = 0.107 \text{ M}$$

11-11.

	$(CH_3)_3NH^+$	+	OH⁻	→	$(CH_3)_3N$	+	H₂O
Initial mmol:	1.00		0.40		—		
Final mmol:	0.60		—		0.40		

First, find the ionic strength:

$[(CH_3)_3NH^+] = 0.60 \text{ mmol}/14.0 \text{ mL} = 0.042\,86 \text{ M}$

$[Br^-] = 1.00 \text{ mmol}/14.0 \text{ mL} = 0.071\,43 \text{ M}$

$[Na^+] = 0.40 \text{ mmol}/14.0 \text{ mL} = 0.028\,57 \text{ M}$

$\mu = \frac{1}{2} \Sigma c_i z_i^2 = 0.071 \text{ M}$

$$\text{pH} = pK_a + \log \frac{[B]\gamma_B}{[BH^+]\gamma_{BH^+}}$$

$$\text{pH} = 9.799 + \log \frac{(0.028\,6)(1.00)}{(0.042\,9)(0.80)} = 9.72$$

In the previous calculation, we used the size of $(CH_3)_3NH^+$ (400 pm) and the activity coefficient interpolated from Table 8-1.

11-12. The sketch should look like Figure 11-9. Before base is added, the pH is determined by the base hydrolysis reaction of B with H₂O. Between the initial point and the equivalence point, each mole of H⁺ converts an equivalent quantity of B into BH⁺. The resulting buffer containing B and BH⁺ determines the pH. At the equivalence point, all B has been converted to BH⁺. The pH is controlled by the acid dissociation reaction of BH⁺. After the equivalence point, excess H⁺ is being added to the solution. To a good approximation, the pH is determined just by the concentration of excess H⁺.

11-13. At the equivalence point, the weak base, B, is converted completely to the conjugate acid, BH⁺, which is necessarily acidic.

Acid-Base Titrations

11-14.

V_a	0	1	5	9	9.9	10	10.1	12
pH	11.00	9.95	9.00	8.05	7.00	5.02	3.04	1.75

$(V_e \text{ (mL)})(1.00 \text{ M HClO}_4) = (100.0 \text{ mL})(0.100 \text{ M B}) \Rightarrow V_e = 10.00 \text{ mL}$

Representative calculations:

0 mL: $\underset{0.100-x}{\text{B}} + \text{H}_2\text{O} \rightleftharpoons \underset{x}{\text{BH}^+} + \underset{x}{\text{OH}^-} \quad \dfrac{x^2}{0.100-x} = 10^{-5.00} \text{ M} \Rightarrow x = 9.95 \times 10^{-4} \text{ M}$

$[\text{H}^+] = \dfrac{K_w}{x} \Rightarrow \text{pH} = 11.00$

1 mL: $\text{pH} = pK_{\text{BH}^+} + \log \dfrac{[\text{B}]}{[\text{BH}^+]} = 9.00 + \log \dfrac{9}{1} = 9.95$

10 mL: $\underset{\left(\frac{100}{110}\right)(1.00)-x}{\text{BH}^+} \rightleftharpoons \underset{x}{\text{B}} + \underset{x}{\text{H}^+} \quad \dfrac{x^2}{0.0909-x} = 10^{-9.00} \text{ M} \Rightarrow x = 9.53 \times 10^{-6} \text{ M}$

$[\text{H}^+] = x \Rightarrow \text{pH} = 5.02$

10.1 mL: $[\text{H}^+] = \left(\dfrac{0.1 \text{ mL}}{110.1 \text{ mL}}\right)(1.00 \text{ M}) = 9.08 \times 10^{-4} \text{ M} \Rightarrow \text{pH} = 3.04$

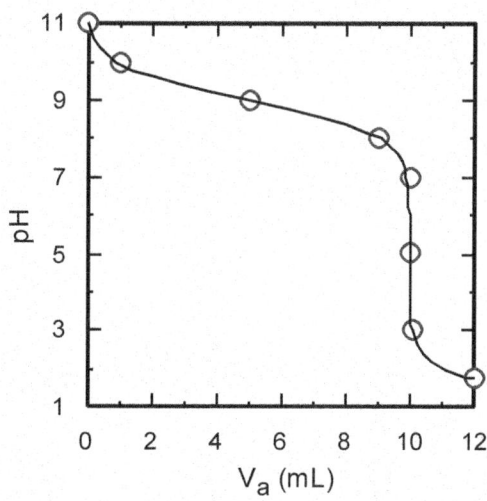

11-15. The maximum buffer capacity is reached when
$V = \tfrac{1}{2} V_e$, at which time $\dfrac{[\text{B}]}{[\text{BH}^+]} = 1$ and pH = pK_a (for BH$^+$).

11-16. $\text{C}_6\text{H}_5\text{—CH}_2\text{NH}_2 + \text{H}^+ \rightleftharpoons \text{C}_6\text{H}_5\text{—CH}_2\overset{+}{\text{NH}}_3$ Reverse of K_a reaction

$K = 1/K_a$ (for $\text{C}_6\text{H}_5\text{CH}_2\text{NH}_3^+$) = 2.2×10^9

11-17. Titration reaction: $B + H^+ \rightarrow BH^+$. To find the equivalence point, we write
$(50.0 \text{ mL})(0.0319 \text{ M}) = (V_e)(0.0500 \text{ M}) \Rightarrow V_e = 31.9 \text{ mL}$.

<u>0 mL</u>: $\underset{0.0319-x}{B} + H_2O \rightleftharpoons \underset{x}{BH^+} + \underset{x}{OH^-}$ $\dfrac{x^2}{0.0319-x} = K_b = \dfrac{K_w}{K_a} = 2.22 \times 10^{-5}$

$\Rightarrow x = 8.31 \times 10^{-4} \text{ M} \Rightarrow \text{pH} = 10.92$

<u>12.0 mL</u>:

	B	+	H⁺	→	BH⁺
Relative initial quantity:	31.9		12.0		—
Relative final quantity:	19.9		—		12.0

$\text{pH} = \text{p}K_a + \log \dfrac{[B]}{[BH^+]} = 9.35 + \log \dfrac{19.9}{12.0} = 9.57$

<u>1/2 V_e</u>: $\text{pH} = \text{p}K_a = 9.35$

<u>30 mL</u>: $\text{pH} = \text{p}K_a + \log \dfrac{1.9}{30.0} = 8.15$

<u>V_e</u>: B has been converted to BH^+ at a concentration of
$\left(\dfrac{50.0 \text{ mL}}{81.9 \text{ mL}}\right)(0.0319 \text{ M}) = 0.0195 \text{ M}$

$\underset{0.0195-x}{BH^+} \rightleftharpoons \underset{x}{B} + \underset{x}{H^+}$ $\dfrac{x^2}{0.0195-x} = K_a \Rightarrow x = 2.96 \times 10^{-6} \text{ M}$

$\Rightarrow \text{pH} = 5.53$

<u>35.0 mL</u>: $[H^+] = \left(\dfrac{3.1 \text{ mL}}{85.0 \text{ mL}}\right)(0.0500 \text{ M}) = 1.82 \times 10^{-3} \text{ M} \Rightarrow \text{pH} = 2.74$

11-18. Titration reaction: $CN^- + H^+ \rightarrow HCN$

At the equivalence point, moles of CN^- = moles of H^+
$(0.100 \text{ M})(50.00 \text{ mL}) = (0.438 \text{ M})(V_e) \Rightarrow V_e = 11.42 \text{ mL}$

(a)

	CN⁻	+	H⁺	→	HCN
Relative initial quantity:	11.42		4.20		—
Relative final quantity:	7.22		—		4.20

$\text{pH} = \text{p}K_a + \log \dfrac{7.22}{4.20} = 9.45$

(b) 11.82 mL is 0.40 mL past the equivalence point.
$[H^+] = \left(\dfrac{0.40 \text{ mL}}{61.82 \text{ mL}}\right)(0.438 \text{ M}) = 2.83 \times 10^{-3} \text{ M} \Rightarrow \text{pH} = 2.55$

(c) At the equivalence point, we have made HCN at a formal concentration of
$\left(\dfrac{50.00 \text{ mL}}{61.42 \text{ mL}}\right)(0.100 \text{ M}) = 0.0814 \text{ M}$.

$$\text{HCN} \rightleftharpoons \text{H}^+ + \text{CN}^- \qquad \frac{x^2}{0.0814-x} = K_a \Rightarrow x = 7.1 \times 10^{-6}$$
$$0.0814-x \qquad x \qquad x$$
$$\Rightarrow \text{pH} = 5.15$$

11-19. 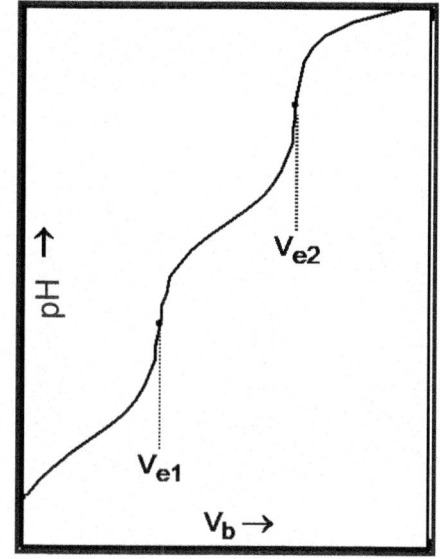 The pH of the initial solution before base is added is determined by the first acid dissociation reaction of H_2A. As base is added, it converts H_2A into an equivalent amount of HA^-. The buffer consisting of H_2A and HA^- governs the pH. At the first equivalence point, we have a solution of "pure" HA^-, the intermediate form of a diprotic acid. The pH is determined by the competitive acid and base reactions of HA^-. Between the two equivalence points there is a mixture of HA^- and A^{2-}, which is another buffer. At the second equivalence point, we have converted all HA^- into A^{2-}, whose base hydrolysis reaction determines the pH. After the second equivalence point, the excess added OH^- is determines the pH, with negligible contribution from A^{2-}.

11-20. The protein has an average charge of 0 at the isoelectric point. The isoionic pH is lower, so the average charge of the protein must be positive at the isoionic point because some basic groups react with H^+ as acid is added to lower the pH.

11-21. The equivalence point could be attained by mixing pure HA plus NaCl. Neglecting effects of ionic strength, the pH is equivalent to that of a solution of pure HA. This is the isoionic pH.

11-22. (a)
$$\begin{array}{ll} \text{HA} \rightleftharpoons \text{A}^- + \text{H}^+ & K_a \\ \text{B} + \text{H}^+ \rightleftharpoons \text{BH}^+ & K = K_b/K_w \\ \hline \text{B} + \text{HA} \rightleftharpoons \text{BH}^+ + \text{A}^- & K = K_aK_b/K_w \end{array}$$
$$= 10^{-2.86} \, 10^{-3.36}/10^{-14.00} = 10^{7.78}$$

(b) In the upper curve, $\frac{3}{2}V_e$ is halfway between the first and second equivalence points. The pH is simply pK_2, since there is a 1:1 mixture of HA^- and A^{2-}. In the lower curve, pK_2 (=pK_{BH^+}) occurs when there is a 1:1 mixture of B and

BH$^+$. To achieve this condition, all of B is first transformed into BH$^+$ by reaction with HA until V_e is reached. Then, at $2V_e$ one more equivalent of B has been added, giving a 1:1 mole ratio B:BH$^+$, so pH = pK_{BH^+}.

11-23.

V_a	0	1	5	9	10	11	15	19	20	22
pH	11.49	10.95	10.00	9.05	8.00	6.95	6.00	5.05	3.54	1.79

Equivalence points:

$(0.100 \text{ M B})(100.0 \text{ mL}) = (1.00 \text{ M HCl})(V_{e1}) \Rightarrow V_{e1} = 10.0 \text{ mL}$

Representative calculations: $V_{e2} = 20.0 \text{ mL}$

<u>0 mL</u>: $\text{B} + \text{H}_2\text{O} \underset{}{\overset{K_{b1}}{\rightleftharpoons}} \text{BH}^+ + \text{OH}^-$ $\dfrac{x^2}{0.100-x} = 10^{-4.00} \Rightarrow x = 3.11 \times 10^{-3}$ M

$0.100-x \qquad x \qquad x$

$$\text{pH} = -\log \dfrac{K_w}{x} = 11.49$$

<u>1 mL</u>: pH = pK_{BH^+} + log $\dfrac{[\text{B}]}{[\text{BH}^+]}$ = 10.00 + log $\dfrac{9}{1}$ = 10.95

<u>10 mL</u>: Predominant form is BH$^+$ with formal concentration $\dfrac{100}{110}(0.100) = 0.0909$ M

$$[\text{H}^+] \approx \sqrt{\dfrac{10^{-6.00}10^{-10.00}(0.0909) + 10^{-6.00}10^{-14.00}}{10^{-6.00} + 0.0909}} = 1.00 \times 10^{-8} \Rightarrow \text{pH} = 8.00$$

<u>11 mL</u>: pH = p$K_{BH_2^{2+}}$ + log $\dfrac{[\text{BH}^+]}{[\text{BH}_2^{2+}]}$ = 6.00 + log $\dfrac{9}{1}$ = 6.95

<u>20 mL</u>: $\text{BH}_2^{2+} \rightleftharpoons \text{BH}^+ + \text{H}^+$ $\dfrac{x^2}{0.0833-x} = 10^{-6.00} \Rightarrow x = 2.88 \times 10^{-4}$

$\dfrac{100}{120}(0.100)-x \qquad x \qquad x$

$\Rightarrow$ pH = 3.54

<u>22 mL</u>: $[\text{H}^+] = \left(\dfrac{2 \text{ mL}}{122 \text{ mL}}\right)(1.00 \text{ M}) = 1.64 \times 10^{-2}$ M $\Rightarrow$ pH = 1.79

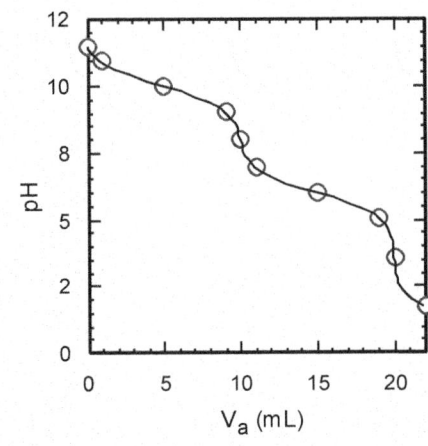

11-24.

V_b	0	1	5	9	10	11	15	19	20	22
pH	2.51	3.05	4.00	4.95	6.00	7.05	8.00	8.95	10.46	12.21

$(0.100 \text{ M H}_2\text{A})(100.0 \text{ mL}) = (1.00 \text{ M NaOH})(V_{e1}) \Rightarrow V_{e1} = 10.0 \text{ mL}$

Representative calculations:

<u>0 mL</u>: $\quad \text{H}_2\text{A} \rightleftharpoons \text{HA}^- + \text{H}^+ \qquad \dfrac{x^2}{0.100-x} = 10^{-4.00} \Rightarrow x = 3.11 \times 10^{-3} \text{ M}$

$\qquad \quad \; 0.100-x \qquad x \qquad x$

$\qquad\qquad\qquad\qquad\qquad\qquad\qquad\qquad\qquad \Rightarrow \text{pH} = 2.51$

<u>1 mL</u>: $\quad \text{pH} = pK_1 + \log \dfrac{[\text{HA}^-]}{[\text{H}_2\text{A}]} = 4.00 + \log \dfrac{1}{9} = 3.05$

<u>10 mL</u>: Predominant form is HA$^-$ with formal concentration $\left(\dfrac{100 \text{ mL}}{110 \text{ mL}}\right)(0.100 \text{ M})$

$= 0.0909 \text{ M}$.

$[\text{H}^+] \approx \sqrt{\dfrac{10^{-4.00}10^{-8.00}(0.0909) + 10^{-4.00}10^{-14.00}}{10^{-4.00} + 0.0909}}$

$= 9.99 \times 10^{-7} \Rightarrow \text{pH} = 6.00$

<u>11 mL</u>: $\quad \text{pH} = pK_2 + \log \dfrac{[\text{A}^{2-}]}{[\text{HA}^-]} = 8.00 + \log \dfrac{1}{9} = 7.05$

<u>20 mL</u>: $\quad \text{A}^{2-} + \text{H}_2\text{O} \rightleftharpoons \text{HA}^- + \text{OH}^- \qquad \dfrac{x^2}{0.0833-x} = \dfrac{K_w}{K_2} \Rightarrow x = 2.88 \times 10^{-4} \text{ M}$

$\qquad \; \left(\dfrac{100}{120}\right)(0.100) - x \qquad x \qquad x$

$\qquad\qquad\qquad\qquad\qquad\qquad\qquad\qquad\qquad \text{pH} = -\log \dfrac{K_w}{x} = 10.46$

<u>22 mL</u>: $[\text{OH}^-] = \left(\dfrac{2 \text{ mL}}{122 \text{ mL}}\right)(1.00 \text{ M}) = 1.64 \times 10^{-2} \text{ M} \Rightarrow \text{pH} = 12.21$

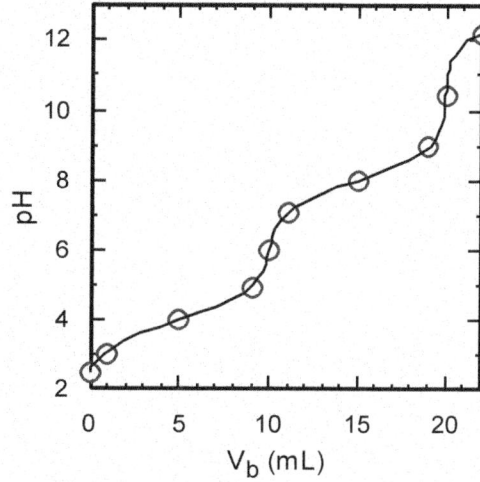

130	Chapter 11

11-25. Titration reactions:

$$HN\!\!\bigcirc\!\!NH + H^+ \longrightarrow H_2\overset{+}{N}\!\!\bigcirc\!\!NH \qquad V_e = 40.0 \text{ mL}$$

$$H_2\overset{+}{N}\!\!\bigcirc\!\!NH + H^+ \longrightarrow H_2\overset{+}{N}\!\!\bigcirc\!\!\overset{+}{N}H_2 \qquad V_e = 80.0 \text{ mL}$$

<u>0 mL</u>: $B + H_2O \rightleftharpoons BH^+ + OH^-$ $\dfrac{x^2}{0.100-x} = K_{b1} = \dfrac{K_w}{K_{a2}} = \dfrac{1.0\times10^{-14}}{1.86\times10^{-10}}$

$\quad\quad 0.100-x \quad\quad\quad\quad x \quad\quad x$

$\Rightarrow x = 2.29 \times 10^{-3}$ M $\Rightarrow$ pH = 11.36

<u>10.0 mL</u>: pH $= pK_2 + \log \dfrac{[B]}{[BH^+]} = 9.731 + \log \dfrac{3}{1} = 10.21$

<u>20.0 mL</u>: pH $= pK_2 = 9.73$

<u>30.0 mL</u>: pH $= pK_2 + \log \dfrac{1}{3} = 9.25$

<u>40.0 mL</u>: B has been converted to BH^+ at a formal concentration of

$$F = \left(\dfrac{40.0 \text{ mL}}{80.0 \text{ mL}}\right)(0.100 \text{ M}) = 0.0500 \text{ M}$$

$$[H^+] = \sqrt{\dfrac{K_1 K_2 F + K_1 K_w}{K_1 + F}}$$

$$= \sqrt{\dfrac{(4.65\times10^{-6})(1.86\times10^{-10})(0.0500) + (4.65\times10^{-6})(1.0\times10^{-14})}{4.65\times10^{-6} + 0.0500}}$$

$\Rightarrow$ pH = 7.53

<u>50.0 mL</u>: pH $= pK_1 + \log \dfrac{[BH^+]}{[BH_2^{2+}]} = 5.333 + \log \dfrac{3}{1} = 5.81$

<u>60.0 mL</u>: pH $= pK_1 = 5.33$

<u>70.0 mL</u>: pH $= pK_1 + \log \dfrac{1}{3} = 4.86$

<u>80.0 mL</u>: B has been converted to BH_2^{2+} at a formal concentration of

$$\left(\dfrac{40.0 \text{ mL}}{120.0 \text{ mL}}\right)(0.100 \text{ M}) = 0.0333 \text{ M}$$

$BH_2^{2+} \rightleftharpoons BH^+ + H^+$ $\dfrac{x^2}{0.0333-x} = K_1 = 4.65 \times 10^{-6} \Rightarrow$

$\quad 0.0333-x \quad\quad x \quad\quad x$

$x = 3.91 \times 10^{-4}$ M $\Rightarrow$ pH = 3.41

<u>90.0 mL</u>: $[H^+] = \left(\dfrac{10.0 \text{ mL}}{130.0 \text{ mL}}\right)(0.100 \text{ M}) \Rightarrow$ pH = 2.11

<u>100.0 mL</u>: $[H^+] = \left(\dfrac{20.0 \text{ mL}}{140.0 \text{ mL}}\right)(0.100 \text{ M}) \Rightarrow$ pH = 1.85

Acid-Base Titrations

11-26.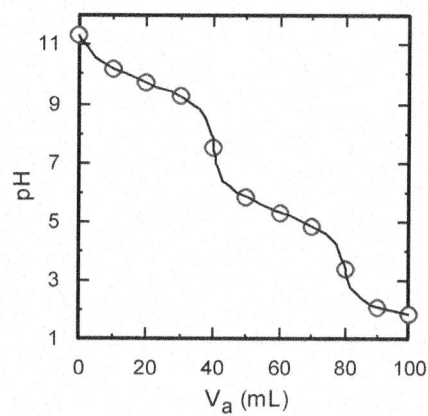

Initial mmol:	0.500	0.164	—
Final mmol:	0.336	—	0.164

$$\text{pH} = pK_1 + \log \frac{[B]}{[BH^+]} = 4.70 + \log \frac{0.336}{0.164} = 5.01$$

11-27. (a) Titration reactions:

$$H_2NCH_2CO_2^- + H^+ \rightarrow H_3^+NCH_2CO_2^- \qquad V_e = 50.0 \text{ mL}$$

$$H_3^+NCH_2CO_2^- + H^+ \rightarrow H_3^+NCH_2CO_2H \qquad V_e = 100.0 \text{ mL}$$

At the second equivalence point, the formal concentration of $H_3^+NCH_2CO_2H$ is $\left(\frac{50.0 \text{ mL}}{150.0 \text{ mL}}\right)(0.100 \text{ M}) = 0.033\ 3 \text{ M}$

$$H_3^+NCH_2CO_2H \rightleftharpoons H_3^+NCH_2CO_2^- + H^+ \qquad \frac{x^2}{0.033\ 3 - x} = K_1 = 0.004\ 47 \Rightarrow$$

$$0.0333 - x \qquad\qquad x \qquad\qquad x$$

$$x = 1.02 \times 10^{-2} \text{ M} \Rightarrow \text{pH} = 1.99$$

(b) At $V_a = 90.0$ mL, the approximation gives $\text{pH} = pK_1 + \log \frac{[HG]}{[H_2G^+]} = 2.35$ $+ \log \frac{1}{4} = 1.75$, which is lower than the correct value at 100.0 mL. The pH calculated with the equation in Table 11-5 is 2.16, which is higher than the equivalence point pH of 1.99.

At $V_a = 101.0$ mL, the approximation gives $[H^+] = \left(\frac{1.0 \text{ mL}}{151.0 \text{ mL}}\right)(0.100 \text{ M}) = 6.62 \times 10^{-4} \text{ M} \Rightarrow \text{pH} = 3.18$, which is higher than the correct value at 100.0 mL. The pH calculated with the equation in Table 11-5 is 1.98, which is lower than the equivalence point pH of 1.99.

132   Chapter 11

11-28. (a)
$$\overset{+}{N}H_3 \text{—}CHCH_2CH_2CO_2H \text{—}CO_2^- + OH^- \rightarrow \overset{+}{N}H_3\text{—}CHCH_2CH_2CO_2^-\text{—}CO_2^- + H_2O$$

(b) V mL of glutamic acid will require $\frac{0.100}{0.025} V = 4.00\ V$ mL of RbOH to reach the equivalence point. The formal concentration of product will be

$$\left(\frac{V}{V+4.00\ V}\right)(0.100\ M) = 0.0200\ M.$$

$$[H^+] = \sqrt{\frac{K_2 K_3 F + K_2 K_w}{K_2 + F}}$$

$$= \sqrt{\frac{(5.0\times 10^{-5})(1.1\times 10^{-10})(0.020\ 0) + (5.0\times 10^{-5})(1.0\times 10^{-14})}{(5.0\times 10^{-5}) + 0.020\ 0}}$$

$$= 7.42 \times 10^{-8}\ M \Rightarrow pH = 7.13$$

11-29.

$$\overset{+}{N}H_3\text{—}CHCH_2\text{—}\bigcirc\text{—}OH + H^+ \rightarrow \overset{+}{N}H_3\text{—}CHCH_2\text{—}\bigcirc\text{—}OH$$
$$\ \ \ \ CO_2^-\ CO_2H$$
$$\ \ \ \ \ \ \ \ \ \ H_2T\ H_3T^+$$

One volume of tyrosine (0.0100 M) requires 2.5 volumes of $HClO_4$ (0.004 00 M).

Formal concentration of tyrosine at equivalence point is $\left(\frac{V}{V+2.5V}\right)(0.010\ 0\ M)$

$= 0.002\ 86\ M$. pH is calculated from acid dissociation of H_3T^+.

$$H_3T^+ \rightleftharpoons H_2T + H^+ \ \ \ \ \ \ \ \frac{x^2}{0.002\ 86 - x} = K_1 = 3.9 \times 10^{-3} \Rightarrow$$
$$0.002\ 86 - x\ \ \ \ \ x\ \ \ \ \ x$$

$$x = 0.001\ 92\ M \Rightarrow pH = 2.72$$

11-30. (a) $C^{2-} + H^+ \rightarrow HC^-$. $V_e = 20.0$ mL. At the equivalence point, the formal concentration of HC^- is $\left(\frac{40.0\ mL}{60.0\ mL}\right)(0.030\ 0\ M) = 0.0200\ M$.

$$[H^+] = \sqrt{\frac{K_2 K_3 F + K_2 K_w}{K_2 + F}}$$

$$= \sqrt{\frac{(4.4\times 10^{-9})(1.82\times 10^{-11})(0.020\ 0) + (4.4\times 10^{-9})(1.0\times 10^{-14})}{(4.4\times 10^{-9}) + 0.020\ 0}}$$

$$= 2.86 \times 10^{-10}\ M \Rightarrow pH = 9.54$$

Acid-Base Titrations

(b) $\quad H_3C^+ \rightleftharpoons H_2C + H^+ \qquad \dfrac{x^2}{0.0500-x} = K_1 = 0.02$

$ 0.0500-x \qquad x \qquad x \qquad \Rightarrow x = 0.023\ M \Rightarrow pH = 1.64$

$$pH = pK_3 + \log\dfrac{[C^{2-}]}{[HC^-]}$$

$$1.64 = 10.74 + \log\dfrac{[C^{2-}]}{[HC^-]} \Rightarrow \dfrac{[C^{2-}]}{[HC^-]} = 7.9 \times 10^{-10}$$

11-31. The two values of pK_a for oxalic acid are 1.250 and 4.266. At a pH of 4.40, the $C_2O_4^{2-}$ has not yet been half-neutralized.

	$C_2O_4^{2-}$	+	H^+	→	$HC_2O_4^-$
Initial mmol:	x		16.0		—
Final mmol:	$x - 16.0$		—		16.0

$$pH = 4.40 = pK_2 + \log\dfrac{[C_2O_4^{2-}]}{[HC_2O_4^-]} = 4.266 + \log\dfrac{x-16.0}{16.0}$$

$\Rightarrow x = 37.8$ mmol of $K_2C_2O_4 = 6.28$ g

11-32. Neutral alanine is designated HA.

	HA	+	OH$^-$	→	A$^-$	+	H$_2$O
Initial mmol:	1.260 5		0.516		—		
Final mmol:	0.744 5		—		0.516		

$$pH = pK_2 + \log\dfrac{[A^-]\gamma_{A^-}}{[HA]\gamma_{HA}} \qquad 9.57 = pK_2 + \log\dfrac{(0.516)(0.77)}{(0.744\ 5)(1)} \Rightarrow pK_2 = 9.84$$

11-33. A Gran plot allows us to find the equivalence point by extrapolating from points measured prior to the equivalence point. Points prior to the equivalence point have less measurement uncertainty than points very close to the equivalence point.

11-34. It is evident from titration data in the problem that the end point is near 23.4 mL, where pH changes most rapidly with small additions of titrant. A graph of $V_b 10^{-pH}$ versus V_b follows. The points from 21.01 to 23.30 mL were fit by the method of least squares to give the equation shown in the graph. The intercept is found by setting $y = 0$ in the equation, giving $x = V_e = 23.39$ mL.

V_b (mL)	$V_b\,10^{-\text{pH}}$	V_b (mL)	$V_b\,10^{-\text{pH}}$	V_b (mL)	$V_b\,10^{-\text{pH}}$
21.01	15.22×10^{-6}	22.10	8.40×10^{-6}	22.97	2.76×10^{-6}
21.10	14.94	22.27	7.37	23.01	2.41
21.13	14.62	22.37	6.60	23.11	1.79
21.20	14.33	22.48	5.91	23.17	1.46
21.30	13.75	22.57	5.29	23.21	1.16
21.41	12.90	22.70	4.53	23.30	0.75
21.51	12.10	22.76	4.14	23.32	0.42
21.61	11.61	22.80	3.78	23.40	0.12
21.77	10.42	22.85	3.46	23.46	0.01
21.93	9.35	22.91	3.16	23.55	0.003

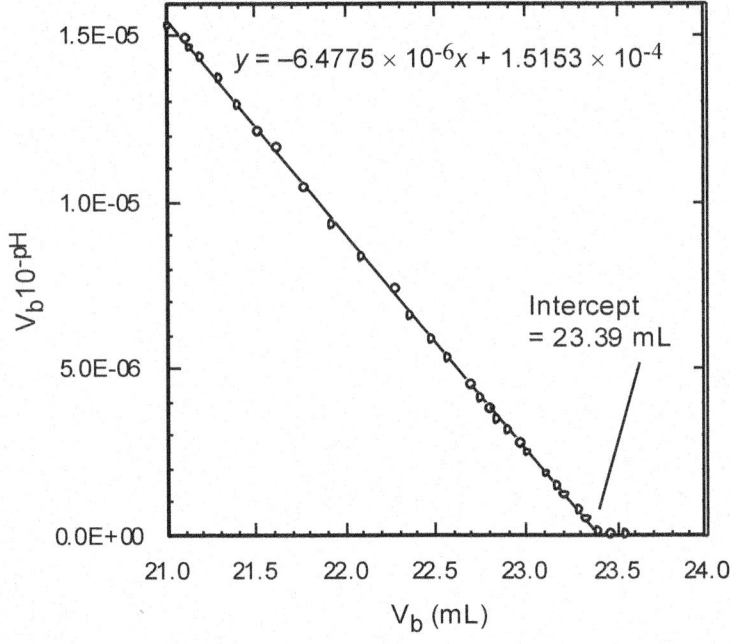

 11-35.

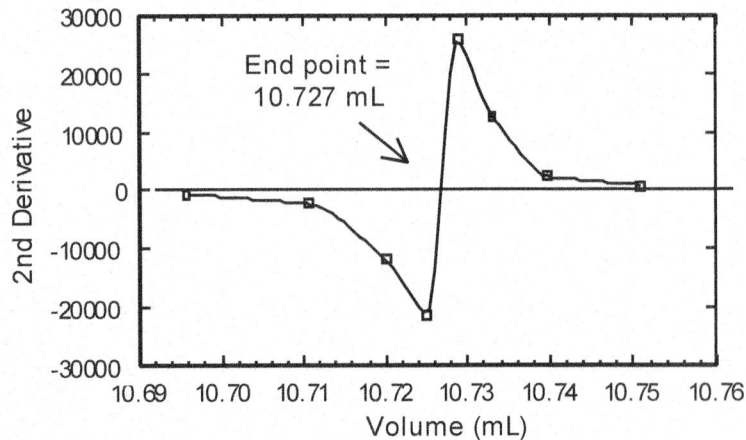

Calculations are shown in the spreadsheet:

Acid-Base Titrations

	A	B	C	D	E	F
1	Derivatives in a titration curve					
2	Data		1st derivative		2nd derivative	
3	mL NaOH	pH	mL	ΔpH/ΔmL	mL	Δ(ΔpH/ΔmL)
4	10.679	7.643				ΔmL
5			10.6875	-11.5		
6	10.696	7.447			10.6960	-553.6
7			10.7045	-20.9		
8	10.713	7.091			10.7108	-2234.7
9			10.7170	-48.9		
10	10.721	6.700			10.7200	-11770.8
11			10.7230	-119.5		
12	10.725	6.222			10.7250	-21375.0
13			10.7270	-205.0		
14	10.729	5.402			10.7290	25687.5
15			10.7310	-102.2		
16	10.733	4.993			10.7333	12411.1
17			10.7355	-46.4		
18	10.738	4.761			10.7398	2351.0
19			10.7440	-26.4		
20	10.750	4.444			10.7508	885.2
21			10.7575	-14.5		
22	10.765	4.227				
23	Representative formulas:					
24	C5 = (A6+A4)/2			E6 = (C7+C5)/2		
25	D5 = (B6-B4)/(A6-A4)			F6 = (D7-D5)/(C7-C5)		

11-36. The quotient [HIn]/[In$^-$] changes from 10:1 when pH = pK_{HIn} − 1 to 1:10 when pH = pK_{HIn} + 1. This change is generally sufficient to cause a complete color change.

11-37. The indicator has its acidic color when pH = pK_{HIn} − 1 because HIn is the dominant species. The indicator has its basic color when pH = pK_{HIn} + 1 because In$^-$ is the dominant species. The color changes from acidic color to intermediate color to basic color as pH rises through the range pK_{HIn} − 1 to pK_{HIn} + 1. If the indicator is chosen correctly for the titration, this indicator pH transition range coincides with the steep part of the titration curve. The color change occurs near the equivalence point, which is the center of the steep portion of the titration curve.

11-38. The Henderson-Hasselbalch equation for the indicator, HIn, is

$$\text{pH} = \text{p}K_{HIn} + \log \frac{[\text{In}^-]}{[\text{HIn}]}.$$ If we know pK_{HIn} and we measure $\frac{[\text{In}^-]}{[\text{HIn}]}$ spectroscopically, then we can calculate the pH.

11-39. Strong acids, such as H_2SO_4, HCl, HNO_3, and $HClO_4$ have pK_a < 0.

11-40. yellow, green, blue

11-41. (a) red (b) orange (c) yellow

11-42. (a) red (b) orange (c) yellow (d) red

11-43. No. When a weak acid is titrated with a strong base, the solution contains A^- at the equivalence point. A solution of A^- must have a pH above 7.

11-44. (a) The titration reaction is $F^- + H^+ \rightarrow HF$.
If V mL of NaF are used, $V_e = \frac{1}{2}V$, since the concentration of $HClO_4$ is twice as great as the concentration of NaF. The formal concentration of HF at the equivalence point is $\left(\dfrac{V}{V + \frac{1}{2}V}\right)(0.030\,0\text{ M}) = 0.020\,0\text{ M}$.

The pH is determined by the acid dissociation of HF.

$$HF \rightleftharpoons H^+ + F^-$$
$$0.0200 - x \quad\quad x \quad\quad x$$

$\dfrac{x^2}{0.020\,0 - x} = K_a \Rightarrow x = 3.36 \times 10^{-3}$

$\Rightarrow pH = 2.47$

(b) The pH is so low that there would not be much (if any) break (inflection) in the titration curve at the equivalence point. A sharp change in indicator color will not be seen.

11-45. (a) violet (red + blue) (b) blue (c) yellow

11-46. (a) $NH_4^+ \rightleftharpoons NH_3 + H^+$
$0.010 - x \quad\quad x \quad\quad x$

$\dfrac{x^2}{0.010 - x} = K_a \Rightarrow x = 2.38 \times 10^{-6}\text{ M}$

$\Rightarrow pH = 5.62$

(b) One possible indicator is methyl red, using the yellow end point.

For a more complete analysis of this problem, we could compute the titration curve for a mixture of HCl and NH_4^+. For simplicity, we consider the mixture to be a "diprotic" acid with $K_1 = 100$ (i.e., a "strong" acid) and $K_2 = 5.7 \times 10^{-10}$ for the ammonium ion. We use the spreadsheet equation in Table 11-5 for titrating H_2A with strong base, taking $C_a = 0.01$ M and $C_b = 0.1$ M. An indicator with a color change in the range ~4.5–7.0 would find the HCl end point without titrating a significant amount of NH_4^+.

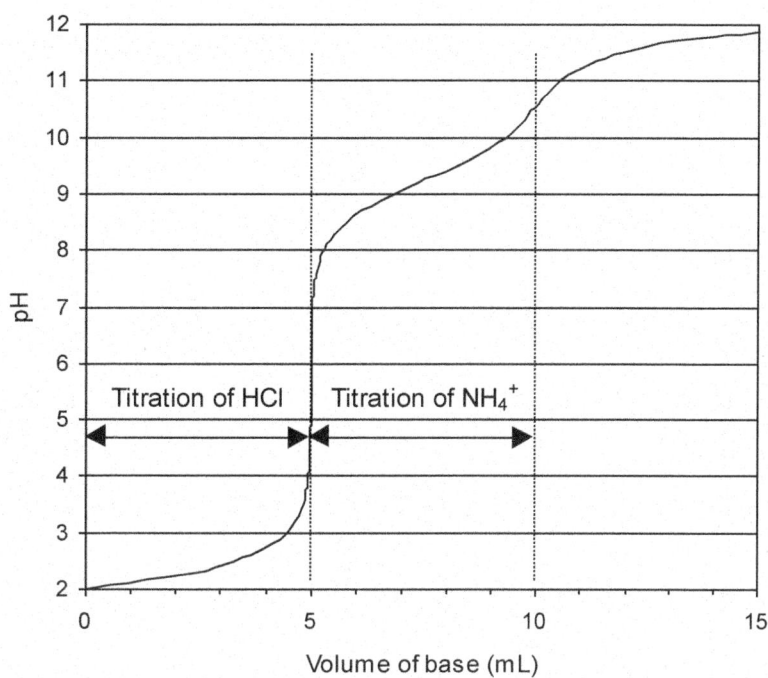

11-47. Grams of cleaner titrated $= \left(\dfrac{4.373 \text{ g}}{10.231 \text{ g} + 39.466 \text{ g}}\right)(10.231 \text{ g}) = 0.9003 \text{ g}$

mol HCl used = mol NH_3 present = $(0.01422 \text{ L})(0.1063 \text{ M}) = 1.512$ mmol

1.512 mmol NH_3 = 25.74 mg NH_3

wt% $NH_3 = \dfrac{2.574 \times 10^{-2} \text{ g}}{0.9003 \text{ g}} \times 100 = 2.859\%$

11-48. *Alkalinity* is the capacity of natural water to react with H^+ to reach pH 4.5, which is the second equivalence point in the titration of carbonate (CO_3^{2-}) with H^+. Alkalinity measures $[OH^-] + [CO_3^{2-}] + [HCO_3^-]$ plus any other bases that are present. Bromocresol green is blue above pH 5.4 and yellow below pH 3.8. It will be at its green end-point color between these two pH values, which approximates pH 4.5.

11-49. Tris(hydroxymethyl) aminomethane ($H_2NC(CH_2OH)_3$), sodium carbonate (Na_2CO_3), and borax ($NaB_4O_7 \cdot 10H_2O$) can be used to standardize HCl. Potassium acid phthalate ($HO_2C-C_6H_4-CO_2^-K^+$), HCl azeotrope, potassium hydrogen iodate ($KH(IO_3)_2$), sulfosalicylic acid double salt ($C_7H_5SO_6K \cdot C_7H_4SO_6K_2$), and sulfamic acid ($^+H_3NSO_3^-$) can be used to standardize NaOH.

11-50. The greater the equivalent mass, the more primary standard is required. There is less relative error in weighing a large mass of reagent than a small mass.

11-51. Potassium acid phthalate is dried at 110°C and weighed accurately into a flask. It is titrated with NaOH, using a pH electrode or phenolphthalein to observe the end point.

11-52. Grams of Tris titrated = $\dfrac{4.963 \text{ g}}{(1.023 \text{ g}+99.367 \text{ g})}(1.023 \text{ g}) = 0.05057 \text{ g} = 0.4175$ mmol

Concentration of HNO_3 = $\dfrac{0.4175 \text{ mmol}}{5.262 \text{ g solution}} = 0.07934$ mol/kg solution

11-53. True mass = $m = \dfrac{(1.023 \text{ g})\left(1-\dfrac{0.0012}{8.0}\right)}{\left(1-\dfrac{0.0012}{1.33}\right)} = 1.023_8$ g

Failure to account for buoyancy introduces a systematic error of
$100 \times (1.023_8 - 1.023)/1.023 = 0.08\%$ in the calculated molarity of HCl.

The true mass is higher than the measured mass of Tris, so the calculated HCl molarity is too low.

11-54. 30 mL of 0.05 M OH^- = 1.5 mmol = 0.31 g of potassium acid phthalate.

11-55. (a) From a graph of weight percent vs pressure (slope = –0.00240000 wt%/Torr, y-intercept = 22.0430 torr), we find HCl = 20.253% when P = 746 Torr.

(b) We need 0.10000 mole of HCl = 3.6458 g

$\dfrac{3.6458 \text{ g HCl}}{0.20253 \text{ g HCl/g solution}} = 18.001_3$ g of solution.

The mass required (weighed in air) is

$m' = \dfrac{(18.001_3)\left(1-\dfrac{0.0012}{1.096}\right)}{\left(1-\dfrac{0.0012}{8.0}\right)} = 17.984$ g

11-56. (a) For a rectangular distribution of uncertainty in atomic mass, divide the uncertainty listed in the periodic table by $\sqrt{3}$ to find the standard uncertainty:

C: $12.011 \pm 0.001/\sqrt{3} = 12.011 \pm 0.000\,5_8$

H: $1.008 \pm 0.001/\sqrt{3} = 1.008 \pm 0.000\,5_8$

O: $15.999 \pm 0.001/\sqrt{3} = 15.999 \pm 0.000\,5_8$

K: $39.098\,3 \pm 0.000\,1/\sqrt{3} = 39.098\,3 \pm 0.000\,05_8$

8C:	$8(12.011 \pm 0.000\,5_8)$	=	$96.088 \pm 0.004\,6$
5H:	$5(1.008 \pm 0.000\,5_8)$	=	$5.040 \pm 0.002\,9$
4O:	$4(15.999 \pm 0.000\,5_8)$	=	$63.996 \pm 0.002\,3$
1K:	$1(39.098\,3 \pm 0.000\,05_8)$	=	$39.098\,3 \pm 0.000\,05_8$
$C_8H_5O_4K$:			$204.222\,3 \pm ?$

Uncertainty = $\sqrt{0.004\,6^2 + 0.002\,9^2 + 0.002\,3^2 + 0.000\,058^2} = 0.005\,9$

Answer: 204.223 ± 0.006 g/mol

(b) For a rectangular distribution, divide the stated uncertainty by $\sqrt{3}$ to find the standard uncertainty. Purity = $1.000\,00 \pm 0.000\,05/\sqrt{3} = 1.000\,00 \pm 0.000\,03$

11-57. 5.00 mL of 0.033 6 M HCl = 0.168 0 mmol. 6.34 mL of 0.010 0 M NaOH = 0.063 4 mmol. HCl consumed by NH_3 = 0.168 0 – 0.063 4 = 0.104 6 mmol = 1.465 mg of nitrogen. 256 µL of protein solution contains 9.702 mg protein. 1.465 mg of N/9.702 mg protein = 15.1 wt%.

11-58. (a) One mol HCl is required for one mol NH_3. 8.28 mL of 0.050 M HCl = 0.414 mmol HCl = 0.414 mmol NH_3 = 0.414 mmol N = 5.80 mg N

wt% N in protein = 100 × (5.80 mg/37.9 mg) = 15.3 wt %.

(b) Mass of $B(OH)_3$ in trapping solution = (5.00 mL)(0.040 0 g $B(OH)_3$/mL) = 0.200 g $B(OH)_3$ = 3.235 mmol $B(OH)_3$. If 0.414 mmol of NH_3 converts an equivalent amount of boric acid to borate, there would be a mixture of 0.414 mmol $B(OH)_4^-$ plus (3.235 – 0.414) = 2.821 mmol $B(OH)_3$.

$$pH = pK_a \text{ (boric acid)} + \log \frac{B(OH)_4^-}{B(OH)_3} = 9.237 + \log \frac{0.414 \text{ mmol}}{2.821 \text{ mmol}} = 8.40$$

(c) $pH = 8.40 = pK_a(NH_4^+) + \log \dfrac{mmol\, NH_3}{mmol\, NH_4^+} = 9.245 + \log \dfrac{mmol\, NH_3}{mmol\, NH_4^+}$

$\Rightarrow \log \dfrac{mmol\, NH_3}{mmol\, NH_4^+} = 8.40 - 9.245 \Rightarrow \dfrac{mmol\, NH_3}{mmol\, NH_4^+} = 10^{-0.845} = 0.143$

Fraction protonated $= \dfrac{mmol\, NH_4^+}{mmol\, NH_4^+ + mmol\, NH_3}$

$= \dfrac{mmol\, NH_4^+}{mmol\, NH_4^+ + 0.143\, mmol\, NH_4^+} = \dfrac{1.00}{1.143} = 87\%$.

Fraction unprotonated = 13%. (Reaction A does not go to completion because its equilibrium constant is ~1. However, formation of polyborates does protonate most of the NH_3.)

(d) $NH_3 + H^+ \rightleftharpoons NH_4^+ \qquad K = 1/K_a = 1/10^{-9.245} = 10^{9.245}$

$B(OH)_3 + H_2O \rightleftharpoons H^+ + B(OH)_4^- \qquad K_a = 10^{-9.237}$

$NH_3 + B(OH)_3 + H_2O \rightleftharpoons NH_4^+ + B(OH)_4^- \qquad K = (10^{9.245})(10^{-9.237}) = 1.02$

11-59. When an acid that is stronger than H_3O^+ is added to H_2O, it reacts to give H_3O^+ and is "leveled" to the strength of H_3O^+. Similarly, bases stronger than OH^- are leveled to the strength of OH^-.

11-60. (a) In acetic acid, strong acids are not leveled to the strength of $CH_3CO_2H_2^+$. Therefore, very weak bases can be titrated in acetic acid.

(b) If tetrabutylammonium hydroxide were added to an acetic acid solution, most of the hydroxide would react with acetic acid instead of analyte. However, OH^- will not react with pyridine, so this solvent would be suitable.

11-61. Sodium amide and phenyl lithium are stronger bases than OH^-. Each reacts with H_2O to give OH^-:

$NH_2^- + H_2O \rightarrow NH_3 + OH^-$

$C_6H_5^- + H_2O \rightarrow C_6H_6 + OH^-$

Acid-Base Titrations   141

11-62. The reaction of pyridine with acid is   [pyridine]-N: + H⁺ ⇌ [pyridine]-NH⁺

Methanol is less polar than water. If methanol is added to the aqueous solution, the neutral pyridine molecule will tend to be favored over the protonated pyridinium cation. It will take a higher concentration of acid (a lower pH) to protonate pyridine in the mixed solvent. pK_a for pyridinium ion is lowered when methanol is added to the solution.

11-63. Titration reaction: $K^+HP^- + Na^+OH^- \rightarrow K^+Na^+P^{2-} + H_2O$

Begin with C_aV_a moles of K^+HP^- and add C_bV_b moles of NaOH

Fraction of titration = $\phi = \dfrac{C_bV_b}{C_aV_a}$

Charge balance: $[H^+] + [Na^+] + [K^+] = [HP^-] + 2[P^{2-}] + [OH^-]$

Substitutions: $[K^+] = \dfrac{C_aV_a}{V_a + V_b}$ $[Na^+] = \dfrac{C_bV_b}{V_a + V_b}$

$[HP^-] = \alpha_{HP^-} \cdot \dfrac{C_aV_a}{V_a + V_b}$ $[P^{2-}] = \alpha_{P^{2-}} \cdot \dfrac{C_aV_a}{V_a + V_b}$

Putting these expressions into the charge balance gives

$$[H^+] + \dfrac{C_bV_b}{V_a + V_b} + \dfrac{C_aV_a}{V_a + V_b} = \alpha_{HP^-} \cdot \dfrac{C_aV_a}{V_a + V_b} + 2\alpha_{P^{2-}} \cdot \dfrac{C_aV_a}{V_a + V_b} + [OH^-]$$

Multiply by $V_a + V_b$ and collect terms:

$$[H^+]V_a + [H^+]V_b + C_bV_b + C_aV_a = \alpha_{HP^-}C_aV_a + 2\alpha_{P^{2-}}C_aV_a + [OH^-]V_a + [OH^-]V_b$$

$$V_a([H^+] + C_a - \alpha_{HP^-}C_a - 2\alpha_{P^{2-}}C_a - [OH^-]) = V_b([OH^-] - [H^+] - C_b)$$

$$\dfrac{V_b}{V_a} = \dfrac{\alpha_{HP^-}C_a + 2\alpha_{P^{2-}}C_a - C_a - [H^+] + [OH^-]}{C_b + [H^+] - [OH^-]}$$

Multiply both sides by $\dfrac{1/C_a}{1/C_b}$:

$$\phi = \dfrac{C_bV_b}{C_aV_a} = \dfrac{\alpha_{HP^-} + 2\alpha_{P^{2-}} - 1 - \dfrac{[H^+] - [OH^-]}{C_a}}{1 + \dfrac{[H^+] - [OH^-]}{C_b}}$$

11-64.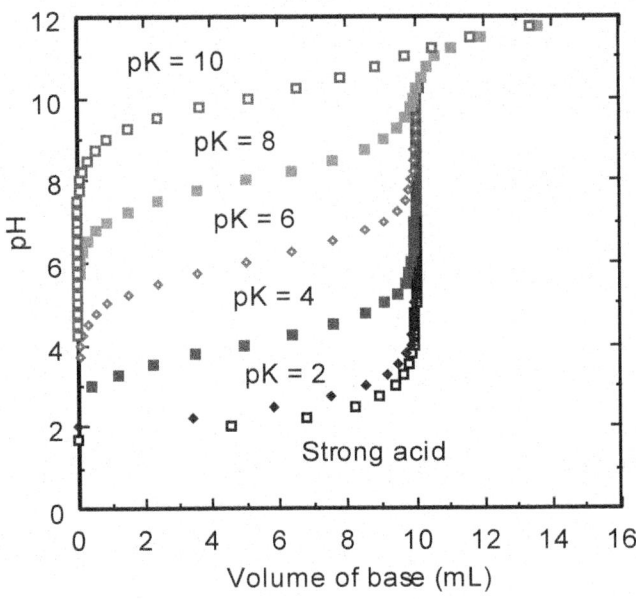

11-65.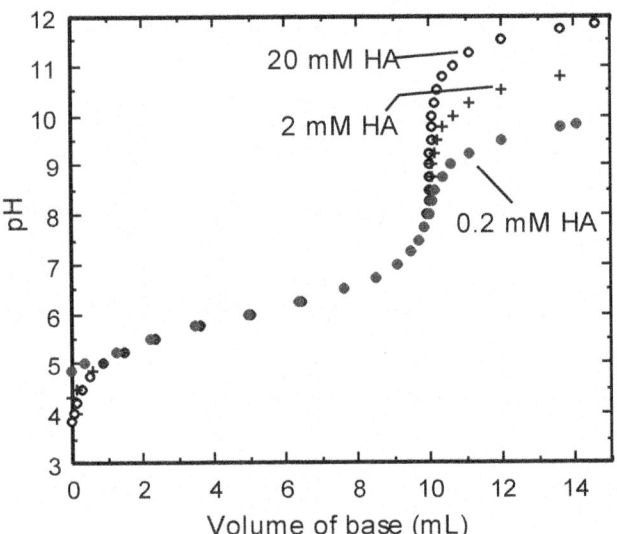

11-66.

	A	B	C	D	E	F	G
1	Effect of pKb in the titration of weak base with strong acid						
2							
3	Ca =	pH	[H+]	[OH-]	Alpha(BH+)	Phi	Va (mL)
4	0.1	2.00	1.00E-02	1.00E-12	9.90E-01	1.66E+00	16.557
5	Cb =	2.90	1.26E-03	7.94E-12	9.26E-01	1.00E+00	10.020
6	0.02	3.50	3.16E-04	3.16E-11	7.60E-01	7.78E-01	7.780
7	Vb =	4.00	1.00E-04	1.00E-10	5.00E-01	5.06E-01	5.055
8	50	4.50	3.16E-05	3.16E-10	2.40E-01	2.42E-01	2.419
9	K(BH+) =	6.00	1.00E-06	1.00E-08	9.90E-03	9.95E-03	0.100
10	1E-04	8.15	7.08E-09	1.41E-06	7.08E-05	5.17E-07	0.000
11	Kw =						
12	1E-14			E4 = C4/(C4+$A$10)			
13			C4 = 10^-B4	F4 = (E4+(C4-D4)/$A$6)/(1-(C4-D4)/$A$4)			
14			D4 = $A$12/C4	G4 = F4*$A$6*$A$8/$A$4			

Acid-Base Titrations

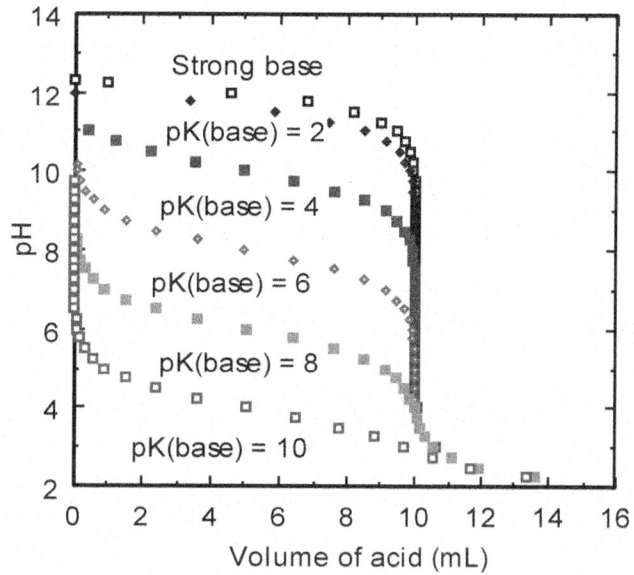

11-67. (a)

	A	B	C	D	E	F	G	H
1	Titrating weak acid with weak base							
2								
3	Cb =	pH	[H+]	[OH-]	Alpha(A-)	Alpha(BH+)	Phi	Vb (mL)
4	0.1	2.86	1.4E-03	7.2E-12	6.76E-02	1.00E+00	-1.4E-03	-0.01
5	Ca =	3.00	1.0E-03	1.0E-11	9.09E-02	1.00E+00	4.1E-02	0.41
6	0.02	4.00	1.0E-04	1.0E-10	5.00E-01	1.00E+00	4.9E-01	4.95
7	Va =	5.00	1.0E-05	1.0E-09	9.09E-01	9.99E-01	9.1E-01	9.09
8	50	6.00	1.0E-06	1.0E-08	9.90E-01	9.90E-01	1.0E+00	10.00
9	Ka =	7.00	1.0E-07	1.0E-07	9.99E-01	9.09E-01	1.1E+00	10.99
10	1E-04	8.00	1.0E-08	1.0E-06	1.00E+00	5.00E-01	2.0E+00	20.00
11	Kw =							
12	1E-14		A16 = $A$12/$A$14		D4 = $A$12/C4			
13	Kb =		C4 = 10^-B4		E4 = $A$10/(C4+$A$10)			
14	1E-06		F4 = C4/(C4+$A$16)					
15	K(BH+) =		G4 = (E4-(C4-D4)/$A$6)/(F4+(C4-D4)/$A$4)					
16	1E-08		H4 = G4*$A$6*$A$8/$A$4					

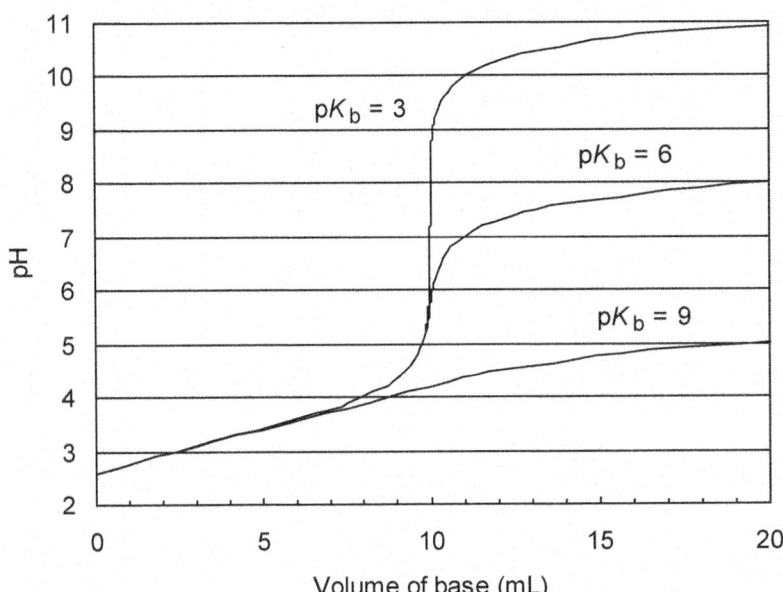

(b)   HA   +   B   ⇌   A⁻ + BH⁺

$K_a = 1.75 \times 10^{-5}$ $K_b = 1.59 \times 10^{-10}$

$V_a = 212$ mL $K_{BH^+} = 6.28 \times 10^{-5}$

$C_a = 0.200$ M $V_b = 325$ mL

                      $C_b = 0.0500$ M

To find the equilibrium constant we write

HA	⇌	A⁻ + H⁺		K_a
H⁺ + B	⇌	BH⁺		$1/K_{BH^+}$
HA + B	⇌	A⁻ + BH⁺		$K = K_a/K_{BH^+} = 0.279$

A <u>pH of 4.16</u> gives $V_b = 325.0$ mL in the following spreadsheet:

	A	B	C	D	E	F	G	H
1	Mixing acetic acid and sodium benzoate							
2								
3	Cb =	pH	[H+]	[OH-]	Alpha(A-)	Alpha(BH+)	Phi	Vb (mL)
4	0.05	4.00	1.0E-04	1.0E-10	1.49E-01	6.14E-01	2.4E-01	204.28
5	Ca =	4.2	6.3E-05	1.6E-10	2.17E-01	5.01E-01	4.3E-01	365.98
6	0.2	4.1	7.9E-05	1.3E-10	1.81E-01	5.58E-01	3.2E-01	272.78
7	Va =	4.15	7.1E-05	1.4E-10	1.98E-01	5.30E-01	3.7E-01	315.79
8	212	4.1598	6.9E-05	1.4E-10	2.02E-01	5.24E-01	3.8E-01	325.03
9	Ka =							
10	1.750E-05		A16 = $A$12/$A$14					
11	Kw =		C4 = 10^-B4					
12	1.E-14		D4 = $A$12/C4					
13	Kb =		E4 = $A$10/(C4+$A$10)					
14	1.592E-10		F4 = C4/(C4+$A$16)					
15	K(BH+) =		G4 = (E4-(C4-D4)/$A$6)/(F4+(C4-D4)/$A$4)					
16	6.281E-05		H4 = G4*$A$6*$A$8/$A$4					

11-68.

	A	B	C	D	E	F	G	H
1	Titrating diprotic acid with strong base							
2								
3	Cb =	pH	[H+]	[OH-]	Alpha(HA-)	Alpha(A2-)	Phi	Vb (mL)
4	0.1	2.865	1.4E-03	7.3E-12	6.83E-02	5.00E-07	5.0E-05	0.000
5	Ca =	4.00	1.0E-04	1.0E-10	5.00E-01	5.00E-05	4.9E-01	4.946
6	0.02	6.00	1.0E-06	1.0E-08	9.80E-01	9.80E-03	1.0E+00	9.999
7	Va =	8.00	1.0E-08	1.0E-06	5.00E-01	5.00E-01	1.5E+00	15.000
8	50	10.0	1.0E-10	1.0E-04	9.90E-03	9.90E-01	2.0E+00	19.971
9	Kw =	12.0	1.0E-12	1.0E-02	1.00E-04	1.00E+00	2.8E+00	27.777
10	1E-14							
11	K1 =		C4 = 10^-B4			D4 = $A$12/C4		
12	1E-4		E4 = C4*$A$12/(C4^2+C4*$A$12+$A$12*$A$14)					
13	K2 =		F4 = $A$12*$A$14/(C4^2+C4*$A$12+$A$12*$A$14)					
14	1.E-08		G4 = (E4+2*F4-(C4-D4)/$A$6)/(1+(C4-D4)/$A$4)					
15			H4 = G4*$A$6*$A$8/$A$4					

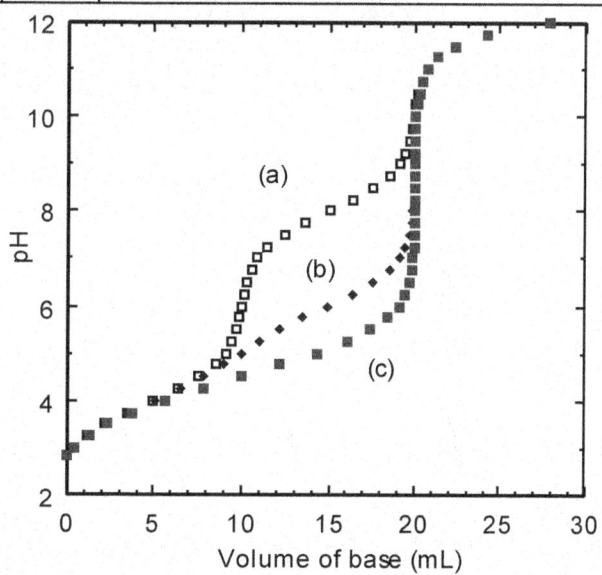

11-69.

	A	B	C	D	E	F	G	H
1	Titrating nicotine with strong acid							
2								
3	Cb =	pH	[H+]	[OH-]	Alpha(BH2)	Alpha(BH)	Phi	Va (mL)
4	0.1	1.75	1.8E-02	5.6E-13	9.62E-01	3.83E-02	2.6E+00	26.023
5	Ca =	2.00	1.0E-02	1.0E-12	9.34E-01	6.61E-02	2.3E+00	22.599
6	0.1	3.00	1.0E-03	1.0E-11	5.86E-01	4.14E-01	1.6E+00	16.117
7	Vb =	4.00	1.0E-04	1.0E-10	1.24E-01	8.76E-01	1.1E+00	11.258
8	10	5.00	1.0E-05	1.0E-09	1.39E-02	9.85E-01	1.0E+00	10.127
9	Kw =	6.00	1.0E-06	1.0E-08	1.39E-03	9.85E-01	9.9E-01	9.875
10	1.E-14	7.00	1.0E-07	1.0E-07	1.24E-04	8.76E-01	8.8E-01	8.764
11	KB1 =	8.00	1.0E-08	1.0E-06	5.86E-06	4.14E-01	4.1E-01	4.145
12	7.079E-7	9.00	1.0E-09	1.0E-05	9.34E-08	6.61E-02	6.6E-02	0.660
13	KB2 =	10.00	1.0E-10	1.0E-04	9.93E-10	7.03E-03	6.0E-03	0.060
14	1.41E-11	10.42	3.8E-11	2.6E-04	1.44E-10	2.68E-03	5.4E-05	0.001
15	KA1 =							
16	7.077E-4		C4 = 10^-B4			D4 = $A$12/C4		
17	KA2 =			E4 = C4*C4/(C4^2+C4*$A$16+$A$16*$A$18)				
18	1.413E-8			F4 = C4*$A$16/(C4^2+C4*$A$16+$A$16*$A$18)				
19				G4 = (F4+2*E4+(C4-D4)/$A$4)/(1-(C4-D4)/$A$6)				
20				H4 = G4*$A$4*$A$8/$A$6				

11-70.

	A	B	C	D	E	F	G	H	I
1	Titrating H$_3$A with NaOH								
2									
3	C$_b$ =	pH	[H$^+$]	[OH$^-$]	Alpha(H$_2$A$^-$)	Alpha(HA^{2-})	Alpha(A^{3-})	Phi	V$_b$ (mL)
4	0.1	1.89	1.29E-02	7.76E-13	6.61E-01	5.50E-05	2.24E-12	1.50E-02	0.150
5	C$_a$ =	2.00	1.00E-02	1.00E-12	7.15E-01	7.66E-05	4.02E-12	1.96E-01	1.958
6	0.02	3.00	1.00E-03	1.00E-11	9.61E-01	1.03E-03	5.40E-10	9.04E-01	9.037
7	V$_a$ =	4.00	1.00E-04	1.00E-10	9.86E-01	1.06E-02	5.54E-08	1.00E+00	10.006
8	50	5.00	1.00E-05	1.00E-09	9.03E-01	9.67E-02	5.08E-06	1.10E+00	10.958
9	K$_w$ =	6.00	1.00E-06	1.00E-08	4.83E-01	5.17E-01	2.71E-04	1.52E+00	15.176
10	1E-14	7.00	1.00E-07	1.00E-07	8.50E-02	9.10E-01	4.78E-03	1.92E+00	19.198
11	K$_1$ =	8.00	1.00E-08	1.00E-06	8.79E-03	9.42E-01	4.94E-02	2.04E+00	20.407
12	2.51E-02	9.00	1.00E-09	1.00E-05	6.12E-04	6.55E-01	3.44E-01	2.34E+00	23.441
13	K$_2$ =	10.00	1.00E-10	1.00E-04	1.49E-05	1.60E-01	8.40E-01	2.85E+00	28.478
14	1.07E-06	11.00	1.00E-11	1.00E-03	1.75E-07	1.87E-02	9.81E-01	3.06E+00	30.619
15	K$_3$ =	12.00	1.00E-12	1.00E-02	1.77E-09	1.90E-03	9.98E-01	3.89E+00	38.868
16	5.25E-10		C4 = 10^-B4						
17	pK$_1$ =		D4 = $A$10/C4						
18	1.60		E4 = $C4^2*$A$12/($C4^3+$C4^2*$A$12+$C4*$A$12*$A$14+$A$12*$A$14*$A$16)						
19	pK$_2$ =		F4 = $C4*$A$12*$A$14/($C4^3+$C4^2*$A$12+$C4*$A$12*$A$14+$A$12*$A$14*$A$16						
20	5.97		G4 = $A$12*$A$14*$A$16/($C4^3+$C4^2*$A$12+$C4*$A$12*$A$14+$A$12*$A$14*$A$16)						
21	pK$_3$ =		H4 = (E4+2*F4+3*G4-(C4-D4)/$A$6)/(1+(C4-D4)/$A$4)						
22	9.28		I4 = H4*$A$6*$A$8/$A$4						

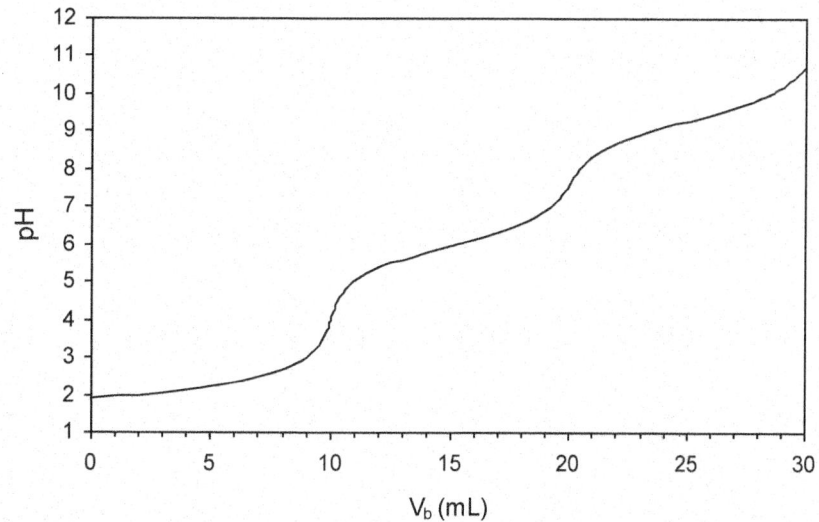

11-71. $\phi = \dfrac{C_a V_a}{C_b V_b} = \dfrac{\alpha_{BH^+} + 2\alpha_{BH_2^{2+}} + 3\alpha_{BH_3^{3+}} + 4\alpha_{BH_4^{4+}} + \dfrac{[H^+]-[OH^-]}{C_b}}{1 - \dfrac{[H^+]-[OH^-]}{C_a}}$

	A	B	C	D	E	F	G	H	I	J
1	Titrating Tetrabasic B with H^+									
2										
3	C_b =	pH	$[H^+]$	$[OH^-]$	$\alpha(BH^+)$	$\alpha(BH_2^{2+})$	$\alpha(BH_3^{3+})$	$\alpha(BH_4^{4+})$	Phi	V_b (mL)
4	0.02	10.9	1.3E-11	7.9E-04	3.1E-02	1.9E-06	4.6E-15	4.6E-25	-0.009	-0.090
5	C_a =	10.0	1.0E-10	1.0E-04	2.0E-01	1.0E-04	1.9E-12	1.5E-21	0.196	1.957
6	0.1	9.0	1.0E-09	1.0E-05	7.1E-01	3.6E-03	6.8E-10	5.4E-18	0.719	7.193
7	V_b =	8.0	1.0E-08	1.0E-06	9.2E-01	4.6E-02	8.8E-08	7.0E-15	1.009	10.094
8	50	7.0	1.0E-07	1.0E-07	6.6E-01	3.3E-01	6.3E-06	5.0E-12	1.330	13.303
9	K_w =	6.0	1.0E-06	1.0E-08	1.7E-01	8.3E-01	1.6E-04	1.3E-09	1.834	18.338
10	1E-14	5.0	1.0E-05	1.0E-09	2.0E-02	9.8E-01	1.9E-03	1.5E-07	1.983	19.830
11	K_{a1} =	4.0	1.0E-04	1.0E-10	2.0E-03	9.8E-01	1.9E-02	1.5E-05	2.024	20.238
12	1.26E-01	3.0	1.0E-03	1.0E-11	1.7E-04	8.4E-01	1.6E-01	1.3E-03	2.235	22.345
13	K_{a2} =	2.0	1.0E-02	1.0E-12	6.5E-06	3.3E-01	6.2E-01	5.0E-02	3.580	35.804
14	5.25E-03	1.7	2.0E-02	5.0E-13	1.9E-06	1.9E-01	7.0E-01	1.1E-01	4.902	49.022
15	K_{a3} =									
16	2.00E-07		C12 = 10^-A20							
17	K_{a4} =		C4 = 10^-B4		D4 = \$A\$10/C4					
18	3.98E-10		Denominator = (\$C4^4+\$C4^3*\$A\$12+\$C4^2*\$A\$12*\$A\$14							
19	pK_1 =			+\$C4*\$A\$12*\$A\$14*\$A\$16+\$A\$12*\$A\$14*\$A\$16*\$A\$18)						
20	0.90		E4 = \$C4*\$A\$12*\$A\$14*\$A\$16/Denominator							
21	pK_2 =		F4 = \$C4^2*\$A\$12*\$A\$14/Denominator							
22	2.28		G4 = \$C4^3*\$A\$12//Denominator							
23	pK_3 =		H4 = \$C4^4/Denominator							
24	6.70		I4 = (E4+2*F4+3*G4+4*H4+(C4-D4)/\$A\$4)/(1-(C4-D4)/\$A\$6)							
25	pK_4 =		J4 = I4*\$A\$4*\$A\$8/\$A\$6							
26	9.40									

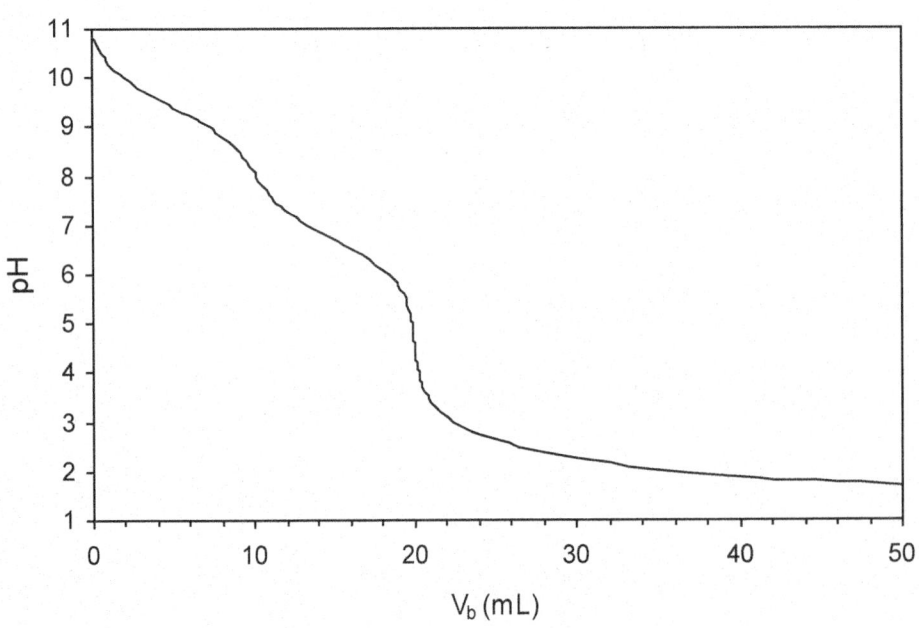

Acid-Base Titrations

11-72. $A_{604} = \varepsilon_{In^-}[In^-](1.00) \Rightarrow [In^-] = \dfrac{0.118}{4.97 \times 10^4} = 2.37 \times 10^{-6}$ M

Since the indicator was diluted with KOH solution, the formal concentration of indicator is 0.700×10^{-5} M.

$[HIn] = 7.00 \times 10^{-6} - 2.37 \times 10^{-6} = 4.63 \times 10^{-6}$ M

$pH = pK_{In} + \log\dfrac{[In^-]}{[HIn]} = 7.10 + \log\dfrac{2.37}{4.63} = 6.81$

Call benzene-1,2,3-tricarboxylic acid H_3A, with $pK_1 = 2.86$, $pK_2 = 4.30$, and $pK_3 = 6.28$. Since the pH is 6.81, the main species is A^{3-} and the second main species is HA^{2-}. Enough KOH to react with H_3A and H_2A^- must have been added, and there is enough KOH to react with part of the HA^{2-}.

	HA^{2-}	+	OH^-	$\rightarrow$	A^{3-}	+	H_2O
Initial mmol:	1.00		x		—		
Final mmol:	$1.00 - x$		—		x		

$pH = pK_3 + \log\dfrac{[A^{3-}]}{[HA^{2-}]}$

$6.81 = 6.28 + \log\dfrac{x}{1.00 - x} \Rightarrow x = 0.77_2$ mmol of OH^-

The total KOH added is 2.77_2 mmol. The molarity is $\dfrac{2.77_2 \text{ mmol}}{20.0 \text{ mL}} = 0.139$ M.

11-73. The pH of the solution is 7.50 and the total concentration of indicator after mixing with buffer is 5.00×10^{-5} M after diluting indicator with buffer. At pH 7.50, there is a negligible amount of H_2In, since $pK_1 = 1.00$. We can write

$[HIn^-] + [In^{2-}] = 5.0 \times 10^{-5}$

$pH = pK_2 + \log\dfrac{[In^{2-}]}{[HIn^-]}$

$7.50 = 7.95 + \log\dfrac{[In^{2-}]}{5.00 \times 10^{-5} - [In^{2-}]} \Rightarrow [In^{2-}] = 1.31 \times 10^{-5}$ M

$[HIn] = 3.69 \times 10^{-5}$ M

$A_{435} = \varepsilon_{435}[HIn^-] + \varepsilon_{435}[In^{2-}]$

$= (1.80 \times 10^4)(3.69 \times 10^{-5}) + (1.15 \times 10^4)(1.31 \times 10^{-5}) = 0.815$

CHAPTER 12
EDTA TITRATIONS

12-1. The chelate effect is the observation that multidentate ligands form more stable metal complexes than do similar, monodentate ligands.

12-2. $\alpha_{Y^{4-}}$ gives the fraction of all free EDTA in the form Y^{4-}.

(a) At pH 3.50:
$$\alpha_{Y^{4-}} = \frac{10^{-0.0}10^{-1.5}10^{-2.00}10^{-2.69}10^{-6.13}10^{-10.37}}{(10^{-3.50})^6 + (10^{-3.50})^5 10^{-0.0} + \cdots + 10^{-0.0}10^{-1.5}\cdots 10^{-10.37}} = 2.7 \times 10^{-10}$$

(b) At pH 10.50:
$$\alpha_{Y^{4-}} = \frac{10^{-0.0}10^{-1.5}10^{-2.00}10^{-2.69}10^{-6.13}10^{-10.37}}{(10^{-10.50})^6 + (10^{-10.50})^5 10^{-0.0} + \cdots + 10^{-0.0}10^{-1.5}\cdots 10^{-10.37}} = 0.57$$

12-3. (a) $K_f' = \alpha_{Y^{4-}} K_f = 0.041 \times 10^{8.79} = 2.5 \times 10^7$

(b) $\quad Mg^{2+} + EDTA \rightleftharpoons MgY^{2-}$
$\qquad\quad x \qquad\quad x \qquad\quad 0.050 - x$

$$\frac{0.050 - x}{x^2} = 2.5 \times 10^7 \Rightarrow [Mg^{2+}] = 4.5 \times 10^{-5} \text{ M}$$

12-4. $[Ca^{2+}] = 10^{-9.00}$ M, so essentially all calcium in solution is CaY^{2-}.

$$[CaY^{2-}] = \frac{1.95 \text{ g}}{(200.12 \text{ g/mol})(0.500 \text{ L})} = 0.019\,49 \text{ M}$$

$$K_f' = (0.041)(10^{10.65}) = \frac{[CaY^{2-}]}{[EDTA][Ca^{2+}]} = \frac{(1.949 \times 10^{-2})}{[EDTA](10^{-9.00})}$$

$\Rightarrow [EDTA] = 0.010\,6$ M

Total EDTA needed = mol CaY^{2-} + mol free EDTA
$\qquad\qquad\qquad\qquad = (0.019\,49$ M$)(0.500$ L$) + (0.010\,6$ M$)(0.500$ L$) = 0.015\,04$ mol
$\qquad\qquad\qquad\qquad = 5.60$ g $Na_2EDTA \cdot 2 H_2O$

12-5.

[Structure of H₅DTPA showing HO₂C—CH₂—N⁺H—CH₂—CH₂—N⁺H(CH₂CO₂⁻)—CH₂—CH₂—N⁺H—CH₂—CO₂H with additional ⁻O₂C—CH₂— group]

Neutral H_5DTPA has 2 carboxylic acid protons and 3 ammonium protons. We are not given pK_a values, but, by analogy with EDTA, we expect carboxyl pK_a to be below ~3 and ammonium pK_a to be above ~6. At pH 14, we expect all acidic protons of DTPA to be dissociated, so the predominant species is DTPA^{5-}. At pH 3–4, nitrogen should be protonated, but carboxyl groups should be deprotonated. The predominant species is probably H$_3$DTPA^{2-}.

For HSO$_4^-$, pK_a = 2.0. At pH 14 and at pH 3, sulfate is in the form SO$_4^{2-}$.

At pH 14, DTPA^{5-} is apparently a strong enough ligand to chelate Ba^{2+} and dissolve BaSO$_4$(s). At pH 3–4, H$_3$DTPA^{2-} is not a strong enough ligand to dissolve BaSO$_4$(s). An equivalent statement is that H$^+$ at a concentration of 10^{-3}–10^{-4} M competes with Ba^{2+} for binding sites on DTPA, but H$^+$ at a concentration of 10^{-14} M does not compete with Ba^{2+} for binding sites on DTPA.

Now that you have seen my reasoning, I'll provide some more information. The pK_a values for DTPA, beginning with the fully protonated H$_8$DTPA^{3+}, are

H$_8$DTPA^{3+}	pK_1 = –0.1	CO$_2$H	H$_4$DTPA$^-$	pK_5 = 2.7	CO$_2$H
H$_7$DTPA^{2+}	pK_2 = 0.7	CO$_2$H	H$_3$DTPA^{2-}	pK_6 = 4.3	NH$^+$
H$_6$DTPA$^+$	pK_3 = 1.6	CO$_2$H	H$_2$DTPA^{3-}	pK_7 = 8.6	NH$^+$
H$_5$DTPA	pK_4 = 2.0	CO$_2$H	HDTPA^{4-}	pK_8 = 10.5	NH$^+$

As pH is lowered from 14, the three nitrogen atoms are 50% protonated at pH 10.5, 8.6, and 4.3. The third nitrogen atom is not quite fully protonated at pH 3–4. The predominant species is H$_3$DTPA^{2-}, as I guessed correctly. The species H$_4$DTPA$^-$ and H$_2$DTPA^{3-} are also found in the pH range 3–4.

12-6. (a) mmol EDTA = mmol M^{n+}

$(V_e)(0.0500 \text{ M}) = (100.0 \text{ mL})(0.0500 \text{ M}) \Rightarrow V_e = 100.0 \text{ mL}$

(b) $[\text{M}^{n+}] = \underbrace{\left(\frac{1}{2}\right)}_{\text{fraction remaining}} \cdot \underbrace{(0.0500 \text{ M})}_{\text{original concentration}} \cdot \underbrace{\left(\frac{100 \text{ mL}}{150 \text{ mL}}\right)}_{\text{dilution factor}} = 0.0167 \text{ M}$

(c) 0.041 (Table 12-1)

(d) $K_f' = (0.041)(10^{12.00}) = 4.1 \times 10^{10}$

(e) $[\text{MY}^{n-4}] = (0.0500 \text{ M})\left(\frac{100 \text{ mL}}{200 \text{ mL}}\right) = 0.0250 \text{ M}$

$\dfrac{[\text{MY}^{n-4}]}{[\text{M}^{n+}][\text{EDTA}]} = \dfrac{0.0250 - x}{x^2} = 4.1 \times 10^{10} \Rightarrow x = [\text{M}^{n+}] = 7.8 \times 10^{-7}$ M

(f) $[\text{EDTA}] = (0.0500 \text{ M})\left(\dfrac{10.0 \text{ mL}}{210.0 \text{ mL}}\right) = 2.38 \times 10^{-3}$ M

$$[MY^{n-4}] = (0.0500\text{ M})\left(\frac{100.0\text{ mL}}{210.0\text{ mL}}\right) = 2.38 \times 10^{-2}\text{ M}$$

$$\frac{[MY^{n-4}]}{[M^{n+}][EDTA]} = \frac{(2.38 \times 10^{-2})}{[M^{n+}](2.38 \times 10^{-3})} = 4.1 \times 10^{10} \Rightarrow [M^{n+}] = 2.4 \times 10^{-10}\text{ M}$$

12-7. $Co^{2+} + EDTA \rightleftharpoons CoY^{2-}$ $\alpha_{Y^{4-}} K_f = (1.8 \times 10^{-5})(10^{16.45}) = 5.1 \times 10^{11}$

$$V_e = (25.00)\left(\frac{0.02026\text{ M}}{0.03855\text{ M}}\right) = 13.14\text{ mL}$$

(a) <u>12.00 mL</u>: $[Co^{2+}] = \left(\frac{13.14\text{ mL} - 12.00\text{ mL}}{13.14\text{ mL}}\right)(0.02026\text{ M})\left(\frac{25.00\text{ mL}}{37.00\text{ mL}}\right)$

$$= 1.19 \times 10^{-3}\text{ M} \Rightarrow pCo^{2+} = 2.93$$

(b) <u>V_e</u>: Formal concentration of $CoY^{2-} = \left(\frac{25.00\text{ mL}}{38.14\text{ mL}}\right)(0.02026\text{ M})$

$$= 1.33 \times 10^{-2}\text{ M}$$

$$\begin{array}{cccc} Co^{2+} & + & EDTA & \rightleftharpoons & CoY^{2-} \\ x & & x & & 1.33 \times 10^{-2} - x \end{array}$$

$$\frac{1.33 \times 10^{-2} - x}{x^2} = \alpha_{Y^{4-}} K_f \Rightarrow x = 1.6 \times 10^{-7}\text{ M} \Rightarrow pCo^{2+} = 6.79$$

(c) <u>14.00 mL</u>: Formal concentration of CoY^{2-} is $\left(\frac{25.00\text{ mL}}{39.00\text{ mL}}\right)(0.02026\text{ M})$

$$= 1.30 \times 10^{-2}\text{ M}$$

$$[EDTA] = \left(\frac{14.0\text{ mL} - 13.14\text{ mL}}{39.00\text{ mL}}\right)(0.03855\text{ M}) = 8.50 \times 10^{-4}\text{ M}$$

$$[Co^{2+}] = \frac{[CoY^{2-}]}{[EDTA]K_f'} = \frac{1.30 \times 10^{-2}}{8.50 \times 10^{-4}(5.1 \times 10^{11})} = 3.0 \times 10^{-11}\text{ M}$$

$$\Rightarrow pCo^{2+} = 10.52$$

12-8. Titration reaction: $Mn^{2+} + EDTA \rightleftharpoons MnY^{2-}$

$$K_f' = \alpha_{Y^{4-}} K_f = (4.2 \times 10^{-3})(10^{13.89}) = 3.3 \times 10^{11}$$

The equivalence point is 50.0 mL. Sample calculations:

<u>20.0 mL</u>: The fraction of Mn^{2+} that has reacted is 2/5 and the fraction remaining is 3/5.

$$[Mn^{2+}] = \left(\frac{30.0\text{ mL}}{50.0\text{ mL}}\right)(0.0200\text{ M})\left(\frac{25.0\text{ mL}}{45.0\text{ mL}}\right) = 6.67 \times 10^{-3}\text{ M}$$

$$\Rightarrow pMn^{2+} = 2.18$$

EDTA Titrations 153

<u>50.0 mL</u>: $[MnY^{2-}] = \left(\frac{25.0 \text{ mL}}{75.0 \text{ mL}}\right)(0.020\ 0\ M) = 0.006\ 67\ M$

$$Mn^{2+} + EDTA \rightleftharpoons MnY^{2-}$$
$$\phantom{Mn^{2+} +\ }x x 0.006\ 67 - x$$

$\frac{0.006\ 67 - x}{x^2} = \alpha_{Y^{4-}} K_f \Rightarrow x = 1.4 \times 10^{-7} \Rightarrow pMn^{2+} = 6.85$

<u>60.0 mL</u>: There are 10.0 mL of excess EDTA.

$[EDTA] = \left(\frac{10.0 \text{ mL}}{85.0 \text{ mL}}\right)(0.010\ 0\ M) = 1.176 \times 10^{-3}\ M$

$[MnY^{2-}] = \left(\frac{25.0 \text{ mL}}{85.0 \text{ mL}}\right)(0.020\ 0\ M) = 5.88 \times 10^{-3}\ M$

$[Mn^{2+}] = \frac{[MnY^{2-}]}{[EDTA]K_f'} = 1.5 \times 10^{-11} \Rightarrow pMn^{2+} = 10.82$

Volume (mL)	pMn^{2+}	Volume	pMn^{2+}	Volume	pMn^{2+}
0	1.70	49.0	3.87	50.1	8.82
20.0	2.18	49.9	4.87	55.0	10.51
40.0	2.81	50.0	6.85	60.0	10.82

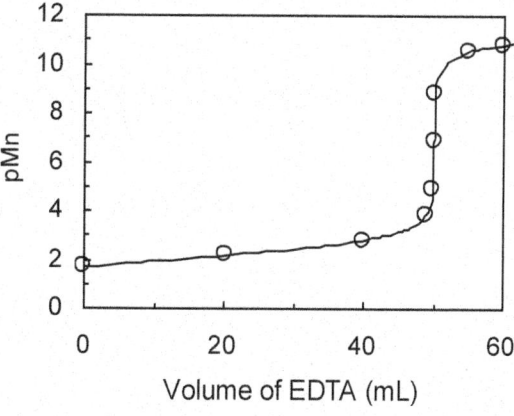

12-9. Titration reaction: $Ca^{2+} + EDTA \rightleftharpoons CaY^{2-}$

$K_f' = \alpha_{Y^{4-}} K_f = (0.30)(10^{10.65}) = 1.3_4 \times 10^{10}$

The equivalence point is 50.0 mL. Sample calculations:

<u>20.0 mL</u>: The fraction of EDTA consumed is 2/5.

$[EDTA] = \left(\frac{30.0 \text{ mL}}{50.0 \text{ mL}}\right)(0.020\ 0\ M)\left(\frac{25.0 \text{ mL}}{45.0 \text{ mL}}\right) = 0.006\ 67\ M$

$[CaY^{2-}] = \left(\frac{20.0 \text{ mL}}{50.0 \text{ mL}}\right)(0.020\ 0\ M)\left(\frac{25.0 \text{ mL}}{45.0 \text{ mL}}\right) = 0.004\ 44\ M$

$[Ca^{2+}] = \frac{[CaY^{2-}]}{[EDTA]K_f'} = 4.9_7 \times 10^{-11} \Rightarrow pCa^{2+} = 10.30$

50.0 mL: $[CaY^{2-}] = \left(\dfrac{25.0 \text{ mL}}{75.0 \text{ mL}}\right)(0.020\ 0 \text{ M}) = 0.006\ 67 \text{ M}$

$$Ca^{2+} + EDTA \rightleftharpoons CaY^{2-}$$
$$\phantom{Ca^{2+} + ED}x \ x \phantom{CaY^{2-}}\ 0.006\ 67 - x$$

$\dfrac{0.006\ 67 - x}{x^2} = \alpha_{Y^{4-}} K_f \Rightarrow x = 7.0_5 \times 10^{-7}$ M $\Rightarrow$ pCa^{2+} = 6.15

50.1 mL: There is an excess of 0.1 mL of Ca^{2+}.

$[Ca^{2+}] = \left(\dfrac{0.1 \text{ mL}}{75.1 \text{ mL}}\right)(0.010\ 0 \text{ M}) = 1.33 \times 10^{-5}$ M $\Rightarrow$ pCa^{2+} = 4.88

Volume (mL)	pCa^{2+}	Volume	pCa^{2+}	Volume	pCa^{2+}
0	(∞)	49.0	8.44	50.1	4.88
20.0	10.30	49.9	7.43	55.0	3.20
40.0	9.52	50.0	6.15	60.0	2.93

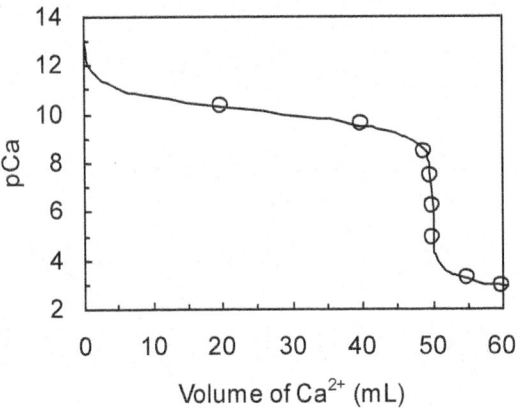

12-10. There is more VO^{2+} than EDTA in this solution.

$[VO^{2+}] = \left(\dfrac{0.10 \text{ mL}}{29.9 \text{ mL}}\right)(0.010\ 0 \text{ M}) = 3.34 \times 10^{-5}$ M

$[VOY^{2-}] = \left(\dfrac{9.90 \text{ mL}}{29.90 \text{ mL}}\right)(0.010\ 0 \text{ M}) = 3.31 \times 10^{-3}$ M

K_f for VOY^{2-} = $10^{18.7}$; pK_6 for H$_6$Y^{2+} = 10.37; pH = 4.00

$[Y^{4-}] = \dfrac{[VOY^{2-}]}{[VO^{2+}]K_f} = 1.98 \times 10^{-17}$ M

$[HY^{3-}] = \dfrac{[H^+][Y^{4-}]}{K_6} = 4.6 \times 10^{-11}$ M

12-11.

	A	B	C	D	E	F	G
1	Titration of V_M mL of C_M M Cu^{2+} with C(ligand) M EDTA						
2							
3	C_M =	pM	M	Phi	V(EDTA)		
4	0.001	3.0	1.00E-03	0.000	0.000		
5	V_M =	4.0	1.00E-04	0.891	0.891		
6	10	5.0	1.00E-05	0.989	0.989		
7	C(ligand) =	6.0	1.00E-06	0.999	0.999		
8	0.01	7.0	1.00E-07	1.000	1.000		
9	K_f' =	8.0	1.00E-08	1.000	1.000		
10	1.75E+12	9.0	1.00E-09	1.001	1.001		
11	$\alpha(Y^{4-})$=	10.0	1.00E-10	1.006	1.006		
12	2.90E-07	11.0	1.00E-11	1.057	1.057		
13	K_f =	12.0	1.00E-12	1.572	1.572		
14	6.0256E+18	12.3	5.01E-13	2.142	2.142		
15							
16	A10 = A12*A14						
17	C4 = 10^-B4						
18	D4 = (1+$A$10*C4-(C4+C4*C4*$A$10)/$A$4)/(C4*$A$10+(C4+C4*C4*$A$10)/$A$8)						
19	E4 = D4*$A$4*$A$6/$A$8						

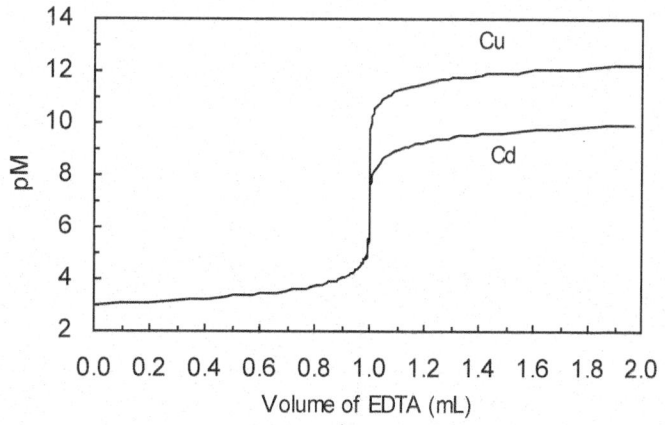

12-12. The spreadsheet below gives representative calculations for the pH 7.

	A	B	C	D	E	F
1	Titration of 10 mL of 1 mM Ca^{2+} with 1 mM EDTA vs pH					
2	pH 7					
3	C_M =	pM	M	Phi	V(ligand)	
4	0.001	3.000	1.00E-03	0.000	0.000	
5	V_M =	3.250	5.62E-04	0.280	2.801	
6	10	3.500	3.16E-04	0.520	5.196	
7	C(ligand) =	3.750	1.78E-04	0.698	6.982	
8	0.001	4.000	1.00E-04	0.819	8.186	
9	K_f' =	4.500	3.16E-05	0.940	9.404	
10	1.70E+07	5.000	1.00E-05	0.986	9.859	
11	$\alpha(Y^{4-})$ =	5.500	3.16E-06	1.012	10.121	
12	3.80E-04	6.000	1.00E-06	1.057	10.567	
13	K_f =	6.500	3.16E-07	1.185	11.855	
14	4.4668E+10	7.000	1.00E-07	1.589	15.887	
15	A10 = A12*A14					
16	C4 = 10^-B4					
17	D4 = (1+A$10*C4-(C4+C4*C4*A$10)/A$4)/(C4*A$10+(C4+C4*C4*A$10)/A$8)					
18	E4 = D4*A$4*A$6/A$8					

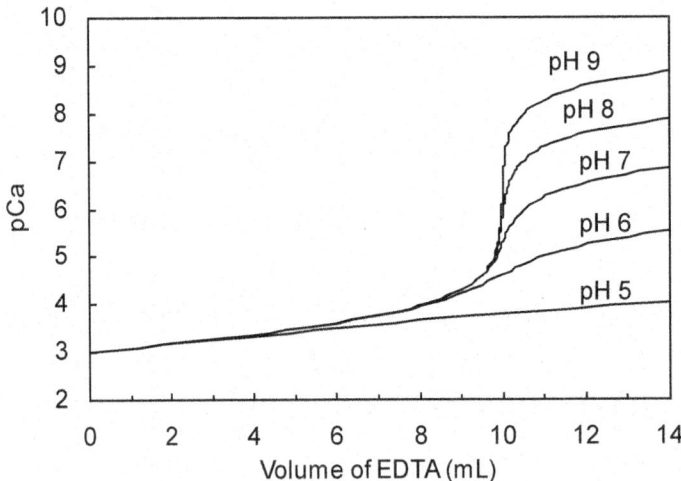

The only change in the spreadsheet for different values of pH is the value of $\alpha_{Y^{4-}}$ in cell A12.

12-13.

	A	B	C	D	E	F	G
1	Titration of EDTA with metal						
2							
3	C_M =	pM	M	Phi	V_M		
4	0.08	14.640	2.29E-15	0.004	0.100		
5	V(ligand) =	12.844	1.43E-13	0.200	5.004		
6	50	12.418	3.82E-13	0.400	10.007		
7	C(ligand) =	12.066	8.59E-13	0.600	15.004		
8	0.04	11.640	2.29E-12	0.800	20.003		
9	K_f' =	10.860	1.38E-11	0.960	24.005		
10	1.75E+12	6.910	1.23E-07	1.000	25.000		
11	$\alpha(Y^{4-})$	2.978	1.05E-03	1.040	25.999		
12	2.90E-07	2.301	5.00E-03	1.200	30.000		
13	K_f =						
14	6.0256E+18						
15							
16	A10 = A12*A14						
17	C4 = 10^-B4						
18	D4 = (C4*$A$10+(C4+C4*C4*$A$10)/$A$8)/(1+$A$10*C4-(C4+C4*C4*$A$10)/$A$4)						
19	E4 = D4*$A$8*$A$6/$A$4						

12-14. An auxiliary complexing agent forms a weak complex with analyte ion, thereby keeping it in solution without interfering with the EDTA titration. For example, NH_3 keeps Zn^{2+} in solution at high pH, but is easily displaced by EDTA.

12-15. (a) $\beta_2 = K_1K_2 = \beta_1 K_2 \Rightarrow K_2 = \beta_2/\beta_1 = 10^{3.63}/10^{2.23} = 10^{1.40} = 25$

(b) $\alpha_{Cu^{2+}} = \dfrac{1}{1+\beta_1[L]+\beta_2[L]^2} = \dfrac{1}{1+10^{2.23}(0.100)+10^{3.63}(0.100)^2} = 0.016$

12-16. $Cu^{2+} + Y^{4-} \rightleftharpoons CuY^{2-}$ $K_f = 10^{18.78} = 6.03 \times 10^{18}$

$\alpha_{Y^{4-}} = 0.81$ at pH 11.00 (Table 12-1)

For Cu^{2+} and NH_3, Appendix I gives log β_1 = 3.99, log β_2 = 7.33, log β_3 = 10.06, and log β_4 = 12.03. Therefore, $\beta_1 = 9.8 \times 10^3$, $\beta_2 = 2.1 \times 10^7$, $\beta_3 = 1.15 \times 10^{10}$, and $\beta_4 = 1.07 \times 10^{12}$.

$\alpha_{Cu^{2+}} = \dfrac{1}{1+\beta_1(1.00)+\beta_2(1.00)^2+\beta_3(1.00)^3+\beta_4(1.00)^4} = 9.2_3 \times 10^{-13}$

$K_f' = \alpha_{Y^{4-}} K_f = 4.8_8 \times 10^{18}$

$K_f'' = \alpha_{Y^{4-}} \alpha_{Cu^{2+}} K_f = 4.5_1 \times 10^6$

Equivalence point = 50.00 mL

(a) At 0 mL, the total concentration of copper is $C_{Cu^{2+}} = 0.001\,00$ M and
$[Cu^{2+}] = \alpha_{Cu^{2+}} C_{Cu^{2+}} = 9.2_3 \times 10^{-16}$ M $\Rightarrow$ pCu^{2+} = 15.03

(b) 1.00 mL: $C_{Cu^{2+}} = \left(\dfrac{49.00\text{ mL}}{50.00\text{ mL}}\right)(0.001\,00\text{ M})\left(\dfrac{50.00\text{ mL}}{51.00\text{ mL}}\right) = 9.61 \times 10^{-4}$ M

        fraction   original   dilution
        remaining concentration factor

$[Cu^{2+}] = \alpha_{Cu^{2+}} C_{Cu^{2+}} = 8.8_7 \times 10^{-16}$ M $\Rightarrow$ pCu^{2+} = 15.05

(c) 45.00 mL: $C_{Cu^{2+}} = \left(\dfrac{5.00\text{ mL}}{50.00\text{ mL}}\right)(0.001\,00\text{ M})\left(\dfrac{50.00\text{ mL}}{95.00\text{ mL}}\right) = 5.26 \times 10^{-5}$ M

$[Cu^{2+}] = \alpha_{Cu^{2+}} C_{Cu^{2+}} = 4.8_6 \times 10^{-17}$ M $\Rightarrow$ pCu^{2+} = 16.31

(d) Equivalence point:

$$C_{Cu^{2+}} + \text{EDTA} \rightleftharpoons \text{CuY}^{2-}$$

   x x $\left(\dfrac{50.00\text{ mL}}{100.00\text{ mL}}\right)(0.001\,00) - x$

$\dfrac{0.000\,500 - x}{x^2} = K_f'' = 4.5_1 \times 10^6 \Rightarrow x = C_{Cu^{2+}} = 1.04 \times 10^{-5}$ M

$[Cu^{2+}] = \alpha_{Cu^{2+}} C_{Cu^{2+}} = 9.6_2 \times 10^{-18}$ M $\Rightarrow$ pCu^{2+} = 17.02

(e) Past the equivalence point at 55.00 mL:

$[\text{EDTA}] = \left(\dfrac{5.00\text{ mL}}{105.00\text{ mL}}\right)(0.001\,00\text{ M}) = 4.76 \times 10^{-5}$ M

$[\text{CuY}^{2-}] = \left(\dfrac{50.00\text{ mL}}{105.00\text{ mL}}\right)(0.001\,00\text{ M}) = 4.76 \times 10^{-4}$ M

$K_f' = \dfrac{[\text{CuY}^{2-}]}{[Cu^{2+}][\text{EDTA}]} = \dfrac{(4.76 \times 10^{-4})}{[Cu^{2+}](4.76 \times 10^{-5})}$

$\Rightarrow [Cu^{2+}] = 2.05 \times 10^{-18}$ M $\Rightarrow$ pCu^{2+} = 17.69

12-17. (a) $\alpha_{ML} = \dfrac{[ML]}{C_M} = \dfrac{\beta_1[M][L]}{[M]\{1 + \beta_1[L] + \beta_2[L]^2\}} = \dfrac{\beta_1[L]}{1 + \beta_1[L] + \beta_2[L]^2}$

$\alpha_{ML_2} = \dfrac{[ML_2]}{C_M} = \dfrac{\beta_2[M][L]^2}{[M]\{1 + \beta_1[L] + \beta_2[L]^2\}} = \dfrac{\beta_2[L]^2}{1 + \beta_1[L] + \beta_2[L]^2}$

(b) For [L] = 0.100 M, $\beta_1 = 1.7 \times 10^2$, and $\beta_2 = 4.3 \times 10^3$, we get $\alpha_{ML} = 0.28$ and $\alpha_{ML_2} = 0.70$.

EDTA Titrations

12-18. In place of Equation 12-8, we write

$$M_{free} + EDTA \rightleftharpoons M(EDTA) \qquad K_f'' = \frac{[M(EDTA)]}{[M]_{free}[EDTA]}$$

where $[M]_{free}$ is the concentration of all metal not bound to EDTA. [EDTA] is the concentration of all EDTA not bound to metal. The mass balances are

Metal: $\quad [M]_{free} + [M(EDTA)] = \dfrac{C_M V_M}{V_M + V_{EDTA}}$

EDTA: $\quad [EDTA] + [M(EDTA)] = \dfrac{C_{EDTA} V_{EDTA}}{V_M + V_{EDTA}}$

These equations have the same form as the first three equations in Section 12-4, with K_f replaced by K_f'', [M] replaced by $[M]_{free}$, and [L] replaced by [EDTA]. The derivation therefore leads to Equation 12-11, with K_f replaced by K_f'', [M] replaced by $[M]_{free}$, and C_L replaced by C_{EDTA}.

12-19. (a)

	A	B	C	D	E	F
1	Titration of 50 mL of 0.001 M Zn^{2+} with 0.001 M EDTA/pH 10 with NH_3					
2						
3	C_M =	pM	M	$[M]_{tot}$	ϕ	V_{EDTA}
4	0.001	8.115	7.67E-09	4.29E-04	0.400	19.9814
5	V_M =	12.014	9.68E-13	5.41E-08	1.000	50.0000
6	50	15.278	5.27E-16	2.95E-11	1.200	59.9965
7	C_{EDTA} =					
8	0.001					
9	K_f'' =	A10 = A12*A16*10^A14				
10	1.70E+11	A12 = 1/(1+A20*A18+B20*A18^2+C20*A18^3+D20*A18^4)				
11	$\alpha(Zn^{2+})$ =					
12	1.79E-05	C4 = 10^-B4				
13	log K_f =	D4 = C4/$A$12				
14	16.5	E4 = (1+$A$10*D4-(D4+D4^2*$A$10)/$A$4)/				
15	$\alpha(Y^{4-})$ =		(D4*$A$10+(D4+D4^2*$A$10)/$A$8)			
16	0.30	F4 = E4*$A$4*$A$6/$A$8				
17	$[NH_3]$ =					
18	0.1					
19	β_1 =	β_2 =	β_3 =	β_4 =		
20	1.51E+02	2.69E+04	5.50E+06	5.01E+08		

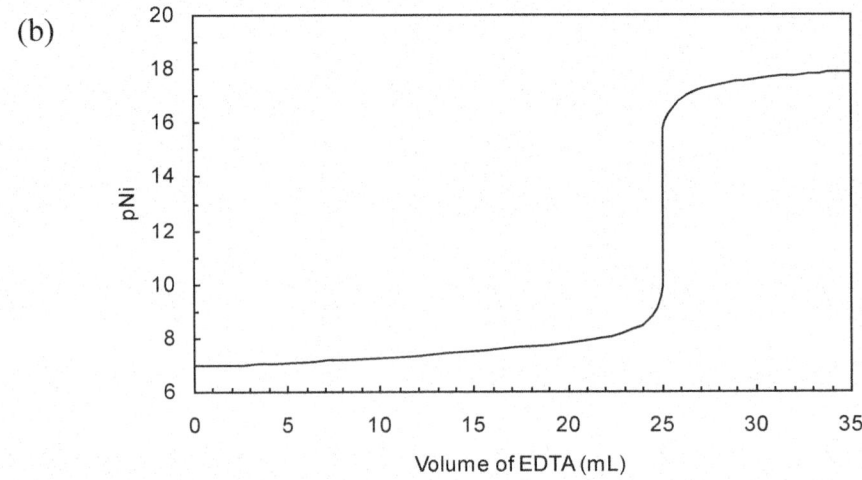
(b)

	A	B	C	D	E	F
1	Titration of 50 mL of 0.05 M Ni^{2+} with 0.1 M EDTA / pH 11 / 0.1 M Oxalate					
2						
3	C_M =	pM	M	$[M]_{tot}$	ϕ	V_{EDTA}
4	0.005	6.97	1.07E-07	4.94E-03	0.008	0.210
5	V_M =	7.00	1.00E-07	4.61E-03	0.054	1.342
6	50	7.20	6.31E-08	2.91E-03	0.324	8.106
7	C_{EDTA} =	7.50	3.16E-08	1.46E-03	0.618	15.461
8	0.01	8.00	1.00E-08	4.61E-04	0.868	21.696
9	K_f'' =	8.40	3.98E-09	1.83E-04	0.946	23.649
10	4.42E+13	8.80	1.58E-09	7.30E-05	0.978	24.456
11	$\alpha(Ni^{2+})$ =	9.50	3.16E-10	1.46E-05	0.996	24.891
12	2.17E-05	10.50	3.16E-11	1.46E-06	1.000	24.989
13	log K_f =	12.80	1.58E-13	7.30E-09	1.000	25.000
14	18.4	14.00	1.00E-14	4.61E-10	1.000	25.001
15	$\alpha(Y^{4-})$ =	15.00	1.00E-15	4.61E-11	1.000	25.012
16	0.81	16.00	1.00E-16	4.61E-12	1.005	25.123
17	[Oxalate^{2-}] =	17.00	1.00E-17	4.61E-13	1.049	26.229
18	0.1	17.40	3.98E-18	1.83E-13	1.123	28.086
19	β_1 =	17.60	2.51E-18	1.16E-13	1.196	29.892
20	1.45E+05	17.80	1.58E-18	7.30E-14	1.310	32.753
21	β_2 =	17.90	1.26E-18	5.80E-14	1.390	34.760
22	3.16E+06	18.00	1.00E-18	4.61E-14	1.491	37.287
23						
24	A10 = A16*A12*10^A14					
25	A12 = 1/(1+A20*A18+A22*A18^2)					
26	C4 = 10^-B4					
27	D4 = C4/$A$12					
28	E4 = (1+$A$10*D4-(D4+D4*D4*$A$10)/$A$4)/					
29	(D4*$A$10+(D4+D4*D4*$A$10)/$A$8)					
30	F4 = E4*$A$4*$A$6/$A$8					

12-20. $[L] + [ML] + 2[ML_2] = \dfrac{C_L V_L}{V_M + V_L}$

$[L] + \alpha_{ML} \dfrac{C_L V_L}{V_M + V_L} + 2\alpha_{ML_2} \dfrac{C_M V_M}{V_M + V_L} = \dfrac{C_L V_L}{V_M + V_L}$

Multiply both sides by $V_M + V_L$:

$[L]V_M + [L]V_L + \alpha_{ML} C_M V_M + 2\alpha_{ML_2} C_M V_M = C_L V_L$

Collect terms $\quad V_L ([L] - C_L) = V_M (-[L] - \alpha_{ML} C_M - 2\alpha_{ML_2} C_M)$

$\dfrac{V_L}{V_M} = \dfrac{[L] + \alpha_{ML} C_M + 2\alpha_{ML_2} C_M}{C_L - [L]}$

Divide the denominator by C_L and divide the numerator by C_M to obtain ϕ, the fraction of the way to the equivalence point:

$\phi = \dfrac{C_L V_L}{C_M V_M} = \dfrac{\dfrac{[L]}{C_M} + \alpha_{ML} + 2\alpha_{ML_2}}{1 - \dfrac{[L]}{C_L}}$

12-21.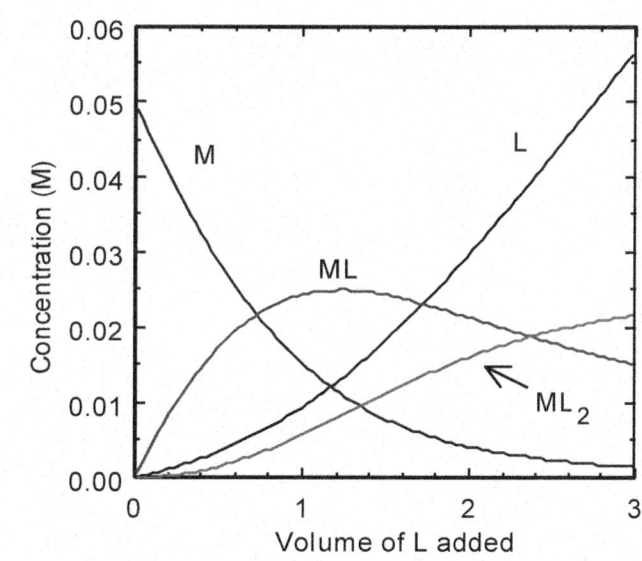

	A	B	C	D	E	F	G	H	I	J	K
1	Copper-acetate complexes ML and ML$_2$										
2											
3	C$_M$ =	pL	[L]	α$_M$	α$_{ML}$	α$_{ML2}$	φ	V(ligand)	[M]	[ML]	[ML$_2$]
4	0.05	4.00	0.0001	0.983	0.017	0.000	0.019	0.019	0.0491	0.0008	0.0000
5	V$_M$ =	3.00	0.0010	0.852	0.145	0.004	0.172	0.172	0.0419	0.0071	0.0002
6	10	2.80	0.0016	0.781	0.210	0.008	0.260	0.260	0.0381	0.0103	0.0004
7	C(ligand)	2.60	0.0025	0.688	0.294	0.019	0.383	0.383	0.0331	0.0141	0.0009
8	0.5	2.40	0.0040	0.573	0.388	0.039	0.550	0.550	0.0272	0.0184	0.0019
9	β$_1$ =	2.20	0.0063	0.446	0.478	0.076	0.766	0.766	0.0207	0.0222	0.0035
10	170	2.00	0.0100	0.319	0.543	0.137	1.039	1.039	0.0145	0.0246	0.0062
11	β$_2$ =	1.90	0.0126	0.262	0.560	0.178	1.199	1.199	0.0117	0.0250	0.0080
12	4300	1.80	0.0158	0.209	0.564	0.226	1.377	1.377	0.0092	0.0248	0.0099
13		1.70	0.0200	0.164	0.556	0.280	1.579	1.579	0.0071	0.0240	0.0121
14		1.60	0.0251	0.125	0.535	0.340	1.808	1.808	0.0053	0.0226	0.0144
15		1.50	0.0316	0.094	0.504	0.403	2.073	2.073	0.0039	0.0209	0.0167
16		1.40	0.0398	0.069	0.464	0.467	2.385	2.385	0.0028	0.0187	0.0189
17		1.30	0.0501	0.049	0.419	0.532	2.761	2.761	0.0019	0.0164	0.0208
18		1.25	0.0562	0.041	0.396	0.563	2.981	2.981	0.0016	0.0152	0.0217
19											
20	C4 = 10^-B4						H4 = G4*$A$4*$A$6/$A$8				
21	D4 = 1/(1+$A$10*C4+$A$12*C4*C4)						I4 = D4*$A$4*$A$6/($A$6+H4)				
22	E4 = $A$10*C4/(1+$A$10*C4+$A$12*C4*C4)						J4 = E4*$A$4*$A$6/($A$6+H4)				
23	F4 = $A$12*C4*C4/(1+$A$10*C4+$A$12*C4*C4)						K4 = F4*$A$4*$A$6/($A$6+H4)				
24	G4 = (C4/$A$4+E4+2*F4)/(1-C4/$A$8)										

12-22. Only a small amount of indicator is employed. Most of the Mg^{2+} is not bound to the indicator. Free Mg^{2+} reacts with EDTA before MgIn reacts. Therefore, the concentration of MgIn is constant until free Mg^{2+} has been consumed. Only when MgIn begins to react does the color change.

12-23. HIn^{2-}, wine red, blue

12-24. Buffer (i) (pH 6–7) will give a yellow → blue color change that will be easier to observe than the violet → blue change expected with the other buffers.

12-25. A back titration is necessary if the analyte precipitates in the absence of EDTA, if it reacts too slowly with EDTA, or if it blocks the indicator.

12-26. In a displacement titration, analyte displaces a metal ion from a complex. The displaced metal ion is then titrated. An example is the liberation of Ni^{2+} from Ni(CN)$_4^{2-}$ by the analyte Ag$^+$. Liberated Ni^{2+} is then titrated by EDTA to find out how much Ag$^+$ was present.

EDTA Titrations

12-27. The Mg^{2+} in a solution of Mg^{2+} and Fe^{3+} can be titrated by EDTA if the Fe^{3+} is masked with CN^- to form $Fe(CN)_6^{3-}$, which does not react with EDTA.

12-28. Hardness refers to the total concentration of alkaline earth cations in water, which normally means $[Ca^{2+}] + [Mg^{2+}]$. Hardness gets its name from the reaction of these cations with soap to form insoluble curds. Temporary hardness, due to $Ca(HCO_3)_2$, is lost by precipitation of $CaCO_3(s)$ upon heating. Permanent hardness derived from other salts, such as $CaSO_4$, is not affected by heat.

12-29. $(50.0 \text{ mL})(0.0100 \text{ mmol/mL}) = 0.500 \text{ mmol } Ca^{2+}$, which requires 0.500 mmol EDTA = 10.0 mL EDTA.

0.500 mmol Al^{3+} requires the same amount of EDTA, 10.0 mL.

12-30. mmol EDTA = mmol Ni^{2+} + mmol Zn^{2+}

$1.250 = x + 0.250 \Rightarrow 1.000$ mmol Ni^{2+} in 50.0 mL = 0.0200 M

12-31. The formula mass of $MgSO_4$ is 120.36. The 50.0 mL aliquot contains

$$\left(\frac{50.0 \text{ mL}}{500 \text{ mL}}\right)\left(\frac{0.450 \text{ g}}{120.36 \text{ g/mol}}\right) = 0.3739 \text{ mmol of } Mg^{2+}$$

37.6 mL of EDTA reacts with this much Mg^{2+}, so the EDTA solution contains 0.3739 mmol/37.6 mL = 9.944×10^{-3} mmol/mL. The formula mass of $CaCO_3$ is 100.09. 1.00 mL of EDTA will react with 9.943×10^{-3} mmol of $CaCO_3$ = 0.995 mg.

12-32. 30.10 mL Ni^{2+} reacted with 39.35 mL 0.01307 M EDTA, so Ni^{2+} molarity is

$$[Ni^{2+}] = \frac{(39.35 \text{ mL})(0.01307 \text{ mol/L})}{30.10 \text{ mL}} = 0.01709 \text{ M}.$$

25.00 mL Ni^{2+} contains 0.4272 mmol Ni^{2+}. 10.15 mL EDTA = 0.1327 mmol EDTA. The Ni^{2+} which must have reacted with CN^- was 0.4272 − 0.1327 = 0.2945 mmol. Cyanide reacting with Ni^{2+} must have been (4)(0.2945 mmol) = 1.178 mmol. Original $[CN^-]$ = 1.178 mmol/12.73 mL = 0.09254 M.

12-33. For 1.00 mL of unknown:

$$\begin{array}{rl}
25.00 \text{ mL of EDTA} &= 0.9680 \text{ mmol} \\
- \quad 23.54 \text{ mL of Zn}^{2+} &= 0.5007 \text{ mmol} \\
\hline
\text{Co}^{2+} + \text{Ni}^{2+} &= 0.4673 \text{ mmol}
\end{array}$$

For 2.000 mL of unknown:

$$\begin{array}{rl}
25.00 \text{ mL of EDTA} &= 0.9680 \text{ mmol} \\
- \quad 25.63 \text{ mL of Zn}^{2+} &= 0.5452 \text{ mmol} \\
\hline
\text{Ni}^{2+} \text{ in 2.000 mL} &= 0.4228 \text{ mmol}
\end{array}$$

Co^{2+} in 2.000 mL of unknown = $2(0.4673) - 0.4228 = 0.5118$ mmol. The Co^{2+} will react with 0.5118 mmol of EDTA, leaving $0.9680 - 0.5118 = 0.4562$ mmol EDTA.

$$\text{mL Zn needed} = \frac{0.4562 \text{ mmol}}{0.02127 \text{ mmol/mL}} = 21.45 \text{ mL}$$

12-34.
$$\begin{array}{rll}
\text{Total EDTA} &= (25.0 \text{ mL})(0.0452 \text{ M}) &= 1.130 \text{ mmol} \\
- \quad \text{Mg}^{2+} \text{ required} &= (12.4 \text{ mL})(0.0123 \text{ M}) &= 0.153 \text{ mmol} \\
\hline
\text{Ni}^{2+} + \text{Zn}^{2+} & &= 0.977 \text{ mmol}
\end{array}$$

Zn^{2+} = EDTA displaced by 2,3-dimercapto-1-propanol
$= (29.2 \text{ mL})(0.0123 \text{ M}) = 0.359$ mmol

$\Rightarrow \text{Ni}^{2+} = 0.977 - 0.359 = 0.618$ mmol; $[\text{Ni}^{2+}] = \dfrac{0.618 \text{ mmol}}{50.0 \text{ mL}} = 0.0124 \text{ M}$

$[\text{Zn}^{2+}] = \dfrac{0.359 \text{ mmol}}{50.0 \text{ mL}} = 0.00718 \text{ M}$

12-35. The precipitation reaction is $\text{Cu}^{2+} + \text{S}^{2-} \rightarrow \text{CuS}(s)$.

$$\begin{array}{rll}
\text{Total Cu}^{2+} \text{ used} &= (25.00 \text{ mL})(0.04332 \text{ M}) &= 1.0830 \text{ mmol} \\
- \quad \text{Excess Cu}^{2+} &= (12.11 \text{ mL})(0.03927 \text{ M}) &= 0.4756 \text{ mmol} \\
\hline
\text{mmol of S}^{2-} & &= 0.6074 \text{ mmol}
\end{array}$$

$[\text{S}^{2-}] = 0.6074 \text{ mmol}/25.00 \text{ mL} = 0.02430 \text{ M}$

12-36. mmol Bi in reaction = (25.00 mL)(0.086 40 M) = 2.160 mmol
EDTA required = (14.24 mL)(0.043 7 M) = 0.622 mmol
mmol Bi that reacted with Cs = 2.160 − 0.622 = 1.538 mmol
Since 2 mol Bi react with 3 mol Cs to give $Cs_3Bi_2I_9$,

mmol Cs^+ in unknown = $\frac{3}{2}$(1.538 mmol) = 2.307 mmol

$$[Cs^+] = \frac{2.307 \text{ mmol}}{25.00 \text{ mL}} = 0.092\,28 \text{ M}.$$

12-37. Total standard Ba^{2+} + Zn^{2+} added to sulfate was (5.000 mL)(0.014 63 M $BaCl_2$) + (1.000 mL)(0.010 00 M $ZnCl_2$) = 0.083 15 mmol. Total EDTA required was (2.39 mL)(0.009 63 M) = 0.023 0$_2$ mmol. Therefore, the original solid must have contained 0.083 15 − 0.023 0$_2$ = 0.060 1$_3$ mmol sulfur (which made 0.060 1$_3$ mmol sulfate that precipitated 0.060 1$_3$ mmol Ba^{2+}). The mass of sulfur was (0.060 1$_3$ mmol)(32.06 mg/mmol) = 1.92$_8$ mg.
wt% S = 100 × (1.92$_8$ mg S/5.89 mg sphalerite) = 32.7 wt%.
Theoretical wt% S in pure ZnS = 100 × (32.06 g S/97.44 g ZnS) = 32.90 wt%.

CHAPTER 13
ADVANCED TOPICS IN EQUILIBRIUM

13-1. As pH is lowered, [H$^+$] increases. H$^+$ reacts with basic anions to increase the solubility of their salts. Dissolution of minerals such as galena and cerussite increases the concentration of Pb^{2+} in the environment.

Galena: $\quad$ PbS(s) + H$^+$ $\rightleftharpoons$ Pb^{2+} + HS$^-$

Cerussite: $\quad$ PbCO$_3$(s) + H$^+$ $\rightleftharpoons$ Pb^{2+} + HCO$_3^-$

13-2. (a) Hydroxybenzene = HA with pK_{HA} = 9.997

Mixture contains 0.010 0 mol HA and 0.005 0 mol KOH in 1.00 L.
Chemical reactions:

$$HA \rightleftharpoons A^- + H^+ \qquad K_{HA} = \frac{[H^+][A^-]}{[HA]} = 10^{-9.997}$$

$$H_2O \rightleftharpoons H^+ + OH^- \qquad K_w = [H^+][OH^-] = 10^{-14.00}$$

Charge balance:

$$[H^+] + [K^+] = [OH^-] + [A^-]$$

Mass balances:

$$[K^+] = 0.005\ 0\ M$$
$$[HA] + [A^-] = 0.010\ 0\ M = F_A$$

We have 5 equations and 5 chemical species.
Fractional composition equations:

$$[HA] = \alpha_{HA} F_A = \frac{[H^+]F_A}{[H^+]+K_{HA}}$$

$$[A^-] = \alpha_{A^-} F_A = \frac{K_{HA}F_A}{[H^+]+K_{HA}}$$

Substitute concentration expressions into the charge balance:

$$[H^+] + [0.005\ 0] = K_w/[H^+] + \alpha_{A^-} F_A \qquad (A)$$

We could solve Equation A for [H$^+$] by using the solution to a quadratic equation. Instead, we will use Solver in the following spreadsheet, with an initial guess of pH = 10 in cell H9. Select Solver and choose Options. In Excel 2016, in the All Methods tab, set Constraint Precision to 1e-15. In the GRG Nonlinear tab Convergence can be left at its default value and Derivatives should be Central. Click OK. In the Solver Parameters window, Set Objective E12 Equal To Value Of 0 By Changing Variable Cells H9. Click Solve and Solver finds pH = 9.980 in cell H9, giving a net charge near 0 in cell E12.

Advanced Topics in Equilibrium   167

	A	B	C	D	E	F	G	H	I
1	Mixture of 0.010 M HA and 0.005 M KOH								
2									
3	F_A =	0.010		$[K^+]$ =	0.005				
4	pK_{HA} =	9.997		pK_w =	14.000				
5	K_{HA} =	1.01E-10		K_w =	1.00E-14				
6									
7	Species in charge balance:						Other concentrations:		
8	$[H^+]$ =	1.05E-10		$[A^-]$ =	4.90E-03		$[HA]$ =	5.10E-03	
9	$[K^+]$ =	5.00E-03		$[OH^-]$ =	9.56E-05		pH =	9.980	
10							↑ initial value is a guess		
11									
12	Positive charge minus negative charge				7.81E-18		= B8+B9-E8-E9		
13	Formulas:								
14	B5 = 10^-B4			B8 = 10^-H9					
15	E5 = 10^-E4			B9 = E3					
16	E8 = B5*B3/(B8+B5)			E9 = E5/B8					
17	H8 = B8*B3/(B8+B5)								

(b) From previous knowledge, we would have said that there is enough KOH to neutralize half of the HA. Therefore, [HA] = [A⁻].

$pH = pK_a + \log([A^-]/[HA]) = pK_a + \log(1) = pK_a = 10.00$. The systematic treatment of equilibrium in the spreadsheet gave pH = 9.98.

(c) If we dilute HA to 0.000 10 M and KOH to 0.000 050 in cells B3 and E3, then Solver finds a pH of 9.45. It makes qualitative sense that as the solution becomes more dilute, the pH must move toward 7.

13-3. Use effective equilibrium constants, K', defined as follows:

$$K_{HA} = \frac{[H^+]\gamma_{H^+}[A^-]\gamma_{A^-}}{[HA]\gamma_{HA}} \Rightarrow K'_{HA} = \frac{[H^+][A^-]}{[HA]} = K_{HA}\frac{\gamma_{HA}}{\gamma_{H^+}\gamma_{A^-}}$$

$$K_w = [H^+]\gamma_{H^+}[OH^-]\gamma_{OH^-} = 10^{-13.995}$$

$$K'_w = \frac{K_w}{\gamma_{H^+}\gamma_{OH^-}} = [H^+][OH^-] \Rightarrow [OH^-] = K'_w/[H^+]$$

$$pH = -\log([H^+]\gamma_{H^+})$$

In Excel 2016, enable circular definitions by clicking on the File menu. Select Options and then select Formulas. In Calculation options, check "Enable iterative calculation." Set Maximum Change = 1E-15. Click OK. Then use Solver to find the pH in cell H13 that gives a net charge near 0 in cell E16. Results are pH =

9.947 and ionic strength = 0.005 0 in cell C17.

	A	B	C	D	E	F	G	H	I
1	Mixture of 0.010 M HA and 0.005 M KOH with activity coefficients								
2									
3	F_A =	0.010		$[K^+]$ =	0.005				
4	pK_{HA} =	9.997		pK_w =	13.995				
5	K_{HA}' =	1.17E-10		K_w' =	1.18E-14				
6									
7	Activity coefficients:								
8	H^+ =	0.93		A^-	0.93				
9	OH^- =	0.93		HA	1.00				
10									
11	Species in charge balance:						Other concentrations:		
12	$[H^+]$ =	1.22E-10		$[A^-]$ =	4.90E-03		[HA] =	5.10E-03	
13	$[K^+]$ =	0.005		$[OH^-]$ =	9.67E-05		pH =	9.947	
14							↑ initial value is a guess		
15									
16	Positive charge minus negative charge =				-7.08E-17		= B12+B13-E12-E13		
17	Ionic strength =		0.005000		= 0.5*(B12+B13+E12+E13)				
18									
19	Formulas:								
20	B5 = (10^-B4)*E9/(B8*E8)				E9 = 1		B13 = E3		
21	E5 = (10^-E4)/(B8*B9)								
22	B8 = B9 = E8 = 10^(-0.51*1^2*(SQRT($C$17)/(1+SQRT($C$17))-0.3*$C$17))								
23	B12 = (10^-H13)/B8				E13 = E5/B12				
24	E12 = B5*B3/(B12+B5))				H12 = B12*B3/(B12+B5)				

13-4. Abbreviating the protonated form of glycine as H_2G^+, we write

$$H_2G^+ \rightleftharpoons HG + H^+ \qquad K_1 = \frac{[HG]\gamma_{HG}[H^+]\gamma_{H^+}}{[H_2G^+]\gamma_{H_2G^+}} \qquad pK_1 = 2.350$$

$$HG \rightleftharpoons G^- + H^+ \qquad K_2 = \frac{[G^-]\gamma_{G^-}[H^+]\gamma_{H^+}}{[HG]\gamma_{HG}} \qquad pK_2 = 9.778$$

At $\mu = 0.1$ M, the activity coefficient of a monovalent ion is 0.78. The activity coefficient of a neutral molecule is 1. Putting these coefficients into the expressions for K_1 and K_2 gives K_1' and K_2' at $\mu = 0.1$ M:

$$K_1' = \frac{[HG][H^+]}{[H_2G^+]} = K_1 \frac{\gamma_{H_2G^+}}{\gamma_{HG}\gamma_{H^+}} = 10^{-2.350} \frac{0.78}{(1)(0.78)} = 10^{-2.350}$$

$$K_2' = \frac{[G^-][H^+]}{[HG]} = K_2 \frac{\gamma_{HG}}{\gamma_{G^-}\gamma_{H^+}} = 10^{-9.778} \frac{1}{(0.78)(0.78)} = 10^{-9.562}$$

Predicted values are $pK_1' = 2.350$ and $pK_2' = 9.562$. Values from fitting data in the spreadsheet are 2.312 and 9.625. The change from pK_1 to pK_1' is expected to be zero and it is observed to be –0.038. The change from pK_2 to pK_2' is expected to be –0.216 and it is observed to be –0.153.

13-5. Ethylenediamine = B from diprotic H_2B^{2+} $pK_1 = 6.848$ $pK_2 = 9.928$

Mixture contains 0.100 mol B and 0.035 mol HBr in 1.00 L.

Chemical reactions:

$H_2B^{2+} \rightleftharpoons HB^+ + H^+$ $K_1 = 10^{-6.848}$

$HB^+ \rightleftharpoons B + H^+$ $K_2 = 10^{-9.928}$

$H_2O \rightleftharpoons H^+ + OH^-$ $K_w = 10^{-14.00}$

Charge balance:

$[H^+] + 2[H_2B^{2+}] + [HB^+] = [OH^-] + [Br^-]$

Mass balances:

$[Br^-] = 0.035\ M$; $[H_2B^+] + [HB] + [B] = 0.100\ M = F_B$

We have 6 equations and 6 chemical species, so there is enough information.

Fractional composition equations:

$$[H_2B^{2+}] = \alpha_{H_2B^{2+}} F_B = \frac{[H^+]^2 F_B}{[H^+]^2 + [H^+]K_1 + K_1 K_2}$$

$$[HB^+] = \alpha_{HB^+} F_B = \frac{K_1[H^+]F_B}{[H^+]^2 + [H^+]K_1 + K_1 K_2}$$

$$[B] = \alpha_B F_B = \frac{K_1 K_2 F_B}{[H^+]^2 + [H^+]K_1 + K_1 K_2}$$

Substitute into charge balance:

$$[H^+] + 2\alpha_{H_2B^{2+}} F_B + \alpha_{HB^+} F_B = K_w/[H^+] + [0.035\ M] \tag{A}$$

We solve Equation A for $[H^+]$ by using Solver in the following spreadsheet, with an initial guess of pH = 10 in cell H11. Select Solver and choose Options. In Excel 2016, in the All Methods tab, set Constraint Precision to 1e-15. In the GRG Nonlinear tab, Derivatives should be Central. Click OK. In the Solver Parameters window, Set Objective E14 Equal To Value Of 0 By Changing Variable Cells H11. Click Solve and Solver finds pH = 10.194 in cell H11, giving a net charge near 0 in cell E14.

	A	B	C	D	E	F	G	H	I
1	Mixture of 0.100 M B and 0.035 M HBr								
2									
3	F_B =	0.100		[Br⁻] =	0.035				
4	pK_1 =	6.848		pK_w =	14.000				
5	pK_2 =	9.928		K_w =	1.00E-14				
6	K_1 =	1.42E-07							
7	K_2 =	1.18E-10							
8									
9	Concentrations:								
10	[H⁺] =	6.39E-11		[H₂B²⁺] =	1.58E-05		[B] =	6.49E-02	
11	[Br⁻] =	3.50E-02		[HB⁺] =	3.51E-02		pH =	10.194	
12	[OH⁻] =	1.56E-04					↑ initial value is a guess		
13									
14	Positive charge minus negative charge				-1.70E-17				
15	Formulas:								
16	B6 = 10^-B4			B7 = 10^-B5		E5 = 10^-E4			
17	B10 = 10^-H11			B11 = E3		B12 = E5/B10			
18	E10 = B10^2*$B$3/(B10^2+B10*$B$6+$B$6*$B$7)								
19	E11 = B10*$B$6*$B$3/(B10^2+B10*$B$6+$B$6*$B$7)								
20	H10 = $B$6*$B$7*$B$3/(B10^2+B10*$B$6+$B$6*$B$7)								

In your earlier life, you would have solved this problem by noting that 0.035 mol HBr converts 0.035 mol B into 0.035 mol HB⁺, leaving (0.100 – 0.035) mol B.

$$pH = pK_2 + \log \frac{[B]}{[HB^+]} = 9.928 + \log \frac{0.065}{0.035} = 10.197 \text{ (near spreadsheet answer)}$$

13-6. Benzene-1,2,3-tricarboxylic acid = H_3A with pK_1 = 2.86, pK_2 = 4.30, pK_3 = 6.28. Imidazole = HB from diprotic H_2B^+ with pK_1 = 6.993, pK_2 = 14.5. Mixture contains 0.040 mol H_3A, 0.030 mol HB, and 0.035 mol NaOH in 1.00 L. Charge balance:

$[H^+] + [H_2B^+] + [Na^+] = [OH^-] + [H_2A^-] + 2[HA^{2-}] + 3[A^{3-}] + [B^-]$

Substitute fractional composition equations into charge balance:

$[H^+] + \alpha_{H_2B^+} F_B + [0.035]$
$\quad = K_w/[H^+] + \alpha_{H_2A^-} F_A + 2\alpha_{H_2A^{2-}} F_A + 3\alpha_{HA^{3-}} F_A + \alpha_{B^-} F_B$

Solve for [H⁺] with the following spreadsheet, with an initial guess of pH = 7 in cell H14. Select Solver and choose Options. In Excel 2016, in All Methods, set Constraint Precision to 1e-15. In the GRG Nonlinear, Derivatives should be Central. Click OK. In Solver Parameters, Set Objective E16 Equal To Value Of 0 By Changing Variable Cells H14. Click Solve to find pH = 4.516 in cell H14, giving a net charge near 0 in cell E16.

Advanced Topics in Equilibrium

	A	B	C	D	E	F	G	H	I
1	Mixture of 0.040 M H$_3$A, 0.030 M HB, and 0.035 M NaOH								
2									
3	F$_A$ =	0.040		F$_B$ =	0.030		[Na$^+$] =	0.035	
4	pK$_1$ =	2.86		pK$_{H2B}$ =	6.993		pK$_w$ =	14.000	
5	pK$_2$ =	4.30		pK$_{HB}$ =	14.5		K$_w$ =	1.00E-14	
6	pK$_3$ =	6.28		K$_{H2B}$ =	1.02E-07				
7	K$_1$ =	1.4E-03		K$_{HB}$ =	3.16E-15				
8	K$_2$ =	5.0E-05							
9	K$_3$ =	5.2E-07							
10									
11	Concentrations								
12	[H$^+$] =	3.05E-05		[A^{3-}] =	4.21E-04		[OH$^-$] =	3.28E-10	
13	[H$_3$A] =	3.27E-04		[H$_2$B$^+$] =	2.99E-02		[Na$^+$] =	0.035	
14	[H$_2$A$^-$] =	1.48E-02		[HB] =	9.98E-05		pH =	4.516	
15	[HA^{2-}] =	2.44E-02		[B$^-$] =	1.04E-14		↑ initial value is a guess		
16	Positive charge minus negative charge =				-2.42E-17				
17									
18	Formulas:								
19	B7 = 10^-B4			B8 = 10^-B5			B9 = 10^-B6		
20	E6 = 10^-E4			E7 = 10^-E5					
21	B12 = 10^-H14			H12 = H5/B12			H13 = H3		
22	B13 = $B$12^3*$B$3/($B$12^3+$B$12^2*$B$7+$B$12*$B$7*$B$8+$B$7*$B$8*$B$9)								
23	B14 = $B$12^2*$B$7*$B$3/($B$12^3+$B$12^2*$B$7+$B$12*$B$7*$B$8+$B$7*$B$8*$B$9)								
24	B15 = $B$12*$B$7*$B$8*$B$3/($B$12^3+$B$12^2*$B$7+$B$12*$B$7*$B$8+$B$7*$B$8*B								
25	E12 = B7*B8*B9*B3/(B12^3+B12^2*B$7+B$12*B$7*B$8+B$7*B$8*B$9)								
26	E13 = $B$12^2*$E$3/($B$12^2+$B$12*$E$6+$E$6*$E$7)								
27	E14 = $B$12*$E$6*$E$3/($B$12^2+$B$12*$E$6+$E$6*$E$7)								
28	E15 = $E$6*$E$7*$E$3/($B$12^2+$B$12*$E$6+$E$6*$E$7)								

13-7. Arginine = HA from H$_3$A^{2+} with pK_1 = 1.823, pK_2 = 8.991, pK_3 = 12.1.

Glutamic acid = H$_2$G from H$_3$G$^+$ with pK_1 = 2.160, pK_2 = 4.30, pK_3 = 9.96.

Mixture contains 0.020 mol arginine, 0.030 mol glutamic acid, and 0.005 mol KOH in 1.00 L. F$_A$ = 0.020 M and F$_G$ = 0.030 M.

Charge balance:

$$[H^+] + 2[H_3A^{2+}] + [H_2A^+] + [H_3G^+] + [K^+] = [OH^-] + [A^-] + [HG^-] + 2[G^{2-}]$$

Substitute fractional composition equations into charge balance:

$$[H^+] + 2\alpha_{H_3A^{2+}} F_A + \alpha_{H_2A^+} F_A + \alpha_{H_3G^+} F_G + [0.005]$$
$$= K_w/[H^+] + \alpha_{A^-} F_A + \alpha_{HG^-} F_G + 2\alpha_{G^{2-}} F_G$$

	A	B	C	D	E	F	G	H	I	J
1	Mixture of 0.020 M arginine, 0.030 M glutamic acid, and 0.005 M KOH									
2										
3	$F_A =$	0.020		$F_B =$	0.030		$[K^+] =$	0.005		
4	$pK_1 =$	1.823		$pK_{H3G} =$	2.160		$pK_w =$	14.00		
5	$pK_2 =$	8.991		$pK_{H2G} =$	4.30		$K_w =$	1.00E-14		
6	$pK_3 =$	12.1		$pK_{HG} =$	9.96					
7	$K_1 =$	1.5E-02		$K_{H3G} =$	6.9E-03					
8	$K_2 =$	1.0E-09		$K_{H2G} =$	5.0E-05					
9	$K_3 =$	7.9E-13		$K_{HG} =$	1.1E-10					
10										
11	Concentrations									
12	$[H_3A^{2+}] =$	1.32E-05		$[H_3G^+] =$	7.14E-06		$[H^+] =$	9.94E-06		
13	$[H_2A^+] =$	2.00E-02		$[H_2G] =$	4.96E-03		$[OH^-] =$	1.01E-09		
14	$[HA] =$	2.05E-06		$[HG^-] =$	2.50E-02		$[K^+] =$	0.005		
15	$[A^-] =$	1.64E-13		$[G^{2-}] =$	2.76E-07		pH =	5.003		
16							↑ initial value is a guess			
17	Positive charge minus negative charge =				8.80E-17					
18					= 2*B12+B13+E12+H12+H14-B15-E14-2*E15-H13					
19	Formulas:									
20	B7 = 10^-B4			B8 = 10^-B5			B9 = 10^-B6			
21	E7 = 10^-E4			E8 = 10^-E5			E9 = 10^-E6			
22	H12 = 10^-H15			H13 = H5/H12			H14 = H3			
23	B12 = $H$12^3*$B$3/($H$12^3+$H$12^2*$B$7+$H$12*$B$7*$B$8+$B$7*$B$8*$B$9)									
24	B13 = $H$12^2*$B$7*$B$3/($H$12^3+$H$12^2*$B$7+$H$12*$B$7*$B$8+$B$7*$B$8*$B$9)									
25	B14 = $H$12*$B$7*$B$8*$B$3/($H$12^3+$H$12^2*$B$7+$H$12*$B$7*$B$8+$B$7*$B$8*$B$9)									
26	B15 = $B$7*$B$8*$B$9*$B$3/($H$12^3+$H$12^2*$B$7+$H$12*$B$7*$B$8+$B$7*$B$8*$B$9)									
27	E12 = $H$12^3*$E$3/($H$12^3+$H$12^2*$E$7+$H$12*$E$7*$E$8+$E$7*$E$8*$E$9)									
28	E13 = $H$12^2*$E$7*$E$3/($H$12^3+$H$12^2*$E$7+$H$12*$E$7*$E$8+$E$7*$E$8*$E$9)									
29	E14 = $H$12*$E$7*$E$8*$E$3/($H$12^3+$H$12^2*$E$7+$H$12*$E$7*$E$8+$E$7*$E$8*$E$9)									
30	E15 = $E$7*$E$8*$E$9*$E$3/($H$12^3+$H$12^2*$E$7+$H$12*$E$7*$E$8+$E$7*$E$8*$E$9)									

We solve for [H$^+$] with the spreadsheet, with an initial guess of pH = 3 in cell H15. Select Solver and choose Options. In Excel 2016, in All Methods, set Constraint Precision to 1e-15. In the GRG Nonlinear, Derivatives should be Central. Click OK. In Solver Parameters, Set Objective E17 Equal To Value of 0 By Changing Variable Cells H15. Click Solve and Solver finds pH = 5.003 in cell H15, giving a net charge near 0 in cell E17.

13-8. $H_3A^{2+} \rightleftharpoons H_2A^+ + H^+$ $\quad K_1 = \dfrac{[H_2A^+]\gamma_{H_2A^+}[H^+]\gamma_{H^+}}{[H_3A^{2+}]\gamma_{H_3A^{2+}}} = 10^{-1.823}$

$H_2A^+ \rightleftharpoons HA + H^+$ $\quad K_2 = \dfrac{[HA]\gamma_{HA}[H^+]\gamma_{H^+}}{[H_2A^+]\gamma_{H_2A^+}} = 10^{-8.991}$

$$HA \rightleftharpoons A^- + H^+ \qquad K_3 = \frac{[A^-]\gamma_{A^-}[H^+]\gamma_{H^+}}{[HA]\gamma_{HA}} = 10^{-12.1}$$

$$K_1' = K_1\left(\frac{\gamma_{H_3A^{2+}}}{\gamma_{H_2A^+}\gamma_{H^+}}\right) = \frac{[H_2A^+][H^+]}{[H_3A^{2+}]} \qquad K' = K_2\left(\frac{\gamma_{H_2A^+}}{\gamma_{HA}\gamma_{H^+}}\right) = \frac{[HA][H^+]}{[H_2A^+]}$$

$$K_3' = K_3\left(\frac{\gamma_{HA}}{\gamma_{A^-}\gamma_{H^+}}\right) = \frac{[A^-][H^+]}{[HA]}$$

$$H_3G^+ \rightleftharpoons H_2G + H^+ \qquad K_{H_3G} = \frac{[H_2G]\gamma_{H_2G}[H^+]\gamma_{H^+}}{[H_3G^+]\gamma_{H_3G^+}} = 10^{-2.160}$$

$$H_2G \rightleftharpoons HG^- + H^+ \qquad K_{H_2G} = \frac{[HG^-]\gamma_{HG^-}[H^+]\gamma_{H^+}}{[H_2G]\gamma_{H_2G}} = 10^{-4.30}$$

$$HG^- \rightleftharpoons G^{2-} + H^+ \qquad K_{HG} = \frac{[G^{2-}]\gamma_{G^{2-}}[H^+]\gamma_{H^+}}{[HG^-]\gamma_{HG^-}} = 10^{-9.96}$$

$$K_{H_3G}' = K_{H_3G}\left(\frac{\gamma_{H_3G^+}}{\gamma_{H_2G}\gamma_{H^+}}\right) = \frac{[H_2G][H^+]}{[H_3G^+]}$$

$$K_{H_2G}' = K_{H_2G}\left(\frac{\gamma_{H_2G}}{\gamma_{HG^-}\gamma_{H^+}}\right) = \frac{[HG^-][H^+]}{[H_2G]}$$

$$K_{HG}' = K_{HG}\left(\frac{\gamma_{HG^-}}{\gamma_{G^{2-}}\gamma_{H^+}}\right) = \frac{[G^{2-}][H^+]}{[HG^-]}$$

In the spreadsheet, $[H^+]$ in cell H18 is computed from $(10^{-pH})/\gamma_{H^+}$. Don't forget that activity coefficient! Enable circular definitions in Excel Options. Ionic strength is computed in cell E23 and activity coefficients are computed in cells A13:H15. From the activity coefficients, effective equilibrium constants are computed in cells H4:H10. pH = 3 is *guessed* in cell H21. From pH and the K' values, concentrations are computed in cells A18:H21. Use Solver to vary pH in cell H21 until the net charge in cell H23 is near 0. pH computed with activities is 4.939 and μ = 0.025 1 M in cell D24. When we found pH without activities in the previous problem, the pH was 5.00.

	A	B	C	D	E	F	G	H	I
1	Mixture of 0.020 M arginine, 0.030 M glutamic acid, and 0.005 M KOH								
2	Solved with Davies activity coefficients								
3	$F_A =$	0.020		$F_B =$	0.030		$[K^+] =$	0.005	
4	$pK_1 =$	1.823		$pK_{H3G} =$	2.160		$K_1' =$	1.1E-02	
5	$pK_2 =$	8.991		$pK_{H2G} =$	4.30		$K_2' =$	1.0E-09	
6	$pK_3 =$	12.1		$pK_{HG} =$	9.96		$K_3' =$	1.1E-12	
7	$K_1 =$	1.5E-02		$K_{H3G} =$	6.9E-03		$K_{H3G}' =$	6.9E-03	
8	$K_2 =$	1.0E-09		$K_{H2G} =$	5.0E-05		$K_{H2G}' =$	6.8E-05	
9	$K_3 =$	7.9E-13		$K_{HG} =$	1.1E-10		$K_{HG}' =$	2.01E-10	
10	$pK_w =$	13.995		$K_w =$	1.01E-14		$K_w' =$	1.37E-14	
11									
12	Davies activity coefficients								
13	$H_3A^{2+} =$	0.55		$H_3G^+ =$	0.86				
14	$H_2A^+ =$	0.86		$HG^- =$	0.86		$H^+ =$	0.86	
15	$A^- =$	0.86		$G^{2-} =$	0.55		$OH^- =$	0.86	
16									
17	Concentrations								
18	$[H_3A^{2+}] =$	2.41E-05		$[H_3G^+] =$	9.58E-06		$[H^+] =$	1.34E-05	
19	$[H_2A^+] =$	2.00E-02		$[H_2G] =$	4.95E-03		$[OH^-] =$	1.02E-09	
20	$[HA] =$	1.52E-06		$[HG^-] =$	2.50E-02		$[K^+] =$	0.005	
21	$[A^-] =$	1.22E-13		$[G^{2-}] =$	3.76E-07		$pH =$	4.939	
22							↑ initial value is a guess		
23	Positive charge minus negative charge =			1.27E-17					
24		Ionic strength =		0.0251	=0.5*(4*B18+B19+B21+E18				
25					+E20+4*E21+H18+H19+H20)				
26									
27	Formulas								
28	K_1' = B7*B13/(B14*H14)			K_2' = B8*B14/(H14)			K_3' = B9/(B15*H14)		
29	K_{H3B}' = E7*E13/H14			K_{H2B}' =E8/(E14*H14)			K_{HB}' = E9*E14/(E15*H14)		
30	K_w' = E10/(H14*H15)			[H+] = (10^-H21)/H14			[OH-] = H10/H18		
31	Activity coefficient = 10^(-0.51*charge^2*(SQRT($D$24)/(1+SQRT($D$24))-0.3*$D$24))								
32	Denom1 = ($H$18^3+$H$18^2*$H$4+$H$18*$H$4*$H$5+$H$4*$H$5*$H$6)								
33	$[H_3A^{2+}]$ = $H$18^3*$B$3/Denom1				$[H_2A^+]$ = $H$18^2*$H$4*$B$3/Denom1				
34	[HA] = $H$18*$H$4*$H$5*$B$3/Denom1				$[A^-]$ = $H$4*$H$5*$H$6*$B$3/Denom1				
35	Denom2 = ($H$18^3+$H$18^2*$H$7+$H$18*$H$7*$H$8+$H$7*$H$8*$H$9)								
36	$[H_3G^+]$ = $H$18^3*$E$3/Denom2				$[H_2G]$ = $H$18^2*$H$7*$E$3/Denom2				
37	$[HG^-]$ = $H$18*$H$7*$H$8*$E$3/Denom2				$[G^{2-}]$ = $H$7*$H$8*$H$9*$E$3/Denom2				
38	E23 = 2*B18+B19+E18+H18+H20-B21-E20-2*E21-H19								

13-9. (a) This is the same problem that was worked in Section 13-2, but with [KH$_2$PO$_4$] = 0.008 695 m and [Na$_2$HPO$_4$] = 0.030 43 m in cells B3 and B4 of the spreadsheet in Figure 13-3. First enable circular references. In Excel 2016, select the File tab and select Options. In the Options window, select Formulas. In Calculation Options, check "Enable iterative calculation" and set Maximum Change to 1e-15. Click OK. Then use Solver to find the pH in cell H15 that reduces the charge balance in cell E18 to near 0. The pH is found to be 7.420 and the ionic strength in cell E19 is 0.100 m.

(b) To use the Debye-Hückel equation, compute activity coefficients with ion size parameters from Table 8-1 or use activity coefficients from Table 8-1 for μ = 0.1 M:

	A	B	C	D	E	F	G	H
8	Activity coefficients from table in textbook:							
9	H$^+$ =	0.83		H$_3$P =	1.00	(fixed at 1)	HP^{2-} =	0.36
10	OH$^-$ =	0.76		H$_2$P$^-$ =	0.78		P^{3-} =	0.10

These activity coefficients produce a pH of 7.403 after executing Solver to reduce the net charge to 0 in cell E18.

13-10. EDTA = H$_4$A from hexaprotic H$_6$A^{2+} with pK_1 = 0.0, pK_2 = 1.5, pK_3 = 2.00, pK_4 = 2.69, pK_5 = 6.13, pK_6 = 10.37; Lysine = HL from triprotic H$_3$L^{2+} with pK_1 = 1.77, pK_2 = 9.07, pK_3 = 10.82

Mixture contains 0.040 mol H$_4$A, 0.030 mol HL, and 0.050 mol NaOH in 1.00 L. Charge balance:
[H$^+$] + 2[H$_6$A^{2+}] + [H$_5$A$^+$] + 2[H$_3$L^{2+}] + [H$_2$L$^+$] + [Na$^+$]
= [OH$^-$] + [H$_3$A$^-$] + 2[H$_2$A^{2-}] + 3[HA^{3-}] + 4[A^{4-}] + [L$^-$]

Substitute fractional composition equations into the charge balance:
[H$^+$] + 2$\alpha_{H_6A^{2+}}$ F$_A$ + $\alpha_{H_5A^+}$ F$_A$ + 2$\alpha_{H_3L^{2+}}$ F$_L$ + $\alpha_{H_2L^+}$ F$_L$ + [0.050]
= K_w/[H$^+$] + $\alpha_{H_3A^-}$ F$_A$ + 2$\alpha_{H_2A^{2-}}$ F$_A$ + 3$\alpha_{HA^{3-}}$ F$_A$ + 4$\alpha_{A^{4-}}$ F$_A$ + α_{L^-} F$_L$

Select Solver and choose Options. In Excel 2016, in All Methods, set Constraint Precision to 1e-15. In the GRG Nonlinear, Derivatives should be Central. Click OK. In Solver Parameters, Set Objective E19 Equal To Value of 0 By Changing Variable Cells H18. Click Solve and Solver finds pH = 4.441 in cell H18, giving a net charge near 0 in cell E19.

	A	B	C	D	E	F	G	H	I
1	Mixture of 0.040 M H_4A, 0.030 M HL, and 0.05 M NaOH								
2									
3	F_A =	0.040		F_L =	0.030		$[Na^+]$ =	0.050	
4	pK_1 =	0.0		pK_{L1} =	1.77				
5	pK_2 =	1.5		pK_{L2} =	9.07				
6	pK_3 =	2.0		pK_{L3} =	10.82		pK_w =	14.00	
7	pK_4 =	2.69		K_{L1} =	1.70E-02		K_w =	1.00E-14	
8	pK_5 =	6.13		K_{L2} =	8.51E-10				
9	pK_6 =	10.37		K_{L3} =	1.51E-11				
10	K_1 =	1.00E+00		K_3 =	1.00E-02		K_5 =	7.41E-07	
11	K_2 =	3.16E-02		K_4 =	2.04E-03		K_6 =	4.27E-11	
12									
13	Species in charge balance:								
14	$[H^+]$ =	3.62E-05		$[HA^{3-}]$ =	7.88E-04		$[OH^-]$ =	2.76E-10	
15	$[H_6A^{2+}]$ =	1.03E-13		$[A^{4-}]$ =	9.28E-10		$[Na^+]$ =	0.050	
16	$[H_5A^+]$ =	2.84E-09		$[H_3L^{2+}]$ =	6.39E-05		$[H_4A]$ =	2.48E-06	
17	$[H_3A^-]$ =	6.84E-04		$[H_2L^+]$ =	2.99E-02		$[HL]$ =	7.03E-07	
18	$[H_2A^{2-}]$ =	3.85E-02		$[L^-]$ =	2.94E-13		pH =	4.441	← initial value
19	Positive charge minus negative charge =				-9.77E-17				is a guess
20	Formulas:								
21	B10 = 10^-B4 with analogous formulas for K_2 - K_6								
22	E7 = 10-E4 with analogous formulas for K_{L2} and K_{L3}								
23	B14 = 10^-H18			H14 = H7/B14			H15 = H3		
24	Denom1 = $B$14^6+$B$14^5*$B$10+$B$14^4*$B$10*$B$11+$B$14^3*$B$10*$B$11*$E$10								
25	+$B$14^2*$B$10*$B$11*$E$10*$E$11+$B$14*$B$10*$B$11*$E$10*$E$11*$H$10								
26	+$B$10*$B$11*$E$10*$E$11*$H$10*$H$11								
27	Denom2 = $B$14^3+$B$14^2*$E$7+$B$14*$E$7*$E$8+$E$7*$E$8*$E$9								
28	B15 = $B$14^6*$B$3/Denom1								
29	B16 = $B$14^5*$B$10*$B$3/Denom1								
30	B17 = $B$14^3*$B$10*$B$11*$E$10*$B$3/Denom1								
31	B18 = $B$14^2*$B$10*$B$11*$E$10*$E$11*$B$3/Denom1								
32	E14 = $B$14*$B$10*$B$11*$E$10*$E$11*$H$10*$B$3/Denom1								
33	E15 = $B$10*$B$11*$E$10*$E$11*$H$10*$H$11*$B$3/Denom1								
34	H16 = $B$14^4*$B$10*$B$11*$B$3/Denom1								
35	E16 = $B$14^3*$E$3/Denom2								
36	E17 = $B$14^2*$E$7*$E$3/Denom2								
37	E18 = $E$7*$E$8*$E$9*$E$3/Denom2								
38	H17 = $B$14*$E$7*$E$8*$E$3/Denom2								
39	E19 = B14+2*B15+B16+2*E16+E17+H15-B17-2*B18-3*E14-4*E15-E18-H14								

Advanced Topics in Equilibrium

13-11. (a) $Fe^{3+} + SCN^- \rightleftharpoons Fe(SCN)^{2+}$ $[Fe(SCN)^{2+}] = \beta_1' [Fe^{3+}][SCN^-]$

$Fe^{3+} + 2SCN^- \rightleftharpoons Fe(SCN)_2^+$ $[Fe(SCN)_2^+] = \beta_2' [Fe^{3+}][SCN^-]^2$

$Fe^{3+} + H_2O \rightleftharpoons FeOH^{2+} + H^+$ $[FeOH^{2+}] = K_a' [Fe^{3+}]/[H^+]$

$H_2O \rightleftharpoons H^+ + OH^-$ $[OH^-] = K_w'/[H^+]$

$\beta_1' = \beta_1 \dfrac{\gamma_{Fe^{3+}} \gamma_{SCN^-}}{\gamma_{Fe(SCN)^{2+}}}$ $\beta_2' = \beta_2 \dfrac{\gamma_{Fe^{3+}} \gamma_{SCN^-}^2}{\gamma_{Fe(SCN)_2^+}}$ $K_a' = K_a \dfrac{\gamma_{Fe^{3+}}}{\gamma_{FeOH^{2+}} \gamma_{H^+}}$

$K_w' = \dfrac{K_w}{\gamma_{H^+} \gamma_{OH^-}}$ $\beta_1 = 10^{3.03}$ $\beta_2 = 10^{4.6}$ $K_a = 10^{-2.195}$

(b) Charge balance: $[OH^-] + [SCN^-] + [NO_3^-] =$
$[H^+] + 3[Fe^{3+}] + 2[Fe(SCN)^{2+}] + [Fe(SCN)_2^+] + 2[FeOH^{2+}] + [Na^+]$

(c) Mass balances:
Total iron $\equiv F_{Fe} = 5.0$ mM $= [Fe^{3+}] + [Fe(SCN)^{2+}] + [Fe(SCN)_2^+] + [FeOH^{2+}]$

Total thiocyanate $\equiv F_{SCN} = 5.0$ μM $= [Fe(SCN)^{2+}] + 2[Fe(SCN)_2^+] + [SCN^-]$

$[Na^+] = 5.0$ μM

$[NO_3^-] = 3(5.0 \text{ mM}) + 15.0 \text{ mM} = 30.0$ mM

(d) The spreadsheet uses Solver to find all concentrations beginning with the guesses pFe = 2.3, pSCN = 5.3, and pH = 1.8 in cells B10:B12. Results are

$[Fe^{3+}] = 4.20$ mM $[SCN^-] = 2.03$ μM $[H^+] = 15.8$ mM

$[Fe(SCN)^{2+}] = 2.97$ μM $[Fe(SCN)_2^+] = 106$ pM $[FeOH^{2+}] = 0.802$ mM

$[OH^-] = 0.920$ pM $\mu = 0.043\ 4$ M

(e) Hydrolysis of Fe(III) produces 0.000 8 M H$^+$ ($Fe^{3+} + H_2O \rightleftharpoons FeOH^{2+} + H^+$)

(f) $\dfrac{[Fe(SCN)^{2+}]}{\{[Fe^{3+}]+[FeOH^{2+}]\}[SCN^-]} = 293$ (The graph in the textbook gives 270.)

	A	B	C	D	E	F	G	H	I	J
1	Iron thiocyanate equilibria with Davies equation for activity coefficients								F_{Fe} =	5.00E-03
2	1. *Estimate* values in cells B10, B11, and B12								F_{SCN} =	5.00E-06
3	2. Use Solver to adjust B10, B11, and B12 to minimize sum in cell J23								$[Na^+]$ =	5.00E-06
4									$[NO_3^-]$ =	3.00E-02
5	Ionic								$[K^+]$ =	0.00E+00
6	strength			μ = 0.5*(D10^2*C10+D11^2*C11+D12^2*C12+D13^2*C13+						
7	μ =	0.04339		D14^2*C14+D15^2*C15+D16^2*C16+J3+J4+J5)						
8					Davies	Activity				
9	Species	pC	C (M)	Charge	log γ	coefficient, γ	F10 = 10^E10		Check:	
10	Fe^{3+}	2.37722	4.195E-03	3	-0.7316	1.855E-01	C10 = 10^-B10		Total Fe =	
11	SCN^-	5.69241	2.030E-06	-1	-0.0813	8.293E-01	C11=10^-B11		5.000E-03	
12	H^+	1.80130	1.580E-02	1	-0.0813	8.293E-01	C12=10^-B12		=C10+C13+C14+C15	
13	$Fe(SCN)^{2+}$		2.969E-06	2	-0.3251	4.730E-01	C13=D19*C10*C11			
14	$Fe(SCN)_2^+$		1.060E-10	1	-0.0813	8.293E-01	C14=D20*C10*C11^2		Total SCN^- =	
15	$FeOH^{2+}$		8.016E-04	2	-0.3251	4.730E-01	C15=D21*C10/C12		5.000E-06	
16	OH^-		9.202E-13	-1	-0.0813	8.293E-01	C16=D22/C12		=C11+C13+2*C14	
17										
18			K' (with activity coefficients)					Mass and charge balances:		$10^6 * b_i$
19	$p\beta_1$ =	-3.03		β_1' =	3.49E+02		$b_1 = F_{Fe} - [Fe^{3+}]-[FeSCN^{2+}]-[Fe(SCN)_2^+]-[FeOH^{2+}]$ =			4.14E-07
20	$p\beta_2$ =	-4.6		β_2' =	6.13E+03		$b_2 = F_{SCN} - [FeSCN^{2+}] - 2[Fe(SCN)_2^+] - [SCN^-]$ =			-2.22E-07
21	pK_a =	2.195		K_a' =	3.02E-03		$b_3 = [OH^-] + [SCN^-] + [NO_3^-] - [H^+] - 3[Fe^{3+}]$			
22	pK_w =	14.00		K_w' =	1.45E-14		$-2[FeSCN^{2+}]-[Fe(SCN)_2^+]-2[FeOH^{2+}]-[Na^+]$ =			3.07E-07
23									$\Sigma(10^6*b_i)^2$ =	3.15E-13
24	Initial values:								J19 = 1e6*(J1-C10-C13-C14-C15)	
25	pFe = 2.3								J20 = 1e6*(J2-C13-2*C14-C11)	
26	pSCN = 5.3						J22 = 1e6*(C16+C11+J4-C12-3*C10-2*C13-C14-2*C15-J3)			
27	pH = 1.8								J23 = J19^2 + J20^2 +J22^2	
28						E10 = -0.51*D10^2*(SQRT($B$7)/(1+SQRT($B$7)) - 0.3*$B$7)				
29									D19 = (10^-B19)*F10*F11/F13	
30									D20 = (10^-B20)*F10*F11^2/F14	
31									D21 = (10^-B21)*F10/(F15*F12)	
32									D22 = (10^-B22)/(F12*F16)	

Spreadsheet for no added KNO_3

(g) With 0.200 M, KNO_3, $[K^+]$ = 0.2 M in cell J5 and $[NO_3^-]$ = 0.23 M in cell J4.

$[K^+]$ has to be added to the charge balance in cell J22. Results are

$[Fe^{3+}]$ = 4.45 mM $\qquad$ $[SCN^-]$ = 2.81 μM $\qquad$ $[H^+]$ = 15.6 mM

$[Fe(SCN)^{2+}]$ = 2.19 μM $\qquad$ $[Fe(SCN)_2^+]$ = 68.2 pM $\qquad$ $[FeOH^{2+}]$ = 0.546 mM

$[OH^-]$ = 1.18 pM $\qquad$ μ = 0.244 M

$$\frac{[Fe(SCN)^{2+}]}{\{[Fe^{3+}]+[FeOH^{2+}]\}[SCN^-]} = 156 \quad \text{(The graph in the textbook gives 150.)}$$

Advanced Topics in Equilibrium

	A	B	C	D	E	F	G	H	I	J
1	Iron thiocyanate equilibria with Davies equation for activity coefficients								F_{Fe} =	5.00E-03
2	1. *Estimate* values in cells B10, B11, and B12								F_{SCN} =	5.00E-06
3	2. Use Solver to adjust B10, B11, and B12 to minimize sum in cell J23								$[Na^+]$ =	5.00E-06
4									$[NO_3^-]$ =	2.30E-01
5	Ionic								$[K^+]$ =	2.00E-01
6	strength		μ = 0.5*(D10^2*C10+D11^2*C11+D12^2*C12+D13^2*C13+							
7	μ =	0.24391	D14^2*C14+D15^2*C15+D16^2*C16+J3+J4+J5)							
8						Davies	Activity			
9	Species	pC	C (M)	Charge	log γ	coefficient, γ	F10 = 10^E10		Check:	
10	Fe^{3+}	2.35142	4.452E-03	3	-1.1816	6.583E-02	C10=10^-B10		Total Fe =	
11	SCN^-	5.55092	2.812E-06	-1	-0.1313	7.391E-01	C11=10^-B11		5.000E-03	
12	H^+	1.80839	1.555E-02	1	-0.1313	7.391E-01	C12=10^-B12		=C10+C13+C14+C15	
13	$Fe(SCN)^{2+}$		2.187E-06	2	-0.5251	2.984E-01	C13=D19*C10*C11			
14	$Fe(SCN)_2^+$		6.821E-11	1	-0.1313	7.391E-01	C14=D20*C10*C11^2		Total SCN^- =	
15	$FeOH^{2+}$		5.455E-04	2	-0.5251	2.984E-01	C15=D21*C10/C12		5.000E-06	
16	OH^-		1.178E-12	-1	-0.1313	7.391E-01	C16=D22/C12		=C11+C13+2*C14	
17										
18			K' (with activity coefficients)				Mass and charge balances:			$10^6 \cdot b_i$
19	$p\beta_1$ =	-3.03	β_1' =	1.75E+02		$b_1 = F_{Fe} - [Fe^{3+}] - [FeSCN^{2+}] - [Fe(SCN)_2^+] - [FeOH^{2+}]$				7.25E-08
20	$p\beta_2$ =	-4.6	β_2' =	1.94E+03		$b_2 = F_{SCN} - [FeSCN^{2+}] - 2[Fe(SCN)_2^+] - [SCN^-]$				-3.90E-07
21	pK_a =	2.195	K_a' =	1.90E-03		charge: $b_3 = [OH^-] + [SCN^-] + [NO_3^-] - [H^+] - 3[Fe^{3+}]$				
22	pK_w =	14.00	K_w' =	1.83E-14		$-2[FeSCN^{2+}] - [Fe(SCN)_2^+] - 2[FeOH^{2+}] - [Na^+] - [K^+]$				2.28E-07
23									Σb_i^2 =	2.09E-13
24	Initial values:								J19 = 1e6*(J1-C10-C13-C14-C15)	
25	pFe = 2.3								J20 = 1e6*(J2-C13-2*C14-C11)	
26	pSCN = 5.3						J22 = 1e6*(C16+C11+J4-C12-3*C10-2*C13-C14-2*C15-J3-J5)			
27	pH = 1.8								J23 = J19^2 + J20^2 + J22^2	
28					E10 = -0.51*D10^2*(SQRT($B$7)/(1+SQRT($B$7)) - 0.3*$B$7)					
29									D19 = (10^-B19)*F10*F11/F13	
30									D20 = (10^-B20)*F10*F11^2/F14	
31									D21 = (10^-B21)*F10/(F15*F12)	
32									D22 = (10^-B22)/(F12*F16)	

Spreadsheet for 0.20 M KNO_3

13-12.
$$Fe^{2+} + G^- \rightleftharpoons FeG^+ \qquad [FeG^+] = \beta_1'[Fe^{2+}][G^-]$$

$$Fe^{2+} + 2G^- \rightleftharpoons FeG_2(aq) \qquad [FeG_2(aq)] = \beta_2'[Fe^{2+}][G^-]^2$$

$$Fe^{2+} + 3G^- \rightleftharpoons FeG_3^- \qquad [FeG_3^-] = \beta_3'[Fe^{2+}][G^-]^3$$

$$Fe^{2+} + H_2O \rightleftharpoons FeOH^+ + H^+ \qquad [FeOH^+] = K_a'[Fe^{2+}]/[H^+]$$

$$HG \rightleftharpoons G^- + H^+ \qquad [HG] = [G^-][H^+]/K_2'$$

$$H_2G^+ \rightleftharpoons HG + H^+ \qquad [H_2G^+] = [HG][H^+]/K_1'$$

$$H_2O \rightleftharpoons H^+ + OH^- \qquad [OH^-] = K_w'/[H^+]$$

$$\beta_1' = \beta_1 \frac{\gamma_{Fe^{2+}}\gamma_{G^-}}{\gamma_{FeG^+}} \qquad \beta_2' = \beta_2 \frac{\gamma_{Fe^{2+}}\gamma_{G^-}^2}{\gamma_{FeG_2}} \qquad \beta_3' = \beta_3 \frac{\gamma_{Fe^{2+}}\gamma_{G^-}^3}{\gamma_{FeG_3^-}}$$

$$K_a' = K_a \frac{\gamma_{Fe^{2+}}}{\gamma_{H^+}} \qquad K_2' = K_2 \frac{\gamma_{HG}}{\gamma_{G^-}\gamma_{H^+}} \qquad K_1' = K_1 \frac{\gamma_{H_2G^+}}{\gamma_{HG}\gamma_{H^+}} \qquad K_w' = \frac{K_w}{\gamma_{H^+}\gamma_{OH^-}}$$

$\beta_1 = 10^{4.31}$ $\qquad\qquad$ $\beta_2 = 10^{7.65}$ $\qquad\qquad$ $\beta_3 = 10^{8.87}$

$K_a = 10^{-9.4}$ $\qquad\qquad$ $K_1 = 10^{-2.350}$ $\qquad\qquad$ $K_2 = 10^{-9.778}$

Charge balance:

$$[OH^-] + [G^-] + [FeG_3^-] + [Cl^-] = [H^+] + 2[Fe^{2+}] + [FeOH^+] + [FeG^+] + [H_2G^+] \quad (A)$$

Mass balances:

$$F_{Fe} = 0.050\ M = [Fe^{2+}] + [FeG^+] + [FeG_2] + [FeG_3^-] + [FeOH^+] \quad (B)$$

$$F_G = 0.100\ M = [G^-] + [HG] + [H_2G^+] + [FeG^+] + 2[FeG_2] + 3[FeG_3^-] \quad (C)$$

There are 11 concentrations to be found and there are 7 equilibria. In the absence of other information, we would need to solve for 4 independent concentrations. However, $[H^+]$ is known from the pH, so we need to find 3 independent variables. The amount of HCl added to the solution to achieve the pH is not known, so $[Cl^-]$ is one of the independent variables to be found by Solver. For the other two independent variables, select $[Fe^{2+}]$ and $[G^-]$ because all other concentrations can be expressed in terms of $[Fe^{2+}]$, $[G^-]$, and $[H^+]$. In the spreadsheet, we tried a few different initial values of pFe, pG, and pCl to find the set pFe = 3, pG = 3, and pCl = 2 that roughly satisfied the mass and charge balances. With these initial values, Solver was able to solve the problem in a single iteration.

Fractions of iron and glycine in each form are computed from concentrations in the spreadsheet:

Fraction of iron		Fraction of glycine	
Fe^{2+}	3.49%	G^-	0.95%
FeG^+	37.40%	FeG^+	18.70%
FeG_2	57.95%	FeG_2	57.95%
FeG_3^-	0.92%	FeG_3^-	1.38%
$FeOH^+$	0.24%	HG	21.02%
		H_2G^+	0.00%

Advanced Topics in Equilibrium

The chemistry: We dissolved FeG_2 and found that the principal species are FeG^+, FeG_2, and HG. The chemistry that requires 20.9 mmol HCl to be added to 50 mmol FeG_2 to obtain pH 8.5 is $FeG_2 \rightleftharpoons FeG^+ + G^-$ followed by $G^- + H^+ \rightleftharpoons HG$. The base G^- released when FeG_2 dissolves requires HCl for neutralization.

	A	B	C	D	E	F	G	H	I
1	Fe(II)-glycine equilibria with fixed pH							pH =	8.50
2	1. *Estimate* values in cells B8, B9, B18							F_{Fe} =	0.050
3	2. Use Solver to adjust B8,B9,B18 to minimize sum in cell I24							F_G =	0.100
4	Ionic strength		μ=0.5*(D8^2*C8+D9^2*C9+D10^2*C10+D11^2*C11+D12^2*C12+D13^2*C13						
5	μ =	0.024051	+D14^2*C14+D15^2*C15+D16^2*C16+D17^2*C17+D18^2*C18)						
6					Davies	Activity			check mass
7	Species	pC	C (M)	Charge	log γ	coefficient, γ			balances:
8	Fe^{2+}	2.758698	1.74E-03	2	-0.2592	5.506E-01	C8 = 10^-B8		Fe_{total} =
9	G^-	3.020251	9.54E-04	-1	-0.0648	8.614E-01	C9 = 10^-B9		5.00E-02
10	H^+	$10^{-8.50}$	3.67E-09	1	-0.0648	8.614E-01	C10 = 10^-I1/F10		G_{total} =
11	FeG^+		1.87E-02	1	-0.0648	8.614E-01	C11 = D20*C8*C9		1.00E-01
12	FeG_2		2.90E-02	0	0.0000	1.000E+00	C12 = D21*C8*C9^2		
13	FeG_3^-		4.59E-04	-1	-0.0648	8.614E-01	C13 = D22*C8*C9^3		
14	$FeOH^+$		1.21E-04	1	-0.0648	8.614E-01	C14 = D23*C8/C10		
15	HG		2.10E-02	0	0.0000	1.000E+00	C15 = C9*C10/D25		
16	H_2G^+		1.28E-08	1	-0.0648	8.614E-01	C16 = C15*C10/D24		
17	OH^-	known	3.67E-06	-1	-0.0648	8.614E-01	C17 = D26/C10		
18	Cl^-	1.680037	2.09E-02	-1	-0.0648	8.614E-01	C18 = 10^-B18		
19			K' (with activity coefficients)				Mass and charge balances:		$10^6 * b_i$
20	$p\beta_1$ =	-4.31	β_1' =	1.12E+04		$b_1 = F_{Fe} - [Fe^{2+}] - [FeG^+] - [FeG_2] - [FeG_3^-] - [FeOH^+]$			-1.52E-06
21	$p\beta_2$ =	-7.65	β_2' =	1.82E+07		$b_2 = F_G - [G^-] - [HG] - [H_2G^+] - [FeG^+] - 2[FeG_2] - 3[FeG_3^-]$			1.30E-06
22	$p\beta_3$ =	-8.9	β_3' =	3.03E+08		$b_3 = [OH^-] + [G^-] + [FeG_3^-] + [Cl^-]$			
23	pK_a =	9.40	K_a' =	2.54E-10		$- [H^+] - 2[Fe^{2+}] - [FeOH^+] - [FeG^+] - [H_2G^+]$ =			-1.30E-05
24	pK_1 =	2.35	K_1' =	6.02E-03				Σb_i^2 =	1.73E-10
25	pK_2 =	9.78	K_2' =	1.67E-10		I20 = 1e6*(I2-C8-C11-C12-C13-C14)			
26	pK_w =	14.00	K_w' =	1.35E-14		I21 = 1e6*(I3-C9-C15-C16-C11-2*C12-3*C13)			
27						I23 = 1e6*(C17+C9+C13+C18-C10-2*C8-C14-C11-C16)			
28	Initial values:					I24 = I20^2 + I21^2 + I23^2			
29	pFe = 3	pG = 3	pCl = 2			E8 = -0.51*D8^2*(SQRT($B$5)/(1+SQRT($B$5)) - 0.3*$B$5)			
30						D20 = 10^-B20*F8*F9/F11			
31						D21 = 10^-B21*F8*F9^2/F12			
32						D22 = 10^-B22*F8*F9^3/F13			
33						D23 = 10^-B23*F8/F10			
34	Optimize pFe, pG, and pCl together					D24 = 10^-B24*F15/(F9*F10)			
35	Then optimize one or two p values while holding other(s) constant					D25 = 10^-B25*F16/(F15*F10)			
36	Continue optimization as long as Σb_i^2 keeps getting smaller					D26 = 10^-B26/(F10*F17)			

13-13. (a) A range of initial values of pK_1 and pK_2, such as 6 and 6 or 10 and 10, converge to the correct solution. Even choosing a ridiculous value for pK'_w, such as 10, converges to the correct value after executing Solver more than once.

(b) Fixing pK'_w at 13.797 and using Solver gives the optimized values of pK_1 and $pK_2 = 2.312$ and 9.630, which are hardly different from the values obtained when pK'_w is allowed to vary. However, inspection of the following curves shows that $\bar{n}_H$ (measured) deviates systematically from $\bar{n}_H$ (theoretical) at the end of the titration when $\bar{n}_H$ (measured) should approach 0.

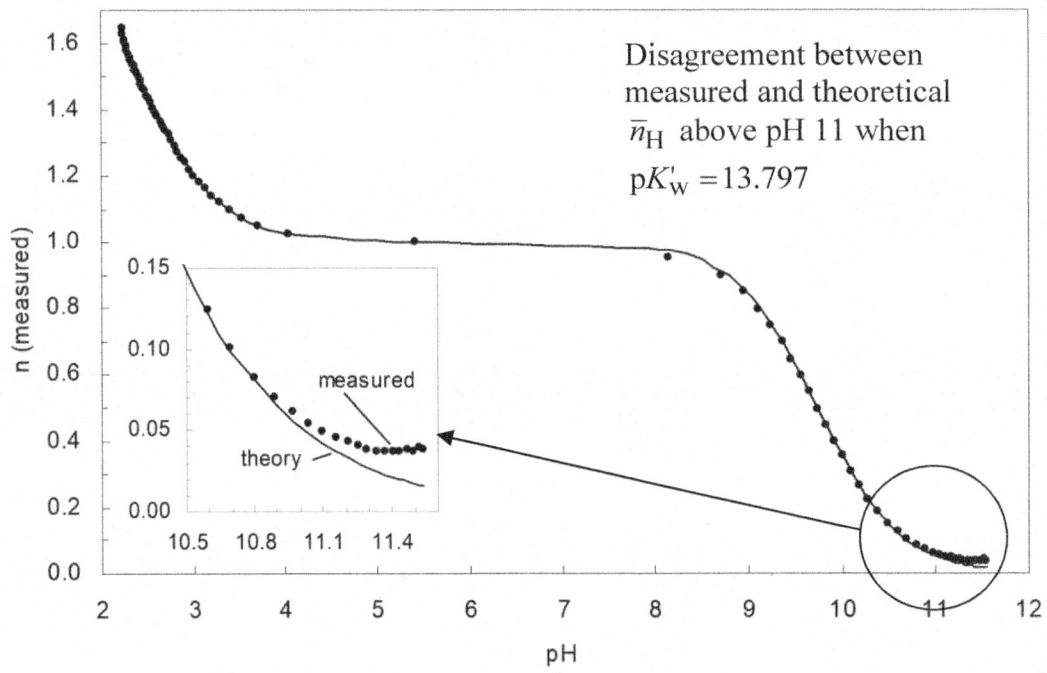

CHAPTER 14
FUNDAMENTALS OF ELECTROCHEMISTRY

14-1. Electric charge (coulombs) refers to the quantity of positive or negative particles. Current (amperes) is the quantity of charge moving past a point in a circuit each second. The electric potential difference (volts) between two points measures the work that can be done by (or must be done to) each coulomb of charge as it moves from one point to another.

14-2. (a) $1/e = 1/(1.602\,176\,634 \times 10^{-19}\text{ C/e}^-) = 6.241\,509\,074 \times 10^{18}\text{ e}^-/\text{C}$

(b) $F = N_A \cdot e = (6.022\,140\,76 \times 10^{23}\text{ mol}^{-1})(1.602\,176\,634 \times 10^{-19}\text{ C})$
$= 96\,485.332\,12\text{ C/mol}$

14-3. (a) $I = \dfrac{6.00\text{ V}}{2.0 \times 10^3\text{ W}} = 3.00\text{ mA} = 3.00 \times 10^{-3}\text{ C/s}$

$\left(\dfrac{3.00 \times 10^{-3}\text{ C/s}}{9.649 \times 10^4\text{ C/mol}}\right)(6.022 \times 10^{23}\text{ e}^-/\text{mole}) = 1.87 \times 10^{16}\text{ e}^-/\text{s}$

(b) $P = E \cdot I = (6.00\text{ V})(3.00 \times 10^{-3}\text{ A}) = 1.80 \times 10^{-2}\text{ W}$

The power appears as heat in the resistor.

Heat per electron $= \dfrac{1.80 \times 10^{-2}\text{ J/s}}{1.87 \times 10^{16}\text{ e}^-/\text{s}} = 9.63 \times 10^{-19}\text{ J/e}^-$

(c) $(1.87 \times 10^{16}\text{ e}^-/\text{s})(1\,800\text{ s}) = 3.37 \times 10^{19}\text{ electrons} = 5.60 \times 10^{-5}\text{ mol}$

(d) $P = EI = E(E/R) = E^2/R \Rightarrow E = \sqrt{PR} = \sqrt{(100\text{ W})(2.00 \times 10^3\text{ W})}$
$= 447\text{ V}$

14-4. (a) I = coulombs/s. Every mol of O_2 accepts 4 mol of e^-. 16 mol O_2/day
= 64 mol e^-/day = 7.41×10^{-4} mol e^-/s = 71.5 C/s = 71.5 A

(b) I = power/E = 500 W/115 V = 4.35 A. The resting human uses 16 times as much current as the refrigerator.

(c) Power = $E \cdot I$ = (1.1 V)(71.5 A) = 79 W

14-5. (a) $I_2 + 2e^- \rightleftharpoons 2I^-$ (I_2 is the oxidant)

(b) $2S_2O_3^{2-} \rightleftharpoons S_4O_6^{2-} + 2e^-$ ($S_2O_3^{2-}$ is the reductant)

(c) 1.00 g $S_2O_3^{2-}$ / (112.12 g/mol) = 8.92 mmol $S_2O_3^{2-}$ = 8.92 mmol e^-
$(8.92 \times 10^{-3}\text{ mol})(9.649 \times 10^4\text{ C/mol}) = 861\text{ C}$

(d) Current (A) = coulombs/s = 861 C/60 s = 14.3 A

14-6. (a) Oxidation numbers of reactants: N (in NH_4^+) Cl (in ClO_4^-) Al
  -3 $+7$ 0

Oxidation numbers of products: N (in N_2) Cl (in HCl) Al (in Al_2O_3)
  0 -1 $+3$

NH_4^+ and Al are reducing agents and ClO_4^- is the oxidizing agent.

(b) Formula mass of reactants = 6(FM NH_4ClO_4) + 10(FM Al) = 974.7

Heat released per gram = 9 334 kJ/974.7 g = 9.576 kJ/g

14-7. In a galvanic cell, two half-reactions are physically separated from each other. At the anode, oxidation generates electrons that can flow through the electric circuit to reach the cathode, where a reduction occurs. The favorable free energy change for the net reaction provides the driving force for electrons to flow through the circuit. There must be a connector (such as a salt bridge) between the anode and cathode compartments to allow ions to flow to maintain electroneutrality.

14-8. (a) Fe(s) | FeO(s) | KOH(aq) | Ag_2O(s) | Ag(s)

$FeO(s) + H_2O + 2e^- \rightleftharpoons Fe(s) + 2OH^-$

$Ag_2O + H_2O + 2e^- \rightleftharpoons 2Ag(s) + 2OH^-$

(b) Pb(s) | $PbSO_4$(s) | K_2SO_4(aq) || H_2SO_4(aq) | $PbSO_4$(s) | PbO_2(s) | Pb(s)

$PbSO_4(s) + 2e^- \rightleftharpoons Pb(s) + SO_4^{2-}$

$PbO_2 + 4H^+ + SO_4^{2-} + 2e^- \rightleftharpoons PbSO_4(s) + 2H_2O$

14-9. $Fe^{3+} + e^- \rightleftharpoons Fe^{2+}$

$Cr_2O_7^{2-} + 14H^+ + 6e^- \rightleftharpoons 2Cr^{3+} + 7H_2O$

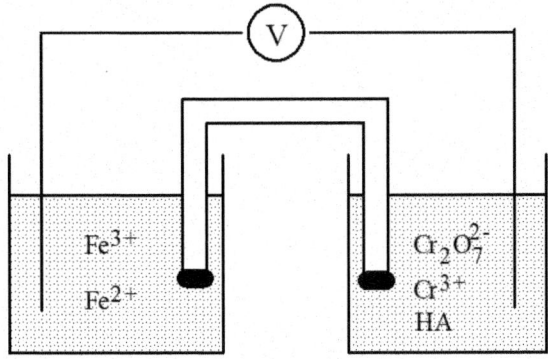

Fundamentals of Electrochemistry 185

14-10. (a) $Zn^{2+} + 2e^- \rightleftharpoons Zn(s)$ $E° = -0.762$ V

$Cl_2(l) + 2e^- \rightleftharpoons 2Cl^-$ $E° \approx 1.4$ V

The Appendix lists standard reduction potentials for $Cl_2(g)$ and $Cl_2(aq)$, but not for $Cl_2(l)$. Both listed potentials are close to 1.4 V, so the potential for $Cl_2(l)$ is probably also close to 1.4 V. Electrons flow from the more negative electrode (Zn) through the circuit to the more positive electrode (C).

(b) One mol of Cl_2 requires 2 mol of e^-.

Moles of Cl_2 consumed in 1.00 h = $\frac{1}{2}$(mol of e^-/hr) =

$\left[\frac{1}{2}\left(1.00 \times 10^3 \frac{C}{s}\right) / (9.64 \times 10^4 \text{ C/mol})\right](3\,600 \text{ s/h}) = 18.7 \text{ mol } Cl_2 = 1.32 \text{ kg}.$

14-11. (a) anode: $C_6Li \rightleftharpoons C_6 + Li^+ + e^-$

cathode: $2Li_{0.5}CoO_2 + Li^+ + e^- \rightleftharpoons 2LiCoO_2$

Formula mass of reactants = 188.80 g ($2Li_{0.5}CoO_2$) + 79.01 g (LiC_6) = 267.81 g

(b) $1 \text{ A} = 1 \frac{C}{s}$. $1 \text{ h} = 3\,600 \text{ s}$; $1 \text{ A·h} = \left(1 \frac{C}{s}\right)(3\,600 \text{ s}) = 3\,600 \text{ C}$

(3 600 C)/(96 485 C/mol) = 0.037 311 mol e^- in 1 A·h

(c) 0.267 81 kg of reactants ($2Li_{0.5}CoO_2 + LiC_6$) provides 1 mol e^-.

(1 mol e^-)/(0.267 81 kg) = 3.734 0 mol e^-/kg

$\left(3.7340 \frac{\text{mol } e^-}{\text{kg}}\right) \div \left(0.037\,311 \frac{\text{mol } e^-}{\text{A·h}}\right) = 100.1 \text{ A·h/kg}$

(d) Power (W) = $E \cdot I$ = volts × amperes.   (3.7 V)(100 A·h/kg) = 370 W·h/kg

14-12. (a) Energy in battery = 3.6×10^9 J = $E \cdot q$ = (1.0 V)(q coulombs)
$\Rightarrow q = 3.6 \times 10^9$ C.

(b) A molecule of 2,6-dihydroxyanthraquinone stores 2e^- in the reaction
$Q^{2-} + 2e^- \rightarrow Q^{4-}$

0.5 M 2,6-dihydroxyanthraquinone stores (1.0 mol e^-/L)(96 485 C/mol) = 9.65×10^4 C/L

Volume require to store 3.6×10^9 C = (3.6×10^9 C)/(9.65×10^4 C/L) = 3.7×10^4 L.

(c) (3.7×10^4 L)/(200 L/drum) = 187 55-gallon drums of liquid.

(d) 0.4 M $K_4Fe(CN)_6$ provides (0.4 mol e^-/L)(96 485 C/mol) = 3.86×10^4 C/L

Volume require to provide 3.6×10^9 C = $(3.6 \times 10^9$ C$)/(3.86 \times 10^5$ C/L$)$ = 9.3×10^4 L or 466 55-gallon drums of liquid.

Total drums to store 1 MW·h of energy = 187 + 466 ≈ 650 drums.

14-13. Cl_2 is strongest because it has the most positive reduction potential.

14-14. (a) Since it becomes harder to reduce Fe(III) to Fe(II) in the presence of CN^-, Fe(III) is stabilized more than Fe(II).

(b) Since it becomes easier to reduce Fe(III) to Fe(II) in the presence of phenanthroline, Fe(II) is stabilized more than Fe(III) by phenanthroline.

14-15. $E°$ applies when activities of reactants and products are unity. E applies to whatever activities exist. At equilibrium, E goes to zero. $E°$ is constant.

14-16. (a) $Zn(s) | Zn^{2+}(0.1 \text{ M}) \| Cu^{2+}(0.1 \text{ M}) | Cu(s)$

right half-cell: $Cu^{2+} + 2e^- \rightleftharpoons Cu(s)$ $E°_+ = 0.339$ V

left half-cell: $Zn^{2+} + 2e^- \rightleftharpoons Zn(s)$ $E°_- = -0.762$ V

$$E = \left(0.339 - \frac{0.059\ 16}{2} \log \frac{1}{0.1}\right) - \left(-0.762 - \frac{0.059\ 16}{2} \log \frac{1}{0.1}\right) = 1.101 \text{ V}$$

The Cu electrode is more positive than the Zn electrode. Electrons move from Zn to Cu. The net reaction is $Cu^{2+} + Zn(s) \rightleftharpoons Cu(s) + Zn^{2+}$.

(b) Since Cu^{2+} ions are consumed in the right half-cell, Zn^{2+} ions must migrate from the left half-cell into the salt bridge to help balance charge. I hope you like Zn^{2+}, because that is what your body will take up.

14-17. $E = -0.238 - \dfrac{0.059\ 16}{3} \log \dfrac{P_{AsH_3}}{[H^+]^3}$

$E = -0.238 - \dfrac{0.059\ 16}{3} \log \dfrac{1.0 \times 10^{-3}}{(10^{-3.00})^3} = -0.356$ V

14-18. (a) $Pt(s) | Br_2(l) | HBr(aq, 0.10 \text{ M}) \| Al(NO_3)_3(aq, 0.010 \text{ M}) | Al(s)$

(b) right half-cell: $Al^{3+} + 3e^- \rightleftharpoons Al(s)$ $E°_+ = -1.677$ V

left half-cell: $Br_2(l) + 2e^- \rightleftharpoons 2Br^-$ $E°_- = 1.078$ V

right half-cell: $E_+ = \left(-1.677 - \dfrac{0.059\ 16}{3} \log \dfrac{1}{[0.010]}\right) = -1.716_4$ V

Fundamentals of Electrochemistry 187

left half-cell: $E_- = \left(1.078 - \dfrac{0.059\,16}{2}\log[0.10]^2\right) = 1.137_2$ V

$E = E_+ - E_- = -1.716_4 - 1.137_2 = -2.854$ V. The Pt electrode is more positive, so electrons flow from Al to Pt. Br_2 is reduced at the Pt electrode:

$$\tfrac{3}{2}Br_2(l) + Al(s) \rightleftharpoons 3Br^- + Al^{3+}$$

The species reacting at Pt is $Br_2(aq)$, which is in equilibrium with $Br_2(l)$.

(c) 14.3 mL of Br_2 = 44.6 g = 0.279 mol of Br_2. 12.0 g of Al = 0.445 mol of Al. The reaction requires 3/2 mol of Br_2 for every mol of Al. The Br_2 will be used up first.

(d) 0.231 mL of Br_2 = 0.721 g of Br_2 = 4.51×10^{-3} mol Br_2 = 9.02×10^{-3} mol e^- = 870 C. Work = $E \cdot q$ = (1.50 V)(870 C) = 1.31 kJ.

(e) $I = \sqrt{P/R} = \sqrt{(1.00\times 10^{-4}\text{ J/s})/(1.20\times 10^{3}\text{ }\Omega)} = 2.89 \times 10^{-4}$ A
 $= 2.99 \times 10^{-9}$ mol e^-/s = 9.97×10^{-10} mol Al/s = 2.69×10^{-8} g/s

14-19. The activities of the solid reagents do not change until they are used up. The only aqueous species, OH^-, is created at the cathode and consumed in equal amounts at the anode, so its concentration remains constant in the cell. Therefore, none of the activities change during the life cycle of the cell until something is used up.

14-20. (a) $\tfrac{1}{2}O_2(g) + 2H^+ + 2e^- \rightleftharpoons H_2O(l)$ $E° = 1.229$ V at 25°C

$$E_{O_2} = \left(1.23 - \dfrac{0.059\,16}{2}\log\dfrac{1}{P_{O_2}^{1/2}[H^+]^2}\right)$$

$$= \left(1.23 - \dfrac{0.059\,16}{2}\log\dfrac{1}{(0.2)^{1/2}[0.5]^2}\right) = 1.201 \text{ V}$$

$2H^+ + 2e^- \rightleftharpoons H_2(g)$ $E° = 0$

$$E_{H_2} = \left(0 - \dfrac{0.059\,16}{2}\log\dfrac{P_{H_2}}{[H^+]^2}\right)$$

$$= \left(0 - \dfrac{0.059\,16}{2}\log\dfrac{(1.0)}{[0.5]^2}\right) = -0.018 \text{ V}$$

$E = E_{O_2} - E_{H_2} = 1.219$ V

(b) Power is the product of current × voltage ($P = IE$).
Current = $I = P/E$ = 20 kW/220 V = 90.9 A.

e^- flow = (90.9 C/s)/(96 485 C/mol) = 0.942 mmol/s

For each mol $H_2(g)$ consumed at the cathode, $2e^-$ are produced.

Therefore, $H_2(g)$ consumed if cell is 100% efficient = ½ rate of e^- production

= (0.942 mmol/s)/2 = 0.471 mmol/s. If the cell is only 70% efficient, the consumption of $H_2(g)$ is (0.471 mmol/s)/0.70 = 0.673 mmol/s.

$H_2(g)$ consumption in 1 h = (0.673 mmol/s)(3 600 s/h) = 2.42 mol/h

= 4.88 g/h

(c) $(20\ 000\ \text{W}) \left(\dfrac{1\ \text{horsepower}}{745.7\ \text{W}}\right)$ = 26.8 horsepower

14-21. (a) right half-cell: $E_+ = \left(0.222 - \dfrac{0.059\ 16}{2} \log[\text{Cl}^-]^2\right) = 0.281_2$ V

left half-cell: $E_- = \left(-0.350 - \dfrac{0.059\ 16}{2} \log[\text{F}^-]^2\right) = -0.290_8$ V

$E = E_+ - E_- = 0.281_2 - (-0.290_8) = 0.572$ V

(b) Electrons flow from the left half-cell ($E = -0.290_8$ V) to the right half-cell ($E = 0.281_2$ V).

(c) $[\text{Pb}^{2+}] = K_{sp}$ (for PbF_2)/$[\text{F}^-]^2 = (3.6 \times 10^{-8})/(0.10)^2 = 3.6 \times 10^{-6}$ M

$[\text{Ag}^+] = K_{sp}$ (for AgCl)/$[\text{Cl}^-] = (1.8 \times 10^{-10})/(0.10) = 1.8 \times 10^{-9}$ M

right half-cell: $E_+ = \left(0.799 - \dfrac{0.059\ 16}{2} \log \dfrac{1}{[\text{Ag}^+]^2}\right) = 0.281_2$ V

left half-cell: $E_- = \left(-0.126 - \dfrac{0.059\ 16}{2} \log \dfrac{1}{[\text{Pb}^{2+}]}\right) = -0.287_0$ V

$E = E_+ - E_- = 0.281_2 - (-0.287_0) = 0.568$ V

The agreement between the two calculations is reasonable.

14-22. A hydrogen pressure of 727.2 Torr corresponds to (727.2 Torr)/(760 Torr/atm) = 0.956 8 atm. (0.956 8 atm)(1.013 25 bar/atm) = 0.969 5 bar.

$0.798\ 3 = E°_{\text{Ag}^+|\text{Ag}} - 0.059\ 16 \log \dfrac{[0.010\ 00](0.914)}{(0.969\ 5)^{1/2}[0.010\ 00](0.898)}$

$\Rightarrow E°_{\text{Ag}^+|\text{Ag}} = 0.799_2$ V

14-23. Balanced reaction: $HOBr + H^+ + 2e^- \rightleftharpoons Br^- + H_2O$

$HOBr + H^+ + e^- \rightarrow \frac{1}{2}Br_2(aq) + H_2O \qquad \Delta G_1° = -1F(1.564)$

$\frac{1}{2}Br_2(aq) + e^- \rightarrow Br^- \qquad \Delta G_2° = -1F(1.098)$

$\overline{HOBr + H^+ + 2e^- \rightarrow Br^- + H_2O} \qquad \Delta G_3° = \Delta G_1° + \Delta G_2° = -2FE_3°$

$E_3° = \dfrac{-1F(1.564) - 1F(1.098)}{-2F} = 1.331 \text{ V}$

14-24.
$2X^+ + 2e^- \rightleftharpoons 2X(s) \qquad E_+° = E_2°$

$-\ X^{3+} + 2e^- \rightleftharpoons X^+ \qquad E_-° = E_1°$

$\overline{3X^+ \rightleftharpoons X^{3+} + 2X(s)} \qquad E_3° = E_2° - E_1°$

If $E_2° > E_1°$, then $E_3° > 0$ and disproportionation is spontaneous.

14-25. right half-cell: $Cu^{2+} + 2e^- \rightleftharpoons Cu(s) \qquad E_+° = 0.339 \text{ V}$

left half-cell: $Ni^{2+} + 2e^- \rightleftharpoons Ni(s) \qquad E_-° = -0.236 \text{ V}$

The ionic strength of the right half-cell is 0.009 0 M, and the ionic strength of the left half-cell is 0.008 0 M. At $\mu = 0.009\,0$ M, $\gamma_{Cu^{2+}} = 0.690$.
At $\mu = 0.008\,0$ M, $\gamma_{Ni^{2+}} = 0.705$.

$E_+ = E_+° - \dfrac{0.059\,16}{2} \log \dfrac{1}{[Cu^{2+}]\gamma_{Cu^{2+}}}$

$= 0.339 - \dfrac{0.059\,16}{2} \log \dfrac{1}{(0.003\,0)(0.690)} = 0.259_6 \text{ V}$

$E_- = E_-° - \dfrac{0.059\,16}{2} \log \dfrac{1}{[Ni^{2+}]\gamma_{Ni^{2+}}}$

$= -0.236 - \dfrac{0.059\,16}{2} \log \dfrac{1}{(0.002\,0)(0.705)} = -0.320_3 \text{ V}$

$E = E_+ - E_- = 0.580 \text{ V}$

Electrons flow from Ni ($E = -0.320_3$ V) to Cu ($E = 0.259_6$ V).

14-26. (a) $Pb(s) | PbSO_4(s) | H_2SO_4(aq) | PbSO_4(s) | PbO_2(s) | Pb(s)$

(b) Cathode: $PbO_2(s) + SO_4^{2-} + 4H^+ + 2e^- \rightleftharpoons PbSO_4(s) + 2H_2O \quad E° = 1.685$ V

Anode: $PbSO_4(s) + 2e^- \rightleftharpoons Pb(s) + SO_4^{2-} \quad E° = -0.355$ V

Net reaction: $Pb(s) + PbO_2(s) + 2SO_4^{2-} + 4H^+ \rightleftharpoons 2PbSO_4(s) + 2H_2O$

$E° = 1.685 - (-0.355) = 2.040$ V

(c) $E_{cathode} = 1.685 - \dfrac{0.05916}{2} \log \dfrac{m_{H_2O}^2 \gamma_{H_2O}^2}{m_{SO_4^{2-}} \gamma_{SO_4^{2-}} m_{H^+}^4 \gamma_{H^+}^4}$

$E_{anode} = -0.355 - \dfrac{0.05916}{2} \log m_{SO_4^{2-}} \gamma_{SO_4^{2-}}$

$E_{net\ reaction} = E_{cathode} - E_{anode} =$

$[1.685 - (-0.355)] - \left(\dfrac{0.05916}{2} \log \dfrac{m_{H_2O}^2 \gamma_{H_2O}^2}{m_{SO_4^{2-}} \gamma_{SO_4^{2-}} m_{H^+}^4 \gamma_{H^+}^4} \right)$

$- \left(-\dfrac{0.05916}{2} \log m_{SO_4^{2-}} \gamma_{SO_4^{2-}} \right)$

(d) $E_{net\ reaction} = 2.040 - \dfrac{0.05916}{2} \log \dfrac{\mathcal{A}_{H_2O}^2}{m_{H^+}^4 m_{SO_4^{2-}}^2 \gamma_{H^+}^4 \gamma_{SO_4^{2-}}^2}$

(e) $\left(\gamma_{H^+}^2 \gamma_{SO_4^{2-}} \right) = \gamma_\pm^3$ and $\left(\gamma_{H^+}^4 \gamma_{SO_4^{2-}}^2 \right) = \gamma_\pm^6$, so

$E_{net\ reaction} = 2.040 - \dfrac{0.05916}{2} \log \dfrac{\mathcal{A}_{H_2O}^2}{m_{H^+}^4 m_{SO_4^{2-}}^2 \gamma_\pm^6}$

$E_{net\ reaction} = 2.040 - \dfrac{0.05916}{2} \log \dfrac{(0.66)^2}{(11.0)^4 (5.5)^2 (0.22)^6} = 2.101$ V

14-27. (a) $E° = \dfrac{-\Delta G°}{nF} = \dfrac{(+257 \times 10^3 \text{ J/mol})}{(2)(9.6485 \times 10^4 \text{ C/mol})} = 1.33_2$ V

(b) $K = 10^{nE°/0.05916} = 10^{(2)(1.33_2 \text{ V})/(0.05916 \text{ V})} = 1 \times 10^{45}$

Fundamentals of Electrochemistry

14-28. (a) $4[Co^{3+} + e^- \rightleftharpoons Co^{2+}]$ $\qquad E_+^\circ = 1.92$ V

$\underline{-2[\tfrac{1}{2}O_2 + 2H^+ + 2e^- \rightleftharpoons H_2O]\qquad E_-^\circ = 1.229\text{ V}}$

$4Co^{3+} + 2H_2O \rightleftharpoons 4Co^{2+} + O_2 + 4H^+ \qquad E^\circ = 0.69_1$ V

$\Delta G^\circ = -4FE^\circ = -2.7 \times 10^5$ J $\qquad K = 10^{4E^\circ/0.059\,16} = 10^{47}$

(b) $Ag(S_2O_3)_2^{3-} + e^- \rightleftharpoons Ag(s) + 2S_2O_3^{2-} \qquad E_+^\circ = 0.017$ V

$\underline{-Fe(CN)_6^{3-} + e^- \rightleftharpoons Fe(CN)_6^{4-} \qquad\qquad E_-^\circ = 0.356\text{ V}}$

$Ag(S_2O_3)_2^{3-} + Fe(CN)_6^{4-} \rightleftharpoons Ag(s) + 2S_2O_3^{2-} + Fe(CN)_6^{3-} \quad E^\circ = -0.339$ V

$\Delta G^\circ = -1FE^\circ = 32.7$ kJ $\qquad K = 10^{1E^\circ/0.059\,16} = 1.9 \times 10^{-6}$

14-29. (a) $5Ce^{4+} + 5e^- \rightleftharpoons 5Ce^{3+} \qquad E_+^\circ = 1.70$ V

$\underline{-MnO_4^- + 8H^+ + 5e^- \rightleftharpoons Mn^{2+} + 4H_2O \qquad E_-^\circ = 1.507\text{ V}}$

$5Ce^{4+} + Mn^{2+} + 4H_2O \rightleftharpoons 5Ce^{3+} + MnO_4^- + 8H^+ \qquad E^\circ = 0.19_3$ V

(b) $\Delta G^\circ = -5FE^\circ = -93._1$ kJ $\qquad K = 10^{5E^\circ/0.059\,16} = 2 \times 10^{16}$

(c) $E = \left(1.70 - \dfrac{0.059\,16}{5}\log\dfrac{[Ce^{3+}]^5}{[Ce^{4+}]^5}\right) - \left(1.507 - \dfrac{0.059\,16}{5}\log\dfrac{[Mn^{2+}]}{[MnO_4^-][H^+]^8}\right)$

$= 1.52_{23} - 1.54_{25} = -0.02_0$ V

(d) $\Delta G = -5FE = +10$ kJ

(e) At equilibrium, $E = 0 \Rightarrow E^\circ = \dfrac{0.059\,16}{5}\log\dfrac{[Ce^{3+}]^5[MnO_4^-][H^+]^8}{[Ce^{4+}]^5[Mn^{2+}]}$

$\Rightarrow [H^+] = 0.62 \Rightarrow pH = 0.21$

14-30. right half-cell: $\qquad\qquad Sn^{4+} + 2e^- \rightleftharpoons Sn^{2+}$

left half-cell: $\underline{\qquad -2VO^{2+} + 4H^+ + 2e^- \rightleftharpoons 2V^{3+} + 2H_2O}$

net reaction: $\qquad Sn^{4+} + 2V^{3+} + 2H_2O \rightleftharpoons Sn^{2+} + 2VO^{2+} + 4H^+$

$E = E^\circ - \dfrac{0.059\,16}{2}\log\dfrac{[VO^{2+}]^2[H^+]^4[Sn^{2+}]}{[V^{3+}]^2[Sn^{4+}]}$

$-0.289 = E^\circ - \dfrac{0.059\,16}{2}\log\dfrac{(0.116)^2(1.57)^4(0.031\,8)}{(0.116)^2(0.031\,8)} \Rightarrow E^\circ = -0.266$ V

$\Rightarrow K = 10^{2E^\circ/0.059\,16} = 1.0 \times 10^{-9}$

14-31. $\quad$ Pd(OH)$_2$(s) + 2e$^-$ $\rightleftharpoons$ Pd(s) + 2OH$^-$ $\qquad E_+^\circ$

$\underline{-\text{ Pd}^{2+} + 2e^- \rightleftharpoons \text{Pd}(s) \qquad\qquad\qquad\qquad E_-^\circ = 0.915 \text{ V}}$

$\qquad\qquad\qquad K_{sp} \qquad\qquad\qquad\qquad\qquad\qquad\qquad E^\circ = E_+^\circ - 0.915$

$\qquad\text{Pd(OH)}_2 \rightleftharpoons \text{Pd}^{2+} + 2\text{OH}^-$

But $K_{sp} = 3 \times 10^{-28} \Rightarrow E^\circ = \dfrac{0.059\ 16}{2} \log K_{sp} = -0.814 \text{ V}$

$-0.814 = E_1^\circ - 0.915 \Rightarrow E_1^\circ = 0.101 \text{ V}$

14-32. $\quad$ Br$_2$(l) + 2e$^-$ $\rightleftharpoons$ 2Br$^-$ $\qquad\qquad E_+^\circ = 1.078$ V

$\underline{-\text{ Br}_2(aq) + 2e^- \rightleftharpoons 2\text{Br}^- \qquad\qquad\qquad E_-^\circ = 1.098 \text{ V}}$

$\qquad\qquad K \qquad\qquad\qquad\qquad\qquad\qquad\qquad E^\circ = -0.020 \text{ V}$

$\qquad\text{Br}_2(l) \rightleftharpoons \text{Br}_2(aq)$

At equilibrium, $E = 0$. Therefore, $0 = -0.020 - \dfrac{0.059\ 16}{2} \log \dfrac{[\text{Br}_2(aq)]}{[\text{Br}_2(l)]}$

$\Rightarrow K = \dfrac{[\text{Br}_2(aq)]}{[\text{Br}_2(l)]} = 0.21_1 \text{ M}.$

That is, the solubility of Br$_2$ in water is 0.21_1 M = 34 g/L.

14-33. (a) $\quad$ Cl$_2$(g) + 2e$^-$ $\rightleftharpoons$ 2Cl$^-$ $\qquad E^\circ = 1.360\ 4$ V $\qquad$ (A)

$\underline{-\text{ Cl}_2(aq) + 2e^- \rightleftharpoons 2\text{Cl}^- \qquad\qquad E^\circ = 1.396 \text{ V} \qquad (B)}$

$\qquad\text{Cl}_2(g) \rightleftharpoons \text{Cl}_2(aq) \qquad\qquad E^\circ = 1.360\ 4 - 1.396 = -0.035_6$ V

$K_h = 10^{nE^\circ/0.059\ 16} = 10^{2(-0.035_6)/0.059\ 16} = 0.062_6$ M/bar

When $P_{Cl_2} = 1.00$ bar, $[\text{Cl}_2(aq)] = 0.063$ M

E° and the constant 0.059 16 V both apply at 298.15 K

(b) For reaction A, $dE^\circ/dT = -1.248$ mV/K. For an increase in temperature of 25 K (from 298.15 to 323.15 K), the change in E° is

$E^\circ(\text{at } 323.15 \text{ K}) = E^\circ(\text{at } 298.15 \text{ K}) + (dE^\circ/dT)\Delta T$

$\qquad = 1.360\ 4 \text{ V} + (-1.248 \times 10^{-3} \text{ V/K})(25 \text{ K}) = 1.329_2 \text{ V}$

For reaction B, $dE^\circ/dT = -0.72$ mV/K, so

$\qquad E^\circ(\text{at } 323.15 \text{ K}) = 1.396 \text{ V} + (-0.72 \times 10^{-3} \text{ V/K})(25 \text{ K}) = 1.378 \text{ V}$

$E^\circ(\text{net reaction at } 323.15 \text{ K}) = 1.329_2 - 1.378 = -0.048_8$ V.

The number 0.059 16 V in the Nernst equation is $(RT \ln 10)/F$, which changes with temperature. At $T = 323.15$ K,

$(RT \ln 10)/F = (8.314\ 5 \text{ J/[mol·K]})(323.15 \text{ K})(\ln 10)/(96\ 485.33 \text{ C/mol})$

$\qquad = 0.064 1_2 \text{ J/C} = 0.064 1_2 \text{ V}$

Fundamentals of Electrochemistry   193

$$K_h = \frac{[Cl_2(aq)]}{P_{Cl_2}} = 10^{nE°/0.064\,12} = 10^{2(-0.048_8)/0.064\,12} = 0.030_1 \text{ M/bar}$$

The solubility of Cl(g) decreases when the temperature is raised, which is typical of most gases.

14-34. $FeY^- + e^- \rightleftharpoons FeY^{2-}$ $E_+^°$

$$\underline{-FeY^- + e^- \rightleftharpoons Fe^{2+} + Y^{4-} \qquad\qquad E_-^° = -0.730 \text{ V}}$$

$Fe^{2+} + Y^{4-} \rightleftharpoons FeY^{2-}$ $E° = E_+^° + 0.730$

But $E° = 0.059\,16 \log [K_f \text{ (for FeY}^{2-})] = 0.847$ V

$\Rightarrow E_+^° = E° + E_-^° = 0.847 - 0.730 = 0.117$ V.

14-35. $E°(T) = E° + \dfrac{dE°}{dT} \Delta T$

$E°(50°C) = -1.677 \text{ V} + (0.533 \times 10^{-3} \text{ V/K})(25 \text{ K}) = -1.664$ V

14-36.
1. $2Cu(s) + 2I^- \rightleftharpoons 2CuI(s) + 2e^-$ $\Delta G_1^° = +2F(-0.185) = -35.7_0$ kJ
2. $2Cu^{2+} + 4e^- \rightleftharpoons 2Cu(s)$ $\Delta G_2^° = -4F(0.339) = -130.8_3$ kJ
3. $\text{hydro} \rightleftharpoons \text{quinone} + 2H^+ + 2e^-$ $\Delta G_3^° = +2F(0.700) = 135.0_8$ kJ
4. (1)+(2)+(3): $2Cu^{2+} + 2I^- + \text{hydro} \rightleftharpoons$ $\Delta G_4^° = \Delta G_1^° + \Delta G_2^° + \Delta G_3^°$
    $2CuI(s) + \text{quinone} + 2H^+$ $= -31.4$ kJ

Since 2e⁻ are transferred in the net reaction, $E_4^° = \dfrac{-\Delta G_4^°}{2F} = +0.163$ V

$K = 10^{2(0.163 \text{ V})/(0.059\,16 \text{ V})} = 3.2 \times 10^5$.

14-37. $E^{°\prime}$ is the effective reduction potential for a half-reaction at pH 7, instead of pH 0. Since living systems tend to have a pH much closer to 7 than to 0, $E^{°\prime}$ provides a better indication of redox behavior in an organism.

14-38. (a) $E = 0.731 - \dfrac{0.059\,16}{2} \log \dfrac{P_{C_2H_4}}{P_{C_2H_4}[H^+]^2}$

(b) $E = \underbrace{0.731 + 0.059\,16 \log [H^+]}_{\text{This is } E^{°\prime} \text{ when pH = 7}} - \dfrac{0.059\,16}{2} \log \dfrac{P_{C_2H_4}}{P_{C_2H_2}}$

(c) $E^{°\prime} = 0.731 + 0.059\,16 \log(10^{-7.00}) = 0.317$ V

14-39. $E = E° - \dfrac{0.05916}{2} \log \dfrac{[HCN]^2}{P_{(CN)_2}[H^+]^2}$

Substituting $[HCN] = \dfrac{[H^+] F_{HCN}}{[H^+] + K_a}$ into the Nernst equation gives

$E = 0.373 - \dfrac{0.05916}{2} \log \dfrac{[H^+]^2 F_{HCN}^2}{([H^+]+K_a)^2 P_{(CN)_2}[H^+]^2}$

$E = \underbrace{0.373 + 0.05916 \log ([H^+] + K_a)}_{\text{This is } E°' \text{ when pH} = 7} - \dfrac{0.05916}{2} \log \dfrac{F_{HCN}^2}{P_{(CN)_2}}$

Inserting $K_a = 6.2 \times 10^{-10}$ for HCN and $[H^+] = 10^{-7.00}$ gives

$E°' = 0.373 + 0.05916 \log (10^{-7.00} + 6.2 \times 10^{-10}) = -0.041$ V.

14-40. $H_2C_2O_4 + 2H^+ + 2e^- \rightleftharpoons 2HCO_2H$

$E = 0.204 - \dfrac{0.05916}{2} \log \dfrac{[HCO_2H]^2}{[H_2C_2O_4][H^+]^2}$

But $[HCO_2H] = \dfrac{[H^+] F_{HCO_2H}}{[H^+] + K_a}$ and $[H_2C_2O_4] = \dfrac{[H^+]^2 F_{H_2C_2O_4}}{[H^+]^2 + K_1[H^+] + K_1K_2}$.

Putting these expressions into the Nernst equation gives

$E = 0.204 - \dfrac{0.05916}{2} \log \dfrac{[H^+]^2 F_{HCO_2H}^2 ([H^+]^2 + K_1[H^+] + K_1K_2)}{([H^+]+K_a)^2 [H^+]^2 F_{H_2C_2O_4}[H^+]^2}$

$E = \underbrace{0.204 - \dfrac{0.05916}{2} \log \dfrac{[H^+]^2 + K_1[H^+] + K_1K_2}{([H^+]+K_a)^2 [H^+]^2}}_{\text{This is } E°' \text{ when pH} = 7} - \dfrac{0.05916}{2} \log \dfrac{F_{HCO_2H}^2}{F_{H_2C_2O_4}}$.

Putting in $[H^+] = 10^{-7.00}$ M, $K_a = 1.80 \times 10^{-4}$, $K_1 = 5.62 \times 10^{-2}$, and $K_2 = 5.42 \times 10^{-5}$ gives $E°' = -0.268$ V.

14-41. $E = E° - 0.05916 \log \dfrac{[H_2Red^-]}{[HOx]}$

But $[HOx] = \dfrac{[H^+] F_{HOx}}{[H^+] + K_a}$ and $[H_2Red^-] = \dfrac{[H^+]^2 F_{H_2Red^-}}{[H^+]^2 + [H^+]K_1 + K_1K_2}$.

Putting these values into the Nernst equation gives

$$E = E° - 0.05916 \log \frac{[H^+]^2 F_{H_2Red^-}([H^+]+K_a)}{[H^+]F_{HOx}([H^+]^2+[H^+]K_1+K_1K_2)}$$

$$E = \underbrace{E° - 0.05916 \log \frac{[H^+]([H^+]+K_a)}{[H^+]^2+[H^+]K_1+K_1K_2}}_{E°'} - 0.05916 \log \frac{F_{H_2Red^-}}{F_{HOx}}.$$

Since $E°' = 0.062$ V, we find $E° = -0.036$ V.

14-42. $E = E° - \dfrac{0.05916}{2} \log \dfrac{[HNO_2]}{[NO_3^-][H^+]^3}$

But $[HNO_2] = \dfrac{[H^+]F_{HNO_2}}{[H^+]+K_a}$ and $[NO_3^-] = F_{NO_3^-}$.

Putting these values into the Nernst equation gives

$$E = \underbrace{E° - \frac{0.05916}{2} \log \frac{1}{([H^+]+K_a)[H^+]^2}}_{E°'} - \frac{0.05916}{2} \log \frac{F_{HNO_2}}{F_{NO_3^-}}$$

$$E°' = 0.433 = 0.940 - \frac{0.05916}{2} \log \frac{1}{(10^{-7}+K_a)(10^{-7})^2} \Rightarrow K_a = 7.2 \times 10^{-4}.$$

14-43.
$$\begin{array}{ll} H_2PO_4^- + H^+ + 2e^- \rightleftharpoons HPO_3^{2-} + H_2O & E_+° \\ -\ HPO_4^{2-} + 2H^+ + 2e^- \rightleftharpoons HPO_3^{2-} + H_2O & E_-° = -0.234 \text{ V} \\ \hline H_2PO_4^- \rightleftharpoons HPO_4^{2-} + H^+ & E° = E_+° - E_-° \end{array}$$

$$E° = \frac{0.05916}{2} \log K_{a2} \text{ (for } H_3PO_4\text{)} = \frac{0.05916}{2} \log(6.34 \times 10^{-8}) = -0.213 \text{ V}$$

$$E_+° = E° - 0.234 \text{ V} = -0.447 \text{ V}$$

14-44. (a) $A = 0.500 =$
$(1.12 \times 10^4 \text{ M}^{-1}\text{cm}^{-1})[\text{Ox}](1.00 \text{ cm}) + (3.82 \times 10^3 \text{ M}^{-1}\text{cm}^{-1})[\text{Red}](1.00 \text{ cm})$
But $[\text{Ox}] = 5.70 \times 10^{-5} - [\text{Red}]$.

Combining these two equations gives
$[\text{Ox}] = 3.82 \times 10^{-5}$ M and $[\text{Red}] = 1.88 \times 10^{-5}$ M.

(b) $[\text{S}^-] = [\text{Ox}] = 3.82 \times 10^{-5}$ M and $[\text{S}] = \text{Red} = 1.88 \times 10^{-5}$ M.

(c) $\quad \text{S} + e^- \rightleftharpoons \text{S}^- \qquad\qquad E_+^{\circ\prime} = ?$
$\underline{- \text{Ox} + e^- \rightleftharpoons \text{Red} \qquad\qquad E_-^{\circ\prime} = -0.128 \text{ V}}$
$\text{Red} + \text{S} \rightleftharpoons \text{Ox} + \text{S}^- \qquad E^{\circ\prime} = 0.059\,16 \log \dfrac{[\text{Ox}][\text{S}^-]}{[\text{Red}][\text{S}]} = 0.036 \text{ V}$

$E_+^{\circ\prime} = E^{\circ\prime} + E_-^{\circ\prime} = -0.092 \text{ V}$

CHAPTER 15
ELECTRODES AND POTENTIOMETRY

15-1. (a) $AgCl(s) + e^- \rightleftharpoons Ag(s) + Cl^-$

$Hg_2Cl_2(s) + 2e^- \rightleftharpoons 2Hg(l) + 2Cl^-$

(b) $E = E_+ - E_- = 0.241 - 0.197 = 0.044$ V

15-2. Diagram for (a):

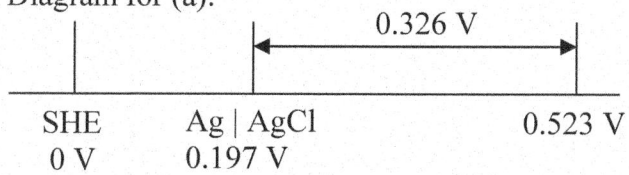

SHE	Ag | AgCl		0.523 V
0 V	0.197 V		

(a) 0.326 V (b) 0.086 V (c) 0.019 V (d) –0.021 V (e) 0.021 V

15-3. $E = E_+ - E_-$

$E = \left(0.771 - 0.05916 \log \dfrac{[Fe^{2+}]}{[Fe^{3+}]}\right) - (0.241) = 0.684$ V

15-4. $E = E° - 0.05916 \log \mathcal{A}_{Cl^-}$

$0.280 = 0.268 - 0.05916 \log \mathcal{A}_{Cl^-} \Rightarrow \mathcal{A}_{Cl^-} = 0.627$

15-5. For the saturated Ag-AgCl electrode, we can write: $E = E° - 0.05916 \log \mathcal{A}_{Cl^-}$. Putting in $E = 0.197$ and $E° = 0.222$ V gives $\mathcal{A}_{Cl^-} = 2.6_5$. For the S.C.E., we can write: $E = E° - 0.05916 \log \mathcal{A}_{Cl^-} = 0.268$ V $- 0.05916 \log 2.6_5$ $= 0.243$ V.

15-6. (a) $Cu^{2+} + 2e^- \rightleftharpoons Cu(s) \qquad E° = 0.339$ V

(b) $E_+ = 0.339 - \dfrac{0.05916}{2} \log \dfrac{1}{[Cu^{2+}]} = 0.309$ V

(c) $E = E_+ - E_- = 0.309 - 0.241 = 0.068$ V

15-7. A silver electrode serves as an indicator for Ag^+ by virtue of the equilibrium $Ag^+ + e^- \rightleftharpoons Ag(s)$ that occurs at its surface. If the solution is saturated with silver halide, then $[Ag^+]$ is affected by changes in halide concentration ($[Ag^+] = K_{sp}/[\text{halide}]$). Therefore, the electrode is also an indicator for halide.

15-8. $V_e = 20.0$ mL. $Ag^+ + e^- \rightleftharpoons Ag(s) \Rightarrow E_+ = 0.799 - 0.05916 \log \dfrac{1}{[Ag^+]}$

0.1 mL: $[Ag^+] = \underbrace{\left(\dfrac{19.9 \text{ mL}}{20.0 \text{ mL}}\right)}_{\substack{\text{Fraction} \\ \text{remaining}}} \underbrace{(0.050\,0\text{M})}_{\substack{\text{Original} \\ \text{concentration}}} \underbrace{\left(\dfrac{10.0 \text{ mL}}{10.1 \text{ mL}}\right)}_{\substack{\text{Dilution} \\ \text{factir}}} = 0.049\,3 \text{ M}$

$E = E_+ - E_- = \left(0.799 - 0.059\,16 \log \dfrac{1}{0.049\,3}\right) - 0.241 = 0.481 \text{ V}$

30.0 mL: This is 10.0 mL past $V_e \Rightarrow [Br^-] = \left(\dfrac{10.0 \text{ mL}}{40.0 \text{ mL}}\right)(0.025\,0 \text{ M}) = 0.006\,25 \text{ M}$

$[Ag^+] = K_{sp}/[Br^-] = (5.0 \times 10^{-13})/0.006\,25 = 8.0 \times 10^{-11}$ M

$E = E_+ - E_- = \left(0.799 - 0.059\,16 \log \dfrac{1}{8.0 \times 10^{-11}}\right) - 0.241 = -0.039$ V.

15-9. The reaction in the right half-cell is $Hg^{2+} + 2e^- \rightleftharpoons Hg(l)$

$E = E_+ - E_-$

$-0.027 = \left(0.852 - \dfrac{0.059\,16}{2} \log \dfrac{1}{[Hg^{2+}]}\right) - (0.241)$

$\Rightarrow [Hg^{2+}] = 2.7 \times 10^{-22}$ M.

The cell contains 5.00 mmol EDTA (in all forms) and 1.00 mmol Hg(II) in 100 mL. 1.00 mmol EDTA reacts with 1.00 mmol Hg(II), leaving 4.00 mmol EDTA.

$K_f = \dfrac{[HgY^{2-}]}{[Hg^{2+}][Y^{4-}]} = \dfrac{[HgY^{2-}]}{[Hg^{2+}]\alpha_{Y^{4-}}[EDTA]}$

$K_f = \dfrac{(1.00 \text{ mmol}/100 \text{ mL})}{(2.7 \times 10^{-22})(0.30)(4.00 \text{ mmol}/100 \text{ mL})} = 3._1 \times 10^{21}$

15-10. (a) $Fe^{3+} + e^- \rightleftharpoons Fe^{2+}$ $E° = 0.771$ V

(b) $E = E_+ - E_-$

$-0.126 = 0.771 - 0.059\,16 \log \dfrac{[Fe^{2+}]}{[Fe^{3+}]} - 0.241 \Rightarrow \dfrac{[Fe^{2+}]}{[Fe^{3+}]} = 1._2 \times 10^{11}$

(c) The solution contains 2.00 mmol Fe(II), 1.00 mmol Fe(III), and 4.00 mmol EDTA. After reaction, concentrations will be $[FeEDTA^{2-}] = 2.00$ mM, $[FeEDTA^-] = 1.00$ mM and unbound [EDTA] = 1.00 mM.

Electrodes and Potentiometry

(d) $\dfrac{K_f(\text{FeEDTA}^-)}{K_f(\text{FeEDTA}^{2-})} = \dfrac{[\text{FeEDTA}^-]}{[\text{Fe}^{3+}][\text{EDTA}^{4-}]} \div \dfrac{[\text{FeEDTA}^{2-}]}{[\text{Fe}^{2+}][\text{EDTA}^{4-}]}$

$= \left(\dfrac{[\text{FeEDTA}^-]}{[\text{FeEDTA}^{2-}]}\right)\left(\dfrac{[\text{Fe}^{2+}]}{[\text{Fe}^{3+}]}\right) = \left(\dfrac{1.00\ \text{mM}}{2.00\ \text{mM}}\right)(1.2 \times 10^{11}) = 6 \times 10^{10}$

15-11. $E = E_+ - E_- = -0.429 - 0.059\ 16 \log \dfrac{[\text{CN}^-]^2}{[\text{Cu(CN)}_2^-]} - 0.197$

Putting in $E = -0.440$ V and $[\text{Cu(CN)}_2^-] = 1.00$ mM gives $[\text{CN}^-] = 0.847$ mM.

$\text{pH} = \text{p}K_a(\text{HCN}) + \log \dfrac{[\text{CN}^-]}{[\text{HCN}]} = 9.21 + \log \dfrac{8.47 \times 10^{-4}}{1.00 \times 10^{-3} - 8.47 \times 10^{-4}} = 9.95_4$

Now we use the pH to see how much HA reacted with KOH:

	HA	+	OH$^-$	→	A$^-$	+	H$_2$O
initial mmol	10.0		x		—		
final mmol	10.0 − x		—		x		

$\text{pH} = \text{p}K_a(\text{HA}) + \log \dfrac{[\text{A}^-]}{[\text{HA}]}$

$9.95_4 = 9.50 + \log \dfrac{x}{10.0 - x} \Rightarrow x = 7.4_0\ \text{mmol of OH}^-$

$[\text{KOH}] = \dfrac{7.4_0\ \text{mmol}}{25.0\ \text{mL}} = 0.29_6\ \text{M}$

15-12. Junction potential arises because different ions diffuse at different rates across a liquid junction, leading to a separation of charge. The resulting electric field retards fast-moving ions and accelerates the slow-moving ions until a steady-state junction potential is reached. This limits the accuracy of a potentiometric measurement, because we do not know what part of a measured cell voltage is due to the process of interest and what is due to junction potential. The cell in Figure 14-5 has no junction potential because there are no liquid junctions.

15-13. H$^+$ has greater mobility than K$^+$. The HCl side of the HCl | KCl junction will be negative because H$^+$ diffuses into KCl faster than K$^+$ diffuses into HCl. K$^+$ has a greater mobility than Na$^+$, so this junction has the opposite sign. The HCl | KCl voltage is larger, because the difference in mobility between H$^+$ and K$^+$ is greater than the difference in mobility between K$^+$ and Na$^+$.

15-14. Relative mobilities:

$$K^+ \rightarrow 7.62 \quad NO_3^- \rightarrow 7.40$$
$$5.19 \leftarrow Na^+ \quad 7.91 \leftarrow Cl^-$$

Both the cation and anion diffusion cause negative charge to build up on the <u>left</u>.

15-15. The junction potentials are dominated by the high mobility of H^+ and OH^-. The high mobility of H^+ makes the right side of 0.1 M HCl | 0.1 M KCl become positive. The high mobility of OH^- makes the right side of 0.1 M NaOH | 0.1 M KCl become negative. There is only a small junction potential at 0.1 M NaOH | KCl (saturated) because diffusion is dominated by the high concentrations of K^+ and Cl^-, which have nearly equal mobility.

15-16. Both half-cells contain saturated KCl. Therefore it makes sense to use saturated KCl in the salt bridge. The Ag | AgCl compartment also contains a very low concentration of Ag^+ and the calomel compartment contains a very low concentration of Hg_2^{2+}. With $[Cl^-] \approx 4.2$ M,

$$[Ag^+] \approx K_{sp}(AgCl)/[Cl^-] = (1.8 \times 10^{-10})/(4.2) \approx 10^{-11} \text{ M}$$

$$[Hg_2^{2+}] \approx K_{sp}(Hg_2Cl_2)/[Cl^-]^2 = (1.2 \times 10^{-18})/(4.2)^2 \approx 10^{-19} \text{ M}$$

The Ag^+ and Hg_2^{2+} concentrations are too low to affect the junction potential. Both junctions contain saturated KCl on both sides of each junction. The junction potential should therefore be very close to 0 at both junctions.

15-17. Velocity = mobility × field = $(36.30 \times 10^{-8} \text{ m}^2/(\text{s·V})) \times (7\,800 \text{ V/m}) = 2.83 \times 10^{-3}$ m s^{-1} for H^+ and $(7.40 \times 10^{-8} \text{ m}^2/(\text{s·V}))(7\,800 \text{ V/m}) = 5.77 \times 10^{-4}$ m s^{-1} for NO_3^-. To cover 0.120 m will require $(0.120 \text{ m})/(2.83 \times 10^{-3} \text{ m s}^{-1}) = 42.4$ s for H^+ and $(0.120)/(5.77 \times 10^{-4}) = 208$ s for NO_3^-.

15-18. (a) $E° = 0.799$ V $\Rightarrow K = 10^{nE°/0.059\,16} = 10^{(1)(0.799)/0.059\,16} = 3.20 \times 10^{13}$

(b) $K' = 10^{0.801/0.059\,16} = 3.46 \times 10^{13}$. $K'/K = 1.08$. The increase is 8%.

(c) $K = 10^{0.100/0.059\,16} = 49.0$. $K' = 10^{0.102/0.059\,16} = 53.0$

$K'/K = 1.08$. The change is still 8%.

Electrodes and Potentiometry

15-19. Both half-cell reactions are the same (AgCl + e⁻ ⇌ Ag + Cl⁻) and the concentration of Cl⁻ is the same on both sides. In principle, the voltage of the cell would be 0 if there were no junction potential. The measured voltage can be attributed to the junction potential. In practice, if both sides contained 0.1 M HCl (or 0.1 M KCl), the two electrodes would probably produce a small voltage because no two real cells are identical. This voltage can be measured and subtracted from the voltage measured with the HCl | KCl junction.

15-20. (a) In phase α, we have 0.1 M H⁺ ($u = 36.3 \times 10^{-8}$ m² s⁻¹ V⁻¹) and 0.1 M Cl⁻ ($u = 7.91 \times 10^{-8}$ m² s⁻¹ V⁻¹). In phase β, we have 0.1 M K⁺ ($u = 7.62 \times 10^{-8}$ m² s⁻¹ V⁻¹) and 0.1 M Cl⁻ ($u = 7.91 \times 10^{-8}$ m² s⁻¹ V⁻¹). Substituting into the Henderson equation gives

$$E_j = \frac{(36.3\times10^{-8})[0-0.1]+(7.62\times10^{-8})[0.1-0]-(7.91\times10^{-8})[0.1-0.1]}{(36.3\times10^{-8})[0-0.1]+(7.62\times10^{-8})[0.1-0]+(7.91\times10^{-8})[0.1-0.1]} \times$$

$$0.059\,16 \log \frac{(36.3\times10^{-8})(0.1)+(7.91\times10^{-8})(0.1)}{(7.62\times10^{-8})(0.1)+(7.91\times10^{-8})(0.1)} = 26.9 \text{ mV}$$

(b)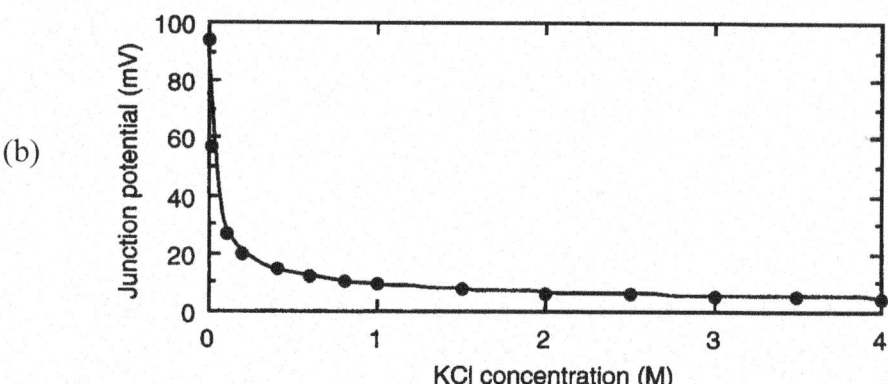

(c)

[HCl]	y M HCl | 1 mM KCl	y M HCl | 4 M KCl
10^{-4} M	9.1 mV	4.6 mV
10^{-3} M	26.9 mV	3.6 mV
10^{-2} M	57.3 mV	3.0 mV
10^{-1} M	93.6 mV	4.7 mV

15-21. The electrode should be calibrated at 37°C using two buffers bracketing the pH of the blood. It is reasonable to use MOPSO and HEPES buffers in Table 15-3 that are recommended for use with physiologic fluids. The pH of these standards at 37°C is 6.695 and 7.370. The standards should be thermostatted to 37°C during calibration, and the blood should also be at 37°C during the measurement.

15-22. Uncertainty in pH of standard buffers, junction potential, junction potential drift, sodium or acid errors at extreme pH values, equilibration time, hydration of glass, temperature of measurement and calibration, and cleaning of electrode

15-23. The error measured in the graph is –0.33 pH units. The electrode will indicate 11.00 – 0.33 = 10.67.

15-24. Saturated potassium hydrogen tartrate and 0.05 m potassium hydrogen phthalate

15-25. If the alkaline solution has a high concentration of Na^+ (as in NaOH), the Na^+ cation competes with H^+ for cation exchange sites on the glass surface. The glass responds as if H^+ were present, and the apparent pH is lower than the actual pH.

15-26. The junction potential changes from –6.4 mV to –0.2 mV. A change of 6.4 – 0.2 = +6.2 mV appears to be a pH change of +6.2/59.16 = +0.10 pH units.

15-27. (a) (4.63)(59.16 mV) = 274 mV. The factor 59.16 mV is the value of $(RT \ln 10)/F$ at 298.15 K.

(b) At 310.15 K (37°C), $(RT \ln 10)/F$
= (8.3145 J mol^{-1} K^{-1})(310.15 K)(ln 10)/(96 485 C mol^{-1}) = 61.54 mV
(4.63)(61.54 mV) = 285 mV.

15-28. pH of 0.025 m KH$_2$PO$_4$/0.025 m Na$_2$HPO$_4$ at 20°C = 6.881

pH of 0.05 m potassium hydrogen phthalate at 20°C = 4.002

$$\frac{E_{unknown} - E_{S1}}{pH_{unknown} - pH_{S1}} = \frac{E_{S2} - E_{S1}}{pH_{S2} - pH_{S1}}$$

$$\frac{E_{unknown} - (-18.3 \text{ mV})}{pH_{unknown} - 6.881} = \frac{(+146.3 \text{ mV}) - (-18.3 \text{ mV})}{4.002 - 6.881} = -57.17_3 \text{ mV/pH unit}$$

$$pH_{unknown} = \frac{E_{unknown} - (-18.3 \text{ mV})}{-57.17_3 \text{ mV/pH unit}} + 6.881 = 5.686$$

Observed slope = –57.17$_3$ mV/pH unit

Electrodes and Potentiometry

$$\text{Theoretical slope} = -\frac{RT \ln 10}{F}$$

$$= -\frac{[8.314\,5\ \text{V}\cdot\text{C/(mol}\cdot\text{K)}][293.15\,\text{K}]\ln 10}{9.648\,53\times 10^4\ \text{C/mol}} = -0.058\,17\ \text{V}$$

$$\beta = \frac{\text{observed slope}}{\text{theoretical slope}} = \frac{-57.17_3\ \text{mV}}{-58.17\ \text{mV}} = 0.983$$

15-29. (a) There is negligible change in concentrations of buffer species when we mix the acid $H_2PO_4^-$ with its conjugate base, HPO_4^{2-}. The ionic strength of $0.025\,0\ m$ KH_2PO_4 (a 1:1 electrolyte) is $0.025\,0\ m$. The ionic strength of $0.025\,0\ m$ Na_2HPO_4 (a 2:1 electrolyte) is $0.075\,0\ m$. Both salts are mixed in one solution, so the total ionic strength is $0.025\,0\ m + 0.075\,0\ m = 0.100\ m$.

(b) $$K_2 = \frac{[H^+]\gamma_{H^+}[HPO_4^{2-}]\gamma_{HPO_4^{2-}}}{[H_2PO_4^-]\gamma_{H_2PO_4^-}}$$

But $K_2 = 10^{-7.198}$ and $[H^+]\gamma_{H^+} = 10^{-\text{pH}} = 10^{-6.865}$

$$\frac{\gamma_{HPO_4^{2-}}}{\gamma_{H_2PO_4^-}} = \frac{K_2[H_2PO_4^-]}{[H^+]\gamma_{H^+}[HPO_4^{2-}]} = \frac{10^{-7.198}[0.025\,0]}{10^{-6.865}[0.025\,0]} = 0.464_5$$

(We can use molality or other concentration units because units cancel in the numerator and denominator.) From the table of activity coefficients, we expect $\gamma_{HPO_4^{2-}}/\gamma_{H_2PO_4^-} = 0.355/0.775 = 0.458$

(c) To get a pH 7.000, increase the concentration of base (HPO_4^{2-}) and decrease the concentration of acid ($H_2PO_4^-$). To maintain a constant ionic strength, we must decrease KH_2PO_4 three times as much as we increase Na_2HPO_4, because Na_2HPO_4 contributes three times as much as KH_2PO_4 to the ionic strength. So let's increase Na_2HPO_4 by x and decrease KH_2PO_4 by $3x$.

$$K_2 = \frac{[H^+]\gamma_{H^+}[HPO_4^{2-}]\gamma_{HPO_4^{2-}}}{[H_2PO_4^-]\gamma_{H_2PO_4^-}}$$

$$\Rightarrow 10^{-7.198} = \frac{10^{-7.000}[0.025\,0+x]}{[0.025\,0-3x]}(0.464_5) \Rightarrow x = 0.001\,8\ m.$$

New concentrations: $Na_2HPO_4 = 0.026\,8\ m$ and $KH_2PO_4 = 0.019\,6\ m$.

15-30. (a)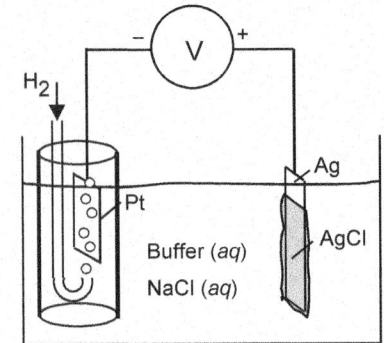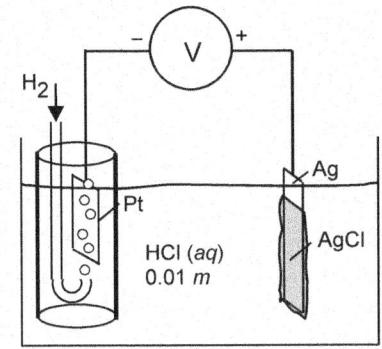

(b) $\log \gamma_{Cl} = \dfrac{-Az^2\sqrt{\mu}}{1+B\alpha\sqrt{\mu}} = \dfrac{(-0.511)(-1)^2\sqrt{0.1}}{1+1.5\sqrt{0.1}} \Rightarrow \gamma_{Cl} = 0.777$

$-\log \mathcal{A}_H \gamma_{Cl} = p(\mathcal{A}_H \gamma_{Cl})^o = 6.972$ from the graph at zero added NaCl

$-\log \mathcal{A}_H(0.777) = 6.972 \Rightarrow \mathcal{A}_H = (10^{-6.972})/0.777 = 1.373 \times 10^{-7}$

$pH = -\log(1.373 \times 10^{-7}) = 6.862$ (value in table = 6.865)

15-31. (a) Analyte ions equilibrate between the outer solution and ligand L in the ion-selective membrane. This equilibrium creates a slight charge imbalance (an electric potential difference) across the interface between the membrane and the analyte solution because other ions in the membrane and outer solution are not free to cross the interface. Changes in analyte ion concentration in the outer solution change the potential difference across the outer boundary of the ion-selective membrane.

(b) Compound electrodes contain a conventional electrode surrounded by a membrane that isolates (or generates) the analyte to which the electrode responds.

15-32. The selectivity coefficient $K_{A,X}^{Pot}$ tells us the relative response of an ion-selective electrode to the ion of interest (A) and an interfering ion (X). The smaller $K_{A,X}^{Pot}$, the more selective is the electrode (smaller response to the interfering ion).

15-33. A ligand dissolved in the membrane liquid phase binds tightly to the ion of interest and weakly to interfering ions.

15-34. A metal ion buffer maintains the desired (small) concentration of metal ion from a large reservoir of metal complex (ML) and free ligand (L). If you just tried to dissolve 10^{-8} M metal ion in most solutions or containers, the metal would probably bind to the container wall or to an impurity in the solution and be lost.

Electrodes and Potentiometry 205

15-35. Electrodes respond to *activity*. If the ionic strength is constant, the activity coefficient of analyte will be constant in all standards and unknowns. In this case, the calibration curve can be written directly in terms of concentration.

15-36. (a) $-0.230 = \text{constant} - 0.05916 \log (1.00 \times 10^{-3}) \Rightarrow \text{constant} = -0.407$ V

(b) $-0.300 = -0.407 - 0.05916 \log x \Rightarrow x = 1.5_5 \times 10^{-2}$ M

(c) $-0.230 = \text{constant} - 0.05916 \log (1.00 \times 10^{-3})$
$-0.300 = \text{constant} - 0.05916 \log x$

subtract: $0.070 = -0.05916 \log \dfrac{1.00 \times 10^{-3}}{x} \Rightarrow x = 1.5_2 \times 10^{-2}$ M

15-37. $E_1 = \text{constant} + \dfrac{0.05916}{2} \log [1.00 \times 10^{-4}]$

(note positive sign for response to cation Mg^{2+})

$E_2 = \text{constant} + \dfrac{0.05916}{2} \log [1.00 \times 10^{-3}]$

$\Delta E = E_2 - E_1 = \dfrac{0.05916}{2} \log \dfrac{1.00 \times 10^{-3}}{1.00 \times 10^{-4}} = +0.0296$ V

15-38. $[F^-]_{\text{Providence}} = 1.00$ mg F^-/L $= 5.26 \times 10^{-5}$ M

$E_{\text{Providence}} = \text{constant} - 0.05916 \log [5.26 \times 10^{-5}]$

$E_{\text{Foxboro}} = \text{constant} - 0.05916 \log [F^-]_{\text{Foxboro}}$

$\Delta E = E_{\text{Foxboro}} - E_{\text{Providence}} = 0.0400$ V

$= -0.05916 \log \dfrac{[F^-]_{\text{Foxboro}}}{5.26 \times 10^{-5}} \Rightarrow [F^-]_{\text{Foxboro}} = 1.11 \times 10^{-5}$ M $= 0.211$ mg/L

15-39. K^+ has the largest selectivity coefficient of Group 1 ions and therefore interferes the most. Sr^{2+} and Ba^{2+} are the worst of the Group 2 ions. Since $\log K^{\text{Pot}}_{Li^+,K^+} \approx -2$, there must be 100 times more K^+ than Li^+ to give equal response.

15-40. $\dfrac{[ML]}{[M][L]} = 4.0 \times 10^8 = \dfrac{0.030 \text{ M}}{[M](0.020 \text{ M})} \Rightarrow [M] = 3.8 \times 10^{-9}$ M

15-41. (a) The least-squares parameters are

$$E = 51.10\ (\pm 0.24) + 28.14\ (\pm 0.08_5) \log [\text{Ca}^{2+}] \quad (s_y = 0.2_7)$$

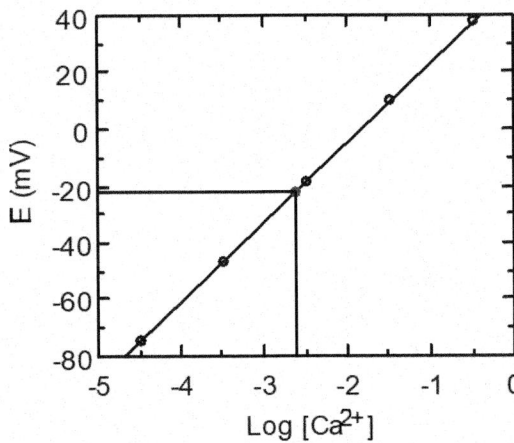

(b) The slope is $0.028\ 14$ V $= \beta(0.059\ 16\ \text{V})/2 \Rightarrow \beta = 0.951$.

(c) If we use Equation 4-27 in a spreadsheet, we find $\log [\text{Ca}^{2+}] = -2.615_3(\pm 0.007_2)$ using $s_y = 0.3$ and $k = 4$.

From Table 3-1, we can write that if $F = 10^x$, $e_F/F = (\ln 10)e_x$.

In this problem, $F = [\text{Ca}^{2+}] = 10^{-2.615_3(\pm 0.007_2)} = 2.43(\pm ?) \times 10^{-3}$ M

$e_F/F = (\ln 10)(0.007_2) = 0.016_6$

$e_F = (0.016_6) F = 4.0 \times 10^{-5} \Rightarrow F = 2.43(\pm 0.04) \times 10^{-3}$ M.

15-42. At pH 7.2 the effect of H^+ will be negligible because $[H^+] \ll [\text{Li}^+]$:

-0.333 V $=$ constant $+ 0.059\ 16 \log [3.44 \times 10^{-4}] \Rightarrow$ constant $= -0.128$ V.

At pH 1.1 ($[H^+] = 0.079$ M), we must include interference by H^+:

$E = -0.128 + 0.059\ 16 \log [3.44 \times 10^{-4} + (4 \times 10^{-4})(0.079)] = -0.331$ V.

15-43. The function to plot on the y-axis is $(V_i + V_s)10^{E/S}$, where $S = -(\beta RT \ln 10)/nF$. (The minus sign comes from the equation for the response of the electrode, which has a minus sign in front of the log term.) Putting in $\beta = 0.933$, $R = 8.314\ 5$ J/(mol·K), $F = 96\ 485$ C/mol, $T = 303.15$ K, and $n = 2$ gives $S = -0.028\ 06$ J/C $= 0.028\ 06$ V. (You can get J/C = V from the equation work(J) = E(V)·q(C).)

V_s (mL)	E (V)	$y = (V_i + V_s)10^{E/S}$
0	0.079 0	0.084 1
0.100	0.072 4	0.144 9
0.200	0.065 3	0.259 9
0.300	0.058 8	0.443 8
0.800	0.050 9	0.856 5

The graph has a slope of $m = 0.989$ and an intercept of $b = 0.080_9$, giving an x-intercept of $-b/m = -0.081_8$ mL. The concentration of original unknown is

$$c_X = -\frac{(x\text{-intercept})c_S}{V_i} = -\frac{(-0.081_8 \text{ mL})(0.020\ 0 \text{ M})}{55.0 \text{ mL}} = 3.0 \times 10^{-5} \text{ M}.$$

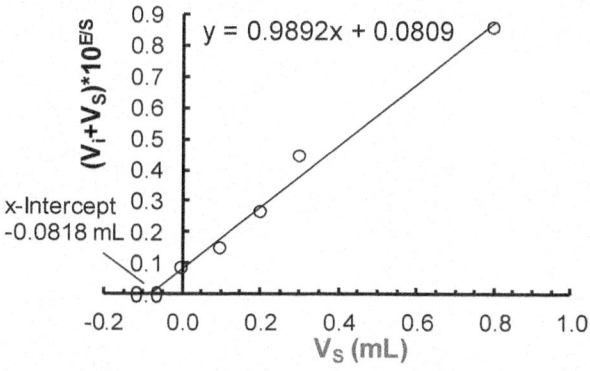

15-44. (a) The spreadsheet and graph are shown below. The x-intercept is at -3.65 mL with a standard deviation in cell B26 of $s = 0.484$ mL. The intercept gives us the concentration of ammonia nitrogen in the volume $V_i = 101.0$ mL:

$$x\text{-intercept} = -3.65 \text{ mL} = -\frac{V_i c_X}{c_S}$$

$$\Rightarrow c_X = \frac{(x\text{-intercept})c_S}{V_i} = \frac{(3.65 \text{ mL})(10.0 \text{ ppm})}{101.0 \text{ mL}} = 0.361_4 \text{ ppm}$$

The concentration of ammonia nitrogen in the original 100.0 mL of seawater, which had been diluted from 100.0 to 101.0 mL, is

$\frac{101.0 \text{ mL}}{100.0 \text{ mL}} (0.361_4 \text{ ppm}) = 0.365$ ppm.

The 95% confidence interval is equal to Student's t times the standard deviation:

95% confidence interval $= \pm t \cdot s = \pm(3.18)(0.484 \text{ mL}) = \pm 1.54$ mL

where t is for 95% confidence and $5 - 2 = 3$ degrees of freedom because there are 5 data points on the standard addition curve. You can find t in the table of Student's t or you can compute it with the statement "=T.INV.2T(0.05,3)" in cell G24 in the spreadsheet. The confidence interval of ± 1.54 mL corresponds to a relative uncertainty of $100 \times \frac{1.54 \text{ mL}}{3.65 \text{ mL}} = 42\%$.

Absolute uncertainty $= (0.42)(0.365 \text{ ppm}) = 0.15$ ppm. The concentration of ammonia nitrogen in the seawater can be expressed as 0.36 ± 0.15 ppm.

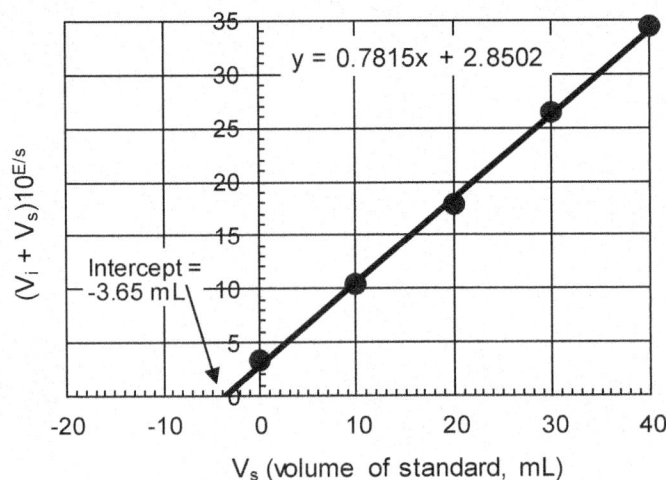

(b) In this experiment, analytical signal is $(V_i + V_S)10^{E/S} = 3.26$ mL for unknown and 34.37 mL for the final standard. Final signal is 10 times as great as the initial signal, which is more than recommended range of 2–5.

	A	B	C	D	E	F	G	H
1	Standard Addition: Ammonia in Seawater							
2								
3	$V_i =$	101	mL					
4	$c_S =$	10	ppm					
5	$s =$	0.0566	V					
6								
7	x			y				
8	Added standard	E = cell	$V_i + V_s$	$(V_i + V_s)10^{E/s}$				
9	(mL)	voltage (V)	(mL)	(mL)				
10	0.00	-0.0844	101.0	3.26				
11	10.00	-0.0581	111.0	10.44				
12	20.00	-0.0469	121.0	17.95				
13	30.00	-0.0394	131.0	26.37				
14	40.00	-0.0347	141.0	34.37				
15								
16		LINEST output:			Highlight cells B17:D19			
17	m	0.7815	2.8502	b	Type '=LINEST(D10:D14,A10:A14,TRUE,TRUE)			
18	u_m	0.0137	0.3364	u_b	Press CTRL+SHIFT+ENTER (on PC)			
19	R^2	0.9991	0.4343	s_y	Press CTRL+SHIFT+RETURN (on Mac)			
20								
21	x-intercept = -b/m =	-3.647	mL		To find 95% confidence interval			
22	n =	5	B22 = COUNT(A10:A14)		we need Student's t for			
23	Mean y =	18.479	B23 = AVERAGE(D10:D14)		3 degrees of freedom			
24	$\Sigma(x_i - \text{mean } x)^2 =$	1000	B24 = DEVSQ(A10:A14)		T.INV.2T(0.05,3) =		3.182446	
25	Std uncertainty of				t*s =		1.541109	
26	x-intercept =	0.484	mL					
27	B26 = (C19/ABS(B17))*SQRT((1/B22) + B23^2/(B17^2*B24))							

Electrodes and Potentiometry

15-45. For the first line of data, with A = Na$^+$ and X = Mg^{2+}

$$\log K_{A,X}^{Pot} = \frac{z_A F(E_X - E_A)}{RT \ln 10} + \log\left(\frac{\mathcal{A}_A}{(\mathcal{A}_X)^{z_A/z_X}}\right)$$

$$= \frac{(+1)(96\,485 \text{ C/mol})(-0.385 \text{ V})}{(8.314\,5 \text{ V}\cdot\text{C/[mol}\cdot\text{K]})(294.65 \text{ K})\ln 10} + \log\left(\frac{10^{-3}}{(10^{-3})^{1/2}}\right)$$

$$= -8.09.$$

For the second line of data, $\log K_{A,X}^{Pot} = -8.15$.

The first and second lines should give the same selectivity coefficient. The difference is experimental error.

For the third line of data, with A = Na$^+$ and X = K$^+$:

$$\log K_{A,X}^{Pot} = \frac{(+1)(96\,485 \text{ C/mol})(-0.285 \text{ V})}{(8.3145 \text{ V}\cdot\text{C/[mol}\cdot\text{K]})(294.65\text{K}) \ln 10} + \log\left(\frac{10^{-3}}{(10^{-3})^{1/1}}\right) = -4.87.$$

For the fourth line of data, $\log K_{A,X}^{Pot} = -4.87$.

15-46. For Na$^+$: % error in $\mathcal{A}_{H^+} = \dfrac{(10^{-8.6})^{1/1}(10^{-2.0})}{(10^{-8.0})^{1/1}} \times 100 = 0.25\%$

For Ca^{2+}: % error in $\mathcal{A}_{H^+} = \dfrac{(10^{-7.8})^{2/1}(10^{-2.0})}{(10^{-8.0})^{2/1}} \times 100 = 2.5\%$

15-47. $E = \underbrace{\text{constant}}_{A} + \underbrace{\dfrac{\beta(0.059\,16)}{2}}_{B} \log\left([\text{Ca}^{2+}] + K_{\text{Ca}^{2+},\text{Mg}^{2+}}^{Pot}[\text{Mg}^{2+}]\right)$

For the first two solutions we can write

$-52.6 \text{ mV} = A + B \log(1.00 \times 10^{-6}) = A - 6B$

$+16.1 \text{ mV} = A + B \log(2.43 \times 10^{-4}) = A - 3.614 B.$

Subtraction gives $68.7 \text{ mV} = 2.386 B \Rightarrow B = 28.80 \text{ mV}$.

Putting this value of B back into the first equation gives $A = 120.2$ mV.

The third set of data now gives the selectivity coefficient:

$-38.0 \text{ mV} = 120.2 + 28.80 \log[10^{-6} + K_{\text{Ca}^{2+},\text{Mg}^{2+}}^{Pot}(3.68 \times 10^{-3})]$

$\Rightarrow K_{\text{Ca}^{2+},\text{Mg}^{2+}}^{Pot} = 6.0 \times 10^{-4}$

$E = 120.2 + 28.80 \log([\text{Ca}^{2+}] + 6.0 \times 10^{-4}[\text{Mg}^{2+}]).$

15-48. There is a large excess of EDTA in the buffer. We expect essentially all lead to be in the form PbY^{2-} (where Y = EDTA).

$$[PbY^{2-}] = \frac{1.0 \text{ mL}}{101.0 \text{ mL}}(0.10 \text{ M}) = 9.9 \times 10^{-4} \text{ M}$$

$$\text{Total EDTA} = \frac{100.0 \text{ mL}}{101.0 \text{ mL}}(0.050 \text{ M}) = 0.049_5 \text{ M}$$

$$\text{Free EDTA} = 0.049_5 \text{ M} - \underbrace{9.9 \times 10^{-4} \text{ M}}_{\text{EDTA bound to } Pb^{2+}} = 0.048_5 \text{ M}$$

$$Pb^{2+} + Y^{4-} \rightleftharpoons PbY^{2-} \quad K_f' = \alpha_{Y^{4-}} K_f = (1.46 \times 10^{-8})(10^{18.0}) = 1.4_6 \times 10^{10}$$

$$K_f' = \frac{[PbY^{2-}]}{[Pb^{2+}][EDTA]}$$

$$\Rightarrow [Pb^{2+}] = \frac{[PbY^{2-}]}{K_f'[EDTA]} = \frac{9.9 \times 10^{-4}}{(1.4_6 \times 10^{10})(0.048_5)} = 1.4 \times 10^{-12} \text{ M}$$

15-49. (a) pK_a for HNO_2 is 3.15. Below pH 4, nitrite reacts with H^+ to make HNO_2, so the concentration of NO_2^- decreases and the electrode potential becomes more positive. The approximate response of the electrode to nitrite is $E = \text{constant} - (59 \text{ mV})\log[NO_2^-]$. We predict that a factor-of-10 decrease in nitrate concentration will increase the electrode potential by 59 mV. At pH 3.15, nitrite is half protonated and its concentration is reduced from 1 µM to 0.5 µM. The electrode potential is expected to change by

$$\Delta E = \{-(59 \text{ mV})\log[0.5 \times 10^{-6}]\} - \{-(59 \text{ mV})\log[1 \times 10^{-6}]\} = +18 \text{ mV}$$

which is approximately equal to the observed change in the graph.

(b) At pH >6, $[NO_2^-]$ is constant at 1 µM. However, $[OH^-]$ increases by a factor of 10 for every unit increase in pH. If OH^- interferes with the measurement of NO_2^-, we expect to see the potential decrease as $[OH^-]$ increases.

15-50. $[Hg^{2+}]$ in the buffers is computed from equilibrium constants for the solubility of HgX_2 and formation of complex ions such as HgX_3^-. Since the data for $HgCl_2$ are not in line with the data for $Hg(NO_3)_2$ and $HgBr_2$, equilibrium constants used for the $HgCl_2$ system could be in error. Whenever we make a buffer by mixing *calculated* quantities of reagents, we are at the mercy of the quality of tabulated equilibrium constants.

Electrodes and Potentiometry   211

15-51. (a) slope $= 29.58 \text{ mV} = \dfrac{E_2 - E_1}{\log \mathcal{A}_2 - \log \mathcal{A}_1} = \dfrac{(-25.90) - 2.06}{\log \mathcal{A}_2 - (-3.000)}$

$\Rightarrow \mathcal{A}_2 = 1.13 \times 10^{-4}$

(b) $\quad \text{Ca}^{2+} \quad + \quad \text{A}^{3-} \quad \rightleftharpoons \quad \text{CaA}^{-}$
$\quad 5.00 \times 10^{-4} - x \quad (0.998)[(5.00 \times 10^{-4}) - x] \quad\quad\quad x$

But $\mathcal{A}_{\text{Ca}^{2+}} = 1.13 \times 10^{-4} = (5.00 \times 10^{-4} - x)(0.405)$
$\quad\quad\quad\quad\quad\quad\quad\quad\quad\quad\quad\quad\quad\uparrow$
$\quad\quad\quad\quad\quad\quad\quad\quad\quad\quad\gamma$ from Table 8-1

$\Rightarrow x = 2.2 \times 10^{-4} \text{ M}$

$K_f = \dfrac{[\text{CaA}^-]\gamma_{\text{CaA}^-}}{[\text{Ca}^{2+}]\gamma_{\text{Ca}^{2+}}[\text{A}^{3-}]\gamma_{\text{A}^{3-}}}$

$K_f = \dfrac{(2.20 \times 10^{-4})(0.79)}{(1.13 \times 10^{-4})[(0.998)(5 \times 10^{-4} - 2.20 \times 10^{-4})](0.115)} = 4.8 \times 10^{4}$
$\quad\quad\quad\quad\quad\quad\quad\quad\quad\quad\quad\quad\uparrow$
$\quad\quad\quad\quad\quad\quad\quad\quad\quad\quad\gamma$ from Table 8-1

15-52. Analyte adsorbed on the surface of the gate changes the electric potential of the gate. This, in turn, changes the current between the source and drain. The potential that must be applied by the external circuit to restore the current to its initial value is a measure of the change in gate potential. Following the Nernst equation, there is close to a 59 mV change in gate potential for each factor-of-10 change in activity of univalent analyte at 25°C. The key to ion-specific response is to have a chemical on the gate that selectively binds one analyte.

15-53. DNA is randomly cleaved into many small fragments. Many copies (~10^5–10^6) of one strand of one fragment are attached to one bead placed into one well of the device. Different beads in each well carry different fragments of DNA. The enzyme and primer required for DNA replication are added to the fragments on each bead. A solution containing a single nucleotide triphosphate is then added to all of the wells. If the next required nucleotide for chain extension was added to DNA in a particular well, it is incorporated and liberates some H$^+$ from the reaction with the existing strand of DNA. The H$^+$-sensitive tantalum oxide at the base of the well regulates the current in a field effect transistor below each well. If the right base was added to extend the DNA in a particular well, H$^+$ is released and there is a voltage change in the transistor associated with that well.

If the wrong base was added, no reaction occurs in that well and no signal is recorded by the transistor. After a few seconds, the wells are washed and fresh solution with a different nucleotide base is added. The sequence is repeated to see which wells respond to which added base. The purpose of each field effect transistor is to respond when the next required base is added to each well.

CHAPTER 16
REDOX TITRATIONS

16-1. (a) $Ce^{4+} + Fe^{2+} \rightarrow Ce^{3+} + Fe^{3+}$

(b) $Fe^{3+} + e^- \rightleftharpoons Fe^{2+}$ $E° = 0.767$ V

$Ce^{4+} + e^- \rightleftharpoons Ce^{3+}$ $E° = 1.70$ V

(c) $E = \left(0.767 - 0.05916 \log \dfrac{[Fe^{2+}]}{[Fe^{3+}]}\right) - (0.241)$ (A)

$E = \left(1.70 - 0.05916 \log \dfrac{[Ce^{3+}]}{[Ce^{4+}]}\right) - (0.241)$ (B)

(d) <u>10.0 mL</u>: Use Eq. (A) with $[Fe^{2+}]/[Fe^{3+}] = 40.0/10.0$, since $V_e = 50.0$ mL $\Rightarrow E = 0.490$ V.

<u>25.0 mL</u>: $[Fe^{2+}]/[Fe^{3+}] = 25.0/25.0 \Rightarrow E = 0.526$ V

<u>49.0 mL</u>: $[Fe^{2+}]/[Fe^{3+}] = 1.0/49.0 \Rightarrow E = 0.626$ V

<u>50.0 mL</u>: This is V_e, where $[Ce^{3+}] = [Fe^{3+}]$ and $[Ce^{4+}] = [Fe^{2+}]$. Eq. 16-11 gives $E_+ = 1.23$ V and $E = 0.99$ V.

<u>51.0 mL</u>: Use Eq. (B) with $[Ce^{3+}]/[Ce^{4+}] = 50.0/1.0 \Rightarrow E = 1.36$ V.

<u>60.0 mL</u>: $[Ce^{3+}]/[Ce^{4+}] = 50.0/10.0 \Rightarrow E = 1.42$ V

<u>100.0 mL</u>: $[Ce^{3+}]/[Ce^{4+}] = 50.0/50.0 \Rightarrow E = 1.46$ V

16-2. (a) $Ce^{4+} + Cu^+ \rightarrow Ce^{3+} + Cu^{2+}$

(b) $Ce^{4+} + e^- \rightleftharpoons Ce^{3+}$ $E° = 1.70$ V

$Cu^{2+} + e^- \rightleftharpoons Cu^+$ $E° = 0.161$ V

(c) $E = \left(1.70 - 0.05916 \log \dfrac{[Ce^{3+}]}{[Ce^{4+}]}\right) - (0.197)$ (A)

$E = \left(0.161 - 0.05916 \log \dfrac{[Cu^+]}{[Cu^{2+}]}\right) - (0.197)$ (B)

(d) <u>1.00 mL</u>: Use Eq. (A) with $[Ce^{3+}]/[Ce^{4+}] = 1.00/24.0$, since $V_e = 25.0$ mL $\Rightarrow E = 1.58$ V.

<u>12.5 mL</u>: $[Ce^{3+}]/[Ce^{4+}] = 12.5/12.5 \Rightarrow E = 1.50$ V

<u>24.5 mL</u>: $[Ce^{3+}]/[Ce^{4+}] = 24.5/0.5 \Rightarrow E = 1.40$ V

25.0 mL: $E_+ = 1.70 - 0.059\,16 \log \dfrac{[\text{Ce}^{3+}]}{[\text{Ce}^{4+}]}$

$E_+ = 0.161 - 0.059\,16 \log \dfrac{[\text{Cu}^+]}{[\text{Cu}^{2+}]}$

$\overline{2E_+ = 1.86_1 - 0.059\,16 \log \dfrac{[\text{Ce}^{3+}]}{[\text{Ce}^{4+}]} \dfrac{[\text{Cu}^+]}{[\text{Cu}^{2+}]}}$

At the equivalence point, $[\text{Ce}^{3+}] = [\text{Cu}^{2+}]$ and $[\text{Ce}^{4+}] = [\text{Cu}^+]$. Therefore, the log term above is zero and $E_+ = 1.86_1/2 = 0.930$ V.

$E = 0.930 - 0.197 = 0.733$ V

25.5 mL: Use Eq. (B) with $[\text{Cu}^+]/[\text{Cu}^{2+}] = 0.5/25.0 \Rightarrow E = 0.065$ V.

30.0 mL: $[\text{Cu}^+]/[\text{Cu}^{2+}] = 5.0/25.0 \Rightarrow E = 0.005$ V

50.0 mL: $[\text{Cu}^+]/[\text{Cu}^{2+}] = 25.0/25.0 \Rightarrow E = -0.036$ V

16-3. (a) $\text{Sn}^{2+} + \text{Tl}^{3+} \rightarrow \text{Sn}^{4+} + \text{Tl}^+$

(b) $\text{Sn}^{4+} + 2e^- \rightleftharpoons \text{Sn}^{2+}$ $\quad E° = 0.139$ V

$\text{Tl}^{3+} + 2e^- \rightleftharpoons \text{Tl}^+$ $\quad E° = 0.77$ V

(c) $E = \left(0.139 - \dfrac{0.059\,16}{2} \log \dfrac{[\text{Sn}^{2+}]}{[\text{Sn}^{4+}]}\right) - (0.241)$ (A)

$E = \left(0.77 - \dfrac{0.059\,16}{2} \log \dfrac{[\text{Tl}^+]}{[\text{Tl}^{3+}]}\right) - (0.241)$ (B)

(d) 1.00 mL: Use Eq. (A) with $[\text{Sn}^{2+}]/[\text{Sn}^{4+}] = 4.00/1.00$, since $V_e = 5.00$ mL $\Rightarrow E = -0.120$ V.

2.50 mL: $[\text{Sn}^{2+}]/[\text{Sn}^{4+}] = 2.50/2.50 \Rightarrow E = -0.102$ V

4.90 mL: $[\text{Sn}^{2+}]/[\text{Sn}^{4+}] = 0.10/4.90 \Rightarrow E = -0.052$ V

5.00 mL: $E_+ = 0.139 - \dfrac{0.059\,16}{2} \log \dfrac{[\text{Sn}^{2+}]}{[\text{Sn}^{4+}]}$

$E_+ = 0.77 - \dfrac{0.059\,16}{2} \log \dfrac{[\text{Tl}^+]}{[\text{Tl}^{3+}]}$

$\overline{2E_+ = 0.90_9 - \dfrac{0.059\,16}{2} \log \dfrac{[\text{Sn}^{2+}][\text{Tl}^+]}{[\text{Sn}^{4+}][\text{Tl}^{3+}]}}$

At the equivalence point, $[\text{Sn}^{4+}] = [\text{Tl}^+]$ and $[\text{Sn}^{2+}] = [\text{Tl}^{3+}]$. Therefore, the log term above is zero and $E_+ = 0.90_9/2 = 0.45_4$ V.

Redox Titrations 215

$$E = 0.45_4 - 0.241 = 0.21 \text{ V}$$

<u>5.10 mL</u>: Use Eq. (B) with $[Tl^+]/[Tl^{3+}] = 5.00/0.10 \Rightarrow E = 0.48$ V

<u>10.0 mL</u>: Use Eq. (B) with $[Tl^+]/[Tl^{3+}] = 5.00/5.00 \Rightarrow E = 0.53$ V

16-4. (a) $2Fe^{3+}$ + ascorbic acid + $H_2O \rightarrow 2Fe^{2+}$ + dehydroascorbic acid + $2H^+$

(b) The equivalence volume is 10.0 mL.

At <u>5.0 mL</u>, half of the Fe^{3+} is titrated and the ratio $[Fe^{2+}]/[Fe^{3+}]$ is 5.0/5.0:

$Fe^{3+} + e^- \rightleftharpoons Fe^{2+}$ $E° = 0.767$ V

$$E = E_+ - E_- = \left(0.767 - 0.059\,16 \log \frac{[Fe^{2+}]}{[Fe^{3+}]}\right) - (0.197)$$

$$= \left(0.767 - 0.059\,16 \log \frac{5.0 \text{ mL}}{5.0 \text{ mL}}\right) - (0.197) = 0.570$$

<u>10.0 mL</u> is the equivalence point. We multiply the ascorbic acid Nernst equation by 2 and add it to the iron Nernst equation:

$$E_+ = 0.767 - 0.059\,16 \log \frac{[Fe^{2+}]}{[Fe^{3+}]}$$

$$2E_+ = 2\left(0.390 - \frac{0.059\,16}{2} \log \frac{[\text{ascorbic acid}]}{[\text{dehydro}][H^+]^2}\right)$$

$$3E_+ = 1.547 - 0.059\,16 \log \frac{[Fe^{2+}][\text{ascorbic acid}]}{[Fe^{3+}][\text{dehydro}][H^+]^2}$$

<u>At the equivalence point</u>, the stoichiometry of the titration reaction tells us that $[Fe^{2+}] = 2[\text{dehydroascorbic acid}]$ and $[Fe^{3+}] = 2[\text{ascorbic acid}]$. Inserting these equalities into the log term just shown gives

$$3E_+ = 1.547 - 0.059\,16 \log \frac{2[\text{dehydro}][\text{ascorbic acid}]}{2[\text{ascorbic acid}][\text{dehydro}][H^+]^2}$$

$$3E_+ = 1.547 - 0.059\,16 \log \frac{1}{[H^+]^2}$$

Using $[H^+] = 10^{-0.30}$ gives $E_+ = 0.504$ V and

$E = 0.504 - 0.197 = 0.307$ V.

At <u>15.0 mL</u>, the ratio [dehydro]/[ascorbic acid] is 10.0 mL/5.0 mL: dehydroascorbic acid + $2H^+ + 2e^- \rightleftharpoons$ ascorbic acid + H_2O

$E° = 0.390$ V

$$E = E_+ - E_- = \left(0.390 - \frac{0.05916}{2}\log\frac{[\text{ascorbic acid}]}{[\text{dehydro}][\text{H}^+]^2}\right) - (0.197)$$

$$= \left(0.390 - \frac{0.05916}{2}\log\frac{(5.0\text{ mL})}{(10.0\text{ mL})[10^{-0.30}]^2}\right) - (0.197) = 0.184\text{ V}$$

16-5. (a) Titration reaction: $\text{Sn}^{2+} + 2\text{Fe}^{3+} \rightarrow \text{Sn}^{4+} + 2\text{Fe}^{2+}$ $\quad V_e = 25.0$ mL

(b) $\text{Fe}^{3+} + e^- \rightleftharpoons \text{Fe}^{2+}$ $\quad E° = 0.732$ V
$\text{Sn}^{4+} + 2e^- \rightleftharpoons \text{Sn}^{2+}$ $\quad E° = 0.139$ V

(c) $E = \left(0.732 - 0.05916\log\frac{\text{Fe}^{2+}}{\text{Fe}^{3+}}\right) - (0.241)$ $\quad$ (A)

$E = \left(0.139 - \frac{0.05916}{2}\log\frac{[\text{Sn}^{2+}]}{[\text{Sn}^{4+}]}\right) - (0.241)$ $\quad$ (B)

(d) Representative calculations:

<u>1.0 mL</u>: $E_+ = 0.139 - \frac{0.05916}{2}\log\frac{[\text{Sn}^{2+}]}{[\text{Sn}^{4+}]}$

initial mol $\text{Sn}^{2+} = (25.0\text{ mL})\left(0.0500\frac{\text{mmol}}{\text{mL}}\right) = 1.25$ mmol

mol Fe^{3+} added $= (1.0\text{ mL})\left(0.100\frac{\text{mmol}}{\text{mL}}\right) = 0.10$ mmol

Note that 1 mol Fe^{3+} consumes ½ mol Sn^{2+} and gives ½ mol Sn^{4+}

$[\text{Sn}^{4+}] = \frac{½(0.10\text{ mmol})}{26.0\text{ mL}} = 1.9_2 \times 10^{-3}$ M

$[\text{Sn}^{2+}] = \frac{1.25 - ½(0.10\text{ mmol})}{26.0\text{ mL}} = 4.62 \times 10^{-2}$ M

$E_+ = 0.139 - \frac{0.05916}{2}\log\frac{[\text{Sn}^{2+}]}{[\text{Sn}^{4+}]}$

$E_+ = 0.139 - \frac{0.05916}{2}\log\frac{[4.62\times10^{-2}\text{M}]}{[1.9_2\times10^{-3}\text{M}]} = 0.098$ V

$E = E_+ - E_- = 0.098 - 0.241 = -0.143$ V

<u>25.0 mL</u>: At the equivalence point, we add the two indicator electrode Nernst equations. To make the factor in front of the log term the same in both equations, we can multiply the $\text{Sn}^{4+}\mid\text{Sn}^{2+}$ equation by 2:

$$E_+ = 0.139 - \frac{0.05916}{2} \log \frac{[Sn^{2+}]}{[Sn^{4+}]}$$

Multiply × 2: $\quad 2E_+ = 0.278 - 0.05916 \log \frac{[Sn^{2+}]}{[Sn^{4+}]}$

$$E_+ = 0.732 - 0.05916 \log \left(\frac{[Fe^{2+}]}{[Fe^{3+}]}\right)$$

Now add the last two equations to get

$$3E_+ = 1.010 - 0.05916 \log \left(\frac{[Sn^{2+}][Fe^{2+}]}{[Sn^{4+}][Fe^{3+}]}\right)$$

But at the equivalence point, $2[Sn^{4+}] = [Fe^{2+}]$ and $2[Sn^{2+}] = [Fe^{3+}]$. Substituting these identities into the log term gives

$$3E_+ = 1.010 - 0.05916 \log \left(\frac{[Sn^{2+}]2[Sn^{4+}]}{[Sn^{4+}]2[Sn^{2+}]}\right)$$

So the log quotient in the log term is 1 and the logarithm is 0. Therefore, $E_+ = 1.010/3 = 0.337$ V and $E = E_+ - E_- = 0.337 - 0.241 = 0.096$ V.

<u>26.0 mL</u>: $E_+ = 0.732 - 0.05916 \log \left(\frac{[Fe^{2+}]}{[Fe^{3+}]}\right)$

There is 1.0 mL of Fe^{3+} beyond the equivalence point.

$$[Fe^{3+}] = \frac{(1.0 \text{ mL})(0.100 \text{ M})}{51.0 \text{ mL}} = 1.9_6 \times 10^{-3} \text{ M}$$

The first 25.0 mL of Fe^{3+} were converted to Fe^{2+}, so

$$[Fe^{2+}] = \frac{(25.0 \text{ mL})(0.100 \text{ M})}{51.0 \text{ mL}} = 4.90 \times 10^{-2} \text{ M}$$

$$E_+ = 0.732 - 0.05916 \log \left(\frac{[Fe^{2+}]}{[Fe^{3+}]}\right)$$

$$E_+ = 0.732 - 0.05916 \log \left(\frac{[4.90 \times 10^{-2}]}{[1.9_6 \times 10^{-3}]}\right) = 0.649 \text{ V}$$

$$E = E_+ - E_- = 0.649 - 0.241 = 0.408 \text{ V}$$

mL	E (V)	mL	E (V)	mL	E (V)
1.0	−0.143	24.0	−0.061	26.0	0.408
12.5	−0.102	25.0	0.096	30.0	0.450

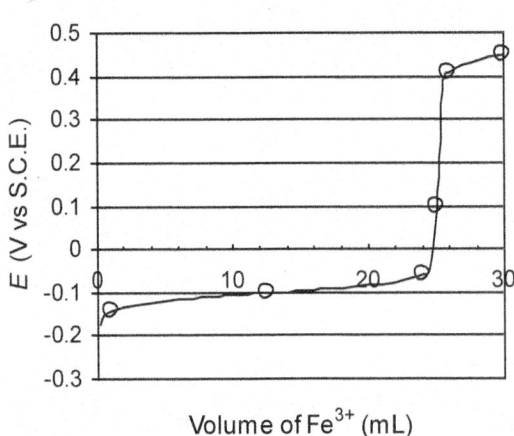

Volume of Fe^{3+} (mL)

16-6. Diphenylamine sulfonic acid: colorless → red-violet
Diphenylbenzidine sulfonic acid: colorless → violet
tris (2,2'-bipyridine) iron: red → pale blue
Ferroin: red → pale blue

16-7. The reduction potentials are
$$Sn^{4+} + 2e^- \rightleftharpoons Sn^{2+} \qquad E° = 0.139 \text{ V}$$
$$Mn(EDTA)^- + e^- \rightleftharpoons Mn(EDTA)^{2-} \qquad E° = 0.825 \text{ V}$$
The end point (versus S.H.E.) is between 0.139 and 0.825 V. Tris(2,2'-bipyridine) iron has too high a reduction potential (1.120 V) to be useful for this titration.

16-8. Preoxidation and prereduction refer to adjusting the oxidation state of analyte to a suitable value for a titration. The preoxidation or prereduction agent must be destroyed so it does not interfere with the titration by reacting with titrant.

16-9. $2S_2O_8^{2-} + 2H_2O \xrightarrow{\text{boiling}} 4SO_4^{2-} + O_2 + 4H^+$

$Ag^{3+} + H_2O \xrightarrow{\text{boiling}} Ag^+ + \frac{1}{2}O_2 + 2H^+$

$2H_2O_2 \xrightarrow{\text{boiling}} O_2 + 2H_2O$

16-10. A Jones reductor is a column packed with zinc granules coated with zinc amalgam. Prereduction is accomplished by passing analyte solution through the column.

Redox Titrations

16-11. Cr^{3+} and TiO^{2+} would interfere if they were reduced to Cr^{2+} and Ti^{3+}. In the Jones reductor, Zn is a strong enough reductant to react with Cr^{3+} and TiO^{2+}.

$E° = -0.764$ for the $Zn^{2+}|Zn$ couple

$E° = -0.42$ for the $Cr^{3+}|Cr^{2+}$ couple

$E° = 0.1$ for the $TiO^{2+}|Ti^{3+}$ couple

In the Walden reductor, Ag is not strong enough to reduce Cr^{3+} and TiO^{2+}:

$E° = 0.222$ for the $AgCl|Ag$ couple

16-12. A weighed amount of the solid mixture is added to a solution containing excess standard Fe^{2+} plus phosphoric acid. Each mol of $(NH_4)_2S_2O_8$ oxidizes 2 mol of Fe^{2+} to Fe^{3+}. Excess Fe^{2+} is then titrated with standard $KMnO_4$ to find out how much Fe^{2+} was consumed by the $(NH_4)_2S_2O_8$. The phosphoric acid masks the yellow color of Fe^{3+}, making the end point easier to see.

16-13. (a) $MnO_4^- + 8H^+ + 5e^- \rightleftharpoons Mn^{2+} + 4H_2O$

(b) $MnO_4^- + 4H^+ + 3e^- \rightleftharpoons MnO_2(s) + 2H_2O$

(c) $MnO_4^- + e^- \rightleftharpoons MnO_4^{2-}$

16-14. $3MnO_4^- + 5Mo^{3+} + 4H^+ \rightarrow 3Mn^{2+} + 5MoO_2^{2+} + 2H_2O$

$(16.43 - 0.04) = 16.39$ mL of 0.01033 M $KMnO_4 = 0.1693$ mmol of MnO_4^-, which will react with $(5/3)(0.1693) = 0.2822$ mmol of Mo^{3+}.

$[Mo^{3+}] = 0.2822$ mmol/25.00 mL $= 0.01129$ M ($=$ original $[MoO_4^{2-}]$).

16-15. $2MnO_4^- + 5H_2O_2 + 6H^+ \rightarrow 2Mn^{2+} + 5O_2 + 8H_2O$

$(27.66 - 0.04) = 27.62$ mL of 0.02123 M $KMnO_4 = 0.58637$ mmol of MnO_4^-, which reacts with $(5/2)(0.58637) = 1.4659$ mmol of H_2O_2, which came from 25.00 mL of diluted solution $\Rightarrow [H_2O_2] = 1.4659$ mmol/25.00 mL $= 0.05864$ M in the dilute solution. The original solution was 10 times more concentrated $= 0.5864$ M.

16-16. (a) Scheme 1:

$$2\,[8H^+ + MnO_4^- + 5e^- \rightarrow Mn^{2+} + 4H_2O]$$
$${+7}{+2}$$
$$5\,[H_2O_2 \rightarrow O_2 + 2e^- + 2H^+]$$
$${-1}{0}$$

$$\overline{6H^+ + 2MnO_4^- + 5H_2O_2 \rightarrow 2Mn^{2+} + 5O_2 + 8H_2O}$$

Scheme 2:

$$2\,[MnO_4^- \rightarrow Mn^{2+} + 2O_2 + 3e^-]$$
$${+7\ -2}{+2}{0}$$
$$3\,[H_2O_2 + 2H^+ + 2e^- \rightarrow 2H_2O]$$
$${-1}{-2}$$

$$\overline{6H^+ + 2MnO_4^- + 3H_2O_2 \rightarrow 2Mn^{2+} + 4O_2 + 6H_2O}$$

(b) $\dfrac{1.023 \text{ g NaBO}_3 \cdot 4\,H_2O}{153.86 \text{ g/mol}} = 6.649 \text{ mmol NaBO}_3$

One tenth of this quantity was titrated = 0.664 9 mmol NaBO$_3$, producing 0.664 9 mmol H$_2$O$_2$ by the reaction $BO_3^- + 2H_2O \rightarrow H_2O_2 + H_2BO_3^-$.

In Scheme 1, 2MnO$_4^-$ react with 5H$_2$O$_2$

$\Rightarrow$ 0.664 9 mmol H$_2$O$_2$ requires $\tfrac{2}{5}(0.664\ 9) = 0.266\ 0$ mmol MnO$_4^-$

$\dfrac{0.266\ 0 \text{ mmol MnO}_4^-}{0.010\ 46 \text{ mmol KMnO}_4/\text{mL}} = 25.43$ mL KMnO$_4$ required

In Scheme 2, 2MnO$_4^-$ react with 3H$_2$O$_2$

$\Rightarrow$ 0.664 9 mmol H$_2$O$_2$ requires $\tfrac{2}{3}(0.664\ 9) = 0.443\ 3$ mmol MnO$_4^-$

$\dfrac{0.443\ 3 \text{ mmol MnO}_4^-}{0.010\ 46 \text{ mmol KMnO}_4/\text{mL}} = 42.38$ mL KMnO$_4$ required

16-17. $2MnO_4^- + 5H_2C_2O_4 + 6H^+ \rightarrow 2Mn^{2+} + 10CO_2 + 8H_2O$

18.04 mL of 0.006 363 M KMnO$_4$ = 0.114 8 mmol of MnO$_4^-$, which reacts with (5/2)(0.114 8) = 0.287 0 mmol of H$_2$C$_2$O$_4$, which came from (2/3)(0.287 0) = 0.191 3 mmol of La^{3+}. [La^{3+}] = 0.191 3 mmol/50.00 mL = 3.826 mM.

16-18. $\underset{\substack{\text{glycerol}\\ \text{(average oxidation}\\ \text{number of C} = -2/3)}}{C_3H_8O_3} + 3H_2O \rightleftharpoons \underset{\substack{\text{formic acid}\\ \text{(oxidation}\\ \text{number of C} = +2)}}{3HCO_2H} + 8e^- + 8H^+$

$8Ce^{4+} + 8e^- \rightleftharpoons 8Ce^{3+}$

$\overline{C_3H_8O_3 + 8Ce^{4+} + 3H_2O \rightleftharpoons 3HCO_2H + 8Ce^{3+} + 8H^+}$

One mole of glycerol requires eight moles of Ce^{4+}.

50.0 mL of 0.083 7 M Ce^{4+} = 4.185 mmol

12.11 mL of 0.044 8 M Fe^{2+} = 0.543 mmol

Ce^{4+} reacting with glycerol = 3.642 mmol

glycerol = (1/8) (3.642) = 0.455_2 mmol = 41.9 mg

⇒ original solution = 41.9 wt% glycerol

16-19. Ferrous ammonium sulfate is $Fe(NH_4)_2(SO_4)_2 \cdot 6H_2O$

50.00 mL of 0.118 6 M Ce^{4+} = 5.930 mmol Ce^{4+}

31.13 mL of 0.042 89 M Fe^{2+} = 1.335 mmol Fe^{2+}

4.595 mmol Ce^{4+} consumed by NO_2^-

Since two moles of Ce^{4+} react with one mole of NO_2^-, there must have been 1/2 (4.595) = 2.298 mmol of $NaNO_2$ = 0.158 5 g in 25.0 mL. In 500.0 mL, there would be $\left(\dfrac{500}{25.0}\right)$(0.158 5) = 3.170 g = 78.67% of the 4.030 g sample.

16-20. Step 2 gives the total Cr content of the crystal, since each Cr^{x+} ion in any oxidation state is oxidized and reacts with $3Fe^{2+}$.

Step 2: $\dfrac{(0.703 \text{ mL})(2.786 \text{ mM})}{0.156\,6 \text{ g of crystal}} = \dfrac{12.51 \text{ μmol Fe}^{2+}}{\text{g of crystal}}$

$\dfrac{\frac{1}{3}(12.51) \text{ μmol Cr}}{\text{g of crystal}} = \dfrac{4.169 \text{ μmol Cr}}{\text{g of crystal}}$

Step 1 tells us how much Cr^{x+} is oxidized above the +3 state. Each Cr^{x+} reacts with $(x - 3)$ Fe^{2+}.

Step 1: $\dfrac{(0.498 \text{ mL})(2.786 \text{ mM})}{0.437\,5 \text{ g of crystal}} = \dfrac{3.171 \text{ μmol Fe}^{2+}}{\text{g of crystal}}$

Since one gram of crystal contains 4.169 μmol of Cr that reacts with 3.171 μm of Fe^{2+}, the average oxidation state of Cr is $3 + \dfrac{3.171}{4.169} = +3.761$.

Total Cr (from Step 2) = 4.169 μmol Cr per gram = 217 μg per gram of crystal.

16-21. (a) Theoretical molarity = (3.214 g/L)/(158.034 g/mol) = $0.020\,33_7$ M.

(b) 25.00 mL of $0.020\,33_7$ M $KMnO_4$ = $0.508\,4_3$ mmol. Two moles of MnO_4^- react with five moles of H_3AsO_3, which comes from $\frac{5}{4}$ moles of As_4O_6.

Moles of As_4O_6 needed to react with $0.508\,4_3$ mmol of MnO_4^- =

$$\left(\frac{5/4 \text{ mol As}_4\text{O}_6}{2 \text{ mol KMnO}_4}\right)(0.508\,4_3 \text{ mmol KMnO}_4) = 0.317\,7_7 \text{ mmol As}_4\text{O}_6 =$$

0.125 7$_4$ g of As$_4$O$_6$.

(c) $\dfrac{0.508\,4_3 \text{ mmol KMnO}_4}{0.125\,7_4 \text{ g As}_2\text{O}_3} = \dfrac{x \text{ mmol KMnO}_4}{0.146\,8 \text{ g As}_4\text{O}_6} \Rightarrow x = 0.593\,6_1 \text{ mmol}$

KMnO$_4$ in (29.98 – 0.03) = 29.95 mL $\Rightarrow$ [KMnO$_4$] = 0.019 82 M.

16-22. I$^-$ reacts with I$_2$ to give I$_3^-$. This reaction increases the solubility of I$_2$ and decreases its volatility.

16-23. Standard triiodide can be prepared from a weighed amount of KIO$_3$ with acid plus excess iodide. Alternatively, triiodide solution can be standardized by reaction with standard S$_2$O$_3^{2-}$ prepared from anhydrous Na$_2$S$_2$O$_3$.

16-24. Starch is not added until just before the end point in iodometry, so it does not irreversibly bind to I$_2$, which is present during the whole titration.

16-25. S$_4$O$_6^{2-}$ + 2e$^-$ $\rightleftharpoons$ 2S$_2$O$_3^{2-}$ or S$_4$O$_6^{2-}$ + 4H$^+$ + 2e$^-$ $\rightleftharpoons$ 2H$_2$SO$_3$

$E°$ for the second half-reaction above is 0.57 V. $E°$ for the half-reaction ½O$_2$(g) + 2H$^+$ + 2e$^-$ $\rightleftharpoons$ H$_2$O is 1.23 V. O$_2$ is a stronger oxidant than tetrathionate.

16-26. (a) 50.00 mL contains exactly 1/10 of the KIO$_3$ = 0.102 2 g = 0.477 5$_7$ mmol KIO$_3$. Each mol of iodate makes 3 mol of triiodide, so I$_3^-$ = 3(0.477 5$_7$) = 1.432$_7$ mmol.

(b) Two moles of thiosulfate react with one mole of I$_3^-$. Therefore, there must have been 2(1.432$_7$) = 2.865$_4$ mmol of thiosulfate in 37.66 mL, so the concentration is (2.865$_4$ mmol)/(37.66 mL) = 0.076 08$_7$ M.

(c) 50.00 mL of KIO$_3$ make 1.432$_7$ mmol I$_3^-$. The unreacted I$_3^-$ requires 14.22 mL of sodium thiosulfate = (14.22 mL)(0.076 08$_7$ M) = 1.082$_0$ mmol, which reacts with $\frac{1}{2}$(1.082$_0$ mmol) = 0.541$_0$ mmol I$_3^-$. The ascorbic acid must have consumed the difference = 1.432$_7$ – 0.541$_0$ = 0.891$_7$ mmol I$_3^-$. Each mole of ascorbic acid consumes one mole of I$_3^-$, so mol ascorbic acid = 0.891$_7$ mmol, which has a mass of (0.891$_7$ × 10^{-3} mol)(176.12 g/mol) = 0.157$_0$ g. Ascorbic acid in the unknown = 100 × (0.157$_0$ g)/(1.223 g) = 12.8 wt%.

Redox Titrations

(d) Starch should not be added until just before the end point because I_3^- is present throughout the titration and will irreversibly bind to starch if the starch is added too early.

16-27. $2Cu^{2+} + 5I^- \rightarrow 2CuI(s) + I_3^-$ $I_3^- + 2S_2O_3^{2-} \rightarrow 3I^- + S_4O_6^{2-}$

23.33 mL of 0.046 68 M $Na_2S_2O_3$ = 1.089$_0$ mmol $S_2O_3^{2-}$ = 0.544 5 mmol I_3^-, which came from 1.089$_0$ mmol Cu^{2+} = 69.20 mg Cu. This much Cu comes from 1/5 of the original solid, which therefore contained 346.0 mg Cu = 11.43 wt%. There is a great deal of I_3^- present at the start of the titration, so starch should not be added until just before the end point.

16-28. (a) 4.0 mL of the 297.6 mL sample were displaced by $MnSO_4$ and alkali solution. After mixing, the resulting 297.6 mL volume contains 293.6 mL of the creek water, or 293.6/297.6 = 98.66%.

(b) After adding 2.0 mL of H_2SO_4, 2.0 mL of solution are displaced from the bottle, which now contains 291.6 mL of creek water in a volume of 297.6 mL. The fraction of creek water in the bottle is 291.6/297.6 = 97.98%.

(c) 200.0 mL that are titrated contain (0.979 8)(200.0 mL) = 196.0 mL of creek water.

(d) $1O_2$ makes $4Mn(OH)_3$, which makes $2I_3^-$.

(e) Thiosulfate required = (14.05 mL)(0.010 22 M) = 0.143 6 mmol $S_2O_3^{2-}$
2 moles of $S_2O_3^{2-}$ react with 1 mole of I_3^-, so mmol I_3^- produced by O_2 = ½(0.143 6 mmol) = 0.071 80 mmol O_2 = 2.297 mg O_2.
Dissolved O_2 concentration (mg/L) = 2.297 mg/0.196 0 L = 11.7 mg/L

(f) Fraction of saturation = (11.7 mg/L)/(14.6 mg/L) = 80%

(g) $2HNO_2 + 2H^+ + 3I^- \rightarrow 2NO + I_3^- + 2H_2O$
(NO could go on to react with even more I^-)

16-29. $H_2S + I_3^- \rightarrow S(s) + 3I^- + 2H^+$ $I_3^- + 2S_2O_3^{2-} \rightarrow 3I^- + S_4O_6^{2-}$

25.00 mL of 0.010 44 M I_3^- = 0.261 0$_0$ mmol I_3^-

14.44 mL of 0.009 336 M $Na_2S_2O_3$ = 0.134 8$_1$ mmol $Na_2S_2O_3$, which would have reacted with ½(0.134 8$_1$ mmol) = 0.067 40$_6$ mmol I_3^-. Therefore, the quantity of I_3^- that reacted with H_2S was $0.261\,0_0 - 0.067\,40_6 = 0.193\,5_9$ mmol. Since 1 mol of I_3^- reacts with 1 mol of H_2S, the H_2S concentration was $0.193\,5_9$ mmol/25.00 mL = 0.007 744 M. I_3^- is present at the start of the titration, so starch should not be added until just before the end point.

16-30. (a)

$$I_2(aq) + 2e^- \rightleftharpoons 2I^- \qquad E° = 0.620 \text{ V}$$
$$3I^- \rightleftharpoons I_3^- + 2e^- \qquad E° = -0.535 \text{ V}$$
$$\overline{I_2(aq) + I^- \rightleftharpoons I_3^-} \qquad E° = 0.085 \text{ V}$$
$$K = 10^{2(0.085)/0.059\,16} = 7 \times 10^2$$

(b)
$$I_2(s) + 2e^- \rightleftharpoons 2I^- \qquad E° = 0.535 \text{ V}$$
$$3I^- \rightleftharpoons I_3^- + 2e^- \qquad E° = -0.535 \text{ V}$$
$$\overline{I_2(s) + I^- \rightleftharpoons I_3^-} \qquad E° = 0.000 \text{ V}$$
$$K = 10^{2(-0.000)/0.059\,16} = 1.0$$

(c)
$$I_2(s) + 2e^- \rightleftharpoons 2I^- \qquad E° = 0.535 \text{ V}$$
$$2I^- \rightleftharpoons I_2(aq) + 2e^- \qquad E° = -0.620 \text{ V}$$
$$\overline{I_2(s) \rightleftharpoons I_2(aq)} \qquad E° = -0.085 \text{ V}$$
$$K = [I_2(aq)] = 10^{2(-0.085)/0.059\,16} = 1.3 \times 10^{-3} \text{ M} = 0.34 \text{ g of } I_2/L$$

16-31. Each mole of NH_3 liberated in the Kjeldahl digestion reacts with 1 mole of H^+ in the standard H_2SO_4 solution. Six moles of H^+ left (3 moles of H_2SO_4) after reaction with NH_3 will react with 1 mole of iodate by Reaction 16-18 to release 3 moles of I_3^-. Two moles of thiosulfate react with 1 mole of I_3^- in Reaction 16-19. Therefore each mole of thiosulfate corresponds to $\frac{1}{2}$ mol of residual H_2SO_4.

$$\text{mol } NH_3 = 2 \text{ (initial mol } H_2SO_4 - \text{final mol } H_2SO_4)$$
$$\text{mol } NH_3 = 2 \text{ (initial mol } H_2SO_4 - \tfrac{1}{2} \times \text{mol thiosulfate)}$$

16-32. (a) $IO_3^- + 8I^- + 6H^+ \rightleftharpoons 3I_3^- + 3H_2O$. The stock solution contained {0.804 3 g KIO_3 (FM 214.00) + 6.0 g KI (FM 166.00)}/100 mL, which translates into 0.037 58 M KIO_3 plus 0.36 M KI, giving a mole ratio KI/KIO_3 = 9.6, which is an excess over the 8:1 ratio required in the reaction. 5.00 mL of the stock solution contain 0.187 9_2 mmol KIO_3 plus

Redox Titrations 225

1.8 mmol KI. 1.0 mL of 6.0 M H_2SO_4 contains 6 mmol H_2SO_4, which is a large excess for the reaction. Neither KI nor H_2SO_4 need to be measured accurately.

(b) $I_3^- + SO_3^{2-} + H_2O \rightarrow 3I^- + SO_4^{2-} + 2H^+$

(c) $0.187\,9_2$ mmol KIO_3 delivered to the wine generates $3 \times 0.187\,9_2 = 0.563\,7_6$ mmol I_3^-. The excess, unreacted I_3^- required 12.86 mL of 0.048 18 M $Na_2S_2O_3 = 0.619\,5_9$ mmol $Na_2S_2O_3$. Each mole of unreacted I_3^- requires 2 moles of $Na_2S_2O_3$, so there must have been $(0.619\,5_9)/2 = 0.309\,8$ mmol I_3^- left over from the reaction with sulfite. Therefore, the I_3^- that reacted with sulfite was $(0.563\,7_6 - 0.309\,8) = 0.254\,0$ mmol I_3^-. One mole of I_3^- reacts with 1 mole of sulfite, so there must have been 0.254 0 mmol SO_3^{2-} in 50.0 mL of wine. $[SO_3^{2-}] = 0.254\,0$ mmol/50.0 mL $= 5.07\,9 \times 10^{-3}$ M. With a formula mass of 80.06 for sulfite, the sulfite content is 406.6 mg/L.

(d) $s_{pooled} = \sqrt{\dfrac{2.2^2(3-1) + 2.1^2(3-1)}{3+3-2}} = 2.15$

$t_{calculated} = \dfrac{|277.7 - 273.2|}{2.15}\sqrt{\dfrac{3 \cdot 3}{3+3}} = 2.56$

$t_{table} = 2.776$ for 95% confidence and $3 + 3 - 2 = 4$ degrees of freedom
$t_{calculated} < t_{table}$, so the difference is not significant at 95% confidence level.

16-33. 25.00 mL of 0.020 00 M $KBrO_3 = 0.500\,0$ mmol of BrO_3^-, which generates 1.500 mmol of Br_2. One mole of excess Br_2 generates one mole of I_2 (from I^-) and one mole of I_2 consumes 2 moles of $S_2O_3^{2-}$. But $S_2O_3^{2-} = (8.83$ mL$)(0.051\,13$ mmol/mL$) = 0.451\,5$ mmol $S_2O_3^{2-}$, so $I_2 = \frac{1}{2}(0.451\,5$ mmol$) = 0.225\,7$ mmol and Br_2 consumed by reaction with 8-hydroxyquinoline $= 1.500 - 0.225\,7 = 1.274$ mmol. But one mol of 8-hydroxyquinoline consumes 2 mol Br_2, so 8-hydroxyquinoline $= \frac{1}{2}(1.274$ mmol$) = 0.637\,1$ mmol and $Al^{3+} = 0.6371/3 = 0.212\,4$ mmol $= 5.730$ mg.

16-34. (a) $YBa_2Cu_3O_7$ contains 1 Cu^{3+} and 2 Cu^{2+}. $YBa_2Cu_3O_{6.5}$ contains no Cu^{3+} and 3 Cu^{2+}. The moles of Cu^{3+} in the formula $YBa_2Cu_3O_{7-z}$ are therefore $1 - 2z$. The moles of superconductor in 1 g of superconductor are

226 Chapter 16

(1 g)/[(666.19 − 15.999 z)g/mol]. The difference between experiments B and A is 5.68 mmol − 4.55 mmol = 1.13 mmol $S_2O_3^{2-}$/g of superconductor.

Since 1 mol of thiosulfate is equivalent to 1 mol of Cu^{3+}, there are 1.13 mmol Cu^{3+}/g of superconductor.

$$\frac{\text{mol } Cu^{3+}}{\text{mol superconductor}} = 1 - 2z = \frac{1.13 \times 10^{-3} \text{mol } Cu^{3+}}{\left(\dfrac{1 \text{ g superconductor}}{(666.19 - 15.999z) \text{ g/mol}}\right)}$$

Solving this equation gives $z = 0.125$. The formula is $YBa_2Cu_3O_{6.875}$.

(b) $1 - 2z = \dfrac{\text{mol } Cu^{3+}}{\text{mol superconductor}} = \dfrac{[5.68(\pm 0.05) - 4.55(\pm 0.10)] \times 10^{-3}}{\left(\dfrac{1}{666.19 - 15.999z}\right)}$

$1 - 2z = \dfrac{1.13\,(\pm 0.112) \times 10^{-3}}{\left(\dfrac{1}{666.19 - 15.999z}\right)}$

$1 - 2z = 0.752\,79\,(\pm 0.074\,61) - 0.018\,079\,(\pm 0.001\,792)\,z$

$0.247\,21\,(\pm 0.074\,61) = 1.981\,21\,(\pm 0.001\,792)\,z$

$z = 0.125 \pm 0.038$. The formula is $YBa_2Cu_3O_{6.875\,\pm 0.038}$.

16-35. A superconductor containing unknown quantities of Cu(I), Cu(II), Cu(III), and peroxide (O_2^{2-}) is dissolved in a known excess of Cu(I) in oxygen-free HCl solution. Possible reactions are

$$Cu^{3+} + Cu^+ \rightarrow 2Cu^{2+}$$
$$H_2O_2 + 2Cu^+ + 2H^+ \rightarrow 2H_2O + 2Cu^{2+}$$

Unreacted Cu(I) is then measured by coulometry to find out how much Cu(I) was consumed by the dissolving superconductor. The amount of Cu(I) consumed is equal to the moles of Cu^{3+} plus 2 times the moles of O_2^{2-} in the superconductor. The coulometry is done under Ar to prevent oxidation of Cu(I) by O_2 from the air. If the superconductor contained Cu(I) (but no Cu(III) or peroxide), then the amount of Cu(I) found by coulometry would be greater than the known amount used in the original solution.

16-36. (a) Initial Fe^{2+} in 5.000 mL = (5.000 mL)(0.100 0 M) = 0.500 0 mmol.
$K_2Cr_2O_7$ required for titration of unreacted Fe^{2+}
= (3.22 8 mL)(0.015 93 M $K_2Cr_2O_7$) = 0.0514 2 mmol.
But 1 mmol $K_2Cr_2O_7$ reacts with 6 mmol Fe^{2+} by the reaction
$K_2Cr_2O_7 + 6Fe^{2+} + 14H^+ \rightarrow 2Cr^{3+} + 6Fe^{3+} + 2K^+ + 14H_2O$.
Therefore, Fe^{2+} left after reaction with $Li_{1+y}CoO_2$
= (0.0514 2 mmol)(6 mmol Fe^{2+}/mmol $K_2Cr_2O_7$) = 0.308 5 mmol.
Fe^{2+} consumed by Co^{3+} = (0.500 0 − 0.308 5) = 0.191 5 mmol.
1 mol Fe^{2+} is consumed by 1 mol Co^{3+}, so Co^{3+} in 25.00 mg solid sample = 0.191 5 mmol.

(b) Co in 25.00 mg solid = (0.564 g Co/g solid)(25.00 g solid) = 14.10 mg
Co in 25.00 mg solid = (14.10 mg)/(58.933 g/mol) = 0.239 3 mmol
From (a), we know that Co^{3+} = 0.191 5 mmol, so
Co^{2+} = 0.239 3 − 0.191 5 = 0.047 8 mmol.
Co oxidation state = $\dfrac{(0.047\ 8\ \text{mmol})(2+) + (0.191\ 5\ \text{mmol})(3+)}{0.239\ 2\ \text{mmol}}$ = 2.80

(c) If Co has an average oxidation number of +2.80 and O has an oxidation number of −2, Li must contribute a charge of 4 − 2.80 = 1.20. Therefore, the formula is $Li_{1.20}CoO_2$ and $y = 0.20$.

(d) Theoretical weight percent for metals in $Li_{1.20}CoO_2$:
Formula mass = 1.20(6.94) + 1(58.933) + 2(15.999) = 99.259
wt% Li = 100 × 1.20(6.94)/99.259 = 8.39%
wt% Co = 100 × 58.933/99.259 = 59.37%
$\dfrac{\text{wt\% Li}}{\text{wt\% Co}}$ = 0.141 3

The observed quotient wt% Li/wt% Co is 0.138 8 ±. 0.000 6, which is not exactly equal to the quotient computed from the oxidation number. The difference represents experimental error between the two methods used find the stoichiometry.

16-37. Denote the average oxidation number of Bi as $3 + b$ and the average oxidation number of Cu as $2 + c$.
$$Bi_2^{3+b}Sr_2^{2+}Ca^{2+}Cu_2^{2+c}O_x$$
Positive charge = $6 + 2b + 4 + 2 + 4 + 2c = 16 + 2b + 2c$
The charge must be balanced by O^{2-} $\Rightarrow$ $x = 8 + b + c$.

The formula mass of the superconductor is $760.37 + 15.999(8 + b + c)$.

One gram contains $1/[760.37 + 15.999(8 + b + c)]$ moles of superconductor

<u>Experiment A</u>: Initial $Cu^+ = 0.2000$ mmol; final $Cu^+ = 0.1085$ mmol.
Therefore, 102.3 mg of superconductor consumed 0.0915 mmol Cu^+.
$2 \times$ mmol Bi^{5+} + mmol Cu^{3+} in 102.3 mg of superconductor $= 0.0915$.

<u>Experiment B</u>: Initial $Fe^{2+} = 0.1000$ mmol; final $Fe^{2+} = 0.0577$ mmol.
Therefore, 94.6 mg of superconductor consumed 0.0423 mmol Fe^{2+}.
$2 \times$ mmol Bi^{5+} in 94.6 mg of superconductor $= 0.0423$.

Normalizing to 1 gram of superconductor gives

  Expt A: $2($mmol $Bi^{5+}) +$ mmol Cu^{3+} in 1 g of superconductor $= 0.89443$

  Expt B: $2($mmol $Bi^{5+})$ in 1 g of superconductor $= 0.44715$

It is easier not to get lost in the arithmetic if we suppose that the oxidized bismuth is Bi^{4+} and equate one mole of Bi^{5+} to two moles of Bi^{4+}. Therefore, we can rewrite the two previous equations as

  mmol Bi^{4+} + mmol Cu^{3+} in 1 g of superconductor $= 0.89443$ (1)

  mmol Bi^{4+} in 1 g of superconductor $= 0.44715$ (2)

Subtracting (2) from (1) gives

  mmol Cu^{3+} in 1 g of superconductor $= 0.44728$ (3)

Equations (2) and (3) tell us that the stoichiometric relationship in the formula of the superconductor is $b/c = 0.44715/0.44728 = 0.9997$.

Since 1 g of superconductor contains 0.44728 mmol Cu^{3+}, we can say

$$\frac{\text{mol } Cu^{3+}}{\text{mol solid}} = 2c$$

$$\frac{\text{mol } Cu^{3+}/\text{mol solid}}{\text{gram solid/mol solid}} = \frac{2c}{760.37 + 15.999(8+b+c)}$$

$$\frac{\text{mol } Cu^{3+}}{\text{gram solid}} = \frac{2c}{760.37 + 15.999(8+b+c)} = 4.4728 \times 10^{-4} \quad (4)$$

Substituting $b = 0.9997c$ in the denominator of (4) allows us to solve for c:

$$\frac{2c}{760.37 + 15.999(8 + 1.9997c)} = 4.4728 \times 10^{-4} \Rightarrow c = 0.200_1$$

$$\Rightarrow b = 0.9997c = 0.200_0$$

Average oxidation numbers are $Bi^{3.200_0+}$ and $Cu^{2.200_1+}$. Oxygen stoichiometry is $x = 8 + b + c$, so formula is $Bi_2Sr_2CaCu_2O_{8.400}$

CHAPTER 17
ELECTROANALYTICAL TECHNIQUES

17-1. charge (C) = $q = n(\text{mol})F$, where, $n = 1$ unit charge per e^-, (mol) is the number of moles of electrons passing through the circuit, and F = Faraday constant.
q = (1 charge/e^-)(0.100 mol e^-)(96 485 C/mol) = 9.648×10^3 C
1 ampere = 1 C/s, so the time required for 9.648×10^3 C is
$(9.648 \times 10^3 \text{ C})/(1.00 \text{ C/s}) = 9.648 \times 10^3$ s = 2.68 h

17-2. $\Delta G° = -nFE°$
where n = number of electrons in balanced half-reaction, F = Faraday constant, and $E°$ = standard voltage. For the reaction $H_2O(l) \rightleftharpoons H_2(g) + \frac{1}{2}O_2(g)$, $n = 2$ electrons per mole of H_2 or per half mole of O_2.
$E° = -\Delta G°/nF = (-237.13 \times 10^3 \text{ J/mol})/[(2)(964\,85 \text{ C/mol})]$
$= -1.228\,8$ J/C $= -1.228\,8$ V

17-3. (a) $E = E(\text{cathode}) - E(\text{anode})$
$= \left\{ E°(\text{cathode}) - 0.059\,16 \log P_{H_2}^{1/2}[\text{OH}^-] \right\}$
$\quad - \left\{ E°(\text{anode}) - 0.059\,16 \log [\text{Br}^-] \right\}$
(remember to write both reactions as reductions)
$= \left\{ -0.828 - 0.059\,16 \log (1.0)^{1/2}[0.10] \right\}$
$\quad - \left\{ 1.078 - 0.059\,16 \log [0.10] \right\} = -1.906$ V

(b) Ohmic potential = $I \cdot R$ = (0.100 A)(2.0 Ω) = 0.20 V

(c) Overpotentials always increase the magnitude of the voltage that must be applied to drive the reaction:
$E = E(\text{cathode}) - E(\text{anode}) - I \cdot R - \text{Overpotentials}$
$= -1.906 - 0.20 - (0.20 + 0.40) = -2.71$ V

(d) $E(\text{cathode}) = E°(\text{cathode}) - 0.059\,16 \log P_{H_2}^{1/2}[\text{OH}^-]_s$
$= -0.828 - 0.059\,16 \log (1.0)^{1/2}[1.0] = -0.828$ V
$E(\text{anode}) = E°(\text{anode}) - 0.059\,16 \log [\text{Br}^-]_s$
$= 1.078 - 0.059\,16 \log [0.010] = 1.196$ V
$E = E(\text{cathode}) - E(\text{anode}) - I \cdot R - \text{Overpotentials}$
$= -0.828 - 1.196 - 0.20 - (0.20 + 0.40) = -2.82$ V

17-4. (a) $H^+ + e^- \rightleftharpoons \frac{1}{2}H_2(g)$ $\quad E = E° - 0.059\ 16 \log P_{H_2}^{1/2}/[H^+]$

$E = 0 - 0.059\ 16 \log (1)^{1/2}/[10^{-14}] = -0.828\ 2$ V

The appendix gives $E° = -0.828\ 0$ for $H_2O + e^- \rightleftharpoons \frac{1}{2}H_2(g) + OH^-$. The values $-0.828\ 0$ V and $-0.828\ 2$ V are equal within a roundoff error from the four decimal places in the number 0.059 16 V. Standard conditions (neglecting activity coefficients) for the reaction $H_2O + e^- \rightleftharpoons \frac{1}{2}H_2(g) + OH^-$ are $P_{H_2} = 1$ bar and $[OH^-] = 1$ M, for which the standard voltage is $E° = -0.828\ 0$ V. These are the same conditions for the half-reaction $H^+ + e^- \rightleftharpoons \frac{1}{2}H_2(g)$ in 1 M NaOH. Therefore, E for the half-reaction $H^+ + e^- \rightleftharpoons \frac{1}{2}H_2(g)$ in 1 M NaOH is the same as $E°$ for the half-reaction $H_2O + e^- \rightleftharpoons \frac{1}{2}H_2(g) + OH^-$.

(b) For a Pt alloy electrode, we might expect close to no overpotential and therefore the reduction should commence near –0.83 V versus S.H.E. In the figure, there is negligible current flowing at –0.8 V and reduction commences within a few hundreds of a volt of –0.83 V. For the Ti electrode, reduction commences at roughly –1.05 to –1.1 V. The difference of ~0.2 V between the start of reduction on the two electrodes is the overpotential (the activation energy) for reduction of water on the Ti electrode.

17-5. (a) V_2 is the voltage between the working and reference electrodes, which is held constant. Working: Reference: Auxiliary:

(b) The objective is to measure the potential difference between the reference electrode and a point in solution in an electrochemical cell. Negligible current flows between the opening of the Luggin capillary and the reference electrode in the reservoir connected to the Luggin capillary. Therefore, there is negligible ohmic loss between the opening of the capillary and the surface of the reference electrode. There is also negligible overpotential and negligible concentration polarization at the reference electrode because there is negligible current at the reference electrode. With negligible potential losses between the reference electrode and the opening of the Luggin capillary, the potential inside the capillary is the potential of the reference electrode. The potential outside the capillary is that of the electrochemical system.

17-6. (a) Copper electrode: $Cu(s) \rightleftharpoons Cu^{2+} + 2e^-$

$E_+ = 0.339 - \dfrac{0.059\,16}{2} \log[Cu^{2+}] = 0.339$ V if $[Cu^{2+}] = 1.0$ M

Silver-silver chloride electrode: $E_- = 0.197$ V

Predicted equilibrium voltage = $E_+ - E_- = 0.339 - 0.197 = 0.142$ V

(observed = 0.109 mV)

The difference between observed value of 109 mV and predicted 142 mV is due to neglect of the activity coefficient of Cu^{2+} and to the difference between the reference electrode potential with saturated KCl or 3 M KCl.

(b) Increasing the imposed potential difference above 1.000 V would increase the current flowing between the Cu electrodes. The overpotential at each electrode increases as current density (current per unit area) increases. The anode intercept would rise above 122 mV and the cathode intercept would go below 85 mV.

17-7. (a) For every mole of Hg produced, one mole of electrons flows because Hg(I) in Hg_2SO_4 is reduced to Hg(0).

1.00 mL Hg = 13.53 g Hg = 0.067 45 mol Hg = 0.067 45 mol e^-.
(0.067 45 mol) (96 485 C/mol) = 6 508 C.
Work = $q \cdot E$ = (6 508 C) (1.02 V) = 6.64×10^3 J.

(b) The power is 0.209 J/min = 0.003 48 J/s.
$P = I^2 R \Rightarrow I = \sqrt{P/R} = \sqrt{(0.003\,48\text{ W})/(100\,\Omega)} = 5.902$ mA.

In 1 h the total charge flowing through the circuit is
$(5.902 \times 10^{-3}$ C/s$)(3\,600$ s/h$) = 21.25$ C/h
$(21.25$ C/h$)/(96\,485$ C/mol$) = 2.202 \times 10^{-4}$ mol of e^-/h
$= 1.101 \times 10^{-4}$ mol of Cd/h $= 0.012\,4$ g Cd/h.

17-8. Hydroxide generated at the cathode and Cl^- in the anode compartment cannot cross the Nafion membrane. Na^+ from seawater crosses from anode to cathode to preserve charge balance. Therefore, NaOH can be formed free from Cl^-.

17-9. $E = E(\text{cathode}) - E(\text{anode}) - IR - \text{overpotentials}$

Suppose that the open-circuit voltage of each cell is $E(\text{cathode}) - E(\text{anode}) = 2.2$ V when no current flows. Ohmic loss and overpotentials for the two half-reactions decrease the output of the cell by 0.2 V, giving a net cell voltage of 2.0 V when the cell is delivering current. The cell can be recharged at very low current flow by applying just over 2.2 V in the opposite direction to reverse the cell chemistry. To charge at a significant rate requires additional voltage to overcome ohmic loss and overpotentials. The recharge requires ~0.2 V more than open-circuit voltage, or ~2.4 V. Electrical losses [IR, overpotentials, and concentration polarization in the terms E(cathode) and E(anode)] always decrease the magnitude of the voltage that can be delivered by a cell and increase the magnitude of the voltage required to reverse the spontaneous cell reaction.

17-10. $Pb(lactate)_2 + 2H_2O \rightarrow PbO_2(s) + 2\,lactate^- + 4H^+ + 2e^-$
  Pb^{2+} $\qquad\qquad\qquad\qquad\quad Pb^{4+}$

Lead lactate is oxidized to PbO_2 at the anode.
The mass of lead lactate (FM 385.3) giving 0.111 1 g of PbO_2
(FM = 239.2) is [(385.3 g/mol)/(239.2 g/mol)](0.111 1 g) = 0.179 0 g.
wt% Pb = $\dfrac{0.179\,0\text{ g}}{0.326\,8\text{ g}} \times 100 = 54.77\%$

17-11. Cathode: $Sn^{2+} + 2e^- \rightleftharpoons Sn(s)$ $E° = -0.141$ V

E(cathode, vs S.H.E.) = $-0.141 - \dfrac{0.059\,16}{2} \log \dfrac{1}{1.0 \times 10^{-8}} = -0.378$ V

E(cathode, vs S.C.E.) = $-0.378 - 0.241 = -0.619$ V

The voltage will be more negative if concentration polarization occurs. Concentration polarization means that $[Sn^{2+}]_s < 1.0 \times 10^{-8}$ M.

17-12. When 99.99% of Cd(II) is reduced, the formal concentration will be 1.0×10^{-5} M, and the predominant form is $Cd(NH_3)_4^{2+}$.

$\beta_4 = \dfrac{[Cd(NH_3)_4^{2+}]}{[Cd^{2+}][NH_3]^4} = \dfrac{(1.0 \times 10^{-5})}{[Cd^{2+}](1.0)^4} \Rightarrow [Cd^{2+}] = 2.8 \times 10^{-12}$ M

$Cd^{2+} + 2e^- \rightleftharpoons Cd(s)$ $E° = -0.402$

E(cathode) = $-0.402 - \dfrac{0.059\,16}{2} \log \dfrac{1}{[Cd^{2+}]} = -0.744$ V

17-13. Ni deposited = (0.479 8 g – 0.477 5 g) = 2.3 mg = 39.19 μmol Ni which would require $(39.19 \times 10^{-6}$ mol Ni$)(2e^-/$Ni$)(96\,485$ C/mol$) = 7.562$ C. Percentage of current going to reduction of Ni^{2+} = $100 \times \dfrac{7.562\text{ C}}{8.082\text{ C}} = 94\%$. The remainder went into reduction of H^+ to H_2.

17-14. When excess Br$_2$ appears in the solution, current flows at a low applied potential difference (0.25 V) in the detector circuit by virtue of the reactions

anode: $2Br^- \rightarrow Br_2 + 2e^-$

cathode: $Br_2 + 2e^- \rightarrow 2Br^-$

Br$_2$ is first generated in the absence of cyclohexene to give a detector current of 20.0 µA. When cyclohexene is added, the current decreases to a tiny value because Br$_2$ is consumed. Br$_2$ is then generated by the coulometric circuit, and the end point is taken when the detector again reaches 20.0 µA.

17-15. A mediator shuttles electrons between analyte and the electrode. After being oxidized or reduced by analyte, the mediator is regenerated at the electrode.

17-16. (a) 0.005 C/s × 1 s = 0.005 C $\dfrac{0.005 \text{ C}}{96\,485 \text{ C/mol}} = 5.2 \times 10^{-8}$ mol e$^-$

(b) A 0.01 M solution of a two-electron reductant delivers 0.02 moles of electrons/liter.

$$\dfrac{5.2 \times 10^{-8} \text{ moles}}{0.02 \text{ moles/liter}} = 2.6 \times 10^{-6} \text{ L} = 0.0026 \text{ mL} = 2.6 \text{ µL}$$

17-17. (a) $\text{mol e}^- = \dfrac{I \cdot t}{nF} = \dfrac{(5.32 \times 10^{-3} \text{ C/s})(964 \text{ s})}{(1 \text{ charge/e}^-)(96\,485 \text{ C/mol})} = 5.32 \times 10^{-5}$ mol

(b) One mol e$^-$ reacts with ½ mol Br$_2$, which reacts with ½ mol cyclohexene, so 5.32 × 10^{-5} mol e$^-$ produces 2.66 × 10^{-5} mol cyclohexene.

(c) 2.66 × 10^{-5} mol/5.00 × 10^{-3} L = 5.32 × 10^{-3} M

17-18. $2I^- \rightarrow I_2 + 2e^- \Rightarrow$ one mole of I$_2$ is created when two moles of electrons flow. $\text{mol e}^- = \dfrac{I \cdot t}{nF} = \dfrac{(52.6 \times 10^{-3} \text{ C/s})(812 \text{ s})}{(1 \text{ charge/e}^-)(96\,485 \text{ C/mol})} = 0.4427$ mmol of e$^-$ = 0.2213 mmol of I$_2$. Therefore, there must have been 0.2213 mmol of H$_2$S (FM 34.08) = 7.542 mg of H$_2$S/50.00 mL = 7.542 × 10^3 µg of H$_2$S/50.00 mL = 151 µg/mL.

17-19. D-fructose → 5-keto-D-fructose + 2H$^+$ + 2e$^-$

n = 2 electrons per mol of fructose

2.00 nmol fructose will liberate 4.00 nmol electrons

q = (4.00 nmol e$^-$)(96 485 C/mol) = 0.386 mC

The asymptotic current in the figure is near 0.38$_7$ mC.

17-20. (a) When one B atom is substituted for one C atom, the crystal lattice contains one less valence electron, making it *p*-type.

(b) Each ferrocene loses one electron. To convert charge to moles, divide by the Faraday constant: $\dfrac{23 \times 10^{-6}\,C}{(0.38\,cm^2)(96\,485\,C/mol)} = 6.2_7 \times 10^{-10}\,\dfrac{mol}{cm^2}$. If there are two ferrocene groups attached to each N atom, the surface density of N atoms is half of the surface density of ferrocene groups $= 3.1_4 \times 10^{-10}\,\dfrac{mol}{cm^2}$.

Surface density of carbon atoms $= \dfrac{1.7 \times 10^{15}\,atoms/cm^2}{6.022 \times 10^{23}\,atoms/mol}$

$= 2.8_2 \times 10^{-9}\,\dfrac{mol}{cm^2}$.

Fraction of C atoms substituted by N $= \dfrac{3.1_4 \times 10^{-10}\,mol/cm^2}{2.8_2 \times 10^{-9}\,mol/cm^2} \approx 11\%$.

17-21. (a) $C_6H_5N{=}NC_6H_5 + 4H^+ + 4e^- \rightarrow 2C_6H_5NH_2$

Electron flow =

$\left(4\,\dfrac{electrons}{C_6H_5N{=}NC_6H_5}\right)\left(25.9\,\dfrac{nmol}{s}\right)\left(96\,485\,\dfrac{C}{mol}\right) = 1.00 \times 10^{-2}\,C/s$

current density $= \dfrac{1.00 \times 10^{-2}\,A}{1.00 \times 10^{-4}\,m^2} = 1.00 \times 10^2\,A/m^2$

$\Rightarrow$ overpotential $= 0.85\,V$

(b) $E(\text{cathode}) = 0.100 - 0.059\,16\,\log \dfrac{[Ti^{3+}]_s}{[TiO^{2+}]_s[H^+]^2}$

$= 0.100 - 0.059\,16\,\log \dfrac{[0.10]}{[0.050][0.10]^2} = -0.036\,V$

(c) $O_2 + 4H^+ + 4e^- \rightleftharpoons 2H_2O \qquad E° = 1.229\,V$

$E(\text{anode}) = 1.229 - \dfrac{0.059\,16}{4}\,\log \dfrac{1}{P_{O_2}[H^+]^4}$

$= 1.229 - \dfrac{0.059\,16}{4}\,\log \dfrac{1}{(0.20)[0.10]^4} = 1.160\,V$

(d) $E = E(\text{cathode}) - E(\text{anode}) - I{\cdot}R - \text{Overpotential}$

$= -0.036 - 1.160 - (1.00 \times 10^{-2}\,A)(52.4\,\Omega) - 0.85 = -2.57\,V$

Electroanalytical Techniques   235

17-22. $F = \dfrac{\text{coulombs}}{\text{mol}} = \dfrac{I \cdot t}{\text{mol}}$

$= \dfrac{[0.203\ 639\ 0\ (\pm 0.000\ 000\ 4)\ \text{A}][18\ 000.075\ (\pm 0.010)\ \text{s}]}{[4.097\ 900\ (\pm 0.000\ 003)\ \text{g}]/[107.868\ 2\ (\pm 0.000\ 2)\ \text{g/mol}]}$

$= \dfrac{[0.203\ 639\ 0\ (\pm 1.96 \times 10^{-4}\%)][18\ 000.075\ (\pm 5.56 \times 10^{-5}\%)]}{[4.097\ 900\ (\pm 7.32 \times 10^{-5}\%)]/[107.868\ 2\ (\pm 1.85 \times 10^{-4}\%)]}$

$= 9.648\ 667 \times 10^4\ (\pm 2.85 \times 10^{-4}\ \%) = 96\ 486.6_7 \pm 0.2_8\ \text{C/mol}$

17-23. (a) $H_2SO_3 \overset{pK_1 = 1.86}{\rightleftharpoons} HSO_3^- \overset{pK_2 = 7.17}{\rightleftharpoons} SO_3^{2-}$

H_2SO_3 is predominant below pH 1.86. HSO_3^- dominates between pH 1.86 and 7.17. SO_3^{2-} is dominant above pH 7.17.

(b) Cathode: $H_2O + e^- \rightarrow \tfrac{1}{2} H_2(g) + OH^-$

Anode: $3I^- \rightarrow I_3^- + 2e^-$

(c) $I_3^- + HSO_3^- + H_2O \rightarrow 3I^- + SO_4^{2-} + 3H^+$

$I_3^- + 2S_2O_3^{2-} \rightleftharpoons 3I^- + $ O=S(O⁻)−S−S−S(O⁻)=O

               Thiosulfate   Tetrathionate

(d) In Step 3, I_3^- was generated by a current of 10.0 mA ($= 10.0 \times 10^{-3}$ C/s) for 4.00 min (= 240 s).

charge = $I \cdot t = (10.0 \times 10^{-3}\ \text{C/s})(240\ \text{s}) = 2.40\ \text{C}$

mol $e^- = I/F = (2.40\ \text{C})/(96\ 485\ \text{C/mol}) = 24.8_7\ \mu\text{mol}\ e^-$

The anode reaction generates ½ mol I_3^- for 1 mol e^-. Therefore, $24.8_7\ \mu$mol e^- will generate $\tfrac{1}{2}(24.8_7) = 12.4_4\ \mu$mol I_3^-.

In Step 5, 0.500 mL of 0.050 7 M thiosulfate = $25.3_5\ \mu$mol $S_2O_3^{2-}$. But 2 mol $S_2O_3^{2-}$ consume 1 mol I_3^-. Therefore, $25.3_5\ \mu$mol $S_2O_3^{2-}$ consume $\tfrac{1}{2}(25.3_5) = 12.6_8\ \mu$mol I_3^-.

We added excess $S_2O_3^{2-}$ in Step 5 and consumed the excess in Step 6. In Step 6, we had to generate I_3^- at 10.0 mA for 131 s to react with excess $S_2O_3^{2-}$.

charge = $I \cdot t = (10.0 \times 10^{-3}\ \text{C/s})(131\ \text{s}) = 1.31\ \text{C}$

mol $e^- = I/F = (1.31\ \text{C})/(96\ 485\ \text{C/mol}) = 13.5_8\ \mu$mol e^-

mol $I_3^- = \tfrac{1}{2}(13.5_8\ \mu\text{mol}\ e^-) = 6.79\ \mu$mol I_3^-

Here is where we are so far:

Step 3: 12.4_4 μmol I_3^- were generated.

Step 4: x μmol I_3^- were consumed by sulfite in wine.

Step 5: We added enough $S_2O_3^{2-}$ to consume 12.6_8 μmol I_3^-.

Step 6: We had to generate 6.79 μmol I_3^- to consume excess $S_2O_3^{2-}$ from Step 5. Therefore, I_3^- left after Step 4 = $12.6_8 - 6.79 = 5.8_9$ μmol I_3^-.

We began with 12.4_4 μmol I_3^- and 5.89 μmol I_3^- were left after reaction with sulfite in wine. Therefore, sulfite in wine consumed $12.4_4 - 5.89 = 6.5_5$ μmol I_3^-. But 1 mol I_3^- reacts with 1 mol sulfite. Therefore, the wine contained 6.5_5 μmol sulfite in the 2.00 mL injected for analysis.

The wine sample prepared in Step 1 consisted of 9.00 mL wine diluted to 10.00 mL. Therefore, the original wine contained 10.00/9.00 of the amount found in the analysis. That is, 2.000 mL of pure wine contains $(10.00/9.00)(6.5_5 \text{ μmol sulfite}) = 7.2_8$ μmol sulfite.

$$\text{sulfite in wine} = \frac{7.2_8 \text{ μmol sulfite}}{2.00 \text{ mL}} = 3.64 \text{ mM}$$

This problem left out a description of the blank titration that should be done in a real analysis. There are components in wine in addition to sulfite that could react with I_3^-. For the blank titration, 1 M formaldehyde is added to the wine to bind all sulfite. The sulfite-formaldehyde adduct is not decomposed in 2 M NaOH and does not react with I_3^-. The blank titration consists of taking this formaldehyde/wine solution through the entire procedure. We subtract I_3^- consumed by the blank from I_3^- consumed by the wine without formaldehyde.

17-24. (a) To consume Ee^- requires $(E/4)O_2$, because each O_2 consumes $4e^-$.

(b) $E = (9.43 \times 10^{-3} \text{ C})/(9.648\,5 \times 10^4 \text{ C/mol}) = 9.77_4 \times 10^{-8}$ mol e^-
mol $O_2 = E/4 = 2.44_3 \times 10^{-8}$ mol

(c) The mass of O_2 in (b) is $(2.44_3 \times 10^{-8} \text{ mol})(32.00 \text{ g/mol}) = 7.81_9 \times 10^{-7}$ g. This much O_2 was required to react with 13.5 μL of sample. The mass of O_2 that would react with 1 L of sample is $(7.81_9 \times 10^{-7} \text{ g})/(13.5 \times 10^{-6} \text{ L}) = 0.057\,9$ g/L = 57.9 mg/L.

(d) The balanced oxidation half-reaction is
$C_9H_6NO_2ClBr_2 + 16H_2O \rightarrow 9CO_2 + 3X^- + NH_3 + 35H^+ + 32e^-$.
The observed number of electrons in the reaction was $9.77_4 \times 10^{-8}$ mol e^-, so there must have been $(9.77_4 \times 10^{-8}$ mol $e^-)/(32$ mol e^-/mol $C_9H_6NO_2ClBr_2) = 3.05_4 \times 10^{-9}$ mol $C_9H_6NO_2ClBr_2$ in 13.5 µL. The molarity of $C_9H_6NO_2ClBr_2$ is $(3.05_4 \times 10^{-9}$ mol$)/(13.5 \times 10^{-6}$ L$) = 2.26 \times 10^{-4}$ M.

17-25. The Clark electrode measures dissolved oxygen by reducing it to H_2O at a gold tip on a platinum electrode held at –0.75 V with respect to Ag|AgCl. The opening of the body of the electrode is filled with a 10- to 40-µm-long plug of silicone rubber that is permeable to O_2. Current is proportional to the concentration of dissolved O_2 in the external medium. The electrode needs to be calibrated in solutions of known O_2 concentration.

17-26. (a) The glucose monitor has a test strip with two carbon indicator electrodes and a silver-silver chloride reference electrode. Indicator electrode 1 is coated with glucose oxidase and a mediator. When a drop of blood is placed on the test strip, glucose from the blood is oxidized near indicator electrode 1 by mediator to gluconolactone and the mediator is reduced. Reduced mediator is re-oxidized at the indicator electrode at a potential of +0.2 V versus Ag|AgCl. Current between indicator electrode 1 and the reference electrode is proportional to the rate of oxidation of mediator, which is proportional to the concentration of glucose plus any interfering reducible species in the blood. Indicator electrode 2 has mediator, but no glucose oxidase. Current measured at indicator electrode 2 is proportional to the concentration of interfering species in the blood. The difference between the two currents is proportional to the concentration of glucose in the blood.

(b) In the absence of mediator, the rate of oxidation of glucose depends on the concentration of O_2 in the blood. If $[O_2]$ is low, current will be low and the monitor will give an incorrect, low reading for the glucose concentration. A mediator such as 1,1'-dimethylferrocene can replace O_2 in the glucose oxidation and be subsequently reduced at the indicator electrode. The concentration of mediator is constant and high enough so that variations in electrode current are due mainly to variations in glucose concentration. Also, by lowering the required electrode potential for oxidation of the mediator, there is less possible interference by other species in the blood.

(c) Glucose oxidase is replaced by glucose dehydrogenase, which does not use O_2 as a reactant. The enzyme oxidizes glucose and reduces PQQ to $PQQH_2$. $PQQH_2$ is oxidized back to PQQ by Os^{3+} bound to the polymer chain. In the process, Os^{3+} is reduced to Os^{2+}. This Os^{2+} can transfer an electron to a nearby Os^{3+}. By moving from Os to Os, electrons eventually reach the carbon electrode. The coulometric sensor measures the total number of electrons needed to oxidize all of the glucose in the small blood sample.

(d) Amperometry measures current during the enzyme-catalyzed oxidation of glucose. Current is proportional to the rate of the oxidation reaction. The rates of most chemical reactions increase with increasing temperature. Therefore, the current will increase with increasing temperature of the blood sample. Coulometry measures the total number of electrons released in the oxidation. Glucose releases 2 electrons per molecule, regardless of temperature. The coulometric signal should have no temperature dependence.

(e) 1.00 g glucose/L = 5.55 mM glucose. A volume of 0.300×10^{-6} L contains 1.665 nmol glucose. Each mole of glucose releases $2e^-$ and $2H^+$ during oxidation. Therefore, 2×1.665 nmol = 3.33 nmol e^- are released. The charge is $Q = (3.33 \times 10^{-9}$ mol$)(96\ 485$ C/mol$) = 321$ µC.

17-27. At low potential ($< \sim 0$ V), the chemistry is $Fe(CN)_6^{3-} + e^- \rightarrow Fe(CN)_6^{4-}$. At high potential ($> \sim 0.5$ V), the chemistry is $Fe(CN)_6^{4-} \rightarrow Fe(CN)_6^{3-} + e^-$. The current density at each plateau depends on the rate at which reactant reaches the electrode which, in turn, depends on the bulk concentration of the reactant. In all cases, the bulk concentration of $Fe(CN)_6^{3-}$ is 10 mM, so the low voltage plateau is at the same current density for all solutions. The bulk concentration of $Fe(CN)_6^{4-}$ varies from 20 to 60 mM, so each curve reaches a different current density at high potential. If the rotation speed were decreased, the thickness of the diffusion layer would increase and analyte would require a longer time to transit the diffusion layer to the electrode. The magnitude of the positive current density and the magnitude of the negative current density on both ends of the curves would decrease.

17-28. ω is the rotation rate in radians per second. We need to convert rpm (revolutions per minute) to radians per second.

$$\left(2.00\times 10^3 \frac{\text{revolutions}}{\text{min}}\right)\left(\frac{1 \text{ min}}{60 \text{ s}}\right)\left(\frac{2\pi \text{ radians}}{\text{revolution}}\right) = 209 \text{ rad/s} = 209 \text{ s}^{-1}$$

(because radian is a dimensionless unit)

$$\delta = 1.61 D^{1/3} \nu^{1/6} \omega^{-1/2}$$
$$= 1.61(2.5 \times 10^{-9} \text{ m}^2/\text{s})^{1/3}(1.1 \times 10^{-6} \text{ m}^2/\text{s})^{1/6}(209 \text{ rad/s})^{-1/2} = 1.5_3 \times 10^{-5} \text{ m}$$

To calculate current density, we need to express the concentration of the species reacting at the electrode in mol/m³ instead of mol/L. Since 1 L is the volume of a 10-cm cube, there are 1 000 L in 1 m³. The concentration of $K_4Fe(CN)_6$ is 50.0 mM = $\left(0.050\ 0 \frac{\text{mol}}{\text{L}}\right)\left(1\ 000 \frac{\text{L}}{\text{m}^3}\right) = 50.0 \text{ mol/m}^3$.

Current density = $0.62nFD^{2/3}\nu^{-1/6}\omega^{1/2}C_o$

$$= 0.62(1)\left(96\ 485\frac{\text{C}}{\text{mol}}\right)\left(2.5\times 10^{-9}\frac{\text{m}^2}{\text{s}}\right)^{2/3}\left(1.1\times 10^{-6}\frac{\text{m}^2}{\text{s}}\right)^{-1/6}\left(209\frac{1}{\text{s}}\right)^{1/2}\left(50.0\frac{\text{mol}}{\text{m}^3}\right)$$

$$= 7.8_4 \times 10^2 \frac{\text{C}}{\text{m}^2 \cdot \text{s}} = 7.8_4 \times 10^2 \frac{\text{A}}{\text{m}^2}$$

17-29. (a)

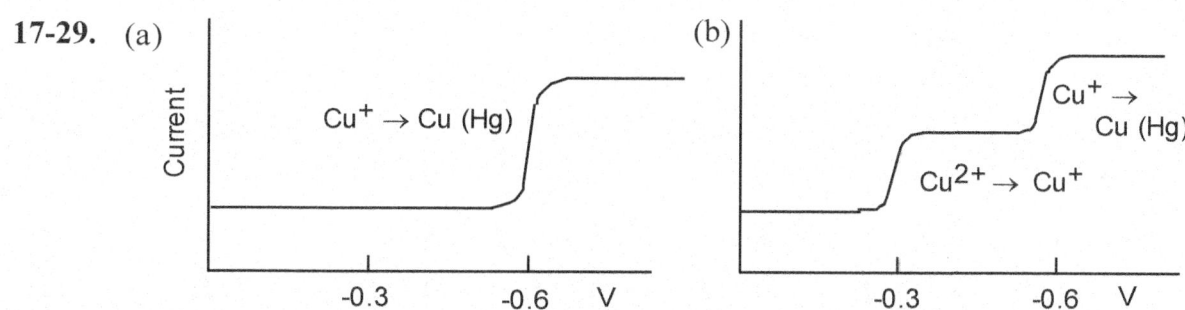

(c) The potential for the reaction Cu(I) → Cu(*in Hg*) will change if Pt is used, since the product obviously cannot be copper amalgam.

17-30. (a) Charging current arises from charging or discharging of the electric double layer at the electrode-solution interface. Faradaic current arises from oxidation or reduction reactions.

(b) Charging current decays more rapidly than faradaic current. By 1 s after a potential step, the charging current decays to near zero and the faradaic current is still significant. The ratio of the desired signal (faradaic current) to the undesired background (charging current) is larger at 1 s than at earlier times. If we wait too long, both signals become too small to measure.

17-31. (a) In sampled current voltammetry, voltage is stepped and then we wait for a fixed time before measuring current. During the waiting period, reactant near the electrode is depleted and ions migrate to charge the electric double layer. When current is measured at the end of the voltage step, reactant concentration near the electrode is less than it was at the beginning of the step, so faradaic current is decreased from its value at the beginning of the step.

In square wave voltammetry, an anodic pulse follows each cathodic pulse. Signal is the difference in current measured for the two pulses. The anodic pulse oxidizes the product of each cathodic pulse, thereby replenishing reactant at the electrode surface for the next pulse. The waiting period after a voltage step before measuring current is just a few milliseconds in square wave voltammetry, so reactants have less time to diffuse away from the electrode and their concentration is higher when current is measured.

(b) The sampled current voltammogram is a graph of current (I) versus potential (E). The square wave voltammogram displays the *difference* ΔI between current 1 and current 2 in Figure 17-27 measured for a small change in potential ΔE. This difference increases in steep parts of the sampled current voltammogram where current rises rapidly for a small change in potential. The square wave voltammogram displays $\Delta I/\Delta E$, which approximates the derivative of the sampled current voltammogram for small values of ΔE.

17-32. (a) In cyclic voltammetry at a macroscopic planar electrode, at sufficiently great potential, reaction at the electrode is fast enough that one-dimensional diffusion is too slow to maintain bulk concentrations near the electrode. Reactant becomes depleted near the electrode, so current decreases.

(b) Reactants and products approach a microscopic electrode from a two-dimensional hemisphere around the electrode. Diffusion from the hemisphere is rapid enough to maintain high concentrations near the electrode. If the voltage scan rate is sufficiently great, there would be a peak, not a plateau, for the microscopic electrode.

(c) A rotating disk electrode actively transports fresh solution to the electrode by inducing convection toward the electrode. Reactant has to diffuse ~10–100 μm through the stagnant diffusion layer to reach the electrode. The constant supply of reactant near the electrode gives a plateau in the voltammogram.

17-33. (a) Reactant is $Fe(CN)_6^{3-}$, which is being reduced at b, c, and d. The middle graph shows b > c > d at distance = 0 from the electrode.

(b) $[Fe(CN)_6^{3-}]$ at the surface of the electrode (distance = 0) is greatest at b, but current is greatest at c because the potential is more negative at c than at b and there is an adequate supply of $[Fe(CN)_6^{3-}]$ diffusing to the electrode at c. The potential is even more negative at d than at c, but current is less at d because $[Fe(CN)_6^{3-}]$ is practically 0 at the electrode and the thickness of the diffusion layer is greater at d than at c.

(c) Reactant is $Fe(CN)_6^{4-}$, which is being oxidized at e, f, and g. The right-hand graph shows e > f > g at distance = 0 from the electrode.

(d) $[Fe(CN)_6^{4-}]$ is less at the electrode surface at point f than at point e. However, the potential is more positive at f than at e. The magnitude of current is greater at f than at e because the potential is more positive at f. $Fe(CN)_6^{4-}$ diffuses to the electrode at a sufficient rate to give greater current at f than at e. Current is not 0 at g because $Fe(CN)_6^{4-}$ is still able to diffuse to the electrode fast enough to maintain some rate of oxidation.

(e) Time increases as e < f < g. Reduction peak at c generates $Fe(CN)_6^{4-}$, which diffuses away from the electrode. Oxidation peak at f consumes $Fe(CN)_6^{4-}$ near the electrode. There is still unreacted $Fe(CN)_6^{4-}$ in solution away from the electrode. As time increases, maximum $[Fe(CN)_6^{4-}]$ recedes from the electrode because $Fe(CN)_6^{4-}$ near the electrode is consumed and $Fe(CN)_6^{4-}$ further from the electrode continues to diffuse out into the solution.

17-34. Electrons flowing in 3.4 min =

$$\frac{(14 \times 10^{-6} \text{ C/s})(60 \text{ s/min})(3.4 \text{ min})}{96\,485 \text{ C/mol}} = 2.9_6 \times 10^{-8} \text{ mol e}^-$$

For the reaction $Cd^{2+} + 2e^- \rightarrow Cd(\text{in Hg})$,

moles of Cd^{2+} = $\frac{1}{2}$ moles of e^- = $1.4_8 \times 10^{-8}$ mol

moles of Cd^{2+} in 25 mL of 0.50 mM solution = 1.25×10^{-5} mol

percentage of Cd^{2+} reduced = $\frac{1.4_8 \times 10^{-8}}{1.25 \times 10^{-5}} \times 100 = 0.11_8 \%$

17-35. $\dfrac{[X]_i}{[S]_f + [X]_f} = \dfrac{I_X}{I_{S+X}}$

$$\dfrac{x(\text{mM})}{(3.00\ \text{mM})\left(\dfrac{2.00\ \text{mL}}{52.00\ \text{mL}}\right) + (x\ \text{mM})\left(\dfrac{50.0\ \text{mL}}{52.0\ \text{mL}}\right)} = \dfrac{0.37\ \mu\text{A}}{0.80\ \mu\text{A}} \Rightarrow x = 0.096\ \text{mM}$$

17-36. In anodic stripping voltammetry, analyte is reduced and concentrated at the working electrode at a controlled potential for a constant time. The potential is then ramped in a positive direction to reoxidize the analyte, during which time current is measured. The height of the oxidation wave is proportional to the original concentration of analyte. Stripping is the most sensitive voltammetric technique because analyte is concentrated from a dilute solution. The longer the period of concentration, the more sensitive is the analysis.

17-37. (a) Concentration (deposition) stage: $Cu^{2+} + 2e^- \rightarrow Cu(s)$

(b) Stripping stage: $Cu(s) \rightarrow Cu^{2+} + 2e^-$

(c) All solutions were made up to the same volume, so $[X]_i = [X]_f \equiv x$. Prepare a graph of I vs. $[S]_f$ using data measured from the figure in the problem. The intercept is at -313 ppb, so the original concentration of Cu^{2+} is 313 ppb.

Added standard (ppb)	Current (μA)
0	0.59_9
100	0.77_4
200	0.94_3
300	1.12_8
400	1.31_4
500	1.54_4

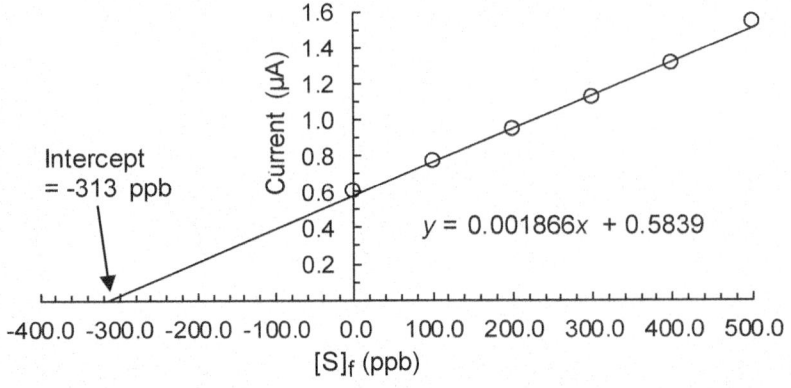

17-38.

	A	B	C	D	E
1	Standard Addition Constant Volume Least-Squares Spreadsheet				
2	x	y			
3	Added Fe(III)	Relative			
4	(pM)	peak height			
5	0	1.00			
6	50	1.56			
7	100	1.98			
8	B10:C12 = LINEST(B5:B7,A5:A7,TRUE,TRUE)				
9		LINEST output:			
10	m	0.00980	1.02333	b	
11	u_m	0.00081	0.05217	u_b	
12	R^2	0.99324	0.05715	s_y	
13	x-intercept = -b/m =	-104.422			
14	n =	3	B14 = COUNT(A5:A7)		
15	Mean y =	1.513	B15 = AVERAGE(B5:B7)		
16	$\Sigma(x_i - \text{mean } x)^2 =$	5000	B16 = DEVSQ(A5:A7)		
17	Std uncertainty of				
18	x-intercept (u_x) =	13.174			
19	B18 = (C12/ABS(B10))*SQRT((1/B14) + B15^2/(B10^2*B16))				

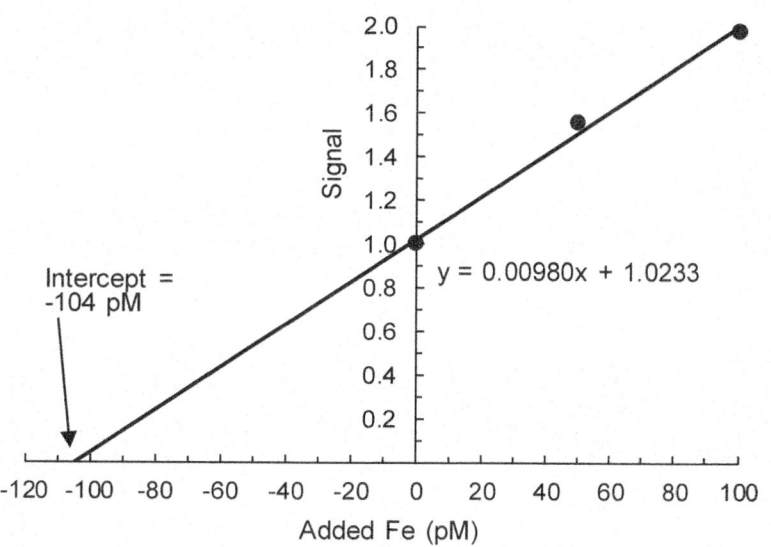

Cells B13 and B18 of the spreadsheet tell us that [Fe(III)] = 104 ± 13 pM.

17-39. In the oxidation step, neutral poly(3-octylthiophene) is oxidized to make it positively charged. One ClO_4^- molecule diffuses into the poly(vinyl chloride) (PVC) membrane to neutralize each positive charge of poly(3-octylthiophene). The number of ClO_4^- molecules diffusing into the PVC membrane in a fixed oxidation time is proportional to the bulk concentration of ClO_4^-. The rate of poly(3-octylthiophene) oxidation cannot exceed the rate at which ClO_4^- diffuses into the PVC. In the cathodic stripping step, electrons are added back to oxidized poly(3-octyl-thiophene) to reduce the charge back to 0. One ClO_4^- diffuses out of the PVC membrane for each electron added to poly(3-octylthiophene) to maintain charge balance. The number of electrons in the cathodic faradaic stripping current equals the number of ClO_4^- molecules that were taken up by the PVC which, in turn, was proportional to the ClO_4^- concentration in bulk solution.

17-40. In the first cathodic scan from 0 to −1.0 V, the reduction is
$RNO_2 + 4e^- + 4H^+ \rightarrow RNHOH + H_2O$
In the reverse anodic scan from −1.0 to +0.9 V, the oxidation is
Peak B: $\quad RNHOH \rightarrow RNO + 2H^+ + 2e^-$
In the second cathodic scan from +0.9 to −0.4 V, the new peak C appears. This peak could be assigned to
Peak C: $\quad RNO + 2H^+ + 2e^- \rightarrow RNHOH$
Peak C was not seen in the first cathodic scan because there was no RNO present in the initial solution. RNO was created by the first anodic scan.

17-41.

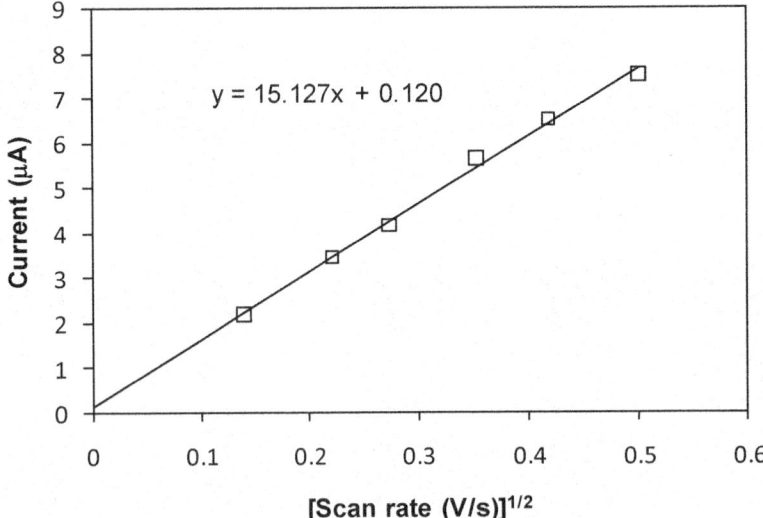

$I_p = (2.69 \times 10^8) n^{3/2} ACD^{1/2} v^{1/2}$

slope $= (2.69 \times 10^8) n^{3/2} ACD^{1/2}$

$\Rightarrow D = \dfrac{\text{slope}^2}{(2.69 \times 10^8)^2 n^3 A^2 C^2}$

$= \dfrac{(15.1 \times 10^{-6}\, A/\sqrt{V/s})^2}{(2.69 \times 10^8)^2 1^3 (0.020\ 1 \times 10^{-4}\, m^2)^2 (1.00 \times 10^{-3}\, M)^2} = 7.8 \times 10^{-10}\, m^2/s$

17-42. Microelectrodes fit into small places, are useful in nonaqueous solution (because of small ohmic losses), and allow rapid voltage scans (because of small capacitance), which permits the study of short-lived species. Low capacitance gives low background charging current, which increases sensitivity to analyte by orders of magnitude.

17-43. Nafion membrane permits neutral and cationic species to pass through to the electrode, but excludes anions. It reduces the background signal from ascorbate anion, which would otherwise swamp the signal from dopamine.

17-44. (a) Upper plateau: $Fe(CN)_6^{4-} \rightarrow Fe(CN)_6^{3-} + e^-$

Lower plateau: $Fe(CN)_6^{3-} + e^- \rightarrow Fe(CN)_6^{4-}$

(b) $r = \dfrac{I_{limit}}{4nFDC} = \dfrac{18.5 \text{ nA}}{4 \cdot 1(96\,485 \text{ C/mol})(9.2 \times 10^{-10} \text{ m}^2/\text{s})(0.010 \text{ M})}$

1 liter = 10^{-3} m^3. Expressing concentration in mol/m^3 gives:

$$= \dfrac{18.5 \times 10^{-9} \text{ C/s}}{4 \cdot 1(96\,485 \text{ C/mol})(9.2 \times 10^{-10} \text{ m}^2/\text{s})(0.010 \text{ mol}/10^{-3} \text{ m}^3)} = 5.2 \text{ μm}$$

Note that the diffusion coefficient cited in Problem 17-28 is almost 3 times greater than the value cited in the present problem. The higher diffusion coefficient gives an electrode radius of r = 1.9 μm.

17-45.

	A	B	C	D	E	F
1	Paired t test of two methods					
2						
3	Sample	Method A	Method B	d_i		
4	A	0.88	0.83	0.05	= B4-C4	
5	B	1.15	1.04	0.11		
6	C	1.22	1.39	-0.17		
7	D	0.93	0.91	0.02		
8	E	1.17	1.08	0.09		
9	F	1.51	1.31	0.20		
10			mean =	0.050	=AVERAGE(D4:D9)	
11			stdev =	0.124	=STDEV.S(D4:D9)	
12			$t_{calculated}$ =	0.987	=D10/D11*SQRT(6)	
13			t_{table} =	2.571	=T.INV.2T(0.05,5)	

$t_{calculated}$ = 0.987 < 2.571 (Student's t for 95% confidence and $n - 1 = 5$ degrees of freedom). The difference is <u>not</u> significant at the 95% confidence level.

17-46. In the following spreadsheet, we find $t_\text{calculated}$ (*t Stat* in cell F10) = 0.99 is less than t_table (*t Critical two-tail* in cell F14) = 2.57. Therefore, the difference between the methods is not significant.

The probability *P(T<=t) two-tail* in cell F13 is 0.37. There is a 37% chance of finding the observed difference between equivalent methods by random variations in results. Probability would have to be ≤0.05 for us to conclude that the methods differ.

	A	B	C	D	E	F	G
1	Paired t test of two methods				t-Test: Paired Two Sample for Means		
2							
3	Sample	Method A	Method B			Variable 1	Variable 2
4	A	0.88	0.83		Mean	1.143333	1.093333
5	B	1.15	1.04		Variance	0.051187	0.048187
6	C	1.22	1.39		Observations	6	6
7	D	0.93	0.91		Pearson Correlation	0.845414	
8	E	1.17	1.08		Hypothesized Mean Difference	0	
9	F	1.51	1.31		df	5	
10					t Stat	0.986928	
11	Absolute value of calculated t Statistic				P(T<=t) one-tail	0.184499	
12	in cell F10 is less than critical t in cell				t Critical one-tail	2.015048	
13	F14. Therefore the difference between				P(T<=t) two-tail	0.368999	
14	methods is not significant.				t Critical two-tail	2.570582	

17-47. $\text{ROH} + \text{SO}_2 + \text{B} \rightarrow \text{BH}^+ + \text{ROSO}_2^-$

$\text{H}_2\text{O} + \text{I}_2 + \text{ROSO}_2^- + 2\text{B} \rightarrow \text{ROSO}_3^- + 2\text{BH}^+\text{I}^-$

One mole of H_2O plus one mole of I_2 are required for the oxidation of the alkyl sulfite to an alkyl sulfate in the second reaction.

17-48. The bipotentiometric detector maintains a constant current (~10 µA) between two detector electrodes, while measuring the voltage needed to sustain that current. Before the equivalence point, the solution contains I^-, but little I_2. To maintain a current of 10 µA, the cathode potential must be negative enough to reduce some component of the solvent system (perhaps $\text{CH}_3\text{OH} + e^- \rightleftharpoons \text{CH}_3\text{O}^- + \frac{1}{2}\text{H}_2(g)$).

At the equivalence point, excess I_2 suddenly appears and current can be carried at low voltage by the reactions below. The abrupt voltage drop marks the end point.

Cathode: $\text{I}_3^- + 2e^- \rightarrow 3\text{I}^-$

Anode: $3\text{I}^- \rightarrow \text{I}_3^- + 2e^-$

CHAPTER 18
FUNDAMENTALS OF SPECTROPHOTOMETRY

18-1. (a) double   (b) halve   (c) double

18-2. (a) $E = h\nu = hc/\lambda = (6.6261 \times 10^{-34}\ \text{J s})(2.9979 \times 10^{8}\ \text{m s}^{-1})/(650 \times 10^{-9}\ \text{m})$
$= 3.06 \times 10^{-19}\ \text{J/photon} = 184\ \text{kJ/mol}$

(b) For $\lambda = 400$ nm, $E = 299$ kJ/mol.

18-3. (a) $\nu = c/\lambda = 2.9979 \times 10^{8}\ \text{m s}^{-1}/(562 \times 10^{-9}\ \text{m}) = 5.33 \times 10^{14}$ Hz

$\tilde{\nu} = 1/\lambda = [1/(562 \times 10^{-9}\ \text{m})]\ (1\ \text{m}/100\ \text{cm}) = 1.78 \times 10^{4}\ \text{cm}^{-1}$

$E = h\nu = (6.6261 \times 10^{-34}\ \text{J s})(5.33 \times 10^{14}\ \text{s}^{-1}) = 3.53 \times 10^{-19}\ \text{J/photon}$
$(3.53 \times 10^{-19}\ \text{J/photon})(6.022 \times 10^{23}\ \text{photon/mol}) = 213\ \text{kJ/mol}$

(b) $\lambda = 1/\tilde{\nu} = [1/(3\,028\ \text{cm}^{-1})]\ (1\ \text{m}/100\ \text{cm}) = 3.303 \times 10^{-6}\ \text{m} = 3.303\ \mu\text{m}$

$\nu = c/\lambda = (2.9979 \times 10^{8}\ \text{m s}^{-1}) / (3.303 \times 10^{-6}\ \text{m}) = 9.076 \times 10^{13}$ Hz

$E = hc\tilde{\nu} = (6.6261 \times 10^{-34}\ \text{J s})(2.9979 \times 10^{8}\ \text{m s}^{-1})(3\,028\ \text{cm}^{-1})(100\ \text{cm/m})$

$= 6.015 \times 10^{-20}\ \text{J/photon}$

$(6.015 \times 10^{-20}\ \text{J/photon})(6.022 \times 10^{23}\ \text{photon/mol}) = 36.22\ \text{kJ/mol}$

18-4. Microwave energies correspond to molecular rotational energy transitions. Infrared energies correspond to vibrational energy level transitions. Visible light can promote electrons to excited electronic states (in colored compounds). Ultraviolet light also promotes electrons and can even break chemical bonds.

18-5. From the definition of index of refraction, we can write

$c_{\text{vacuum}} = n \cdot c_{\text{air}}$ $\lambda_{\text{vacuum}} \cdot \nu = n \cdot \lambda_{\text{air}} \cdot \nu$

$\nu = c_{\text{vacuum}}/\lambda_{\text{vacuum}} = 5.088\,49$ and $5.083\,33 \times 10^{14}$ Hz

$\lambda_{\text{air}} = \lambda_{\text{vacuum}}/n = \lambda_{\text{vacuum}}/1.000\,277 = 588.995$ and 589.593 nm

$\tilde{\nu}_{\text{air}} = 1/\lambda_{\text{air}} = 1.697\,808$ and $1.696\,086 \times 10^{4}\ \text{cm}^{-1}$

18-6. Transmittance (T) is the fraction of incident light transmitted by a substance: $T = P/P_0$, where P_0 is incident irradiance and P is transmitted irradiance. Absorbance is logarithmically related to transmittance: $A = -\log T$. When all light is transmitted, absorbance is zero. When no light is transmitted, absorbance is infinite. Absorbance is proportional to concentration. Molar absorptivity ε is the constant of proportionality between absorbance at a particular wavelength and the product cb, where c is concentration and b is pathlength.

18-7. An absorption spectrum is a graph of absorbance (A) vs. wavelength (λ).

18-8. The color of transmitted light is the complement of the color that is absorbed. If blue-green light is absorbed, red light is transmitted.

18-9.

Curve	Absorption peak (nm)	Predicted color (Table 18-1)	Observed color
A	760	green	green
B	700	green	blue-green
C	600	blue	blue
D	530	violet	violet
E	500	red or purple-red	red
F	410	green-yellow	yellow

18-10. If absorbance is too high, too little light reaches the detector for accurate measurement. If absorbance is too low, there is too little difference between sample and reference for accurate measurement.

18-11. (a) $\varepsilon = A/bc = 0.822/[(1.000 \text{ cm})(2.31 \times 10^{-5} \text{ M})] = 3.56 \times 10^4 \text{ M}^{-1} \text{ cm}^{-1}$

(b) Quartz or cyclic olefin copolymer.

18-12. Violet-blue, based on Table 18-1. Plastic, glass and quartz cuvettes can be used at visible wavelengths. Best precision is with matched glass or quartz cuvettes.

18-13. [Fe] in reference cell = $\left(\frac{10.0 \text{ mL}}{50.0 \text{ mL}}\right)(6.80 \times 10^{-4} \text{ M}) = 1.36 \times 10^{-4} \text{ M}$. Setting the absorbances of sample and reference equal to each other gives $\varepsilon_s b_s c_s = \varepsilon_r b_r c_r$. But $\varepsilon_s = \varepsilon_r$, so $(2.48 \text{ cm})c_s = (1.00 \text{ cm})(1.36 \times 10^{-4} \text{ M})$
$\Rightarrow c_s = 5.48 \times 10^{-5}$ M. This is a 1/4 dilution of runoff, so [Fe] in runoff = 2.19×10^{-4} M.

18-14. (a) $A = -\log T = -\log 0.244 = 0.612_6$

(b) $PV = nRT$. Molar concentration $= \dfrac{\text{moles}}{\text{volume}} = \dfrac{n}{V} = \dfrac{P}{RT}$.

To convert 30.3 μbar = 30.3 × 10^{-6} bar to mol/L, we write

$$\dfrac{n}{V} = \dfrac{P}{RT} = \dfrac{30.3 \times 10^{-6} \text{ bar}}{\left(0.083\,145\,\dfrac{\text{L} \cdot \text{bar}}{\text{mol} \cdot \text{K}}\right)(298\,\text{K})} = 1.22_3 \times 10^{-6}\ \text{M}$$

(c) Molar absorptivity $= \varepsilon = \dfrac{A}{bc}$

Pathlength $b = 3.00$ cm; concentration $c = 1.22_3 \times 10^{-6}$ M.

$$\varepsilon = \dfrac{A}{bc} = \dfrac{0.612_6}{(3.00\ \text{cm})(1.22_3 \times 10^{-6}\ \text{M})} = 1.67 \times 10^5\ \text{M}^{-1}\ \text{cm}^{-1}$$

18-15. (a) Measured from the graph:

$\sigma \approx 40 \times 10^{-19}$ cm^2 at 280 nm $\sigma \approx 0.02 \times 10^{-19}$ cm^2 at 340 nm

at 280 nm: $T = e^{-(8 \times 10^{18}\ \text{cm}^{-3})(40 \times 10^{-19}\ \text{cm}^2)(1\ \text{cm})} = 1.3 \times 10^{-14}$

$A = -\log T = 13.9$

at 340 nm: $T = e^{-(8 \times 10^{18}\ \text{cm}^{-3})(0.02 \times 10^{-19}\ \text{cm}^2)(1\ \text{cm})} = 0.98_4$

$A = -\log T = 0.007$

(b) $T = e^{-n\sigma b}$ $0.14 = e^{-(8 \times 10^{18}\ \text{cm}^{-3})\sigma(1\ \text{cm})}$

$\Rightarrow \sigma = 2.4_{576} \times 10^{-19}$ cm^2

If n is decreased by 1%, $T = e^{-(7.92 \times 10^{18}\ \text{cm}^{-3})(2.457\,6 \times 10^{-19}\ \text{cm}^2)(1\ \text{cm})}$

$= 0.142\,8$

Increase in transmittance is $\dfrac{0.142\,8 - 0.14}{0.14} = 2.0\%$.

Note that the fractional increase in transmittance is greater than the fractional decrease in ozone concentration.

(c) $T_\text{winter} = e^{-(290\ \text{D.U.})(2.69 \times 10^{16}\ \text{molecules/cm}^3/\text{D.U.})(2.5 \times 10^{-19}\ \text{cm}^2)(1\ \text{cm})}$

$= 0.142$

$T_\text{summer} = e^{-(350)(2.69 \times 10^{16})(2.5 \times 10^{-19})(1)} = 0.095$

Fractional increase in transmittance is $(0.142 - 0.095)/(0.095) = 49\%$.

18-16. (a) The calibration curve would have the same slope, but would have a positive intercept equal to the absorbance of the reagent blank.

(b) Molar absorptivity is lower at 600 nm than at 562 nm, so the slope of the calibration curve would be lower. Also, the molar absorptivity changes quickly with wavelength near 600 nm. If the wavelength is not exactly 600 nm each time the monochromator was set, the slope of the calibration would change significantly. If the 600 nm light is not perfectly monochromatic, the calibration may show negative deviation due to some of the polychromatic light not being absorbed as much as expected for 600 nm.

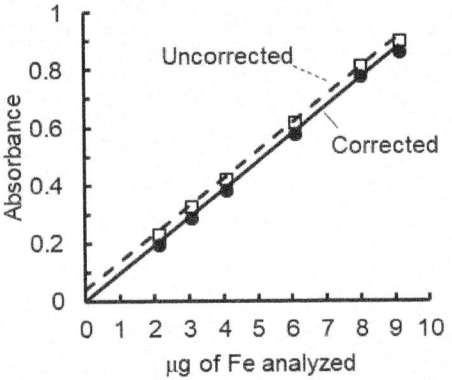
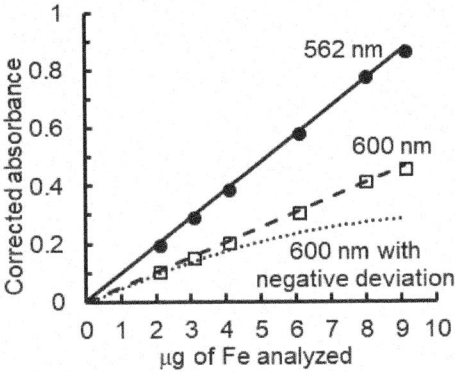

(c) We cannot extrapolate the current calibration line because we do not know if the calibration is linear at absorbances greater than 0.9. The unknown should be diluted 2–4 fold to bring it within the calibrated range.

18-17. (a) The certified reference material is run to check if the procedure is accurate. A certified reference material is sold by a national measurement institute, and is certified to contain known quantities of analyte in a matrix similar to samples being analyzed.

(b) 95% confidence interval = $\bar{x} \pm \dfrac{ts}{\sqrt{n}} = 1.49_3 \pm \dfrac{2.776(0.04_5)}{\sqrt{5}}$

$= 1.49_3 \pm 0.05_6$ µg/mL

The certified value (1.36 µg/mL) is not within the 95% confidence interval. The analysis results are significantly different from the certified value.

(c) Neocuproine reacts with Cu(I) and prevents it from forming a complex with ferrozine that would give a false positive result in the analysis of iron.

Fundamentals of Spectrophotometry

(d) The standard deviation is almost the same for both procedures, and so we use
$$t = \frac{|\bar{x}_1 - \bar{x}_2|}{s_{pooled}} \sqrt{\frac{n_1 n_2}{n_1 + n_2}}.$$

$$s_{pooled} = \sqrt{\frac{s_1^2(n_1-1) + s_2^2(n_2-1)}{n_1 + n_2 - 2}} = \sqrt{\frac{(0.04_5)^2(5-1) + (0.04_3)^2(4-1)}{5+4-2}}$$

$$= 0.04_4 \text{ µg/mL}$$

$$t = \frac{|\bar{x}_1 - \bar{x}_2|}{s_{pooled}} \sqrt{\frac{n_1 n_2}{n_1 + n_2}} = \frac{|1.49_3 - 1.38_7|}{0.04_4} \sqrt{\frac{5 \cdot 4}{5+4}} = 3.591$$

The critical value of t for 95% confidence and $n_1 + n_2 - 2 = 7$ degrees of freedom is 2.365. Because $t_{calculated} > t_{table}$, the difference is significant.

18-18. (a) $c = A/\varepsilon b = 0.427/[(6\,130 \text{ M}^{-1} \text{ cm}^{-1})(1.000 \text{ cm})] = 6.97 \times 10^{-5}$ M

(b) The sample had been diluted $\times 10 \Rightarrow 6.97 \times 10^{-4}$ M.

(c) $\dfrac{x \text{ g}}{(292.16 \text{ g/mol})(5.00 \times 10^{-3} \text{ L})} = 6.97 \times 10^{-4} \text{ M} \Rightarrow x = 1.02$ mg

18-19. Yes

18-20. (a) $\varepsilon = \dfrac{A}{cb} = \dfrac{0.267 - 0.019}{(3.15 \times 10^{-6} \text{ M})(1.000 \text{ cm})} = 7.87 \times 10^4 \text{ M}^{-1} \text{ cm}^{-1}$

(b) $c = \dfrac{A}{\varepsilon b} = \dfrac{0.175 - 0.019}{(7.87 \times 10^4 \text{ M}^{-1} \text{ cm}^{-1})(1.000 \text{ cm})} = 1.98 \times 10^{-6}$ M

18-21. (a)

	A	B	C	D	E	F	G	H
1	Determination of DNA at 260 nm							
2	[DNA]							
3	(µg/mL)	Absorbance	Residual					
4	2.5	0.080	0.0010	C4=B4-(B$12*A4+C$12)				
5	5.0	0.149	-0.0045					
6	10.0	0.307	0.0044					
7	15.0	0.442	-0.0096	Highlight cells B12:C14				
8	20.0	0.620	0.0194	Type				
9	25.0	0.739	-0.0107	"=LINEST(B4:B9,A4:A9,TRUE,TRUE)"				
10				Press CTRL+SHIFT+ENTER on PC				
11		LINEST output		Press CTRL+SHIFT+RETURN on Mac				
12	m	0.02981	0.0045	b				
13	u_m	0.00064	0.0097	u_b				
14	R^2	0.9982	0.0125	s_y				

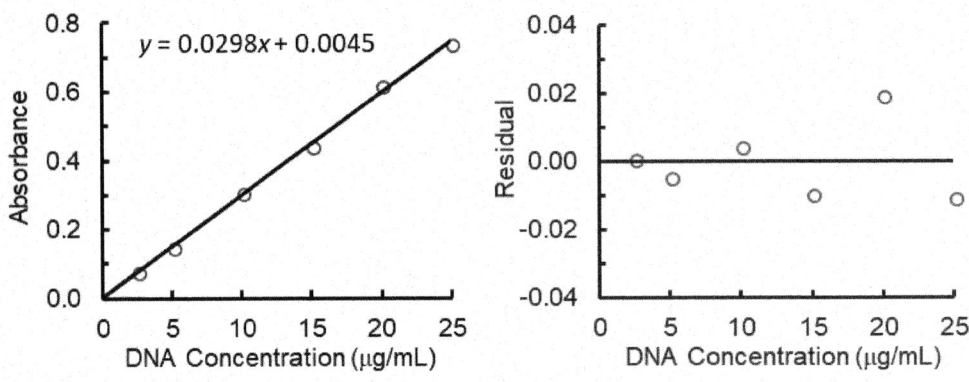

$y = (0.029\,8 \pm 0.000\,6)x + (0.004\,5 \pm 0.009\,7)$

Calibration is linear based on residuals being randomly scattered about the line, intercept is statistically equal to 0, and $R^2 > 0.995$.

(b) The cell pathlength is $b = 1$ cm, so the molar absorptivity (ε) is the slope of the graph of absorbance (A) versus concentration (c) because $A = \varepsilon cb$.

$\varepsilon_{260} = 0.029\,8\ (\mu g/mL)^{-1}cm^{-1} = 0.029\,8\ (mg/L)^{-1}cm^{-1}$

$= \dfrac{0.0298}{(mg/L)(cm)} \cdot \dfrac{1000\ mg}{g} = 29.8\ (g/L)^{-1}cm^{-1}$

$= 29.8\ (g/L)^{-1}cm^{-1} \times 4\,568\ g/mol = 1.36 \times 10^5\ M^{-1}\ cm^{-1}$

(c) $\text{Error} = \left(\dfrac{\text{obs. } A - \text{web } A}{\text{obs. } A}\right) \times 100\% = \left(\dfrac{0.0298 - 0.027}{0.0298}\right) \times 100\% = 9.4\%$

Using $0.027\ (\mu g/mL)^{-1}cm^{-1}$ would result in a 9.4% error. The magnitude of error would depend on the composition of the specific DNA being analyzed.

(d) $F_{\text{calculated}} = 0.154^2/0.017^2 = 82._1 > F_{\text{table}} = 39.00$ (for 2 degrees of freedom in both numerator and denominator). Standard deviations are significantly different at 95% confidence level. Because standard deviations are different, we use the equations for comparison of means.

$t_{\text{calculated}} = \dfrac{|\bar{x}_1 - \bar{x}_2|}{\sqrt{(s_1^2/n_1) + (s_2^2/n_2)}} = \dfrac{|8.385 - 8.264|}{\sqrt{(0.154^2/3) + (0.017^2/3)}} = 1.353$

$\text{Degrees of freedom} = \dfrac{(s_1^2/n_1 + s_2^2/n_2)^2}{\dfrac{(s_1^2/n_1)^2}{n_1 - 1} + \dfrac{(s_2^2/n_2)^2}{n_2 - 1}} = \dfrac{(0.154^2/3 + 0.017^2/3)^2}{\dfrac{(0.154^2/3)^2}{3-1} + \dfrac{(0.017^2/3)^2}{3-1}}$

$= 2.05$, which we round to 2.

$t_{\text{calculated}} = 1.353 < 4.303$ (Student's t for 95% confidence and 2 degrees of freedom). The difference is <u>not</u> significant.

18-22. (a) $\varepsilon_{280\,nm}$ (M^{-1} cm^{-1}) $\approx$ 5 500 n_{Trp} + 1 490 n_{Tyr} + 125 n_{S-S}
= 5 500 (8) + 1 490 (26) + 125 (19) = 8.512 × 10^4 M^{-1} cm^{-1}

(b) 1 wt% protein ≈ 10 g/L if density = 1.00 g/mL
molarity = (10 g/L)/(79 570 g/mol) = 1.257 × 10^{-4} M
$A = \varepsilon bc$ = (8.512 × 10^4 M^{-1} cm^{-1})(1.000 cm)(1.257 × 10^{-4} M) = 10.7

(c) A 1 wt% solution is predicted to have A = 10.7. A solution with A = 1.50 has a concentration of (1.50/10.7)(1 wt%) = 0.140 wt%.
Since 1 wt% = 10 mg/mL, 0.140 wt% = (0.140)(10 mg/mL) = 1.40 mg/mL

18-23. (a) The absorbance due to the colored product from nitrite added to sample C is 0.967 − 0.622 = 0.345. The concentration of colored product due to added nitrite in sample C is $\dfrac{(7.50 \times 10^{-3}\,M)(10.0 \times 10^{-6}\,L)}{0.0540\,L}$ = 1.38$_9$ × 10^{-6} M.

$\varepsilon = A/bc$ = 0.345/[(1.38$_9$ × 10^{-6} M)(5.00 cm)] = 4.97 × 10^4 M^{-1} cm^{-1}

(b) A = 0.622 − 0.153 = 0.469.
$c = A/\varepsilon b$ = 0.469/[(4.97 × 10^4 M^{-1} cm^{-1})(5.00 cm)] = 1.88$_7$ × 10^{-6} M.
mass = ($c \times V$)(46.005 g/mol) = (1.88$_7$ × 10^{-6} M) × 0.0540 L × 46.005 g/mol
= 4.68$_8$ × 10^{-6} g = 4.69 μg NO$_2^-$

(c) c = (4.69 μg NO$_2^-$)/0.0500L = 93.8 mg/L NO$_2^-$
$c = \left(\dfrac{14.007}{46.005}\right)$(4.69 μg NO$_2^-$)/0.0500L = 28.6 mg/L NO$_2^-$ nitrogen

18-24. (a) Mass required = (500 mL)(500 μg Fe/mL) = 2.50 × 10^5 μg Fe = 0.250 g Fe
Mass fraction of Fe in Fe(H$_3$NCH$_2$CH$_2$NH$_3$)(SO$_4$)$_2$ · 4H$_2$O =
(55.845 g Fe)/(382.13 g reagent) = 0.146 14
Mass of Fe(H$_3$NCH$_2$CH$_2$NH$_3$)(SO$_4$)$_2$ · 4H$_2$O required to provide 0.250 g Fe
= (0.250 g Fe) / (0.146 14 Fe/g reagent) = 1.71$_1$ g

(b) g Fe weighed = (0.146 14 g Fe/g reagent)(1.627 g reagent) = 0.237 8 g Fe
[Fe] = $\dfrac{0.2378\text{ g Fe}}{500.0\text{ mL}}$ = 0.475 6 mg Fe/mL = 475.6 μg Fe/mL

(c) To prepare ~ 1 μg Fe/mL, dilute 1.000 mL of stock solution from (b) up to 500.0 mL with 0.1 M H$_2$SO$_4$
[Fe] = $\dfrac{475.6\text{ μg Fe}}{500.0\text{ mL}}$ = 0.9512 μg Fe/mL

In a similar manner, dilute 2.000, 3.00, 4.00, and 5.00 mL of stock solution from (b) up to 500.0 mL with 0.1 M H_2SO_4 to obtain [Fe] = 1.902 µg Fe/mL, 2.85 µg Fe/mL, 3.80 µg Fe/mL, and 4.76 µg Fe/mL.

(d) First make a 1/10 dilution by diluting 5 mL of ~500 µg Fe/mL stock solution up to 50 mL with 0.1 M H_2SO_4. This solution contains ~50 µg Fe/mL. To prepare ~1 µg Fe/mL, dilute 1.000 mL of ~50 µg Fe/mL stock solution up to 50.0 mL with 0.1 M H_2SO_4. In a similar manner, dilute 2.000, 3.00, 4.00, and 5.00 mL of ~50 µg Fe/mL stock solution up to 50.0 mL with 0.1 M H_2SO_4 to obtain [Fe] ≈ 2, 3, 4, and 5 µg Fe/mL.

18-25. (a) Mass required = (500.0 mL)(1 000 µg/mL) = 5.000×10^5 µg = 0.500 0 g Fe

Mass fraction of Fe in $Fe(NH_4)_2(SO_4)_2 \cdot 6H_2O$ =
= (55.845 g Fe) / (392.12 g reagent) = $0.142\,41_8$

Mass of $Fe(NH_4)_2(SO_4)_2 \cdot 6H_2O$ required to provide 0.500 g Fe
= (0.5000 g Fe) / ($0.142\,41_8$ g Fe/g reagent) = 3.511 g

(b) g Fe weighed out = ($0.142\,41_8$ g Fe/g reagent)(3.627 g reagent)
= $0.516\,5_5$ g Fe

[Fe] = ($0.516\,5_5$ g Fe) / 500.0 mL = 1.033_1 mg Fe/mL = 1 033 µg Fe/mL

(c) The table below lists two serial dilutions to prepare each desired standard

First dilution volume (mL) of 1 000 µg/mL to dilute to 250 mL	Resulting [Fe] (µg/mL)	Second dilution volume of first dilution to dilute to 250 mL	Resulting [Fe] (µg/mL)
10	$\left(\frac{10}{250}\right)$(1 033 µg Fe/mL) = 41.32 µg Fe/mL	5	$\left(\frac{5}{250}\right)$(41.32 µg Fe/mL) = 0.826 µg Fe/mL
10	41.32 µg Fe/mL	10	$\left(\frac{10}{250}\right)$(41.32 µg Fe/mL) = 1.653 µg Fe/mL
10	41.32 µg Fe/mL	15 (= 10 + 5)	$\left(\frac{15}{250}\right)$(41.32 µg Fe/mL) = 2.479 µg Fe/mL
15 (= 10 + 5)	$\left(\frac{15}{250}\right)$(1 033 µg Fe/mL) = 61.98 µg Fe/mL	15 (= 10 + 5)	$\left(\frac{15}{250}\right)$(61.98 µg Fe/mL) = 3.719 µg Fe/mL
15 (= 10 + 5)	61.98 µg Fe/mL	(20) (= 10 + 10)	$\left(\frac{20}{250}\right)$(61.98 µg Fe/mL) = 4.958 µg Fe/mL
20 (= 10 + 10)	$\left(\frac{20}{250}\right)$(1 033 µg Fe/mL) = 82.64 µg Fe/mL	(20) (= 10 + 10)	$\left(\frac{20}{250}\right)$(82.64 µg Fe/mL) = 6.611 µg Fe/mL

20 (= 10 + 10)	82.64 μg Fe/mL	(25) (= 10 + 10 + 5)	$\left(\frac{25}{250}\right)$(82.64 μg Fe/mL) = 8.264 μg Fe/mL
20 (= 10 + 10)	82.64 μg Fe/mL	(30) (= 10 + 10 + 10)	$\left(\frac{30}{250}\right)$(82.64 μg Fe/mL) = 9.917 μg Fe/mL

18-26. Uncertainty in 3-mL pipet = ±0.01 mL = 0.333%.

Uncertainty in volume from two 3-mL aliquots =

$\pm\sqrt{0.01^2 + 0.01^2} = \pm 0.014$ mL

Volume delivered from two 3-mL aliquots = 6.00 ± 0.014 mL

= 6.00 ± 0.23$_6$% mL

Uncertainty in volume of 1-L flask = ±0.30 mL = ±0.030%

Fe stock concentration = 1.086 ± 0.002 g Fe/L = 1.086 ± 0.18$_4$% g Fe/L

After diluting two 3-mL volumes of stock solution to 1 L, the concentration is

$[Fe] = \left(\frac{6.00 \pm 0.23_6\% \text{ mL}}{1\,000.0 \pm 0.03\% \text{ mL}}\right)(1.086 \pm 0.18_4\% \text{ g/L})$

$= 0.006\,516 \pm \sqrt{0.23_6\%^2 + 0.03\%^2 + 0.18_4\%^2}$ g/L

$= 0.006\,516 \pm 0.301\%$ g/L $= 0.006\,516 \pm 0.000\,020$ g/L

$= 6.516 \pm 0.020$ mg/L $= 6.516 \pm 0.020$ μg/mL $= 6.516 \pm 0.30\%$ μg Fe/mL

18-27. (a) The orange light emitting diode emission coincides with the maximum of the In^{2-} absorbance, and so it would give the greatest sensitivity.

(b) Orange diode. Beer's law assumes "monochromatic light." Both diodes emit "polychromatic" light—consisting of a range (bandwidth) of wavelengths. Polychromatic light causes greater negative (downward curving) deviation from Beer's law if the absorbance changes significantly over the bandwidth of the light source. The orange diode emission coincides with the maximum of the In^{2-} absorbance, where there is little variation in absorbance with wavelength. The red diode emission is on the side of the In^{2-} absorbance band, where the absorbance changes rapidly with wavelength. So, the orange diode should yield more linear calibration.

(c) Methods for assessing linearity, in preferential order, are: Spike recovery > residual plot > inspecting graph > intercept = 0? > R^2. The spreadsheet plots the calibration curve and a residual plot. LINEST provides the slope (cell B11) and its standard uncertainty (cell B12), intercept (C11) and its standard uncertainty (C12), and the R^2 = 0.996 (B13).

	A	B	C	D
1	Linearity of student-made photometer			
2				
3	Concentration (mM)	Absorbance	Deviation	
4	3.3	0.135	0.0182	
5	10	0.302	-0.0316	
6	16.7	0.567	0.0163	
7	23.3	0.754	-0.0106	
8	29.9	0.986	0.0077	
9			C5=B5-(B$11*A5+C$11)	
10	{B11:C13}=LINEST(B4:B8,A4:A8,TRUE,TRUE)			
11	m	0.0324	0.0097	b
12	u_m	0.0012	0.0221	u_b
13	R^2	0.9962	0.0243	s_y

Inspect of the calibration curve and residual plot show random residuals. Intercept (0.009 7 ± 0.022 1) is statistically equal to 0. Finally, R^2 > 0.995. Based on all of these criteria, the calibration is linear.

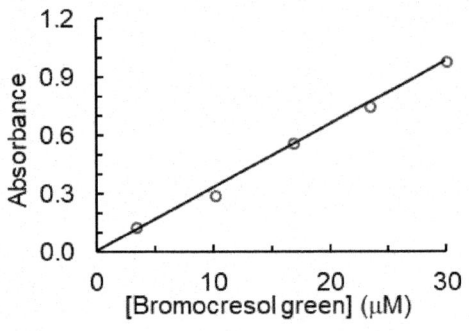 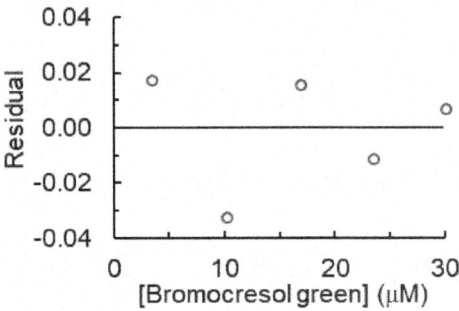

(d) Using LINEST on the commercial spectrophotometer data yields a calibration equation $y = (0.017\ 94 \pm 0.000\ 04)x + (0.001\ 5 \pm 0.000\ 8)$ with R^2 > 0.999 98 with tiny residuals. The slope, 0.017 94, is the product of molar absorptivity times pathlength. The slope for the student photometer is 0.032 4. The greater slope must arise from a longer pathlength for the student photometer:

$$\text{slope} = \frac{A}{c} = b\varepsilon \Rightarrow \frac{b_2}{b_1} = \frac{\text{slope}_2}{\text{slope}_1} \Rightarrow b_2 = \frac{\text{slope}_2}{\text{slope}_1} \cdot b_1$$

$$b_2 = \frac{0.032\ 4}{0.017\ 9} \cdot 1.00\ \text{cm} = 1.81\ \text{cm}$$

18-28. (a) Absorbance of the blue In⁻ form of the indicator is monitored at 600 nm. Before the equivalence point, there is an excess of Mg^{2+} in solution, so all indictor is in the weakly absorbing MgIn form. Near the equivalence point EDTA starts displacing Mg^{2+} from MgIn, resulting in formation of free In⁻ and a rapid rise in absorbance. After the equivalence point, all indicator is in the In⁻ form, and corrected absorbance stays constant.

(b) 0.014 5 9 L of 0.010 8 3 M EDTA = 1.580×10^{-4} mol EDTA

(c) Formula mass of $CaCO_3$ is 100.0 g/mol.

g $CaCO_3$ in sample = $(1.580 \times 10^{-4}$ mol$)(100.0$ g/mol$) = 0.015$ 8 g

mg CaCO3 per L = $(0.015\ 8\ g)\left(\dfrac{1\ 000\ mL/L}{100\ mL}\right)(1\ 000\ mg/g)$ = 158 mg/L

18-29. (a) Use Table 18-1 to predict color observed from wavelength of absorption.

basic form ⇒ In²⁻ ⇒ absorbs 610 nm ⇒ predicted color = blue

acidic form ⇒ HIn⁻ ⇒ absorbs 445 nm ⇒ predicted color = orange/yellow

Table 11-3 states that the colors are blue and yellow.

(b) Equivalence point = (30.0 mL × 0.00845 M) / 0.0496 M = 5.11 mL

Solution is initially in acid HIn⁻ form, so absorbance at 610 nm is initially low. Once all HCl is neutralized, NaOH starts deprotonating HIn⁻ to absorbing In²⁻ form and absorbance increases rapidly. To find end point volume draw a straight line through the plateau before the rise (2–4 mL), and a line through the straight portion of the rise (5.3–5.7 mL). The end point is the volume where these lines intersect.

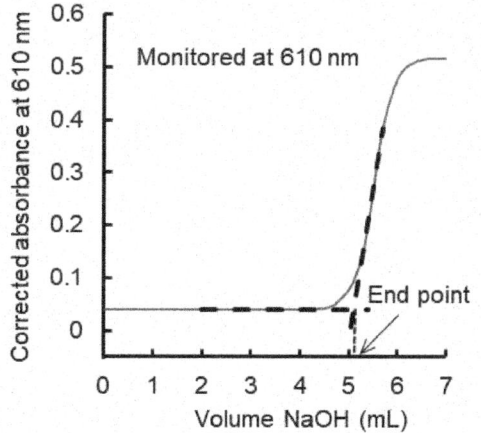
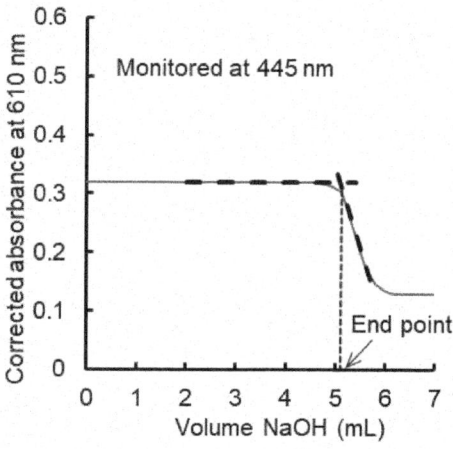

(c) Equivalence point remains at 5.11 mL as in part (b). Absorbance at 445 nm is initially ~0.3 due to HIn in acid. Near the end point all HCl is consumed and further NaOH deprotonates HIn⁻ to weaker absorbing In²⁻ form, so absorbance decreases. To find end point volume, draw a straight line through the plateau before the fall (2–4 mL), and a line through the straight portion of the drop (5.3–5.7 mL). The end point is the intersection. As both the HIn⁻ and In²⁻ forms absorb, the change in absorbance is not as dramatic as in (b), and so precision of extrapolation is poorer and 610 is the preferred wavelength.

18-30. (a) In the graph below, least-squares lines were fit through the first 6 points and the last 3 points. Their intersection is the estimated end point of 21.3 µL. The moles of TCNQ in the titration are (0.700 mL)(1.00 × 10⁻⁴ M TCNQ) = 70.0 nmol TCNQ. This many moles of Au(0) are present in 21.3 µL of nanoparticle solution. The mass of nanoparticles in 21.3 µL is (21.3 µL)(1.43 g/L) = 30.5 µg nanoparticles. To find the moles of Au(0) in 1.00g of nanoparticles, set up a proportion:

$$\frac{70.0\times 10^{-9}\text{ mol Au(0)}}{30.5\times 10^{-6}\text{ g nanoparticles}} = \frac{x \text{ mol Au(0)}}{1.00 \text{ g nanoparticles}} \Rightarrow x = 2.30 \text{ mmol Au(0)}$$

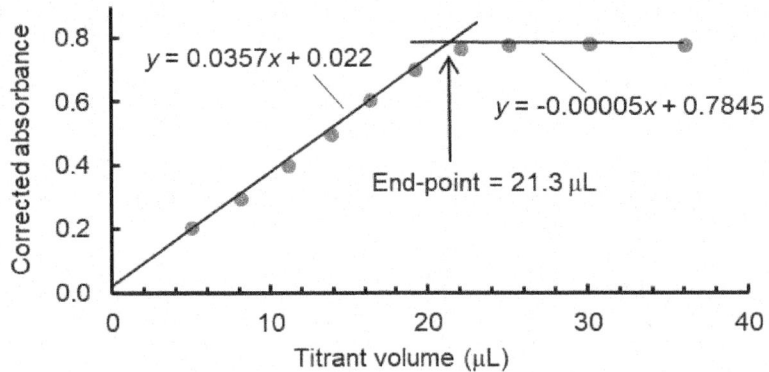

(b) Absorbance of the titration product (TCNQ⁻·) is monitored. Each Au(0) in the nanoparticle titrant reacts to form one TCNQ⁻·, causing a steady rise in the absorbance during the titration. Once all TCNQ has been converted to product, addition of more nanoparticles results in no additional TCNQ⁻· product, and the absorbance is constant.

(c) 1.00 g of nanoparticles is estimated to contain 0.25 g $C_{12}H_{25}S$ (FM 201.39), which is 1.24 mmol $C_{12}H_{25}S$.

(d) From (a), the mass of Au(0) in 1.00 g is (2.30 mmol Au(0)(196.97 g/mol) = 0.453 g. The mass of Au(I) is estimated as the difference 1.00 − 0.453 − 0.25 = 0.29$_7$ g = 1.51 mmol Au(I). The calculated mole ratio Au(I):C$_{12}$H$_{25}$S is 1.51/1.24 = 1.22. Ideally, this mole ratio should be 1.00.

18-31. $n \rightarrow \pi^*(T_1)$: $E = h\dfrac{c}{\lambda} = (6.626 \times 10^{-34}\,\text{J·s})\dfrac{2.997\,9 \times 10^8\,\text{s}^{-1}}{397 \times 10^{-9}\,\text{m}} = 5.00 \times 10^{-19}\,\text{J}$

To convert to J/mol, multiply by Avogadro's number:
5.00×10^{-19} J/molecule × 6.022×10^{23} molecules/mol = 301 kJ/mol.

$n \rightarrow \pi^*(S_1)$:

$E = (6.626 \times 10^{-34}\,\text{J·s})\dfrac{2.997\,9 \times 10^8\,\text{s}^{-1}}{355 \times 10^{-9}\,\text{m}} = 5.60 \times 10^{-19}\,\text{J} = 337$ kJ/mol.

The difference between the T_1 and S_1 states is 337 − 301 = 36 kJ/mol.

18-32. Fluorescence is emission of light with no change in electronic spin state (for example, singlet → singlet). In phosphorescence, electronic spin must change during emission (for example, triplet → singlet). Phosphorescence is less probable, so molecules spend more time in the excited state prior to phosphorescence than to fluorescence. That is, phosphorescence has a longer lifetime than fluorescence. Phosphorescence also comes at lower energy (longer wavelength) than fluorescence, because the triplet excited state is at lower energy than the singlet excited state.

18-33. In Rayleigh scattering, electrons in molecules oscillate at the frequency of incoming radiation and emit that same frequency in all directions. The time scale is ~10^{-15} s for visible light. In Raman scattering, molecules extract a quantum of vibrational energy from incoming light and scatter light with less energy than the incoming light. Again, the time scale is ~10^{-15} s for visible light. Fluorescence occurs in 10^{-9} to 10^{-4} s, which is 10^6 to 10^{11} times longer than scattering.

18-34. Phosphorescence is emitted at longer wavelength than fluorescence. Absorption is at shortest wavelength.

18-35. In an excitation spectrum, the exciting wavelength (λ_{ex}) is varied while the detector wavelength (λ_{em}) is fixed. In an emission spectrum, λ_{ex} is held constant and λ_{em} is varied. The excitation spectrum resembles an absorption spectrum because emission intensity is proportional to absorption of the exciting radiation.

18-36. The spreadsheet computes relative fluorescence given by

$$\text{relative intensity} = \frac{I}{k'P_0} = 10^{-\varepsilon_{ex}b_1c}(1 - 10^{-\varepsilon_{ex}b_2c})10^{-\varepsilon_{em}b_3c}.$$

	A	B	C
1	Anthracene fluorescence response		
2			
3	$\varepsilon_{ex} =$	9000	M⁻¹cm⁻¹
4	$\varepsilon_{em} =$	50	M⁻¹cm⁻¹
5			
6	$b_1 =$	0.30	cm
7	$b_2 =$	0.40	cm
8	$b_3 =$	0.50	cm
9			
10		Relative	
11		fluorescence	
12	c (M)	I/k'P$_o$	Intensity/c
13	1.E-08	8.288E-05	8288
14	2.E-08	0.00017	8288
15	4.E-08	0.00033	8286
16	1.E-07	0.00083	8281
17	2.E-07	0.00165	8272
18	4.E-07	0.00330	8255
19	1.E-06	0.00820	8203
20	2.E-06	0.01624	8118
21	4.E-06	0.03181	7951
22	6.E-06	0.04673	7788
23	8.E-06	0.06102	7628
24	1.E-05	0.07471	7471
25	2.E-05	0.13476	6738
26	4.E-05	0.21957	5489
27	6.E-05	0.26893	4482
28	8.E-05	0.29345	3668
29	1.E-04	0.30087	3009
30	2.E-04	0.23078	1154
31	4.E-04	0.07833	196
32	6.E-04	0.02301	38
33	8.E-04	0.00660	8
34	1.E-03	0.00188	2
35	B13=10^-($B$3*$B$6*A13)		
36	*(1-(10^-($B$3*$B$7*A13)))		
37	*(10^-($B$4*$B$8*A13))		
38	C13=B13/A13		

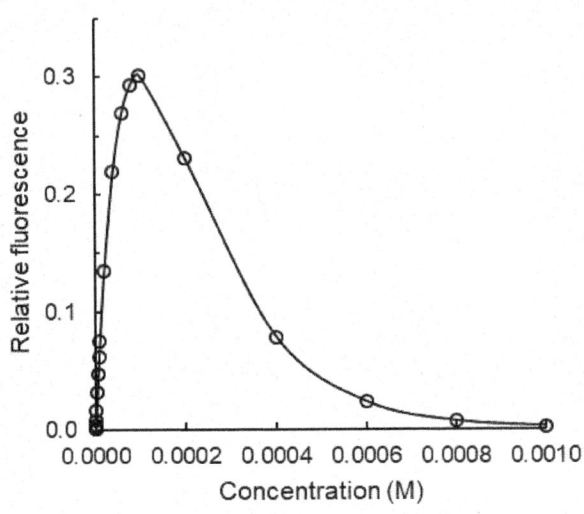

Fluorescence increases with concentration and then decreases because of self-absorption. Most of the absorption occurs at the excitation wavelength, but a small amount of emission at 402 nm is also absorbed. Column C of the spreadsheet gives the relative fluorescence intensity divided by concentration. If intensity were proportional to concentration, then column C would be constant. We see that it is constant at low concentration, and falls by ~5% at ~5 μM. The calibration curve in the text goes up to 0.6 μM, which is in the linear range.

Fundamentals of Spectrophotometry

18-37. The equation for standard addition is

$$\underbrace{I_{X+S}\left(\frac{V_i + V_S}{V_i}\right)}_{\text{Function to plot on } y\text{-axis}} = I_X + \underbrace{\frac{I_X}{[X]_i}[S]_i\left(\frac{V_S}{V_i}\right)}_{\text{Function to plot on } x\text{-axis}}$$

where V_i is the volume of unknown in the cuvette (2.00 mL), V_S is the volume of standard added (0 to 40 µL), $(V_i + V_S)$ is the total volume of unknown plus added standard, $[S]_i$ is the initial concentration of standard (1.40 µg Se/mL), $[X]_i$ is the initial concentration of unknown in the 2.00-mL solution, I_X is the fluorescence intensity from the unknown, and I_{X+S} is the fluorescence intensity from the unknown plus standard addition. We make a graph of $I_{X+S}(V_i + V_S)/V_i$ versus $[S]_i V_S/V_i$ in the following spreadsheet.

	A	B	C	D	E
1	Selenium standard addition with Variable Volume				
2					
3	V_i (mL) =	V_S = mL Se	I_{X+S} =	x-axis function	y-axis function
4	2.00	std added	signal	S_i*V_S/V_i (ppm)	$I_{X+S}*(V_i+V_S)/V_i$
5	$[S]_i$ (µg/mL) =	0.0000	41.4	0.00000	41.4
6	1.40	0.0100	49.2	0.00700	49.4
7		0.0200	56.4	0.01400	57.0
8		0.0300	63.8	0.02100	64.8
9		0.0400	70.3	0.02800	71.7
10			D5=$A$6*B5/$A$4		E5=C5*($A$4+B5)/$A$4
11		Linest output			
12		slope	1084.6	41.670	y-intercept
13		u_m	14.9	0.255	u_b
14		R^2	0.9994	0.330	s_y
15	x-intercept =	-0.0384192	B15=-C12/B12		
16	n =	5	B16=COUNT(B5:B9)		
17	Mean y =	56.8546	B17=AVERAGE(E5:E9)		
18	$\Sigma(x_i-\text{mean } x)^2$=	0.00049	B18=DEVSQ(D5:D9)		
19	u_x =	0.00073222	B19=C14/ABS(B12)*		
20			SQRT(1/B16+B17^2/(B12^2*B18))		

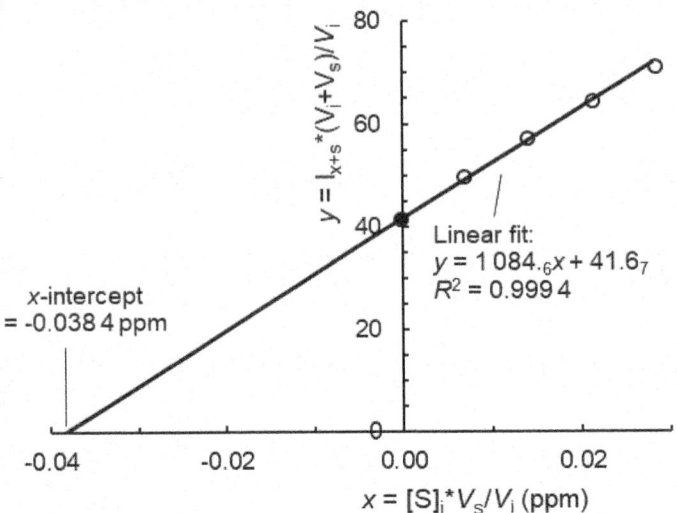

The x-intercept is found from the equation of the straight line by setting $y = 0$: $0 = 1084._6 x + 41.6_7 \Rightarrow x = -0.038\,4_2$. The concentration of Se in the unknown is $0.038\,4_2$ μg/mL. All Se from 0.108 g of Brazil nuts was dissolved in 10.0 mL of solvent, which contained $(10.0\text{ mL})(0.038\,4_2\text{ μg/mL}) = 0.384_2$ μg Se. The wt% Se in the nuts is $100 \times (0.384_2 \times 10^{-6}\text{ g}/0.108\text{ g}) = 3.56 \times 10^{-4}$ wt%.

The standard deviation of the x-intercept is

$$\text{Standard deviation of } x\text{-intercept} = \frac{s_y}{|m|}\sqrt{\frac{1}{n} + \frac{\bar{y}^2}{m^2 \sum(x_i - \bar{x})^2}}$$

where s_y is the standard deviation of y, m is the slope, n is the number of data points (= 5), $\bar{y}$ is the mean value of y for the 5 points, x_i are individual values of x, and $\bar{x}$ is the mean value of x for the 5 points. Cell B19 of the spreadsheet gives a standard deviation of $0.000\,7_3$ for the x-intercept.

The relative uncertainty in the intercept is $0.000\,7_3 / 0.038\,4_2 = 1.9_0\%$. This is the relative uncertainty in wt% Se if other sources of error are insignificant. Uncertainty in wt% = $(0.019_0)(3.56 \times 10^{-4}$ wt%$) = 6._8 \times 10^{-6}$ wt%. Answer = $3.56 (\pm 0.07) \times 10^{-4}$ wt%.

The confidence interval is $\pm t \times$ (standard uncertainty), where t is Student's t (Table 4-2) for $n - 2 = 3$ degrees of freedom. The 95% confidence interval is $\pm(3.182)(0.06_8 \times 10^{-4}$ wt%$) = \pm 0.22 \times 10^{-4}$ wt%. The value $t = 3.182$ is from Table 4-2 for 3 degrees of freedom. Answer = $3.56 (\pm 0.22) \times 10^{-4}$ wt%.

Fundamentals of Spectrophotometry 263

18-38. (a) Linear range is 0.15–12.0 µg/mL. Relative standard deviation is < 0.4%.

(b) Other methods have poor accuracy, interfering species, or prohibitive costs. Second paragraph of *Introduction* mentions titrations, inductively coupled plasma, atomic absorption, X-ray fluorescence, and spectrophotometry.

(c) In *Experimental*, the *Procedure* says alloy samples were dissolved in sulfuric and nitric acid with heating, diluted to volume, oxidized with persulfate with silver (I) catalyst, acidified with sulfuric acid and diluted to volume. An aliquot was mixed with tartrate and reacted with sodium hydrate.

(d) *Interference study* in *Results and discussion* explains that tartrate was used as a masking agent to eliminate interference by Ti(IV) and V(V).

(e) *Absorption Spectra* in *Results and Discussion* and Figure 3 show 278 or 374 nm could be used. 374 nm was used due to its greater sensitivity.

(f) *Accuracy and Precision* within *Results and Discussion* states that because no certified reference material was available, they checked accuracy by comparing their analysis of an alloy sample with that of inductively coupled plasma-optical emission spectroscopy.

18-39. (a) *Abstract* states that 309 ± 93 antibodies were on each nanoparticle. Conventional methods yielded 660 ± 87 antibodies.

(b) *Introduction* on page 3851 mentions: analyzing excess protein using Bradford and bicinchoninic acid (BCA) total protein assays; analyzing excess fluorescent labeled proteins; surface plasmon resonance; and other measurements that compare the amount of adsorbed protein to the saturation value, but cannot provide an absolute quantitation.

(c) In *Experimental*, the *Fluorescence* section states NanoOrange was used, with excitation at 485 nm and emission monitored at 590 nm. The caption for Figure 3 also provides this information.

(d) Yes. Data in Figure 3 are randomly scattered about the line. $R^2 = 0.992$ (figure caption) reflects significant random error, rather than curvature.

(e) *t* test of two methods with comparable standard deviation. Error bars in Figure 4 show the standard deviations of the two methods are comparable, and associate text describes the *t* test and conclusion.

(f) Figure S1 of *Supplementary Materials* shows an intercept of 0.49. The *y*-intercept is *y* value when $x = 0$. In this graph the *y*-axis is positioned at $x = -2$.

CHAPTER 19
APPLICATIONS OF SPECTROPHOTOMETRY

19-1. Putting $b = 0.100$ cm into determinants gives

$$[X] = \frac{\begin{vmatrix} 0.233 & 387 \\ 0.200 & 642 \end{vmatrix}}{\begin{vmatrix} 1\,640 & 387 \\ 399 & 642 \end{vmatrix}} = 8.03 \times 10^{-5} \text{ M} \qquad [Y] = \frac{\begin{vmatrix} 1\,640 & 0.233 \\ 399 & 0.200 \end{vmatrix}}{\begin{vmatrix} 1\,640 & 387 \\ 399 & 642 \end{vmatrix}} = 2.62 \times 10^{-4} \text{ M}$$

The spreadsheet solution looks like this, with answers in column F:

	A	B	C	D	E	F	G
1	Analysis of a mixture by spreadsheet matrix operations						
2							
3	Wavelength	Coefficient matrix		Absorbance		Concentrations	
4	(nm)	(εb)		of unknown		in mixture	
5	272	1640	387	0.233		8.034E-05	<-[X]
6	327	399	642	0.200		2.616E-04	<-[Y]
7		K		A		C	
8							
9	1. Highlight block of blank cells required for solution (F5:F6).						
10	2. Type the formula "=MMULT(MINVERSE(B5:C6),D5:D6)"						
11	3. Press CTRL+SHIFT+ENTER on PC or CTRL+SHIFT+RETURN on Mac.						
12	4. The answer appears in cells F5:F6.						

19-2.

	A	B	C	D	E	F	G	H
1	Analysis of a Mixture When You Have More Data Points that Components of the Mixture							
2				Measured			Calculated	
3		Absorbance		Absorbance	Molar		Absorbance	
4	Wavelength	of Standard		of Mixture	Absorptivity		of Mixture	
5	(nm)	MnO_4^-	$Cr_2O_7^{2-}$	A_m	MnO_4^-	$Cr_2O_7^{2-}$	A_{calc}	$(A_{calc}-A_m)^2$
6	266	0.042	0.410	0.766	420	4100	0.765	1.02E-06
7	288	0.082	0.283	0.571	820	2830	0.572	1.76E-06
8	320	0.168	0.158	0.422	1680	1580	0.422	1.13E-07
9	350	0.125	0.318	0.672	1250	3180	0.671	2.05E-06
10	360	0.056	0.181	0.366	560	1810	0.369	9.11E-06
11						SUM(H6:H10) =		1.40E-05
12	Standards							
13	[Mn] (M) =	1.00E-04		E6 = B6/($B$16*$B$13)	Concentrations in mixture			
14	[Cr] (M) =	1.00E-04		F6 = C6/($B$16*$B$14)	(found by Solver)			
15	Pathlength			G6 = E6*$B$16*$G$15	$[MnO_4^-]$ =		8.356E-05	M
16	(cm) =	1.000		+ F6*$B$16*$G$16	$[Cr_2O_7^{2-}]$ =		1.780E-04	M

Applications of Spectrophotometry

19-3. (a) Quartz and cyclic olefin copolymer are transparent in the ultraviolet. Either could be used.

(b) The spectra are well resolved. These wavelengths are at or near the maximum absorbance of each of the components.

(c) Modern spectrometers are most precise (reproducible) at intermediate levels of absorbance (0.3 to 2). Also, instrumental causes of deviation from Beer's law such as polychromatic light and stray light are most significant at higher absorbances. A range of 0.3 to 1.2 was used for greatest reproducibility and to ensure that calibration curves are linear.

(d) Absorbance spectra (and molar absorptivity at specific wavelengths) of weakly acidic and basic compounds can vary with pH. Acidifying the standards and mixture controlled the pH, eliminating possible chemical deviations from Beer's law.

(e)

	A	B	C	D	E	F	G	H
1	Solving simultaneous linear equations with Excel matrix operations							
2								
3	Wavelength	Coefficient matrix (εb)			Absorbance		Concentrations	
4	(nm)	ASA	ACE	CAF	of unknown		in mixture	
5	228	8170	6820	5380	1.220		5.62E-05	<-[ASA]
6	243	2940	9440	2610	0.951		6.80E-05	<-[ACE]
7	273	1070	2120	9710	0.741		5.53E-05	<-[CAF]
8			K		A		C	
9								
10	1. Highlight block of blank cells for solution (G5:G7)							
11	2. Type formula "=MMULT(MINVERSE(B5:D7),E5:E7)"							
12	3. Press CONTROL+SHIFT+ENTER on PC or CONTROL+SHIFT+RETURN on Mac.							
13	4. The answer appears in cells G5:G7.							

19-4. If the spectra of two compounds with a constant total concentration cross at any wavelength, all mixtures with the same total concentration will go through that same point, called an isosbestic point. The appearance of isosbestic points in a chemical reaction is good evidence that we are observing one main species being converted to one other major species.

19-5. As VO^{2+} is added (traces 1–9), the peak at 439 decreases and a new one near 485 nm develops, with an isosbestic point at 457 nm. When VO^{2+}/xylenol orange > 1, the peak near 485 nm decreases and a new one at 566 nm grows in, with an isosbestic point at 528 nm. This sequence is logically interpreted by the sequence

$$\text{M} + \underset{439 \text{ nm}}{\text{L}} \rightarrow \underset{485 \text{ nm}}{\text{ML}}$$

$$\underset{485 \text{ nm}}{\text{ML}} + \text{M} \rightarrow \underset{566 \text{ nm}}{\text{M}_2\text{L}}$$

where M is vanadyl ion and L is xylenol orange. The structure of xylenol orange in Table 12-3 shows that it has metal-binding groups on both ends of the molecule, and could form an M$_2$L complex.

19-6. Convert T to A ($= -\log T$) and then A to ε ($= A/bc = A/[(0.005 \text{ cm})(0.01 \text{ M})]$)

	Absorbance		ε (M^{-1} cm^{-1})		εb (M^{-1})	
	2 022 cm^{-1}	1 993 cm^{-1}	2 022 cm^{-1}	1 993 cm^{-1}	2 022 cm^{-1}	1 993 cm^{-1}
A	0.508 6	0.098 54	10 170	1 971	50.85	9.855
B	0.011 44	0.699 0	228.8	13 980	1.144	69.90

For the mixture, $A_{2022} = -\log(0.340) = 0.468\ 5$ and $A_{1993} = -\log(0.383) = 0.416\ 8$. Equation 19-5 gives

$$[A] = \frac{\begin{vmatrix} 0.468\ 5 & 1.144 \\ 0.416\ 8 & 69.90 \end{vmatrix}}{\begin{vmatrix} 50.85 & 1.144 \\ 9.855 & 69.90 \end{vmatrix}} = \frac{(0.468\ 5)(69.90) - (1.144)(0.416\ 8)}{(50.85)(69.90) - (1.144)(9.855)} = 9.11 \times 10^{-3} \text{ M}$$

$$[B] = \frac{\begin{vmatrix} 50.85 & 0.468\ 5 \\ 9.855 & 0.416\ 8 \end{vmatrix}}{\begin{vmatrix} 50.85 & 1.144 \\ 9.855 & 69.90 \end{vmatrix}} = \frac{(50.85)(0.416\ 8) - (0.468\ 5)(9.855)}{(50.85)(69.90) - (1.144)(9.855)} = 4.68 \times 10^{-3} \text{ M}$$

19-7.

	A	B	C	D	E	F	G	H
1	Solving simultaneous linear equations with Excel matrix operations							
2								
3	Wavelength	Coefficient matrix (εb)			Absorbance		Concentrations	
4	(nm)	TB	STB	MTB	of unknown		in mixture	
5	455	4800	11100	18900	0.412		1.2194E-05	<-[TB]
6	485	7350	11200	11800	0.350		9.2953E-06	<-[STB]
7	545	36400	13900	4450	0.632		1.3243E-05	<-[MTB]
8			K		A		C	
9								
10	1. Highlight block of blank cells required for solution (G5:G7)							
11	2. Type the formula "=MMULT(MINVERSE(B5:D7),E5:E7)"							
12	3. Press CONTROL+SHIFT+ENTER on a PC or CONTROL+SHIFT+RETURN on a Mac							
13	4. The answer apears in cells G5:G7							

19-8.

	A	B	C	D	E	F	G	H	I
1	Solving 4 simultaneous linear equations with Excel matrix operations								
2	Wave-								
3	length	Coefficient matrix			Ethyl-	Absorbance		Conc. In	
4	(µm)	p-Xylene	m-Xylene	o-Xylene	benzene	of unknown		mixture	
5	12.5	1.5020	0.0514	0.0000	0.0408	0.1013		0.0627	p-xylene
6	13.0	0.0261	1.1516	0.0000	0.0820	0.09943		0.0795	m-xylene
7	13.4	0.0342	0.0355	2.5320	0.2933	0.2194		0.0759	o-xylene
8	14.3	0.0340	0.0684	0.0000	0.3470	0.03396		0.0761	Ethylbz
9			K			A		C	
10									
11	1. Highlight block of blank cells required for solution (H5:H8)								
12	2. Type the formula "=MMULT(MINVERSE(B5:E8),F5:F8)"								
13	3. Press CONTROL+SHIFT+ENTER on a PC or CONTROL+SHIFT+RETURN on a Mac								
14	4. The answer appears in cells H5:H8								

19-9. (a) Cells H16:H17 in the spreadsheet give [In^{2-}] = 3.28 μM and [HIn$^-$] = 6.91 μM. These concentrations add up to 10.1$_9$ instead of 10.0 μM represents 2% experimental error in the quality of the procedure and measurements.

	A	B	C	D	E	F	G	H
1	Analysis of a Mixture When you Have More Data Points than Components of the Mixture							
2				Measured				
3		Absorbance		Absorbance	Molar Absorptivity		Calculated	
4	Wavelength	of Standards		of Mixture	(M^{-1} cm^{-1})		Absorbance	
5	(nm)	Blue In^{2-}	Yellow HIn$^-$	A$_m$	In^{2-}	HIn$^-$	A$_{calc}$	[A$_{calc}$ - A$_m$]^2
6	400	0.182	0.274	0.248	18200	27400	0.2490	9.819E-07
7	432	0.076	0.344	0.265	7600	34400	0.2626	5.533E-06
8	450	0.046	0.323	0.237	4600	32300	0.2383	1.707E-06
9	550	0.356	0.027	0.131	35600	2700	0.1353	1.830E-05
10	570	0.57	0.011	0.193	57000	1100	0.1943	1.742E-06
11	616	0.811	0.005	0.272	81100	500	0.2691	8.302E-06
12							SUM(H6:H11) =	3.656E-05
13	Standards		E6 = B6/($B$17*$B$14)					
14	[In^{2-}] (M) =	1.00E-05	F6 = C6/($B$17*$B$15)				Concentrations in mixture	
15	[HIn$^-$] (M) =	1.00E-05	G6 = E6*$B$17*$H$16+F6*$B$17*$H$17				(to be found by Solver)	
16	Pathlength		H6 = (G6-D6)^2				[In^{2-}] (M) =	3.276E-06
17	(cm) =	1.0					[HIn$^-$] (M) =	6.911E-06

(b) $$\text{pH} = \text{p}K_{\text{HIn}^-} + \log\frac{[\text{In}^{2-}]}{[\text{HIn}^-]} = 7.10 + \log\frac{3.28\ \mu\text{M}}{6.91\ \mu\text{M}} = 6.78$$

(c) Absorbance and absorptivity appear in rows 7 and 11 of the spreadsheet, with b = 1.000 cm. In Equations 19-5, let X = In^{2-}, Y = HIn$^-$, λ' = 432 nm, and λ'' = 616 nm.

$$[\text{In}^{2-}] = \frac{\begin{vmatrix} 0.265 & 34\,400 \\ 0.272 & 500 \end{vmatrix}}{\begin{vmatrix} 7\,600 & 34\,400 \\ 81\,100 & 500 \end{vmatrix}} = 3.31 \times 10^{-6}\ \text{M}$$

$$[\text{HIn}^-] = \frac{\begin{vmatrix} 7\,600 & 0.265 \\ 81\,100 & 0.272 \end{vmatrix}}{\begin{vmatrix} 7\,600 & 34\,400 \\ 81\,100 & 500 \end{vmatrix}} = 6.97 \times 10^{-6}\ \text{M}$$

Applications of Spectrophotometry

The spreadsheet solution looks like this, with answers in column F:

	A	B	C	D	E	F	G
1	Solving simultaneous linear equations with Excel matrix operations						
2							
3	Wavelength	Coefficient matrix		Absorbance		Concentrations	
4	(nm)	(εb)		of unknown		in mixture	
5	432	7600	34400	0.265		3.31E-06	<-[X]
6	616	81100	500	0.272		6.97E-06	<-[Y]
7		K		A		C	
8	1. Enter matrix of coefficients εb in cells B5:C6						
9	2. Enter absorbance of unknown at each wavelength (cells D5:D6)						
10	3. Highlight block of blank cells required for solution (F5 and F6)						
11	4. Type the formula "=MMULT(MINVERSE(B5:C6),D5:D6)"						
12	5. Press CTRL+SHIFT+ENTER on a PC or CTRL+SHIFT+RETURN on a Mac						
13	6. Behold! The answer apears in cells F5 and F6						

The calculation in (a) uses six wavelengths to find [In^{2-}] and [HIn$^-$], whereas (c) uses just two wavelengths. We expect that (a) should be more accurate. Notice that the sum [In^{2-}] + [HIn$^-$] in (a) is 10.1_9 μM and the sum in (c) is 10.2_8 μM. The answer from (a) is higher than the theoretical by 2% and the answer from (c) is higher than the theoretical by 3%.

19-10. The quantity of HIn is small compared to aniline and sulfanilic acid. Calling aniline B and sulfanilic acid HA, we can write

$$B + HA \underset{}{\overset{K}{\rightleftharpoons}} BH^+ + A^-$$

Initial mmol: 2.00   1.500   —   —
Final mmol: $2.00 - x$ $1.500 - x$ x x

$$K = \frac{K_a K_b}{K_w} = \frac{(10^{-3.232})(K_w/10^{-4.601})}{K_w} = 23.39$$

$$\frac{x^2}{(2.00-x)(1.500-x)} = 23.39 \Rightarrow x = 1.372 \text{ mmol}$$

$$\text{pH} = \text{p}K_{BH^+} + \log \frac{[B]}{[BH^+]} = 4.601 + \log \frac{2.00 - 1.372}{1.372} = 4.26$$

For HIn we can write:

absorbance = 0.110 =

$(2.26 \times 10^4 \text{ M}^{-1} \text{ cm}^{-1})(5.00 \text{ cm})$ [HIn] + $(1.53 \times 10^4 \text{ M}^{-1} \text{ cm}^{-1})$ (5.00 cm)[In$^-$]. Substituting [HIn] = 1.23×10^{-6} − [In$^-$] gives [In$^-$] = 7.94×10^{-7} and [HIn] = 4.36×10^{-7}. The Henderson-Hasselbalch equation for HIn is therefore

$$\text{pH} = \text{p}K_{HIn} + \log \frac{[In^-]}{[HIn]} \Rightarrow 4.26 = \text{p}K_{HIn} + \log \frac{7.94 \times 10^{-7}}{4.36 \times 10^{-7}} \Rightarrow \text{p}K_{HIn} = 4.00.$$

Chapter 19

19-11. (a) $A_{620} = \varepsilon_{620}^{\text{HIn}^-} b[\text{HIn}^-] + \varepsilon_{620}^{\text{In}^{2-}} b[\text{In}^{2-}]$

$A_{434} = \varepsilon_{434}^{\text{HIn}^-} b[\text{HIn}^-] + \varepsilon_{434}^{\text{In}^{2-}} b[\text{In}^{2-}]$

The solution of these two equations is given by Equation 19-5 in the text:

$[\text{HIn}^-] = \dfrac{1}{D}\left(A_{620}\varepsilon_{434}^{\text{In}^{2-}} b - A_{434}\varepsilon_{620}^{\text{In}^{2-}} b\right)$

$[\text{In}^{2-}] = \dfrac{1}{D}\left(A_{434}\varepsilon_{620}^{\text{HIn}^-} b - A_{620}\varepsilon_{434}^{\text{HIn}^-} b\right)$

where $D = b^2\left(\varepsilon_{620}^{\text{HIn}^-}\varepsilon_{434}^{\text{In}^{2-}} - \varepsilon_{620}^{\text{In}^{2-}}\varepsilon_{434}^{\text{HIn}^-}\right)$

Dividing the expression for $[\text{In}^{2-}]$ by the expression for $[\text{HIn}^-]$ gives

$\dfrac{[\text{In}^{2-}]}{[\text{HIn}^-]} = \dfrac{A_{434}\varepsilon_{620}^{\text{HIn}^-} - A_{620}\varepsilon_{434}^{\text{HIn}^-}}{A_{620}\varepsilon_{434}^{\text{In}^{2-}} - A_{434}\varepsilon_{620}^{\text{In}^{2-}}}$

Dividing numerator and denominator on the right side by A_{434} gives

$\dfrac{[\text{In}^{2-}]}{[\text{HIn}^-]} = \dfrac{\varepsilon_{620}^{\text{HIn}^-} - R_A\varepsilon_{434}^{\text{HIn}^-}}{R_A\varepsilon_{434}^{\text{In}^{2-}} - \varepsilon_{620}^{\text{In}^{2-}}} = \dfrac{R_A\varepsilon_{434}^{\text{HIn}^-} - \varepsilon_{620}^{\text{HIn}^-}}{\varepsilon_{620}^{\text{In}^{2-}} - R_A\varepsilon_{434}^{\text{In}^{2-}}}$

(b) Mass balance for indicator: $[\text{HIn}^-] + [\text{In}^{2-}] = F_{\text{In}}$

Dividing both sides by $[\text{HIn}^-]$ gives

$\dfrac{[\text{HIn}^-]}{[\text{HIn}^-]} + \dfrac{[\text{In}^{2-}]}{[\text{HIn}^-]} = \dfrac{F_{\text{In}}}{[\text{HIn}^-]} \Rightarrow 1 + R_{\text{In}} = \dfrac{F_{\text{In}}}{[\text{HIn}^-]} \Rightarrow [\text{HIn}^-] = \dfrac{F_{\text{In}}}{R_{\text{In}}+1}$

Acid dissociation constant of indicator:

$K_{\text{In}} = \dfrac{[\text{In}^{2-}][\text{H}^+]}{[\text{HIn}^-]}$

Substituting $F_{\text{In}}/(R_{\text{In}}+1)$ for $[\text{HIn}^-]$ gives

$K_{\text{In}} = \dfrac{[\text{In}^{2-}][\text{H}^+](R_{\text{In}}+1)}{F_{\text{In}}} \Rightarrow [\text{In}^{2-}] = \dfrac{K_{\text{In}} F_{\text{In}}}{[\text{H}^+](R_{\text{In}}+1)}$

(c) Equation A in the problem defines R_{In} as $[\text{In}^{2-}]/[\text{HIn}^-]$. So,

$K_{\text{In}} = \dfrac{[\text{In}^{2-}][\text{H}^+]}{[\text{HIn}^-]} = R_{\text{In}}[\text{H}^+] \Rightarrow [\text{H}^+] = K_{\text{In}}/R_{\text{In}}$

(d) From the acid dissociation reaction of carbonic acid, we can write

$K_1 = \dfrac{[\text{HCO}_3^-][\text{H}^+]}{[\text{CO}_2(aq)]} \Rightarrow [\text{HCO}_3^-] = \dfrac{K_1[\text{CO}_2(aq)]}{[\text{H}^+]}$

From the acid dissociation reaction of bicarbonate, we can write

$$K_2 = \frac{[CO_3^{2-}][H^+]}{[HCO_3^-]} \Rightarrow [CO_3^{2-}] = \frac{K_2[HCO_3^-]}{[H^+]}$$

Substituting in expression for $[HCO_3^-]$ gives $[CO_3^{2-}] = \dfrac{K_1 K_2 [CO_2(aq)]}{[H^+]^2}$

(e) Charge balance:

$[Na^+] + [H^+] = [OH^-] + [HIn^-] + 2[In^{2-}] + [HCO_3^-] + 2[CO_3^{2-}]$

$$F_{Na} + [H^+] = \frac{K_w}{[H^+]} + \frac{F_{In}}{R_{In}+1} + 2\frac{K_{In}F_{In}}{[H^+](R_{In}+1)} + \frac{K_1[CO_2(aq)]}{[H^+]} + 2\frac{K_1 K_2 [CO_2(aq)]}{[H^+]^2}$$

(f) From part (c) we know that $[H^+] = K_{In}/R_{In}$. We calculate R_{In} from part (a):

$$R_{In} = \frac{R_A \varepsilon_{434}^{HIn^-} - \varepsilon_{620}^{HIn^-}}{\varepsilon_{620}^{In^{2-}} - R_A \varepsilon_{434}^{In^{2-}}} = \frac{(2.84)(8.00 \times 10^3) - (0)}{(1.70 \times 10^4) - (2.84)(1.90 \times 10^3)} = 1.95_8$$

$[H^+] = K_{In}/R_{In} = (2.0 \times 10^{-7})/1.95_8 = 1.0_2 \times 10^{-7}$ M

Substituting this value of $[H^+]$ into the mass balance in part (e) produces an equation in which the only unknown is $[CO_2(aq)]$:

$$F_{Na} + [H^+] = \frac{K_w}{[H^+]} + \frac{F_{In}}{R_{In}+1} + 2\frac{K_{In}F_{In}}{[H^+](R_{In}+1)} + \frac{K_1[CO_2(aq)]}{[H^+]} + 2\frac{K_1 K_2 [CO_2(aq)]}{[H^+]^2}$$

$$92.0 \times 10^{-6} + 1.0_2 \times 10^{-7} =$$

$$\frac{(6.7 \times 10^{-15})}{(1.0_2 \times 10^{-7})} + \frac{(50.0 \times 10^{-6})}{1.95_8 + 1} + 2\frac{(2.0 \times 10^{-7})(50.0 \times 10^{-6})}{(1.0_2 \times 10^{-7})(1.95_8 + 1)}$$

$$+ \frac{(3.0 \times 10^{-7})[CO_2(aq)]}{(1.0_2 \times 10^{-7})} + 2\frac{(3.0 \times 10^{-7})(3.3 \times 10^{-11})[CO_2(aq)]}{(1.0_2 \times 10^{-7})^2}$$

$9.21 \times 10^{-5} = 6.56 \times 10^{-8} + 1.69 \times 10^{-5} + 6.62 \times 10^{-5} +$
$\phantom{9.21 \times 10^{-5} =} + 2.94\,[CO_2(aq)] + 0.001\,9\,[CO_2(aq)]$

$\Rightarrow [CO_2(aq)] = 3.0_4 \times 10^{-6}$ M

(g) The ions in solution are Na^+, HIn^-, In^{2-}, HCO_3^-, CO_3^{2-}, H^+, and OH^-. We know that $[Na^+] = 92.0$ μM and $[H^+] = 0.10$ μM. If the total cation charge is 92.1 μM, the total anion charge must be 92.1 μM, and the ionic strength must be ~92 μM ≈ 10^{-4} M. (The ionic strength is not exactly 92.1 μM because some anions have a charge of –2, which will increase the ionic strength from 92.1 μM.) An ionic strength of 10^{-4} M is low enough that the activity coefficients are close to 1.00.

We can calculate the exact ionic strength from the following expressions derived above:

$$[OH^-] = \frac{K_w}{[H^+]} = 0.07 \text{ μM}$$

$$[HIn^-] = \frac{F_{In}}{R_{In}+1} = 16.9 \text{ μM}; \quad [In^{2-}] = \frac{K_{In}F_{In}}{[H^+](R_{In}+1)} = 33.1 \text{ μM}$$

$$[HCO_3^-] = \frac{K_1[CO_2(aq)]}{[H^+]} = 2.94\,[CO_2(aq)] = 8.9 \text{ μM}$$

$$[CO_3^{2-}] = \frac{K_1K_2[CO_2(aq)]}{[H^+]^2} = 0.0019\,[CO_2(aq)] = 0.003 \text{ μM}$$

$$\text{Ionic strength} = \frac{1}{2}\sum_i c_i z_i^2 =$$

$$\frac{1}{2}\{[Na^+]\cdot 1^2 + [H^+]\cdot 1^2 + [OH^-]\cdot 1^2 + [HIn^-]\cdot 1^2 + [In^{2-}]\cdot 2^2 + [HCO_3^-]\cdot 1^2 + [CO_3^{2-}]\cdot 2^2\}$$

$$= 125 \text{ μM}$$

19-12. (a) $\text{Error (xylene)} = 100 \times \dfrac{2.36_4 - 2.43}{2.43} = -2.7\%$

Error (dichloroethane) = 4.5%   Error (toluene) = 2.6%

(b)

	A	B	C	D	E	F	G	H	I	J
1	Analysis of a mixture when you have more data points than components of the mixture									
2										
3					Measured	Molar absorptivity			Calculated	
4	Wavenumber	Absorbance of standards			absorbance	(M^{-1} cm^{-1})			absorbance	
5	(cm^{-1})	Xylene	$C_2H_4Cl_2$	Toluene	of mixture	Xylene	$C_2H_4Cl_2$	Toluene	A_{calc}	$[A_c - A_m]^2$
6	3064.1	0.198	0.046	0.465	0.657	1.172	0.088	2.372	0.654	0.00001
7	3047.4	0.634	0.056	0.374	1.194	3.751	0.107	1.908	1.211	0.00030
8	3028.1	0.541	0.088	1.043	1.593	3.201	0.168	5.321	1.592	0.00000
9	2955.4	0.589	0.795	0.336	1.674	3.485	1.517	1.714	1.672	0.00000
10	2922.6	1.167	0.105	0.537	2.114	6.905	0.200	2.740	2.120	0.00004
11	2971.5	0.570	0.595	0.273	1.443	3.373	1.135	1.393	1.450	0.00004
12	2868.8	0.569	0.107	0.244	1.097	3.367	0.204	1.245	1.062	0.00119
13								sum = SUM(J6:J12) =		0.00159
14	Standards:				F6 = B6/($B$19*$B$15)					
15	[xylene] =	1.69	M		G6 = C6/($B$19*$B$16)			Concentrations in the mixture		
16	[$C_2H_4Cl_2$] =	5.24	M		H6 = D6/($B$19*$B$17)			(found by Solver)		
17	[toluene] =	1.96	M		I6 = ($B$19*F6*$I$17) + ($B$19*			[xylene]=	2.389	M
18					G6*$I$18) + ($B$19*H6*$I$19)			[$C_2H_4Cl_2$]=	3.915	M
19	Pathlength =	0.100	cm		J6 = (I6-E6)^2			[toluene]=	1.431	M

$$\text{Error (xylene)} = 100 \times \frac{2.38_9 - 2.43}{2.43} = -1.7\%$$

Error (dichloroethane) = 3.0%   Error (toluene) = 1.5%

(c) SUM(J6:J12) = 0.00157   [xylene] = 2.36_4 M

[dichloromethane] = 3.97_9 M   [toluene] = 1.44_6 M

$$\text{Error (xylene)} = 100 \times \frac{2.36_4 - 2.43}{2.43} = -2.7\%$$

Error (dichloroethane) = 4.7%   Error (toluene) = 2.6%

(d) Error for all analytes decreased when 2971.5 and 2868.8 cm^{-1} were added in (b). Adding 3124.5 and 2780.6 cm^{-1} in (c) resulted in the same error for xylene and toluene, and increased error for dichloroethane. All analytes absorb strongly at 2971.5 and 2868.8 cm^{-1}. Including these increased signal information in the calibration, and improved accuracy. No analytes absorb at 3124.5 and 2780.6 cm^{-1}. Including such wavenumbers increases the noise (uncertainty), and reduces the accuracy of the multivariate calibration.

19-13. After running Solver, the average value of K in cell E10 is 0.464 and ε in cell A12 is 1.073×10^4 M^{-1}cm^{-1}.

	A	B	C	D	E
1	Equilibrium constant for reaction of I$_2$ with mesitylene				
2					
3	[Mesitylene] (M)	[I$_2$]$_{tot}$ (M)	A	[Complex] =A/εb	K$_{eq}$
4	1.6900	7.817E-05	0.369	3.437E-05	0.46444
5	0.9218	2.558E-04	0.822	7.657E-05	0.46350
6	0.6338	3.224E-04	0.787	7.331E-05	0.46439
7	0.4829	3.573E-04	0.703	6.549E-05	0.46474
8	0.3900	3.788E-04	0.624	5.813E-05	0.46481
9	0.3271	3.934E-04	0.556	5.180E-05	0.46354
10				Average =	0.46424
11	Guess for ε:			Standard dev =	0.00058
12	1.073E+04			Stdev/Average =	0.00125
13					
14	D4 = C4/$A$12				
15	E4 = D4/(A4*(B4-D4)) = [complex]/([Mesitylene][Free I$_2$])				
16	E10 = AVERAGE(E4:E9)				
17	E11 = STDEV.S(E4:E9)				
18	E12 = E11/E10				
19	Use Solver to vary ε (cell A12) until cell E12 is minimized				

19-14. (a) A standard method has validated reliability achieved through collaborative testing, and has received approval by an official agency such as ASTM International or United States Environmental Protection Agency.

(b) Analysis of a blank sample spiked with a known amount of analyte checks accuracy of method in presence of sample matrix.

(c) Particulate matter would scatter light, causing an apparent increase in absorbance that is unrelated to chloride concentration.

19-15. (a) III, 96-well microplate. A microplate can be used to analyze many standards and samples at once, and requires only small quantities of reagents.

(b) IV, flow injection analysis. Flow injection can perform many analyses of one analyte quickly.

(c) II, discrete analyzer. Discrete analyzers can analyze multiple analytes in a single sample by dispensing many aliquots of sample, and reacting each with reagents for a different analyte. While not as fast as flow injection analysis, we have only 20 samples. A microplate might also have been used, but the microplate was most appropriate for part (a).

(d) I, colorimeter. The colorimeter is simple and light weight, allowing analyses to be performed in the field. Only samples from streams with high concentrations of PO_4^{3-} need to be collected and carried. With accuracy of only ±10%, I would collect a few extra samples to ensure that I have the 10 most concentrated.

19-16. (a) $b_{well} = \left(\dfrac{(A_{1000} - A_{900})_{well}}{(A_{1000} - A_{900})_{1\ cm\ cuvet}} \right) \times 1.000\ cm$

$= \left(\dfrac{0.226 - 0.052}{0.243 - 0.058} \right) \times 1.000\ cm = \dfrac{0.174}{0.185} \times 1.000\ cm = 0.940\ cm$

(b) $A_{1\ cm} = \left(\dfrac{A_{well} - A_{blank}}{b_{well}} \right) \times 1.000\ cm = \left(\dfrac{0.634 - 0.016}{0.940\ cm} \right) \times 1.000\ cm = 0.657$

(c) $\varepsilon_{280\ nm}\ (M^{-1}\ cm^{-1}) \approx 5\,500\ n_{Trp} + 1\,490\ n_{Tyr} + 125\ n_{S-S}$

$= 5\,500 \times 3 + 1\,490 \times 10 + 125 \times 1$

$= 31\,525\ M^{-1}\ cm^{-1} = 31\,500\ M^{-1}\ cm^{-1}$

(d) $c = \dfrac{A}{\varepsilon b} = \dfrac{0.657}{31500 \text{ M}^{-1}\text{cm}^{-1}(1.000 \text{ cm})} = 2.09 \times 10^{-5}$ M

$c = 45\,000$ g mol$^{-1} \times (2.09 \times 10^{-5}$ M$) = 0.940$ g/L $= 0.94$ mg/mL

19-17. (a) Late stage: $y = (0.060\,8 \pm 0.000\,4)x + (0.014 \pm 0.005)$; $R^2 = 0.999\,8$

Residuals are small, intercept is near zero, and $R^2 > 0.999$.
The response is linear.

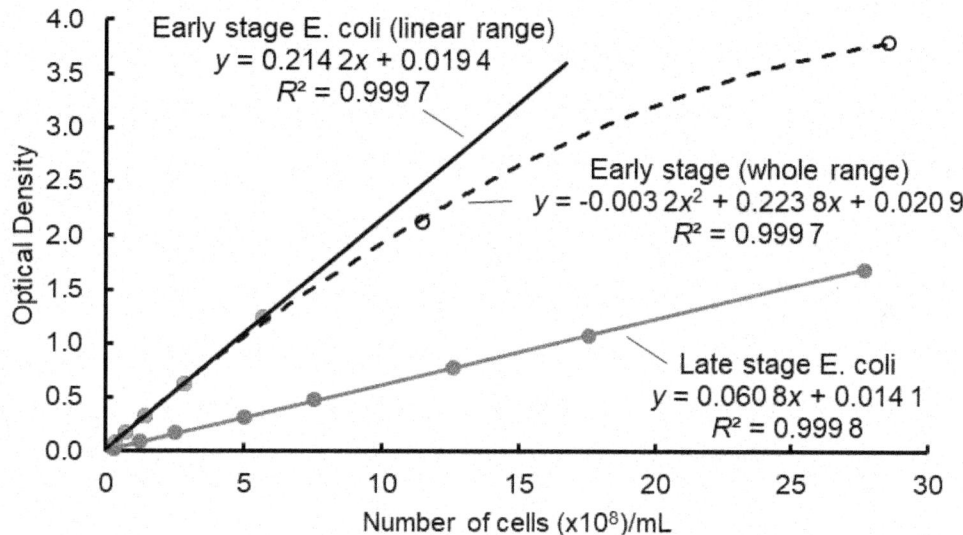

(b) Plot is severely curved. Linear range is $(0.36$ to $5.71) \times 10^8$ cells/mL for which R^2 is $0.999\,7$. A line through $(0.36$ to $11.41) \times 10^8$ cells/mL results in an R^2 of $0.993\,5$, which does not seem too bad, but there is an increased positive intercept (0.07), and systematic deviations between line and data.

(c) $2.50 = -0.003\,2x^2 + 0.223\,8x + 0.020\,9 \Rightarrow 0 = -0.003\,2x^2 + 0.223\,8x - 2.47_9$

$x = \dfrac{-b \pm \sqrt{b^2 - 4ac}}{2a} = \dfrac{-0.2238 \pm \sqrt{(0.2238)^2 - 4(-0.003\,2)(-2.47_9)}}{2(-0.0032)}$

$= 13.8$ or 56.2

Based on graph 13.8×10^8 cells/mL is the correct answer.

(d) Correction factor $= \dfrac{b_{1.000 \text{ cm}}}{b_{\text{well}}} \Rightarrow b_{\text{well}} = \dfrac{1.000 \text{ cm}}{6.37} = 0.157$ cm

Correction factor $= \dfrac{b_{1.000 \text{ cm}}}{b_{\text{well}}} = \dfrac{A_{1.000 \text{ cm}}}{A_{\text{well}}} \Rightarrow A_{\text{well}} = \dfrac{2.50}{6.37} = 0.392$

19-18. In a *discrete analyzer*, a robot dispenses a sample aliquot into an open reaction vessel. Reagents needed for the analysis are dispensed by the robot into the reaction vessel. Once reaction is complete, solution is transferred by the robot to a cuvette where absorbance or fluorescence is measured. Multiple aliquots of sample may be dispensed, and each reacted with reagents for a different analyte. Reaction must go to completion, so analysis time depends on the reaction rate.

In *flow injection analysis*, unknown is injected into a continuously flowing stream that may contain a reagent which reacts with analyte. One or more reagents can be added to the flow downstream of sample injection. Sample and reagents travel through a coil in which mixing and reaction occur. Absorbance or fluorescence is measured in the detector. In general, reactants and products do not reach equilibrium in flow injection analysis. The precision of the result depends on executing the same process in a very repeatable manner. Each flow injection system is dedicated to analysis of a specific analyte. If a different analyte is to be analyzed, a new flow injection system must be assembled.

The principal difference between the two is that the reaction must go to completion in the discrete analyzer. Reaction conditions and timing are highly reproducible in flow injection analysis, and so reproducible signals are achieved even when the reaction is not complete. Not waiting to reach equilibrium allows flow injection analysis to analyze more samples per hour. Discrete analyzers are more flexible, many aliquots of sample may be dispensed, and each reacted with reagents for a different analyte. A flow injection analysis system is dedicated to analysis of a single analyte. The closed nature of flow injection analysis makes it less prone to contamination, and so better suited for trace analysis.

19-19. See the answer to 19-17 for a description of *flow injection analysis*.

In *sequential injection*, unknown and one or more reagents are sequentially injected under computer control into a holding coil, and then flow is stopped. At an appropriate time, flow is reversed and the mixture flows through a detector. As in flow injection, equilibrium is generally not reached and precision depends on repeatability of the sequence of steps.

The principal difference between flow injection and sequential injection is that there is continuous flow in the former and a sequence of steps without continuous flow in the latter. Sequential injection uses less reagent and generates less waste than flow injection. Sequential injection is called "lab-on-a-valve" because sample, reagents, and product all flow through a central multi-port valve.

19-20. (a) Each method is used to determine a number of individual samples, and so a paired t test comparing individual differences is used. Measurements by certified method are in column B, and sequential injection method are in column C. Differences between the two measures for each sample are computed in column D.

	A	B	C	D
1	Validation of sequential injection for International Space Station			
2				
3			Ammonium (mg/L)	
4	Sample	Certified method	Sequential injection	Difference (d_i)
5	PW1	0.33	0.32	-0.01
6	PW2	0.70	0.75	0.05
7	HW1	1.05	1.08	0.03
8	HW2	5.72	5.38	-0.34
9	AC1	3.55	3.28	-0.27
10	AC2	0.34	0.32	-0.02
11	AC3	18.66	19.02	0.36
12			mean difference =	-0.0286
13			standard deviation of difference =	0.230
14	D5 = C5−B5		number of samples =	7
15	D12 = AVERAGE(D5:D11)		t_{calc} =	0.329
16	D13 = STDEV.S(D5:D11)		t_{table} =	2.447
17	D14 = COUNT(D5:D11)			
18	D15 = ABS(D12)*SQRT(D14)/D13	(ABS = absolute value)		
19	D16 = T.INV.2T(0.05,D14-1)			

The t statistic is $t_{calculated} = \dfrac{|\bar{d}|}{s_d}\sqrt{n} = \dfrac{|-0.028\,6|}{0.230}\sqrt{7} = 0.329$

t_{table} for 95% confidence and $7 - 1 = 6$ degrees of freedom is 2.447.

$t_{calculated} < t_{table}$, so there is no significant difference between the methods.

(b) In addition to comparing the new method to a standard method, accuracy can be demonstrated by: (i) analyzing a certified reference material (which the authors also did); (ii) analyzing a blank sample fortified with a known amount of analyte; and (iii) performing standard addition calibration.

19-21. Luminescence is light given off after a molecule absorbs light. Chemiluminescence is light given off by a molecule created in an excited state by a chemical reaction. Bioluminescence is light given off by a living system.

19-22. (a) Thiols react with maleimides such as III.

(b) Carbonyls can be derivatized with hydrazines such as II.

(c) Amines can be derivatized with dialdehydes such as I.

19-23. (a) Both titrations show that 1 mol of streptavidin binds 4 mol of biotin

Addition of BF to SA: Addition of SA to BF:

Equivalence point = 3.99 [BF]:[SA]   Equivalence point = 0.250 [SA]:[BF]

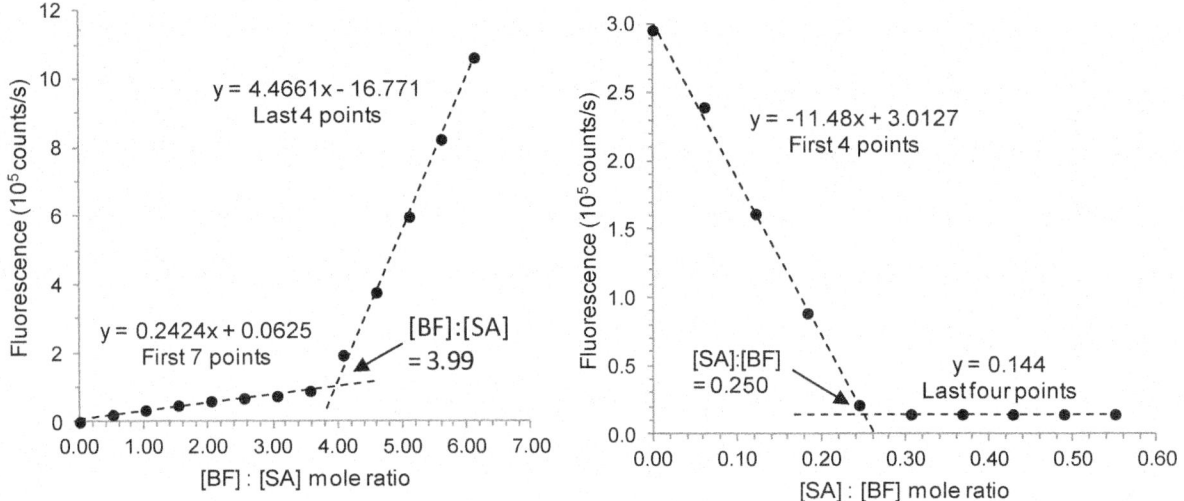

(b) When adding BF to SA, initial fluorescence is 0 with no BF present. As long as added BF binds to SA, fluorescence increases slowly because BF is not highly fluorescent when bound to SA. When [BF]:[SA] > 4, fluorescence increases rapidly because highly fluorescent BF is present.

When adding SA to BF, initial fluorescence of free BF is high. Fluorescence then decreases rapidly as SA is added until enough SA has been added to bind all BF ([SA]:[BF] = 0.25). Beyond this point, fluorescence is low and constant because all BF is bound to SA.

19.24 (a) Oxygen is a collisional quencher. When O_2 collides with an excited molecule M*, energy is transferred from M* to O_2 converting O_2 from the triple state ($^3O_2$) to an excited singlet state ($^1O_2^*$). This energy transfer converts M* to its ground state without emitting a photon.

(b and c) Stern-Volmer plots either the ratio (initial intensity) over (quenched intensity) versus quencher concentration; or (initial lifetime) over (quenched lifetime) versus quencher concentration.

	A	B	C	D	E	F	G
1		Stern-Volmer graph for O_2 quenching of ruthenium complex					
2							
3		Conc O_2	Fluorescence			Fluorescence	
4		(µM)	intensity	I_0/I_Q		lifetime (µs)	τ_0/τ_Q
5		0	590	1.000		1.008	1.000
6		22	574	1.028		0.978	1.031
7		44	557	1.059		0.964	1.046
8		72	552	1.069		0.950	1.061
9		132	518	1.139		0.916	1.100
10		174	503	1.173		0.893	1.129
11		267	478	1.234		0.836	1.206
12				D5=$C$5/C5			G5=$F$5/F5

The Stern-Volmer graph based on intensity (b) is shown with the solid symbol, and that based on lifetime (c) is the open symbols.

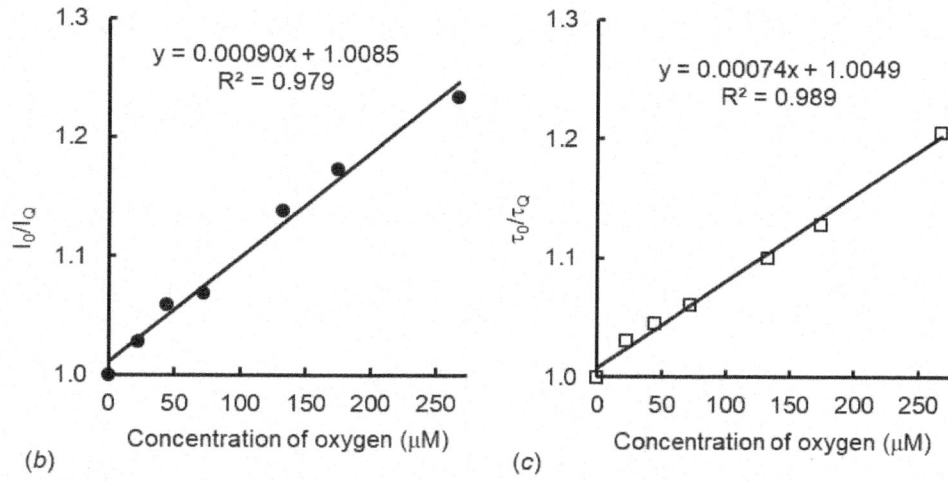

(b) (c)

(b) R^2 is < 0.995, but the intercept is near the theoretical Stern-Volmer value of 1 and data points are randomly scattered about the line. Calibration is linear.

(c) τ_0/τ_Q shows linear dependence on quencher concentration. This indicates collisional quenching is occurring. If static quenching were occurring, τ_0/τ_Q would have a constant value of 1, regardless of the oxygen concentration.

19-25. The graph of K_{SV} versus pH has a plateau at low pH near $K_{SV} \approx 100$ and a plateau at high pH near $K_{SV} \approx 1\,350$. The quencher, 2,6-dimethylphenol, is a weak acid whose pK_a is expected to be near 10. A logical interpretation is that the basic form, A$^-$, is a strong quencher with $K_{SV} \approx 1\,350$, and the acidic form, HA, is a weak quencher with $K_{SV} \approx 100$. We estimate pK_a as the midpoint in the curve at $K_{SV} \approx (1\,350 - 100)/2 = 625$. At this point, pH ≈ 10.8, which is our estimate for pK_a. The literature value is 10.63.

The smooth curve in the graph is a least-squares fit to the equation

$$K_{SV} = K_{SV}^{HA} \underbrace{\left(\frac{[H^+]}{[H^+]+K_a}\right)}_{\text{Fraction in form HA}} + K_{SV}^{A^-} \underbrace{\left(\frac{K_a}{[H^+]+K_a}\right)}_{\text{Fraction in form A}^-}$$

$$\underbrace{\phantom{K_{SV}^{HA} \left(\frac{[H^+]}{[H^+]+K_a}\right)}}_{\text{Quenching by HA}} \quad \underbrace{\phantom{K_{SV}^{A^-}\left(\frac{K_a}{[H^+]+K_a}\right)}}_{\text{Quenching by A}^-}$$

The least-squares fit gave $K_{SV}^{HA} = 69.0$ M^{-1}, $K_{SV}^{A^-} = 1\,370$ M^{-1}, and p$K_a = 10.78$.

19-26. (a) Alexa Fluor 488 would absorb at 488 nm and emit at less than 600 nm if no other dye is close enough for Förster energy transfer. The excitation spectra of Alexa Fluor 555 overlaps with the emission spectrum of Alexa Fluor 488. If the dyes are close, energy absorbed by the 488 dye is transferred to the 555 dye, and emitted as 555 dye fluorescence, which is predominantly above 600 nm.

(b) The 488 dye absorbs at 488 nm and its fluorescence maximum is at 520 nm. So it is the donor. For Förster quenching to occur, the acceptor must absorb at the wavelength emitted by the donor—520 nm in this case—and should not fluoresce. QSY dyes have no fluorescence, and so both could act as quenchers. The absorption spectrum of QSY 9 best overlaps with the emission spectrum of the 488 dye, and so would be the best quencher for this application.

(c) The Alexa Fluor 555 and 647 pair would have these characteristics. The 555 dye absorbs at 532 and would emit at 570 nm in absence of energy transfer. The excitation spectrum of the 647 dye overlaps with the emission of the 555 dye, and emits at 670 nm. If the 555 and 647 dyes were close, Förster resonance energy transfer would result in emission from the 647 dye.

19-27. (a) The graph at the left shows that the Stern-Volmer equation is not obeyed. If it were obeyed, the graph would be linear.

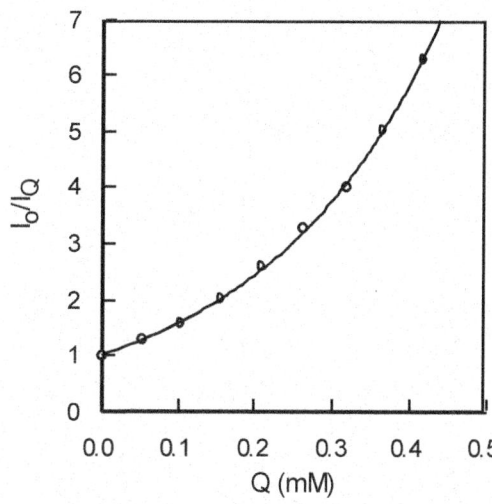
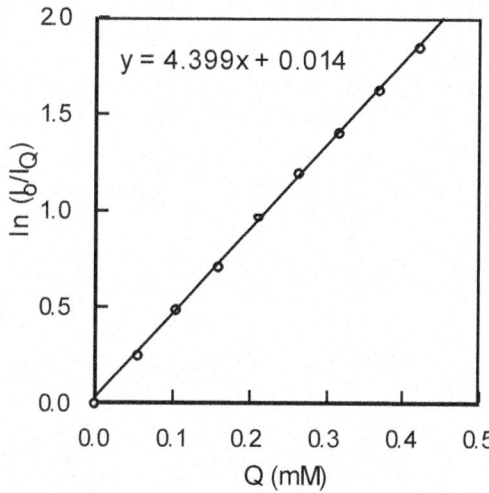

(b) The graph at the right shows that Equation 5 is obeyed. Ideally, the intercept should be zero. The slope of the graph is $N_{av}/([S] - [CMC])$. Given that $[Q]$ was expressed in mM, we will express $[S]$ and $[CMC]$ in mM:
$4.399 = N_{av} / ([20.8] - [8.1]) \rightarrow N_{av} = 55.9$

(c) $[M] = ([S] - [CMC]) / N_{av} = ([20.8] - [8.1]) / 55.9 = 0.227$ mM
$\bar{Q} = [Q]/[M] = 0.200$ mM $/ 0.227$ mM $= 0.881$ molecules per micelle

(d) $P_n = \dfrac{\bar{Q}^n}{n!} e^{-\bar{Q}}$. For $n = 0$, $P_0 = e^{-0.881} = 0.414$

$P_1 = \dfrac{(0.881)^1}{1!} e^{-0.881} = 0.365 \qquad P_2 = \dfrac{(0.881)^2}{2!} e^{-0.881} = 0.161$

19-28. Many biological chromophores absorb visible light and some of them are fluorescent. Stimulation with visible light creates significant background visible fluorescence from the biological matrix, in addition to the desired fluorescence from analyte. Few molecules absorb radiation in the near-infrared range 800 to 1 000 nm, so less background fluorescence occurs. With low background fluorescence, the signal/noise ratio for analyte fluorescence is high.

19-29. (a) Previous studies indicated that a similar dye's fluorescence was quenched by albumin. Studies of the quenching mechanism resulted in synthesis of Pittsburgh Green II, which was quenched more than the original compound.

(b) pH 7 and HEPES. Fluorescence was greatest at pH 7 (Figure 4d). Fluorescence quenching is greatest using HEPES (Figure 2b). The primary role of a buffer is to control pH, but it may affect the analysis in other ways. The type of buffer should always be optimized, in addition to the pH.

(c) Last paragraph of page 2 333, they analyzed a certified reference material. Two co-authors independently performed analysis on the certified material, and obtained results statistically equivalent to the certified value.

(d) Page 2 333, right column, paragraph 2 says 2–10 mg/L. Figure 3a shows the calibration is nonlinear both below 2 mg/L and above 10 mg/L.

(e) Figure caption and page 2 333 right column middle of third paragraph state the free dye was photobleached when exposed repeatedly to excitation light. Photobleaching is irreversible loss of a molecule's ability to fluoresce due to the excited state molecule reacting to form a nonfluorescent product.

(f) Fluorescence is emitted in all directions. If the side walls of the microplate were translucent, fluorescence from one well might cross over into another well. For this reason black or opaque walled microplates are used for luminescence measurements.

19-30. Each molecule of analyte bound to antibody 1 also binds one molecule of antibody 2 that is linked to one enzyme molecule. Each enzyme molecule catalyzes many cycles of reaction in which a colored or fluorescent product is created. Therefore, many product molecules are created for every analyte molecule.

19-31. In time-resolved emission measurements, the short-lived background fluorescence decays to near zero prior to recording emission from the lanthanide ion. By reducing background noise, the signal-to-noise ratio is increased. Also, the wavelength of the lanthanide emission is longer than the wavelength of much of the background emission.

19-32.

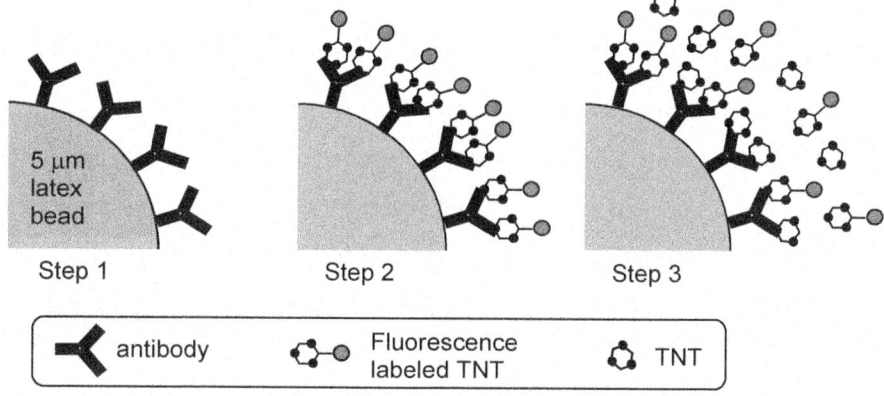

In Step 1, antibodies for TNT are attached to latex beads. In Step 2, the antibody is saturated with a fluorescent derivative of TNT. Excess fluorescent derivative is removed. In Step 3, the beads are incubated with TNT, which displaces some fluorescent derivative from binding sites on the antibodies. The suspension of beads is then injected into the flow cytometer. As each bead passes in front of the detector, it is excited by a laser and its fluorescence is measured. The graph in the textbook shows median bead fluorescence versus TNT concentration in a series of standards. The more TNT in the standard, the less fluorescence remains associated with the beads.

CHAPTER 20
SPECTROPHOTOMETERS

20-1. (a) The light source provides ultraviolet, visible, or infrared radiation. The monochromator selects a narrow band of wavelengths to pass on to the sample. As the experiment progresses, the monochromator scans through a desired range of wavelengths. The sample cuvette holds the sample of interest. The reference cuvette is an identical cell containing pure solvent or a reagent blank. The beam chopper is a rotating mirror that alternately directs light to the sample or reference. The mirror after the reference cuvette and the chopper after the sample cuvette pass both beams of light through to the detector, which could be a photomultiplier tube that generates an electric current proportional to the photon irradiance. The electronics convert the alternating current for sample and reference into an absorbance signal for display.

(b) The beamsplitter directs a portion of monochromatic light to the sample cuvette and another portion to the reference cuvette. Light passing through the sample is directed to a detector, which could be a photodiode that generates an electric current proportional to photon irradiance. Light passing through the reference strikes a second photodiode. The electronics convert the ratio of the sample and reference currents into absorbance for display.

20-2. Deuterium. Silicon carbide globar.

20-3. $M = \sigma T^4 = [5.669\,8 \times 10^{-8}\ \text{W}/(m^2 K^4)]T^4$

T (K)	M (W/m²)
77	1.99
298	447

20-4. (a) $M_\lambda = \dfrac{2\pi hc^2}{\lambda^5}\left(\dfrac{1}{e^{hc/\lambda kT}-1}\right)$

At $T = 1\,000$ K:

λ (μm)	M_λ (W/m³)
2.00	8.79×10^9
10.00	1.164×10^9

(b) $M_\lambda \Delta\lambda = (8.79 \times 10^9\ \text{W/m}^3)(0.02 \times 10^{-6}\ \text{m}) = 1.8 \times 10^2\ \text{W/m}^2$ at 2.00 μm

(c) $M_\lambda \Delta\lambda = (1.164 \times 10^9\ \text{W/m}^3)(0.02 \times 10^{-6}\ \text{m}) = 2.3 \times 10^1\ \text{W/m}^2$ at 10.00 μm

Spectrophotometers 285

(d) At $T = 100$ K:

λ (µm)	M_λ (W/m^3)
2.00	6.69×10^{-19}
10.00	2.111×10^{3}

$$\frac{M_{2.00 \; \mu m}}{M_{10.00 \; \mu m}} = \frac{8.79 \times 10^{9} \; W/m^3}{1.164 \times 10^{9} \; W/m^3} = 7.55 \text{ at } 1\,000 \text{ K}$$

$$\frac{M_{2.00 \; \mu m}}{M_{10.00 \; \mu m}} = \frac{6.69 \times 10^{-19} \; W/m^3}{2.111 \times 10^{3} \; W/m^3} = 3.17 \times 10^{-22} \text{ at } 100 \text{ K}$$

At 100 K, there is virtually no emission at 2.00 µm compared to 10.00 µm, whereas at 1 000 K, there is a great deal of emission at both wavelengths.

20-5. An excited state of the lasing material is pumped to a high population by light, an electric discharge, or other means. Photons emitted when the excited state decays to a less populated lower state stimulate emission from other excited molecules. The stimulated emission has the same energy and phase as the incident photon. In the laser cavity, most light is retained by reflective end mirrors. Some light is allowed to escape from one end. Properties of laser light: monochromatic, bright, collimated, polarized, coherent.

20-6. Resolving power increases in proportion to the number of grooves that are illuminated and to the diffraction order. The number of grooves can be increased with a more finely ruled grating (closer grooves) and with a longer grating. The diffraction order is optimized by appropriate choice of blaze angle of the grating.

Dispersion is proportional to the diffraction order and inversely proportional to the spacing between lines in the grating. The closer the lines, the greater the dispersion.

The optimum wavelength selected by a grating is the one for which the diffraction condition $n\lambda = d(\sin \theta + \sin \phi)$ is satisfied by specular reflection when the angle of incidence is equal to the angle of reflection ($\alpha = \beta$).

20-7. A filter removes higher order diffraction (different wavelengths) at the same angle as the desired diffraction.

20-8. Advantage - increased ability to resolve closely spaced spectral peaks.
Disadvantage - more noise because less light reaches detector.

20-9. (a) $n\lambda = d(\sin\theta + \sin\phi)$

$1 \cdot 600 \times 10^{-9}$ m $= d(\sin 40° + \sin(-30°)) \Rightarrow d = 4.20 \times 10^{-6}$ m

Lines/cm $= 1/(4.20 \times 10^{-4}$ cm$) = 2.38 \times 10^3$ lines/cm

(b) $\lambda = 1/(1\,000$ cm$^{-1}) = 10^{-3}$ cm $\Rightarrow d = 7.00 \times 10^{-3}$ cm $\Rightarrow$ 143 lines/cm

20-10. 10^3 grooves/cm means $d = 10^{-5}$ m $= 10$ μm

$$\text{Dispersion} = \frac{n}{d\cos\phi} = \frac{1}{(10\text{ μm})\cos 10°} = 0.102\,\frac{\text{radians}}{\text{μm}}$$

$$0.102\,\frac{\text{radians}}{\text{μm}} \times \frac{180°}{\pi\text{ radians}} = 5.8°/\text{μm}$$

20-11. (a) Resolving power $= \dfrac{\lambda}{\Delta\lambda} = \dfrac{512.245\text{ nm}}{0.03} = 1.7 \times 10^4$

(b) $\Delta\lambda = \dfrac{\lambda}{10^4} = \dfrac{512.23\text{ nm}}{10^4} = 0.05$ nm

(c) Resolving power $= nN = (4)(8.00$ cm $\times 1\,850$ cm$^{-1}) = 5.9 \times 10^4$

(d) 250 lines/mm $= 4$ μm/line $= d$

$$\frac{\Delta\phi}{\Delta\lambda} = \frac{n}{d\cos\phi} = \frac{1}{(4\text{ μm})\cos 3.0°} = 0.250\,\frac{\text{radians}}{\text{μm}} = 14.3°/\text{μm}$$

For $\Delta\lambda = 0.03$ nm, $\Delta\phi = (14.3°/\text{μm})(3 \times 10^{-5}$ μm$) = 4.3 \times 10^{-4}$ degrees.

For 30th order diffraction, the dispersion will be 30 times greater, or 0.013°.

20-12. Sensitivity (slope of calibration curve) depends on pathlength and molar absorptivity (Sensitivity $= \varepsilon b$). Molar absorptivity is greatest at λ_{max}, and so sensitivity is greatest at λ_{max}.

Polychromatic deviation from Beer's law depends on change in molar absorptivity ($\Delta\varepsilon$) over monochromator bandwidth. At λ_{max} the molar absorptivity is near constant, yielding a more linear calibration. On the side of the absorbance peak, the molar absorptivity changes dramatically with wavelength. Thus $\Delta\varepsilon$ would change dramatically over the monochromator bandwidth, and so the calibration would show negative deviation from Beer's law.

Spectrophotometers

20-13. (a) True transmittance = $10^{-1.500}$ = 0.0316. With 0.50% stray light, the apparent transmittance is

$$\text{Apparent transmittance} = \frac{P+S}{P_0+S} = \frac{0.0316+0.0050}{1+0.0050} = 0.0364$$

The apparent absorbance is $-\log 0.0364 = 1.439$.

(b) Apparent absorbance = 1.999

Apparent transmittance = $10^{-1.999}$ = 0.010 023

$$\text{Apparent transmittance} = \frac{P+S}{P_0+S} = \frac{0.010+S}{1+S} = 0.010\,023$$

$\Rightarrow S = 2.3 \times 10^{-5}$ or 0.002 3%

(c) For true absorbance = 2,

$$\text{Apparent transmittance} = \frac{P+S}{P_0+S} = \frac{0.010+0.000\,000\,5}{1+0.000\,000\,5} = 0.010\,000\,49$$

Apparent absorbance is $-\log T = -\log(0.010\,000\,49) = 1.999\,978$

Absorbance error = $2 - 1.999\,978 = 0.000\,022$

For true absorbance = 3,

$$\text{Apparent transmittance} = \frac{P+S}{P_0+S} = \frac{0.001+0.000\,000\,5}{1+0.000\,000\,5} = 0.001\,000\,50$$

Apparent absorbance is $-\log T = -\log(0.001\,000\,50) = 2.999\,78$

Absorbance error = $3 - 2.999\,78 = 0.000\,22$

20-14. (a) Residuals between least-squares calibration line on next page and readings are small and randomly distributed. Intercepts of both curves are near 0. R^2 values are lower than 0.995, but see part (b). Both calibrations are linear.

(b) The 30.4 mM copper standard is an outlier in both plots based on its large deviation from the lines. Removing the 30.4 mM outlier from the calibration yields R^2 of 0.999 6 for both plots.

(c) Student absorbance = 0.258 $\Rightarrow$ Apparent transmittance = $10^{-0.258}$ = 0.552

Commercial absorbance = 0.440 $\Rightarrow$ True transmittance = $10^{-0.440}$ = 0.363

Student apparent transmittance = $\dfrac{P+S}{P_0+S} \Rightarrow 0.552 = \dfrac{0.363+S}{1+S}$

$S = 0.42 \Rightarrow$ ~42% stray light

	A	B	C	D	E	F	G	H
1	Calibration curves using two spectrophotometers							
2								
3		Copper	Absorbance	Absorbance				
4		conc. (mM)	commercial	student-built				
5		10.2	0.079	0.053				
6		20.3	0.146	0.089				
7		30.4	0.184	0.111				
8		40.6	0.287	0.176				
9		50.8	0.368	0.219				
10		60.9	0.440	0.258				
11		71.1	0.516	0.299				
12		76.1	0.554	0.322				
13								
14	{B15:C17}=LINEST(C5:C12,B5:B12,TRUE,TRUE)							
15	m	0.00737	-0.010	b				
16	u_m	0.00022	0.011	u_b	Highlight cells B15:C17.			
17	R^2	0.9946	0.014	s_y	Type "= LINEST(C5:C12,B5:B12,TRUE,TRUE)"			
18					For PC, press CTRL+SHIFT+ENTER			
19	{B20:C22}=LINEST(D5,D12,B5:B12,TRUE,TRUE)				For Mac, press CTRL+SHIFT+RETURN			
20	m	0.00418	0.0025	b				
21	u_m	0.00014	0.0069	u_b	Highlight cells B20:C22.			
22	R^2	0.9937	0.0086	s_y	Type "= LINEST(D5:D12,B5:B12,TRUE,TRUE)"			

(d) Stray light lowered slope of calibration. Absorbances are low. Nonlinearity due to stray or polychromatic light is most pronounced at high absorbance, *i.e.*, when true transmittance is low.

20-15. $b = \dfrac{30}{2 \cdot 1}\left(\dfrac{1}{1\,906 - 698 \text{ cm}^{-1}}\right) = 0.124\,2 \text{ mm}$

(Air between the plates has refractive index of 1.)

20-16. (a) A photomultiplier tube has a photosensitive cathode that emits an electron when struck by a photon. Electrons from the cathode are accelerated by a positive electric potential toward the first dynode. When an accelerated electron strikes the dynode, more electrons are emitted from the dynode. This multiplication process continues through several successive dynodes until ~10^6 electrons are finally collected at the anode for each photon striking the photocathode. The signal is the current measured at the anode.

(b) Each photodiode in a linear array has *p*-type silicon on a substrate of *n*-type silicon. Reverse bias draws electrons and holes away from the junction, which is a depletion region with few electrons and holes. The junction acts as a capacitor, with charge on either side of the depletion region. At the beginning of a measurement cycle, each diode is fully charged. Free electrons and holes created when radiation is absorbed in the semiconductor migrate to regions of opposite charge, partially discharging the capacitor. Charge left in each capacitor is measured at the end of a collection cycle by measuring the current needed to recharge each capacitor.

(c) A charge coupled device is made from *p*-doped Si on an *n*-doped substrate. The structure is capped with an insulating layer of SiO_2, on top of which is a two-dimensional pattern of conducting Si electrodes. When light is absorbed in the *p*-doped region, an electron is introduced into the conduction band and a hole is left in the valence band. The electron is attracted to the region beneath the positive electrode, where it is stored. The hole migrates to the *n*-doped substrate, where it combines with an electron. Each electrode can store ~10^5 electrons. After the desired observation time, electrons stored in each pixel of the top row of the array are moved into a serial register at the top and then moved, one pixel at a time, to the top right position, where the charge is read out. Then the next row is moved up and read out, and the sequence is repeated until the entire array has been read.

20-17. DTGS has a permanent electric polarization. That is, one face of crystal has a positive charge and opposite face has a negative charge. When temperature of crystal changes by absorption of infrared radiation, polarization (voltage difference between two faces) changes. Change in voltage is detector signal.

20-18. (a) $A = \dfrac{L}{c \ln 10}\left(\dfrac{1}{\tau} - \dfrac{1}{\tau_o}\right)$

$= \dfrac{0.210 \text{ m}}{(3.00 \times 10^8 \text{ m/s}) \ln 10}\left(\dfrac{1}{16.06 \times 10^{-6} \text{s}} - \dfrac{1}{18.52 \times 10^{-6} \text{s}}\right) = 2.51 \times 10^{-6}$

(b) (i) The ordinate is ring-down lifetime with sample in the cavity. The greater the absorbance, the shorter the ring-down time. The spectrum is somewhat analogous to a transmission spectrum in which transmittance is plotted on the ordinate.

(ii) $\tilde{\nu}$ = wavenumber ≡ 1/wavelength = $1/\lambda$ ⇒ $\lambda = 1/\tilde{\nu}$.

$\lambda = 1/(6\,046.954\,6 \text{ cm}^{-1}) = 1.653\,725\,00 \times 10^{-4}$ cm

$(1.653\,725\,00 \times 10^{-4} \text{ cm})(10^7 \text{ nm/cm}) = 1\,653.725\,00$ nm

(iii) A wavelength of 1.7×10^{-6} m is in the infrared region (see Figure 18-2)

20-19. $n_1 \sin \theta_1 = n_2 \sin \theta_2$, where $n_1 = 1.50$ and $n_2 = 1.33$

(a) If $\theta_1 = 30°$, $\theta_2 = 34°$

(b) If $\theta_1 = 0°$, $\theta_2 = 0°$ (no refraction)

20-20. Light inside the fiber strikes the wall at an angle greater than the critical angle for total reflection. Therefore, all light is reflected back into the core and continues to be reflected from wall-to-wall as it moves along the fiber. If the bending angle is not too great, the angle of incidence will still exceed the critical angle and light will not leave the core.

20-21. When traveling from medium 1 into medium 2, the critical angle for total internal reflection is $\sin \theta_{critical} = n_2/n_1$, where n_i is the refractive index of medium i. For solvent/silica interface, $n_1 = 1.50$ and $n_2 = 1.46$, so $\sin \theta_{critical} = 1.46/1.5 = 0.973\ 3$. $\theta_{critical} = \sin^{-1}(0.973\ 3) = 1.339$ radians from the Excel function ASIN(0.9733). Degrees $= 180 \times \dfrac{\text{radians}}{\pi} = 180 \times \dfrac{1.339}{\pi} = 76.7°$.

For silica/air interface, $n_1 = 1.46$ and $n_2 = 1.00$, so $\sin \theta_{critical} = 1.00/1.46 = 0.684\ 9$. $\theta_{critical} = \sin^{-1}(0.684\ 9) = 0.754\ 5$ radians. Degrees $= 180 \times \dfrac{0.754\ 5}{\pi} = 43.2°$.

The angle in photograph is ~55°, which exceeds the critical angle at the silica/air interface but does not exceed the critical angle at the solvent/silica interface. Total internal reflection in photo must be from the silica/air interface.

20-22. Light is transmitted through diamond crystal waveguide by total internal reflection at upper and lower surfaces. The upper surface is in contact with dried blood sample. Downward pressure applied to top of sample ensures good contact between sample and waveguide. Compounds characteristic of malaria parasites absorbs some of the evanescent wave within totally internally reflected radiation, decreasing the radiant power transmitted. The integrated area of the absorption spectrum of transmitted power is proportional to concentration of parasite.

20-23. (a) $\tilde{\nu}$ = wavenumber $\equiv 1/\text{wavelength} = 1/\lambda \Rightarrow \lambda = 1/\tilde{\nu}$.

$\lambda = 1/(1\ 500\text{ cm}^{-1}) = 6.67 \times 10^{-4}\text{ cm} = 6.67\ \mu\text{m}$

$$d_p = \dfrac{\lambda/n_1}{2\pi\sqrt{\sin^2\theta_i - (n_2/n_1)^2}} = \dfrac{6.67\ \mu\text{m}/2.42}{2\pi\sqrt{\sin^2(45°) - (1.5/2.42)^2}} = 1.29\ \mu\text{m}$$

(b) $\lambda = 1/(3\,000\text{ cm}^{-1}) = 3.33 \times 10^{-4}\text{ cm} = 3.33\text{ μm}$

Other terms are constant, so d_p is proportional to the wavelength.

$$d_p = \frac{3.33\text{ μm}}{6.67\text{ μm}} \cdot 1.29\text{ μm} = 0.64\text{ μm}$$

(c) In transmission mode, pathlength is constant—independent of wavenumber. In attenuated total reflectance, pathlength depends on penetration depth which decreases with wavenumber. So in attenuated total reflectance, absorbances at high wavenumber will have weaker relative intensity.

(d) $d_p = \dfrac{\lambda/n_1}{2\pi\sqrt{\sin^2\theta_i - (n_2/n_1)^2}} = \dfrac{0.500\text{ μm}/2.42}{2\pi\sqrt{\sin^2(45°) - (1.5/2.42)^2}} = 0.097\text{ μm}$

Penetration depth for visible light is small and would yield weak signals.

20-24. Sensitivity increases as the number of reflections inside the waveguide increases, because there is some attenuation at each reflection. For a constant angle of incidence, number of reflections increases as thickness of waveguide decreases.

Three reflections

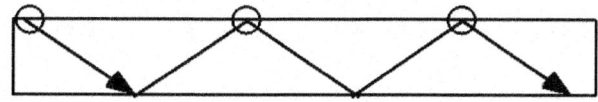

Six reflections in the same length when waveguide thickness is cut in half

20-25. (a) $\theta_i = \theta_c$ (critical angle) when $(n_1/n_2)\sin\theta_c = 1$. For $n_1 = 1.52$ and $n_2 = 1.50$, $\theta_c = 80.7°$. That is, θ_i must be $\geq 80.7°$ for total internal reflection.

(b) $\dfrac{\text{power out}}{\text{power in}} = 10^{-\ell(\text{dB/m})/10} = 10^{-(20.0\text{ m})(0.0100\text{ dB/m})/10} = 0.955$

20-26. (a) $n_{\text{core}}\sin\theta_i = n_{\text{cladding}}\sin\theta_r$

For total reflection, $\sin\theta_r \geq 1 \Rightarrow \sin\theta_i \geq \dfrac{n_{\text{cladding}}}{n_{\text{core}}}$

For $n_{\text{cladding}} = 1.400$ and $n_{\text{core}} = 1.600$, $\sin\theta_i \geq \dfrac{1.400}{1.600} \Rightarrow \theta_i \geq 61.04°$

(b) For $n_{\text{cladding}} = 1.400$ and $n_{\text{core}} = 1.800$, $\theta_i \geq 51.06°$

20-27. Angle of incidence = angle of reflection = 45°. Angle of refraction ≡ θ. $n_{prism} \sin 45° = n_{air} \sin θ$. If total reflection occurs, there is no refracted light. This happens if $\sin θ > 1$, or $\dfrac{n_{prism} \sin 45°}{n_{air}} > 1$. Using $n_{air} = 1$ gives $n_{prism} > \sqrt{2}$. As long as $n_{prism} > \sqrt{2}$, no light will be transmitted through the prism and all light will be reflected.

20-28. (a) The Teflon tube acts as an optical fiber because the internal solution has a higher refractive index (1.33) than the walls (1.29). The tube is a 2.5-m-long sample cell that can be conveniently coiled to fit in a reasonable volume and guide the incident radiation all the way through the tube. The long pathlength allows us to obtain measurable absorbance for a very low concentration of analyte.

(b) $n_{core} \sin θ_i = n_{cladding} \sin θ_r$

For total reflection, $\sin θ_r \geq 1 \Rightarrow \sin θ_i \geq \dfrac{n_{cladding}}{n_{core}}$

For $n_{cladding} = 1.29$ and $n_{core} = 1.33$, $\sin θ_i \geq \dfrac{1.29}{1.33} \Rightarrow θ_i \geq 76°$

(c) Deviations between data and line are very small, intercept is near 0, and R^2 is greater than 0.999. Calibration is linear.

$y = 0.107 = 0.003\,18_3 x + 0.005\,3 \Rightarrow x = \dfrac{0.107 - 0.005_3}{0.003\,18_3} = 32.0$ nM

	A	B	C	D
1	Phosphate analysis with long pathlength cell			
2				
3		Phosphate		
4		(nM)	Absorbance	
5		6.07	0.024	
6		12.13	0.044	
7		24.27	0.083	
8		48.53	0.160	
9		97.07	0.314	
10				
11	{B12:C14}=LINEST(C5:C9,B5:B9,TRUE,TRUE)			
12	m	0.003183	0.00529	b
13	u_m	0.000007	0.00033	u_b
14	R^2	0.99999	0.00049	s_y

Plot: $y = 0.003\,18x + 0.005\,3$, $R^2 = 0.999\,99$

(d) % recovery $= \dfrac{C_{spiked} - C_{unspiked}}{C_{added}} \times 100 = \dfrac{73.0 - 7.7}{63.1} \times 100 = 103\%$

20-29. (a) The following diagram shows the path of a light wave through the waveguide. The length of one bounce, ℓ, satisfies the equation $(0.60\ \mu m)/\ell = \tan 20°$, giving $\ell = 1.648\ \mu m$. The hypotenuse of the triangle, h, satisfies the equation $(0.60\ \mu m)/h = \sin 20°$, giving $h = 1.754\ \mu m$. The number of intervals of length ℓ in 3.0 cm is $(3.0\ cm)/(1.648\ \mu m) = 1.820 \times 10^4$. Therefore, pathlength covered by the light is $h \times (1.820 \times 10^4) = 3.19\ cm$.

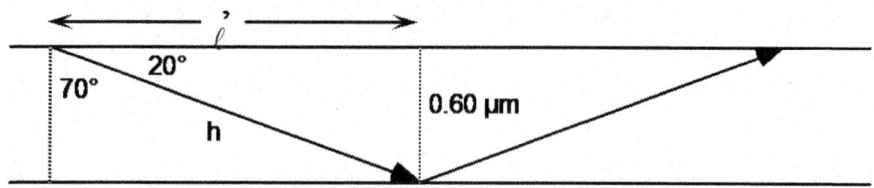

$$\frac{\text{power out}}{\text{power in}} = 10^{-\ell(dB/m)/10} = 10^{-(3.19\ cm)(0.050\ dB/cm)/10} = 0.964$$

(b) Wavelength = λ_o/n, where λ_o is the wavelength in vacuum and n is the refractive index. Wavelength = $(514\ nm)/1.5 = 343\ nm$.

The frequency is unchanged from that in vacuum:
$$\nu = c/\lambda = (2.9979 \times 10^8\ m/s)/(514 \times 10^{-9}\ m) = 5.83 \times 10^{14}\ Hz$$

20-30. (a)

λ (μm)	n	λ (μm)	n
0.2	1.5505	2	1.4381
0.4	1.4701	3	1.4192
0.6	1.4580	4	1.3890
0.8	1.4533	5	1.3405
1	1.4504	6	1.2580

(b) $dn/d\lambda$ is greater for blue light (~400 nm) than for red light (~600 nm)

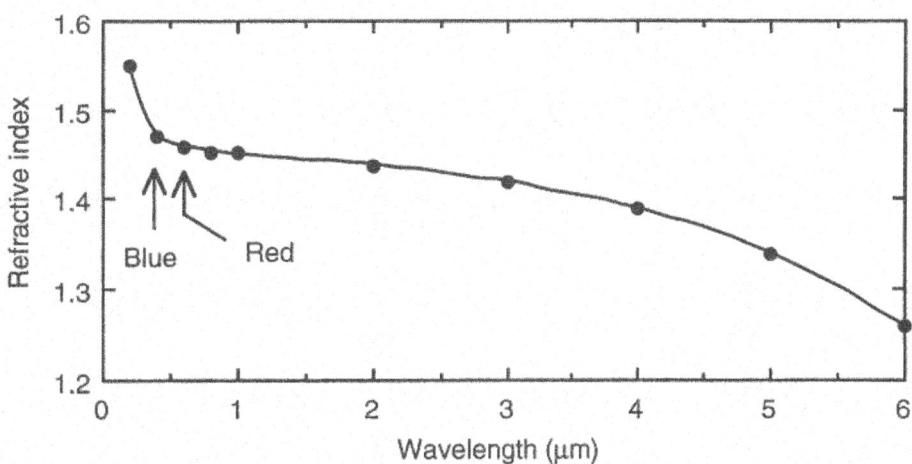

20-31. (a) $\Delta = \pm 2$ cm

(b) Resolution refers to the ability to distinguish closely spaced peaks.

(c) Resolution $\approx 1/\Delta = 0.5$ cm^{-1}

(d) $\delta = 1/(2\Delta\nu) = 1/(2 \times 2\,000$ cm$^{-1}) = 2.5$ μm

20-32. The background transform gives the incident irradiance P_0. The sample transform gives the transmitted irradiance P. Transmittance is P/P_0, not $P - P_0$.

20-33. White noise is independent of frequency. One source is random motion of charge carriers in electronic circuits. $1/f$ noise decreases with increasing frequency. Drift and flicker of the lamp of a spectrophotometer are sources of $1/f$ noise. Line noise results from disturbances at discrete frequencies. The 60 Hz wall frequency is the most common electromagnetic disturbance seen in electronic instruments.

20-34. In beam chopping, the beam is alternately directed through the sample and reference cells. Beam chopping moves the analytical signal from zero frequency to the frequency of the chopper, which can be selected so that $1/f$ noise and line noise are minimal.

20-35. To increase the ratio from 8 to 20 (a factor of $20/8 = 2.5$) requires $2.5^2 = 6.25 \approx 7$ scans.

20-36. (a) $(100 \pm 1) + (100 \pm 1) = 200 \pm \sqrt{2}$, since $e = \sqrt{e_1^2 + e_2^2} = \sqrt{1^2 + 1^2} = \sqrt{2}$

(b) $(100 \pm 1) + (100 \pm 1) + (100 \pm 1) + (100 \pm 1) = 400 \pm 2$, since $e = \sqrt{1^2 + 1^2 + 1^2 + 1^2} = 2$. The signal-to-noise ratio is $400:2 = 200:1$.

(c) The initial measurement has signal/noise = 100/1.
Averaging n measurements gives

average signal $= \dfrac{n \cdot 100}{n} = 100$

average noise $= \dfrac{\sqrt{n}}{n} = 1/\sqrt{n}$

$\dfrac{\text{average signal}}{\text{average noise}} = \dfrac{100}{1/\sqrt{n}} = 100\sqrt{n}$,

which is $\sqrt{n}$ times greater than the original value of signal/noise.

Spectrophotometers 295

20-37. (a) Gridlines are not evenly spaced on a log scale. On *x*-axis, 4 s is 1 gridline to the right of "3." The datum at 4 s is on the horizontal gridline that is 3 above "0.3," meaning the standard uncertainty for 4 s of averaging is 0.6 ppm.

To determine the standard uncertainty for 16 s of signal averaging, we must interpolate 16 on the x-axis. 16 can be written as 1.6×10^1. Log 16 = 1.204, where 1 is the characteristic reflecting the power and 0.204 is the mantissa for 1.6. 16 is then 20.4% of the distance between 10 and 100 along the *x*-axis. The datum at 16 s has a standard uncertainty of 0.3 ppm.

(b) Averaging 4 times as many signals should decrease noise by $\sqrt{4} = 2$. Expected standard uncertainty after 4 s of averaging is $1.18/\sqrt{4} = 0.59$. The observed standard uncertainty is 0.6 ppm.

Averaging 16 times as many signals should decease noise by $\sqrt{16} = 4.0$. Expected standard deviation after 16 s of averaging is $1.18/\sqrt{16} = 0.30$. Observed standard uncertainty is 0.3 ppm. Observed improvement in detection limit follows theoretical prediction at least up to 16 s of averaging.

(c) Averaging 100 times as many signals should decease noise by $\sqrt{100} = 10.0$. Expected standard uncertainty after 100 s of averaging is $1.18/\sqrt{100} = 0.118$. The datum at 100 s is ~9% along the distance between 0.1 and 1. $10^{0.09} = 1.2$, meaning the standard uncertainty is 1.2×0.1 ppm = 0.12 ppm. This is in good agreement with theoretical prediction.

(d) Above 200 s, drift—slow changes in response over time—cause more uncertainty in the averaged signal than $\sqrt{n}$ improvement provided by signal averaging. As a result, standard uncertainty grows when averaging signal over longer times.

20-38. The theoretical signal-to-noise (S/N) ratio should increase in proportion to the square root of the number of cycles that are averaged.

Number of cycles = n	$\sqrt{n}$	Predicted relative S/N ratio
1	1	$\equiv 1$
100	10.00	10.00
300	17.32	17.32
1 000	31.62	31.62

If the observed S/N = 60.0 for the average of 1 000 cycles, then the predicted S/N for the other experiments are shown in the following table:

Number of cycles	Predicted S/N ratio	Observed S/N ratio
1 000	60.0 (observed)	60.0
300	$60.0 \left(\frac{17.32}{31.62}\right) = 32.9$	35.9
100	$60.0 \left(\frac{10.00}{31.62}\right) = 19.0$	20.9
1	$60.0 \left(\frac{1}{31.62}\right) = 1.90$	1.95

20-39. (a) For data in the table in the problem:

$$y_{12}(\text{7-pt moving average}) = \frac{y_9 + y_{10} + y_{11} + y_{12} + y_{13} + y_{14} + y_{15}}{7}$$

$$= \frac{6.956 + 5.000 + 6.086 + 14.347 + 14.130 + 13.913 + 11.739}{7} = 10.310$$

$$y_{12}(\text{7-pt cubic smoothing}) = \frac{-2y_9 + 3y_{10} + 6y_{11} + 7y_{12} + 6y_{13} + 3y_{14} - 2y_{15}}{21}$$

$$= \frac{-2(6.956) + 3(5.000) + 6(6.086) + 7(14.347) + 6(14.130) + 3(13.913) - 2(11.739)}{21}$$

$$= 11.480$$

(b) Partial data for 7-point moving average and cubic smoothing are shown in the spreadsheet. Final results are in the graph.

	A	B	C	D	E
1	7-point smoothing				
2				$y_{smoothed}$	$y_{smoothed}$
3	n	x	y	Mvg. avg.	S-G
4	1	6.750	0.000		
5	2	6.775	0.000		
6	3	6.800	0.869		
7	4	6.825	1.086	0.559	0.683
8	5	6.850	0.000	0.838	0.683
9	6	6.875	1.086	1.832	0.300
10	7	6.900	0.869	2.422	1.573
11	8	6.925	1.956	3.136	3.178
12	9	6.950	6.956	5.186	3.830
13	10	6.975	5.000	7.049	6.293
14	11	7.000	6.086	8.913	9.057
15	12	7.025	14.347	10.310	11.480
16		D7 = AVERAGE(C4:C10)			
17		E7 = (-2*C4 + 3*C5 + 6*C6 + 7*C7			
18		+ 6*C8 + 3*C9 -2*C10)/21			

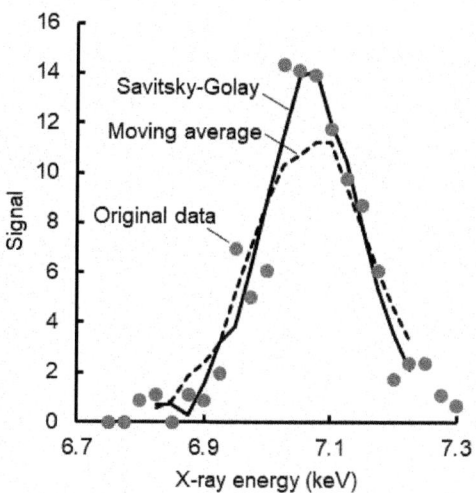

(c) Excel's 7-point moving average produces a smooth curve similar to part (b), but maximum of Excel curve is shifted $(n - 1)/n$ data points to right of true maximum.

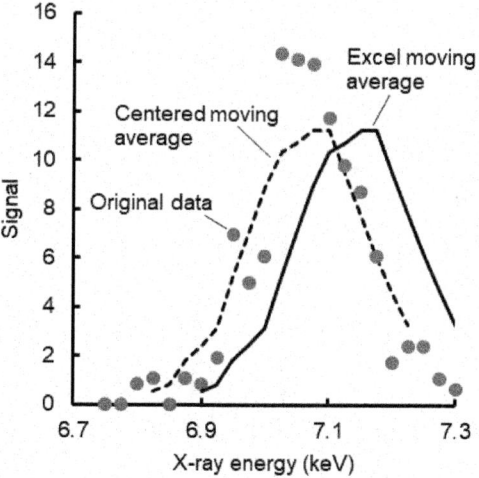

20-40. (a)

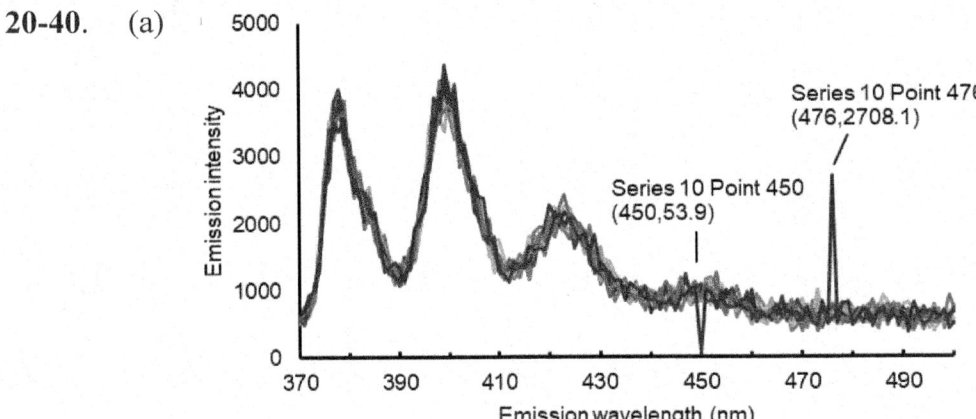

450 and 476 nm of scan 10 are outliers. We will not use scan 10 in further calculations. Alternately outliers could be replaced with a smoothed value.

(b)

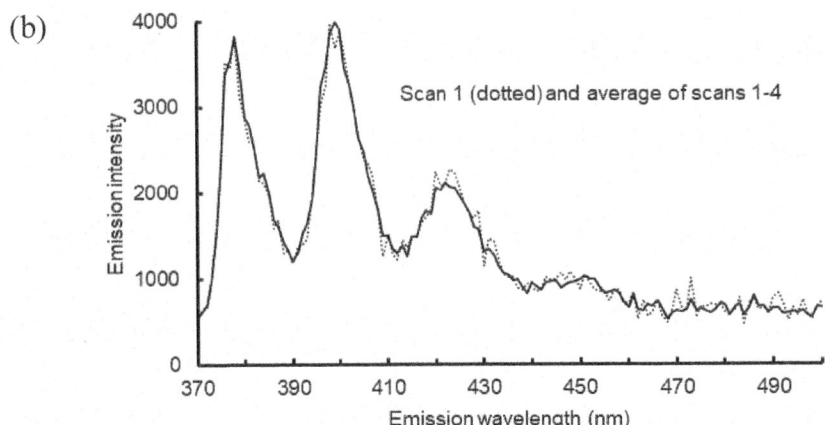

(d)

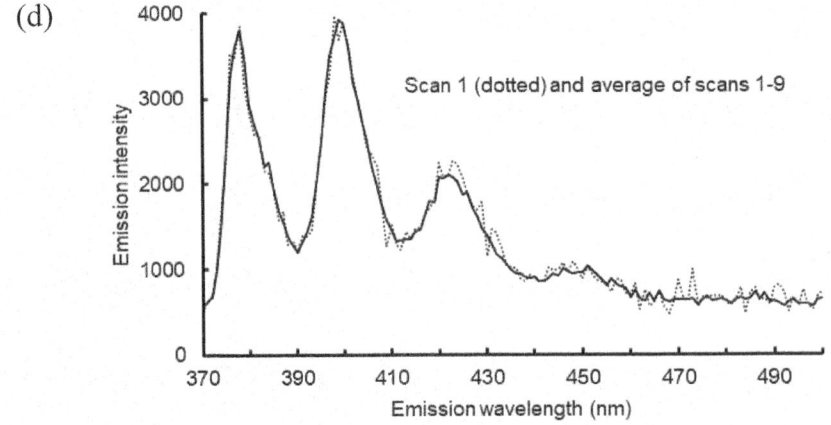

(e) Root-mean-square may be calculated in Excel using STDEV.P which calculates $\sqrt{\left[\sum_i (A_i - \overline{A})^2\right]/n}$.

Scan 1: root-mean-squared 470 to 500 nm = 108.9

Average scans 1–4: root-mean-squared 470 to 500 nm = 52.2

Average scans 1–9: root-mean-squared 470 to 500 nm = 36.9

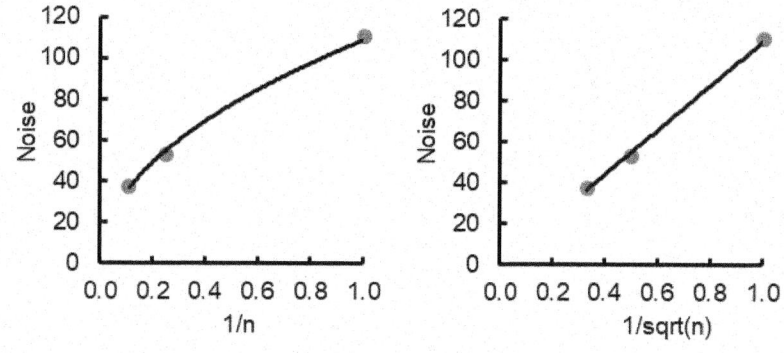

Noise depends inversely on square root of number of scans averaged.

(f) Signal to noise improves with square root of number of scans averaged.

	Signal at 399 nm	Background (470-500 nm)	Corrected signal	Noise (part *e*)	S/N	Relative S/N
Scan 1	3 687.3	669.5	3 017.8	108.9	27.7	1
Average 1–4	3 996.9	637.0	3 359.9	52.2	64.4	2.3
Average 1–9	3 922.8	629.3	3 293.5	36.9	89.3	3.2

CHAPTER 21
ATOMIC SPECTROSCOPY

21-1. Liquid sample containing ionic compound M^+X^- is nebulized to form a fine mist. Solvent evaporates to leave a dry aerosol which is vaporized in the flame. The heat of the flame decomposes gaseous compound MX into gas phase atoms $M(g)$. In atomic absorbance, the absorbance of light from a hollow cathode lamp by $M(g)$ is measured.

In atomic emission spectroscopy, the sample goes through the same steps to form $M(g)$. Heat from the flame or plasma causes $M(g)$ to be excited to $M^*(g)$. $M^*(g)$ relaxes to $M(g)$ by emission of a photon of characteristic wavelength.

21-2. Temperature is more critical in emission spectroscopy, because the small population of the excited state varies substantially as the temperature is changed. The population of the ground state does not vary much.

21-3. Only 5% of a nebulized sample reaches the flame, where analyte resides in the optical path for <1 s. Formation of a volatile species like AsH_3 results in more analyte transported into the heated zone. Also, use of a quartz cell holds atoms in the optical path longer, resulting in greater sensitivity.

21-4. Furnaces give increased sensitivity and require smaller sample volumes but give poorer reproducibility with manual sample introduction. Automated sample introduction gives good precision. Furnaces are more expensive.

21-5. Drying (~20–100°C) removes water from the sample. Ashing (~100–500°C) is intended to remove as much matrix as possible without evaporating analyte. Atomization (~500–2 000°C) vaporizes analyte (and most of the rest of the sample) for the atomic absorption measurement. For some samples, ashing temperatures might be much higher than 500°C to remove more of the matrix, but it must be demonstrated that the ashing does not remove analyte.

21-6. The plasma operates at higher temperature than a flame and the environment is Ar, not combustion gases. The plasma decreases chemical interference (such as oxide formation) and allows emission instead of absorption to be used. Lamps are not required and simultaneous multi-element analysis is possible. Self-absorption is reduced in the plasma because the temperature is more uniform. Disadvantages of the plasma are increased cost of equipment and operation.

21-7. Doppler broadening occurs because an atom moving toward the radiation source sees a higher frequency than one moving away from the source. Increasing temperature gives increased speeds (more broadening) and increased mass gives decreased speeds (less broadening).

21-8. (a) Chopper alternately blocks or exposes lamp to the flame and detector. When lamp is blocked, signal is due to background. When lamp is exposed, signal is due to analyte plus background. Difference is the desired analytical signal.

(b) Flame or furnace is alternately exposed to D_2 lamp and hollow-cathode lamp. Absorbance from D_2 lamp is due to background. Absorbance from hollow-cathode lamp is due to analyte plus background. Difference is desired signal.

(c) When a magnetic field parallel to the viewing direction is applied to the furnace, the analytical signal is split into two components that are separated from the analytical wavelength, and one component at the analytical wavelength. The component at the analytical wavelength is not observed because of its polarization. The other two components have the wrong wavelength to be observed. Analyte is essentially "invisible" to the detector when the magnetic field is applied, and only background is seen. Corrected signal is that observed without a field minus that observed with the field.

21-9. Spectral interference refers to overlap of analyte signal with signals due to other elements or molecules in the sample or with signals due to the flame or furnace. Physical interference occurs when viscosity or density differences of sample alter nebulization or analyte transfer into the flame. Chemical interference occurs when a component of the sample decreases the extent of atomization of analyte through some chemical reaction. Ionization interference refers to a loss of neutral analyte atoms through ionization. Isobaric interference is the overlap of different species with nearly the same mass-to-charge ratio in a mass spectrum.

21-10. (a) $Mg(NO_3)_2$ raises temperature of atomization of Mn by forming a manganese oxide, enabling use of a higher char temperature and more complete removal of matrix.

(b) La^{3+} acts as releasing agent by binding tightly to PO_4^{3-} and freeing Mn^{2+}.

(c) NH$_3$ reacts with polyatomic species such as ^{38}Ar^{16}O^{1}H$^+$ which would otherwise be an isobaric interference for ^{55}Mn$^+$.

21-11. (a) Fast moving ions are slowed by collision with an inert gas like He. Polyatomics are larger than M$^+$ ions and are slowed more. An electrostatic barrier at the cell outlet allows high kinetic energy M$^+$ to pass into the mass spectrometer, but blocks low kinetic-energy interferents.

(b) A collision cell slows polyatomic ^{14}N$_2^+$ and rejects the resultant low kinetic-energy ions. ^{28}Si$^+$ from the sample and background is the same species. The collision cell would not reduce the silicon background.

21-12. (a) The dynamic reaction cell contains reactive gas such as NH$_3$, CH$_4$, N$_2$O, CO, or O$_2$. The electric field is configured to select lower and upper masses of ions to pass through cell. Plasma species that interfere with M$^+$ are reduced by up to 9 orders of magnitude by reactions such as electron transfer (^{40}Ar^{16}O$^+$ + NH$_3$ → NH$_3^+$ + Ar + O) and proton transfer (^{40}ArH$^+$ + NH$_3$ → NH$_4^+$ + Ar). Reactive gas can also be used to shift analyte signal from a position at which interference occurs (for example, ^{40}Ar^{16}O$^+$ interferes with ^{56}Fe$^+$) to one where there is no interference (^{56}Fe$^+$ + N$_2$O → ^{56}Fe^{16}O$^+$ + N$_2$).

(b) When ^{87}Sr$^+$ is converted to ^{87}Sr^{19}F$^+$, which has a mass of 106, it no longer overlaps with ^{87}Rb$^+$ in the mass spectrum.

21-13. (a) No. With no interference removal (without use of collision or reaction cell), apparent vanadium increases with [HCl] concentration, suggesting possible V impurity in the HCl. Use of a collision or dynamic reaction cell reduces or eliminates the V signal. Collision/ dynamic reaction cells do not remove signal due to contamination, but rather remove isobaric interference due to polyatomic species like ^{35}Cl^{16}O$^+$ on the ^{51}V$^+$ signal.

(b) No. Natural abundance of ^{50}V is only 0.25% versus 99.75% for ^{51}V. The relative sensitivity for ^{50}V$^+$ is $\left(0.25/99.75\right) \times 100\% = 0.25\%$ that of ^{51}V$^+$. Using ^{50}V$^+$ would greatly reduce sensitivity.

(c) Trace-metal-grade HNO$_3$ is preferred for inductively coupled plasma–mass spectrometry. HCl, H$_2$SO$_4$, and H$_3$PO$_4$ all create isobaric interferences.

21-14. The extent to which an element is ablated, transported to the plasma, and atomized depends on the matrix. Different elements in a given matrix might not behave in the same manner. The most reliable calibration is for the analyte of interest to be measured in the same matrix as the unknown—if possible.

21-15. For Pb:
$$\left(104 \pm 17 \frac{\text{pg Pb}}{\text{g snow}}\right)\left(11.5 \frac{\text{g snow}}{\text{cm}^2}\right) = 1\,196 \pm 196 \frac{\text{pg Pb}}{\text{cm}^2}$$

$$\left(1\,196 \pm 196 \frac{\text{pg Pb}}{\text{cm}^2}\right)\left(\frac{1 \text{ ng}}{1\,000 \text{ pg}}\right) = 1.2 \pm 0.2 \frac{\text{ng Pb}}{\text{cm}^2}$$

Similarly, we multiply each of the other concentrations by 11.5 g snow/cm² to find Tl: 0.005 ± 0.001; Cd: 0.04 ± 0.01; Zn: 2.0 ± 0.3; Al: $7 (\pm 2) \times 10^1$ ng/cm².

21-16. $\lambda = \dfrac{hc}{\Delta E} = \dfrac{(6.626 \times 10^{-34} \text{ J}\cdot\text{s})(2.998 \times 10^8 \text{ m/s})}{3.371 \times 10^{-19} \text{ J}} = 5.893 \times 10^{-7}$ m = 589.3 nm

21-17. We derive the value for 6 000 K as follows:

$$\Delta E = h\nu = \frac{hc}{\lambda} = \frac{(6.626 \times 10^{-34} \text{ J}\cdot\text{s})(2.998 \times 10^8 \text{ m/s})}{500 \times 10^{-9} \text{ m}} = 3.97 \times 10^{-19} \text{ J}$$

$$\frac{N^*}{N_o} = \frac{g^*}{g_o} e^{-\Delta E/kT} = \frac{g^*}{g_o} e^{-(3.97 \times 10^{-19} \text{ J})/(1.381 \times 10^{-23} \text{ J/K})(6\,000\text{K})}$$

$$= \frac{g^*}{g_o}(8.3 \times 10^{-3})$$

If $g^*/g_o = 3$, then $N^*/N_o = 3\,(8.3 \times 10^{-3}) = 0.025$.

21-18. Doppler linewidth: $\Delta\lambda = \lambda\,(7 \times 10^{-7})\,\sqrt{T/M}$

For $\lambda = 589$ nm, $M = 23$ (sodium) at $T = 2\,000$ K,
$\Delta\lambda = (589 \text{ nm})(7 \times 10^{-7})\sqrt{(2\,000)/23} = 0.003_8$ nm

For $\lambda = 254$ nm, $M = 201$ (mercury) at $T = 2\,000$ K,
$\Delta\lambda = (254 \text{ nm})(7 \times 10^{-7})\sqrt{(2\,000)/201} = 0.000\,5_6$ nm

21-19. (a) $\Delta E = h\nu = \dfrac{hc}{\lambda} = \dfrac{(6.626\,1 \times 10^{-34} \text{ J}\cdot\text{s})(2.997\,9 \times 10^8 \text{ m/s})}{422.7 \times 10^{-9} \text{ m}}$

$= 4.699 \times 10^{-19}$ J/molecule = 283.0 kJ/mol

(b) $\dfrac{N^*}{N_o} = \dfrac{g^*}{g_o} e^{-\Delta E/kT} = 3e^{-(4.699 \times 10^{-19}\,\text{J})/(1.381 \times 10^{-23}\,\text{J/K})(2\,500\,\text{K})}$

$= 3.66_{57} \times 10^{-6}$

(c) At 2 515 K, $N^*/N_o = 3.97_{58} \times 10^{-6} \Rightarrow 8.5\%$ increase from 2 500 to 2 515 K

(d) At 6 000 K, $N^*/N_o = 1.03 \times 10^{-2}$

21-20.

Element:	Na	Cu	Br
Excited state energy (eV):	2.10	3.78	8.04
Wavelength (nm):	591	328	154
Degeneracy ratio (g^*/g_o):	3	3	2/3
N^*/N_o at 2 600 K in flame:	2.6×10^{-4}	1.4×10^{-7}	1.8×10^{-16}
N^*/N_o at 6 000 K in plasma:	5.2×10^{-2}	2.0×10^{-3}	1.2×10^{-7}

Calculations: wavelength $= hc/\Delta E \qquad N^*/N_o = (g^*/g_o)\, e^{-\Delta E/kT}$

Br is not readily observed in atomic absorption, because its lowest excited state requires far-ultraviolet radiation for excitation. N_2 and O_2 in air absorb far-ultraviolet light and have to be excluded from optical path. The excited state is not sufficiently populated to provide adequate intensity for atomic emission.

21-21. Dissociation energy of YC is greater than that of BaC, so BaC + Y $\rightleftharpoons$ Ba + YC is driven to the right, increasing concentration of free Ba atoms in gas phase.

21-22. (a) Least-squares trendline passes through all data, and residuals are negligible. y-Intercept is near 0. $R^2 > 0.999$. Calibration is linear.

Atomic Spectroscopy

	A	B	C	D	E	F	G
1	Least-Squares Spreadsheet						
2		x =	y =				
3	Highlight cells B10:C12	µg/mL	signal				
4	Type "LINEST (C4:C8,	0.00	0				
5	B4:B8, TRUE,TRUE)	5.00	124				
6	For PC, press	10.00	243				
7	CTRL + SHIFT + ENTER	20.00	486				
8	For Mac, press	30.00	712				
9	CTRL + SHIFT + RETURN						
10	LINEST output: m	23.767	4.03	b			
11	u_m	0.226	3.81	u_b			
12	R^2	0.9997	5.43	s_y			
13	n =	5	B13 = COUNT(B4:B8)				
14	Mean y =	313.00	B14 = AVERAGE(C4:C8)				
15	$\Sigma(x_i - \text{mean } x)^2$ =	580.00	B15 = DEVSQ(B4:B8)				
16	Measured y =	417	<= Input				
17	k = Number of replicate measurements of y =	1	<= Input				
18	Derived x =	17.376	B18 = (B16-C10)/B10				
19	Std uncertainty u_x =	0.254	=(C12/ABS(B10))				
20	*SQRT((1/B17)+(1/B13)+((B16-B14)^2)/(B10^2*B15))						

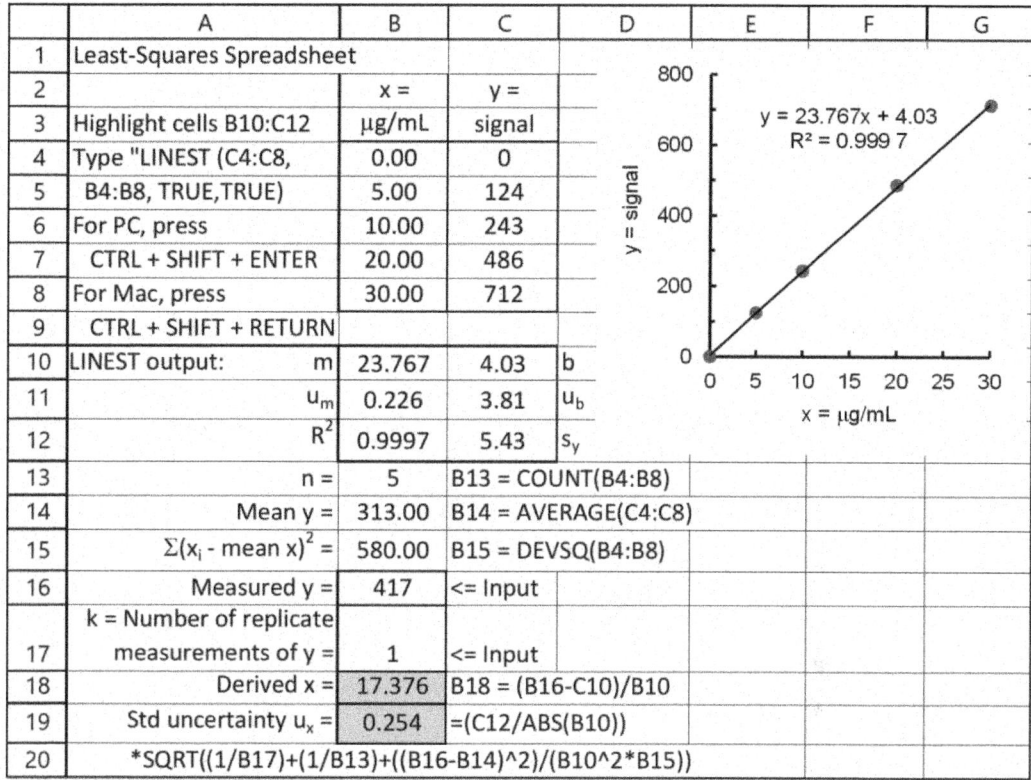

(b) Cells B18 and B19 give us [unknown] = 17.4 ± 0.3 µg/mL for an emission intensity of 417.

21-23. (a) Least-squares trendline passes through all data, and residuals appear negligible. The y-intercept is essentially 0. $R^2 > 0.999$. All characteristics suggest calibration is linear. Blank was not used to determine corrected absorbance and so is included in least squares.

	A	B	C	D	E	F	G
1	Calibration for K by atomic absorbance						
2		x = [Fe]	y = atomic				
3		(ppm)	absorbance				
4		0.00	0.000				
5		0.15	0.049				
6		0.30	0.105				
7		0.45	0.159				
8		0.60	0.212				
9		0.75	0.266				
10	B11:C13 = LINEST(C4:C9,B4:B9,TRUE,TRUE)						
11	m	0.3568	−0.0020	b			
12	u_m	0.0026	0.0012	u_b			
13	R^2	0.9998	0.0016	s_y			
14	n =	6	B14 = COUNT(C4:C9)				
15	Measured y =	0.179	<= Value from problem				
16	Predicted x =	0.507	B16 = (B15−C11)/B11				

Chart: y = 0.356 8x − 0.002, R^2 = 0.999 8. Atomic absorbance vs Potassium concentration (ppm).

(b) $\left[\mathrm{K}^+ \right] = \dfrac{\text{absorbance} - b}{m} = \dfrac{0.179 - (-0.002)}{0.356_8} = 0.507 \text{ ppm}$

(c) Standard uncertainty in $x = u_x = \dfrac{s_y}{|m|} \sqrt{\dfrac{1}{k} + \dfrac{1}{n} + \dfrac{(y - \bar{y})^2}{m^2 \sum (x_i - \bar{x})^2}}$.

$(y - \bar{y})^2$ is smallest, and so u_x is smallest, when measured y is near middle of calibration range.

21-24. (a) Least-squares trendline does not pass through data, and residuals show distinct curvature. The y-intercept is ~10% maximum value. $R^2 = 0.96$. All characteristics suggest calibration is nonlinear.

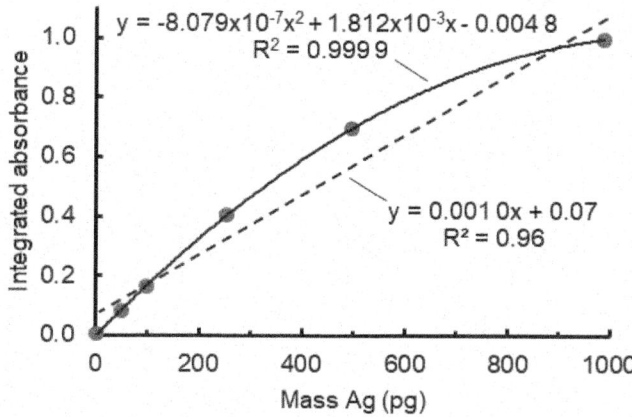

(b) Insert reading into equation:

$$0.598\,2 = -8.079 \times 10^{-7}x^2 + 1.812 \times 10^{-3}x - 0.004\,8$$

Rearranging yields $0 = -8.079 \times 10^{-7}x^2 + 1.812 \times 10^{-3}x - 0.603\,0$

Using $x = \dfrac{-b \pm \sqrt{b^2 - 4ac}}{2a}$ yields $x = 406.4$ and $x = 1\,836.4$

The plot in part (a) tells us invertebrate contains 406 pg Ag.

21-25. Analyte and standard are lost in equal proportions, so their ratio remains constant.

21-26. (a) Least-squares line passes through all data, and residuals appear random. $R^2 > 0.999$. Calibration is linear.

(b) From spreadsheet, x-intercept is $-5.60 \pm 0.16_3$ mL.

(c) Standard [Ca] = 20.0 µg Ca/mL. The intercept corresponds to Ca = $(5.60 \pm 0.16_3 \text{ mL})(20.0 \text{ µg Ca/mL}) = 112.0 \pm 3.2_6$ µg, which is the mass of Ca in 5.00 mL of unknown. The total volume of unknown was 100.0 mL, so mass of Ca in total unknown

$$= (100.0 \text{ mL}/5.00 \text{ mL})(112.0 \pm 3.2_6 \text{ µg Ca}) = 2\,240 \pm 65 \text{ µg Ca}.$$

$$\text{wt\% Ca in cereal} = \frac{100 \times (2\,240 \pm 65 \text{ µg Ca})}{0.521\,6 \text{ g cereal}} = 0.429 \pm 0.012 \text{ wt\%}.$$

	A	B	C	D	E
1	Standard Addition Constant Volume Least-Squares				
2	x	y			
3	Volume added (mL)	Absorbance			
4	0.00	0.151			
5	1.00	0.185			
6	3.00	0.247			
7	5.00	0.300			
8	8.00	0.388			
9	10.00	0.445			
10	15.00	0.572			
11	20.00	0.723			
12	B13:C15=LINEST(B4:B11,A4:A11,TRUE,TRUE)				
13	LINEST output: m	0.0282	0.1579	b	
14	u_m	0.0003	0.0031	u_b	
15	R^2	0.9993	0.0057	s_y	
16	x-intercept = -b/m =	-5.599	B16 = -C13/B13		
17	n =	8	B17=COUNT(A4:A11)		
18	Mean y =	0.376	B18=AVERAGE(B4:B11)		
19	$\Sigma(x_i - \text{mean } x)^2 =$	343.5	B19 = DEVSQ(A4:A11)		
20	Std uncertainty of		B21=(C15/ABS(B13))*SQRT		
21	x-intercept = u_x =	0.1630	((1/B17)+B18^2/(B13^2*B19))		

21-27. (a) All solutions are made to constant volume, so we plot I_{X+S} vs. $[S]_f$. The least-squares trendline in the graph passes through all data, and residuals appear random. The y-intercept should not be 0 for standard addition plots. R^2 is greater than 0.999. The calibration is linear.

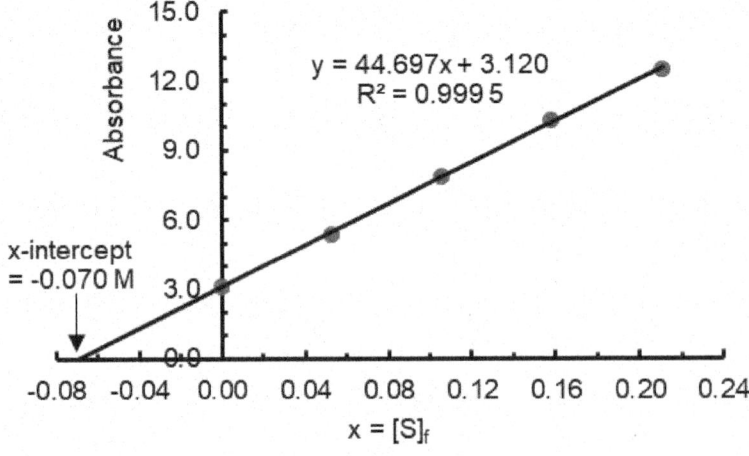

(b) To execute LINEST, highlight cells B13-C15, enter "=LINEST(D6:D10,C6:C10,TRUE,TRUE)", and press CONTROL + SHIFT + ENTER on a PC or CONTROL + SHIFT + RETURN on a Mac.

	A	B	C	D
1	Sodium in serum by standard addition			
2				
3	V_{total} (mL) =	V = mL of	x	y
4	50.00	Na standard	Concentration	I_{X+S} =
5	$[S]_i$ (M) =	added	of added Na $[S]_f$	signal
6	2.64	0.000	0.0000	3.13
7	V_0 (mL) =	1.000	0.0528	5.40
8	25.00	2.000	0.1056	7.89
9		3.000	0.1584	10.30
10		4.000	0.2112	12.48
11	C6 = $A$6*B6/$A$4			
12	B13:C15 = LINEST(D6:D10,C6:C10,TRUE,TRUE)			
13	LINEST output: m	44.70	3.120	b
14	u_m	0.55	0.071	u_b
15	R^2	0.9995	0.092	s_y
16	x-intercept = -b/m =	-0.06980	B16 = -C13/B13	
17	n=	5	B17 = COUNT(B6:B10)	
18	Mean y =	7.84	B18 = AVERAGE(D6:D10)	
19	$\Sigma(x_i - \text{mean } x)^2$ =	0.02788	B19 = DEVSQ(C6:C10)	
20	Std uncertainty of		B20 = (C15/ABS(B13))*SQRT	
21	x-intercept = u_x =	0.00235	((1/B17)+B18^2/(B13^2*B19))	

The intercept in cell B16 is $[X]_f = -0.069\,8$ M. The initial concentration of NaCl is larger by the dilution factor of (50.00 mL/25.00 mL) = 2.000. The initial concentration of NaCl in serum was $2.000 \times 0.069\,8$ M $= 0.139\,6$ M.

(c) The x-intercept is computed in cell B16 and its standard uncertainty is in cell B21. The relative uncertainty is $100 \times (0.002\,35)/(0.069\,8) = 3.37\%$. This uncertainty is much larger than the relative uncertainties in volume measurement, so the uncertainty in the original concentration of Na^+ should be 3.37%. A reasonable expression of $[Na^+]$ in the original serum is 0.140 (±3.37%) M = 0.140 (±0.004$_7$) M.

95% confidence interval = $\pm\, tu_x = \pm\, (3.182)(0.004_7$ M$) = \pm\, 0.015$ M, where t is taken for $5 - 2 = 3$ degrees of freedom.

21-28. (a) [S] in unknown mixture = $(8.24\ \mu g/mL)\left(\dfrac{5.00}{50.0}\right) = 0.824\ \mu g/mL$

Standard mixture has equal concentrations of X and S:
$$\dfrac{A_X}{[X]_f} = F\left(\dfrac{A_S}{[S]_f}\right) \Rightarrow \dfrac{0.930}{[X]_f} = F\dfrac{1.000}{[S]_f} \Rightarrow F = 0.930$$

Unknown mixture:
$$\dfrac{A_X}{[X]_f} = F\left(\dfrac{A_S}{[S]_f}\right) \Rightarrow \dfrac{1.690}{[X]_f} = 0.930\left(\dfrac{1.000}{0.824\ mg/mL}\right)$$

$\Rightarrow [X] = 1.497\ \mu g/mL$

But X was diluted by a factor of (10.00 mL)/(50.0 mL), so the original concentration in the unknown was

$(1.49_7\ \mu g/mL)\left(\dfrac{50.0\ mL}{10.00\ mL}\right) = 7.49\ \mu g/mL.$

(b) Concentration of X is 3.42 times that of S:
$$\dfrac{A_X}{[X]_f} = F\left(\dfrac{A_S}{[S]_f}\right) \Rightarrow \dfrac{0.930}{3.42} = F\left(\dfrac{1.000}{1.00}\right) \Rightarrow F = 0.271_9$$

Unknown mixture:
$$\dfrac{A_X}{[X]_f} = F\left(\dfrac{A_S}{[S]_f}\right) \Rightarrow \dfrac{1.690}{[X]_f} = 0.271_9\left(\dfrac{1.000}{0.824\ \mu g/mL}\right) \Rightarrow [X] = 5.12_2\ \mu g/mL$$

But X was diluted by a factor of (10.00 mL)/(50.0 mL), so the original concentration in the unknown was

$(5.12_2\ \mu g/mL)\left(\dfrac{50.0\ mL}{10.00\ mL}\right) = 25.6\ \mu g/mL.$

21-29. (a) CsCl provides Cs atoms which ionize to $Cs^+ + e^-$ in the plasma. Electrons in the plasma inhibit ionization of Sn. Therefore, emission from atomic Sn is not lost to emission from Sn^+.

(b)

	A	B	C	D	E	F
1	Tin in canned food - *Anal. Bioanal. Chem.* **2002**, *374*, 235					
2	Calibration data for 189.927 nm					
3						
4	Conc (µg/L)	Signal intensity				
5	0	4.0		Output from LINEST		
6	10	8.5	slope (m)	0.781526	0.862272	intercept (b)
7	20	19.6	u_m	0.01844	1.551016	u_b
8	30	23.6	R^2	0.996671	3.201818	s_y
9	40	31.1				
10	60	41.7				
11	100	78.8				
12	200	159.1				
13	Select cells D6:E8					
14	Enter the formula: =LINEST(B5:B12,A5:A12,TRUE,TRUE)					
15	CONTROL+SHIFT+ENTER on PC or CONTROL+SHIFT+RETURN on Mac					

From LINEST in D6:E8, $m = 0.782 \pm 0.018$; $b = 0.86 \pm 1.55$; $R^2 = 0.997$

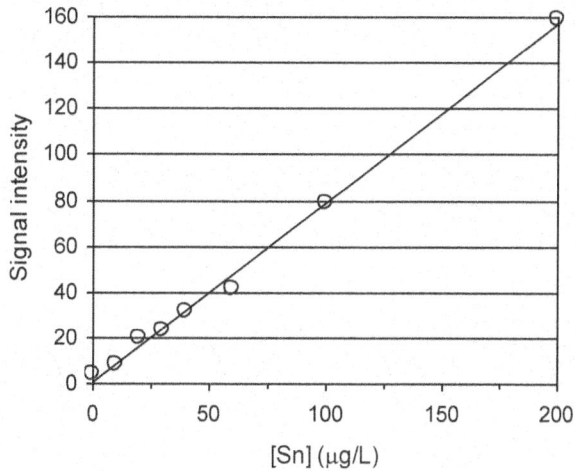

(c) For the 189.927 nm Sn emission line, spike recoveries are all near 100 µg/L, which is near 100%. None of the elements in the table appears to interfere significantly at 189.927 nm. For the 235.485 nm line, interference from an emission line of Fe is so serious that the Sn signal cannot be measured. Several other elements interfere enough to reduce the accuracy of the Sn measurement. These elements include Cu, Mn, Zn, Cr, and, perhaps, Mg. The 189.927 nm line is clearly the better of the two wavelengths for minimizing interference.

(d) Limit of detection = minimum detectable concentration = $3s/m$

where s is the standard deviation of the replicate samples and m is the slope of the calibration curve. Putting in the values $s = 2.4$ units and $m = 0.782$ units per (µg/L) gives

$$\text{limit of detection} = \frac{3s}{m} = \frac{3(2.4 \text{ units})}{0.782 \text{ units}/(\mu g/L)} = 9.2 \ \mu g/L$$

$$\text{limit of quantitation} = \frac{10s}{m} = \frac{10(2.4 \text{ units})}{0.782 \text{ units}/(\mu g/L)} = 30.7 \ \mu g/L$$

It would be reasonable to quote a limit of detection as 9 µg/L and a limit of quantitation as 31 µg/L.

(e) A 2-g food sample ends up in a volume of 50 mL. The limit of quantitation is 30.7 µg Sn/L for the solution. A 50-mL volume with Sn at the limit of quantitation contains (0.050 L)(30.7 µg Sn/L) = 1.54 µg Sn. The quantity of Sn per unit mass of food is

$$\frac{(1.54 \ \mu g \ Sn)(1 \ mg/1\,000 \ \mu g)}{(2.0 \ g \ food)(1 \ kg/1\,000 \ g)} = 0.77 \ \frac{mg \ Sn}{kg \ food} = 0.77 \ ppm$$

21-30. Standard addition graph: plot signal versus Ti or S concentration.

Ti (ppm)	Signal	S (ppm)	Signal
0.00	0.86	0.0	0.0174
3.00	1.10	37.0	0.0221
6.00	1.34	74.0	0.0268
12.00	1.82	148.0	0.0362

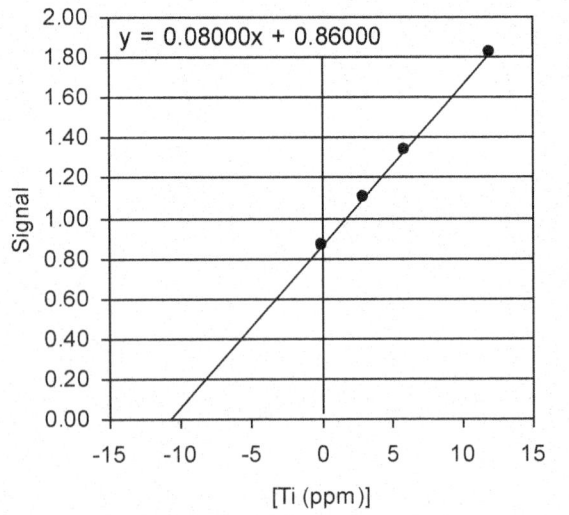
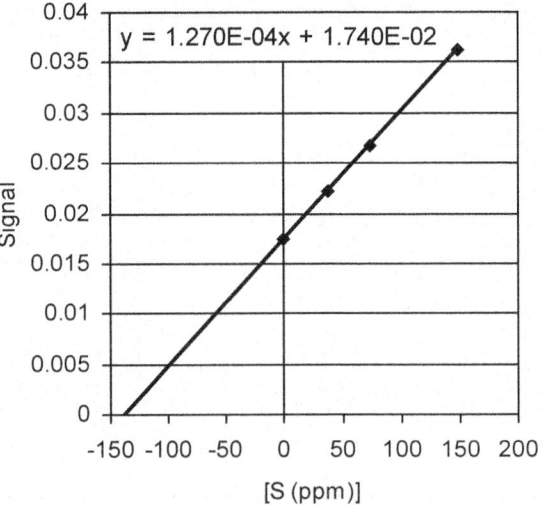

Ti standard addition graph: negative intercept = (0.860)/(0.0800 ppm^{-1})

= 10.75 ppm = 10.75 mg/L

S standard addition graph: negative intercept = (0.0174)/(0.000127 ppm^{-1})

= 137.0 ppm = 137.0 mg/L

Ti atomic mass = 47.867   S atomic mass = 32.06

Atomic Spectroscopy 313

$[Ti] = (10.75 \text{ mg/L})/(47.867 \text{ g/mol}) = 2.246 \times 10^{-4} \text{ M}$

$[S] = (137.0 \text{ mg/L})/(32.06 \text{ g/mol}) = 4.273 \times 10^{-3} \text{ M}$

$[Transferrin] = [S]/39 = 1.096 \times 10^{-4} \text{ M}$

$Ti/transferrin = (2.246 \times 10^{-4} \text{ M})/(1.096 \times 10^{-4} \text{ M}) = 2.05$

21-31. Absorption of X-rays of sufficient energy ionizes electrons from atoms. When an electron is removed from an inner shell of an atom, an electron from an outer shell falls into the vacancy. The excess energy of the electron making the transition is emitted as an X-ray. Electronic energy levels are different for every element, so the signature (energies of emitted X-rays) is different for each element.

21-32. K_β peaks for Ti, Se, and Zr should have energies of 4.93, 12.50, and 17.67 keV, respectively. A weak peak is observed in the spectrum near 4.9 keV, which could be Ti K_β. Se K_β at 12.50 keV comes under the strong Pb L_β peak at 12.61 keV. There is a shoulder at the left side of the base of Pb L_β from Se K_β. Zr K_β at 17.67 keV is observed as a weak peak superimposed on the broad bremsstrahlung radiation.

21-33. We observe Pb L_α = 10.55 and Pb L_β = 12.61 keV in the fluorescence spectrum. Pb K_α (74.97) and K_β (84.94) keV are beyond the working energy range of the handheld analyzer.

21-34. $E = (392 \text{ eV})(1.602 \times 10^{-19} \text{ J/eV}) = 6.28 \times 10^{-17} \text{ J}$ (conversion from Table 1-4)
$(6.28 \times 10^{-17} \text{ J/atom})(6.022 \times 10^{23} \text{ atoms/mol}) = 3.78 \times 10^7 \text{ J/mol} = 3.78 \times 10^4$ kJ/mol. The K_α energy is $(37\,800 \text{ kJ/mol})/(945 \text{ kJ/mol}) = 40$ times greater than the N≡N bond energy.

21-35.

Energy (keV)	Assignment	
6.40	Fe K_α	
7.05	Fe K_β	
7.50	Ni K_α	K_β peak would be at 8.26, but too small to see
8.07	Cu K_α	K_β peak would be at 8.90, but too small to see
8.62	Zn K_α	There is a possible K_β peak at 9.57
10.57	Pb L_α	
12.60	Pb L_β	
~14.1	Sr K_α?	K_β peak would be at 15.84

~14.8	?		
~15.78	Zr K_α?	K_β peak would be at 17.67	
17.50	Mo K_α	Probably X-ray tube anode	
~18.58	?		
19.59	Mo K_β	Probably X-ray tube anode	

21-36.

Energy (keV)	Assignment	
3.70	Ca K_α	
4.01	Ca K_β	
6.40	Fe K_α	
7.06	Fe K_β	
8.73	?	
9.20	?	
9.99	Hg L_α	
10.55	Pb L_α	
11.84	Hg L_β	
12.63	Pb L_β	
13.84	Hg L_γ	L_γ is observed because there is so much Hg in the sample. L_γ is not in the table in the text.
14.20	Sr K_α?	K_β would be at 15.84, where there is weak bump
14.76	Pb L_γ	L_γ is observed because there is so much Pb in the sample. L_γ is not in the table in the text.
15.17	?	
25.25	Sn K_α	
28.48	Sn K_β	

21-37. (a) Quantitative → Few → As low as 100 ppt = graphite furnace atomic absorption. Small sample volume also favors graphite furnace, which is what the authors used.

(b) Page 843 right column, first paragraph. Flame atomic absorbance was the gold standard, but required large amount of sample (1–2 mL).

(c) Glassware was cleaned with detergent, rinsed several times with deionized water, then with EDTA solution, and several times with deionized water. Cleanliness was checked by graphite furnace atomic absorption analysis of blank diluent.

(d) 213.9 nm in Table 1.

(e) Quadratic curve, as stated in Table 1, caption of Figure 2, and in Results and Discussion on page 846 in second paragraph of "Analysis of plasma samples and method validation." R^2 refers to the quadratic fit.

(f) Pd(II) acetate, magnesium nitrate, and Na_2H_2EDTA. Last paragraph of page 844 discusses use of Mg/Pd. Page 845 discusses why EDTA was also used.

(g) Analysis of reference material (page 846, left column, last paragraph), spike recoveries (Table 4), and comparison with flame atomic absorption and inductively coupled plasma–mass spectrometry (page 846, right column, top).

21-38. (a) Quantitative → The need to measure many elements leads to microwave plasma, inductively coupled plasma–atomic emission and inductively coupled plasma-mass spectrometry. Microwave plasma–atomic emission is limited to ~10 simultaneous elements. High dissolved solids favor inductively coupled plasma–atomic emission. However, in this paper flow injection enabled analysis of high dissolved-solid samples with inductively coupled plasma–mass spectrometry.

(b) To minimize blank concentrations arising from airborne sources of analytes.

(c) Plasma: $^{40}Ar^{16}O$, $^{40}Ar^{15}N^{1}H$, $^{38}Ar^{18}O$, $^{38}Ar^{17}O^{1}H$
Sample: $^{40}Ca^{16}O$, $^{37}Cl^{18}O^{1}H$

(d) Page 1199 states that kinetic energy discrimination is preferred for complex matrices. Reactive gases like NH_3 form cluster ions like $M(NH_3)_n^+$, $M(NH)(NH_3)_n^+$, and $M(NH_2)(NH_3)_n^+$ that may cause new interferences.

(e) Table 4 shows the certified concentrations range from 0.047 µg/g Tl to 12 300 µg/g P. Always read the concentration units..

CHAPTER 22
MASS SPECTROMETRY

22-1. Gas-phase ions are created when a ~1 ns pulse from an ultraviolet laser is absorbed by matrix crystals mixed with sample on the sample probe. After a 0.2–2 µs time delay for the plume to expand, the potential on grid 1 is lowered for 10 µs to accelerate ions by 20 kV between the back plate and grid 2. Ions pass through grid 2 into the high-vacuum drift region. Lighter ions travel faster than heavier ions with equal kinetic energy from 20 keV acceleration. Ions reach the detector in order of increasing mass-to-charge ratio (m/z). In fact, there is some spread in the kinetic energy of ions leaving the source region, so there is a spread in arrival time at the detector for ions of equal m/z. The advantage of reflecting ions through 180° with an electrostatic mirror is that ions with greater kinetic energy penetrate into the reflector deeper than ions with less kinetic energy. Faster ions travel a longer path, so all ions of the same m/z reach the detector located beside the ion source at the same time, enabling high-resolution mass spectra to be obtained.

22-2. The parent ion $H_2CO^{+\bullet}$ is observed at m/z 30. The base (most intense) peak at m/z 29 is the fragment HCO^+ that has lost one H atom. A small peak at m/z 31 with ~1% of the parent ion intensity is from the 1% natural abundance of $H_2{}^{13}CO^{+\bullet}$. Weak peaks at m/z 12–14 are the fragments C^+, CH^+, and CH_2^+. The ion at m/z 15 is H_2CO^{2+} with a mass of 30 and two positive charges. Lewis structures show that C^+, CH_2^+, and H_2CO^+ have an unpaired electron.

$:C\!\cdot^+ \qquad H\!:\!\overset{\cdot+}{C}\!: \qquad H\!:\!\overset{+}{C}\!:\!H \qquad \overset{H}{\underset{H}{\!>\!}}\overset{+}{C}\!-\!\overset{..}{\underset{..}{O}}{}^+ \qquad H\!:\!\overset{+}{C}\!=\!\overset{..}{\underset{..}{O}} \qquad \overset{H}{\underset{H}{\!>\!}}C\!=\!\overset{..}{\underset{..}{O}}{}^{\cdot+}$

22-3. For the electron ionization spectrum, pentobarbital is bombarded by electrons with an energy of 70 electron volts. The molecular ion (m/z = 226) produced by the impact has enough energy to break into fragments, so little $M^{+\bullet}$ is observed. Large peaks correspond to the most stable cation fragments. For chemical ionization, pentobarbital reacts with CH_5^+, which is a potent proton donor, but does not have excess kinetic energy. The dominant peak is usually MH^+ (m/z = 227). In the case of pentobarbital, some fragmentation is observed even in the chemical ionization spectrum.

22-4. 1 dalton (Da) ≡ 1/12 of the mass of ^{12}C = $\left(\dfrac{(1/12) \times 12 \text{ g/mol (exactly)}}{6.022\,14 \times 10^{23} \text{ mol}^{-1}} \right)$

= 1.660 54 × 10^{-24} g

[5.03 (±0.14) × 10^{10} Da][1.660 54 × 10^{-24} g/Da] = 8.35 (±0.23) × 10^{-14} g

$\dfrac{8.35\,(\pm 0.23) \times 10^{-14} \text{ g}}{10^{-15} \text{ g/fg}} = 83.5\,(\pm 0.23)$ fg

22-5. Atomic mass is the weighted average of masses of all isotopes of an element. We can estimate the relative abundance of the two major isotopes of Ni from the heights of their mass spectral peaks. The heights of the peaks that I measured from an earlier version of this illustration are 42.6 mm for ^{58}Ni and 17.1 mm for ^{60}Ni. The weighted average is

atomic mass
= (^{58}Ni mass)(% abundance of ^{58}Ni) + (^{60}Ni mass)(% abundance of ^{60}Ni)
= $(57.935\,3) \left(\dfrac{42.6}{42.6 + 17.1} \right) + (59.933\,2) \left(\dfrac{17.1}{42.6 + 17.1} \right) = 58.51$

The atomic mass in the periodic table is 58.69. This main reason for disagreement is that we neglected the existence of ^{61}Ni (1.13% natural abundance), ^{62}Ni (3.59%), and ^{64}Ni (0.90%).

22-6. (a) Resolving power = $\dfrac{m}{m_{1/2}} = \dfrac{268\ m/z}{0.015\ m/z} = 1.8 \times 10^4$ for upper spectrum

Resolving power = $\dfrac{m}{m_{1/2}} = \dfrac{268\ m/z}{0.002\,2\ m/z} = 1.2 \times 10^5$ for lower spectrum.

(b) $C_{15}H_{26}NO_3^+$ exact mass = 268.1907; $^{13}CC_{13}H_{23}N_2O_3^+$ exact mass = 268.1737

22-7. Resolving power = $(m/z)/\Delta(m/z) \approx 31/0.010 \approx 3\,100$.

22-8. Resolving power $(m/z)/(m/z)_{1/2}$ = 906.49/0.000 27 = 3.4 × 10^6

Resolving power $(m/z)/\Delta(m/z)$ = 906.49/0.000 45 = 2.0 × 10^6

The mass of an electron, 0.000 55 Da, is greater than the mass difference between the two compounds. Mass difference of compounds = 82% of electron mass.

318 Chapter 22

22-9. $C_5H_7O^{+\bullet}$ $\quad 5 \times 12.000\,00$
$\quad\quad\quad\quad\quad +7 \times 1.007\,825$
$\quad\quad\quad\quad\quad +1 \times 15.994\,91$
$-e^- \text{ mass } \underline{-1 \times 0.000\,55}$
$\quad\quad\quad\quad\quad\quad 83.049\,14$

$C_6H_{11}^{+\bullet}$ $\quad 6 \times 12.000\,00$
$\quad\quad\quad\quad\quad +11 \times 1.007\,825$
$-e^- \text{ mass } \underline{-1 \times 0.000\,55}$
$\quad\quad\quad\quad\quad\quad 83.085\,52$

$C_6H_{11}^+$ is a closer match than $C_5H_7O^+$ to the observed mass of 83.086 5 Da.

22-10. $^{31}P^+ = {}^{31}P - e^- = 30.973\,76 - 0.000\,55 = 30.973\,21$ (observed: 30.973_5)

To measure m/z, I enlarged the figure and sketched a Gaussian curve over each signal by eye. I then measured the position of the center of the peak with a millimeter scale ruler.

$^{15}N^{16}O^+ = {}^{15}N + {}^{16}O - e^- = 15.000\,11 + 15.994\,91 - 0.000\,55$
$\quad\quad\quad = 30.994\,47$ (observed: 30.994_6)

$^{14}N^{16}OH^+ = {}^{14}N + {}^{16}O + {}^{1}H - e^- =$
$\quad\quad 14.003\,07 + 15.994\,91 + 1.007\,82 - 0.000\,55$
$\quad\quad = 31.005\,25$ (observed: 31.005_6)

22-11. (a) m/z 140: $^{12}C_7{}^1H_{10}{}^{14}N^{16}O_2{}^+$

m/z 141: $^{13}C^{12}C_6{}^1H_{10}{}^{14}N^{16}O_2{}^+$ and $^{12}C_7{}^2H^1H_9{}^{14}N^{16}O_2{}^+$ and
$\quad\quad\quad ^{12}C_7{}^1H_{10}{}^{15}N^{16}O_2{}^+$ and $^{12}C_7{}^1H_{10}{}^{14}N^{17}O^{16}O^+$

Intensity at m/z 141 = $\underbrace{7 \times 1.08\%}_{^{13}C} + \underbrace{10 \times 0.012\%}_{^2H} + \underbrace{1 \times 0.369\%}_{^{15}N} + \underbrace{2 \times 0.038\%}_{^{17}O} = 8.1\%$

(b) m/z 142: $^{12}C_7{}^1H_{10}{}^{14}N^{18}O^{16}O^+$ and $^{13}C_2{}^{12}C_5{}^1H_{10}{}^{14}N^{16}O_2^+$ and
$\quad\quad\quad ^{13}C^{12}C_6{}^1H_{10}{}^{15}N^{16}O_2^+$

22-12. (a) $^{13}C^{12}C_7{}^1H_{11}{}^{15}N^{14}N_3{}^{16}O_2^+$, $^{12}C_8{}^1H_{11}{}^{14}N_4{}^{18}O^{16}O^+$, $^{13}C_2{}^{12}C_6{}^1H_{11}{}^{14}N_4{}^{16}O_2^+$,
and $^{13}C^{12}C_7{}^2H^1H_{10}{}^{14}N_4{}^{16}O_2^+$

(b) $^{12}C_8{}^1H_{11}{}^{14}N_4{}^{18}O^{16}O^+$ has one ^{18}O substitution in the parent ion at m/z 195. The abundance of $^{12}C_8{}^1H_{11}{}^{14}N_4{}^{18}O^{16}O^+$ relative to the parent ion at m/z 195 is expected to be 0.205% × 2 = 0.410% because the natural abundance of

Mass Spectrometry 319

$^{18}O/^{16}O = 0.205/99.757 = 0.205\%$.

$^{13}C_2{}^{12}C_6{}^1H_{11}{}^{14}N_4{}^{16}O_2^+$ has two ^{13}C atoms in place of two ^{12}C atoms in the parent ion at m/z 195. The abundance at m/z 197 relative to the parent ion is predicted to be $(0.005\ 8\%)(8)(7) = 0.325\%$.

Expected relative peak heights = $0.410\%/0.325\% = 1.26$. Measured relative peak heights in experimental spectrum = 1.59. Measured relative peak heights in theoretical spectrum = 1.26.

22-13. If two peaks are separated by one half m/z unit, a likely explanation is that they both have a charge $z = 2$. For example, $^{12}C_6{}^1H_6{}^{2+}$ would appear at m/z 39 and $^{13}C^{12}C_5{}^1H_6{}^{2+}$ would appear at m/z 39.5.

22-14. Intensity at X+2 = $\underbrace{12 \times 11 \times 0.005\ 8\%}_{^{13}C} + \underbrace{8 \times 0.205\%}_{^{18}O} = 2.4\%$

The observed intensity is nearly 100% because the predominant species is $[^{12}C_{12}{}^1H_{18}{}^{16}O_8{}^{35}Cl_2{}^{37}Cl]^-$.

22-15. Mass differences in the bottom row of the spreadsheet are 0.2, 0.6, and 1.2 ppm

	mass	m/z 395 $C_{12}H_{18}O_8{}^{35}Cl_3$	m/z 397 $C_{12}H_{18}O_8{}^{35}Cl_2{}^{37}Cl$	m/z 399 $C_{12}H_{18}O_8{}^{35}Cl{}^{37}Cl_2$	m/z 401 $C_{12}H_{18}O_8{}^{37}Cl_3$
^{12}C	12	144	144	144	144
^1H	1.007825	18.14085	18.14085	18.14085	18.14085
^{16}O	15.99491	127.95928	127.95928	127.95928	127.95928
^{35}Cl	34.96885	104.90655	69.9377	34.96885	
^{37}Cl	36.9659		36.9659	73.9318	110.8977
e^-	0.00055	0.00055	0.00055	0.00055	0.00055
Calculated exact mass =		395.00723	397.00428	399.00133	400.99838
Observed mass =		395.0073	397.0045	399.0018	not measured
Difference (ppm) =		0.2	0.6	1.2	
Difference (ppm) = 10^6*(observed mass - calculated mass)/calculated mass					

22-16. (a) The base peak has a mass loss of $390 - 149 = 241$ from the molecular ion
Nominal mass of fragments that might be lost: OC_8H_{17} 129
C_8H_{17} 113

These two fragments have a total mass of 242. Loss of OC_8H_{17} and C_8H_{17} with retention of one H atom gives a possible structure for the base peak:

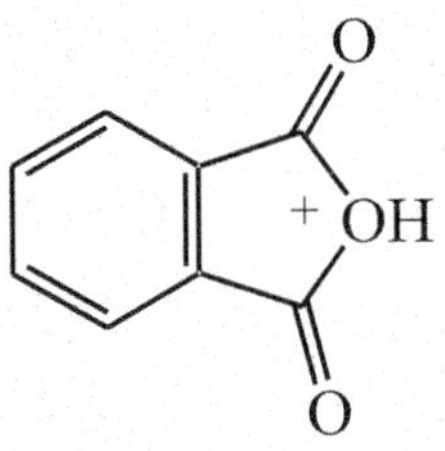

320

[Structure: phthalic anhydride with +OH] m/z 149

The m/z 279 peak corresponds to a mass loss of $390 - 279 = 111$ from $M^{+\cdot}$. Loss of C_8H_{17} with retention of two hydrogen atoms gives a possible structure

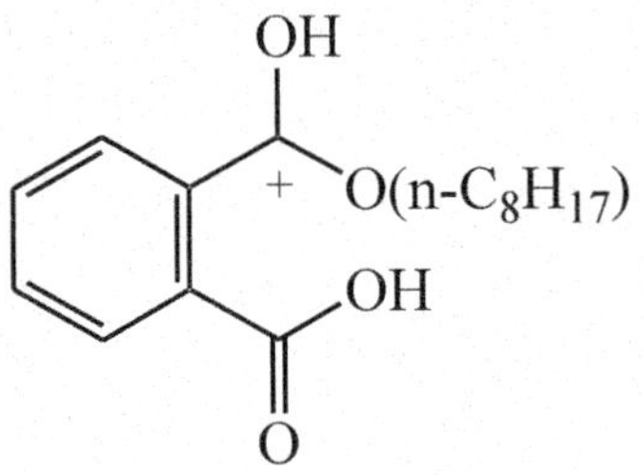

[Structure with OH, $^+O(n\text{-}C_8H_{17})$, and COOH] m/z 279

(b) m/z 277 has a mass loss of $390 - 277 = 113 =$ loss of C_8H_{17} from $M^{-\cdot}$

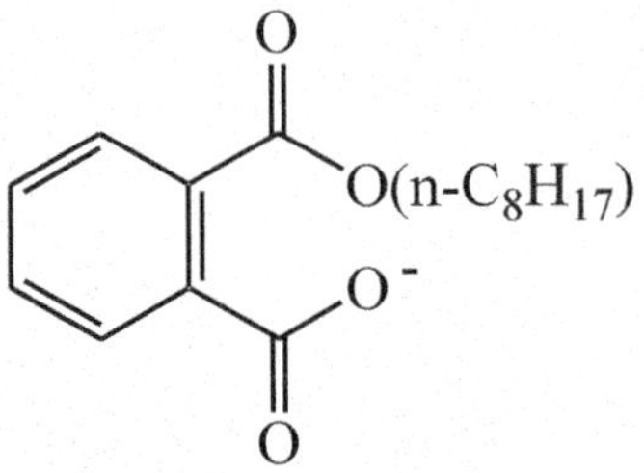

[Structure with $O(n\text{-}C_8H_{17})$ and O^-] m/z 277

m/z 221 has a mass loss of $390 - 221 = 169 = 390 - 113 - 56$
$= 390 - C_8H_{17} - C_4H_8$ from $M^{-\cdot}$. A possible structure is

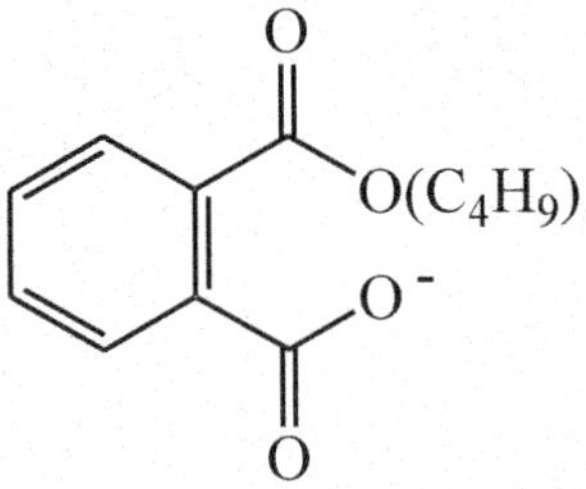

[Structure with $O(C_4H_9)$ and O^-] m/z 221

22-17. ^{79}Br abundance $\equiv a = 0.506\,9$ $\quad\quad$ ^{81}Br abundance $\equiv b = 0.493\,1$

Abundance of $C_2H_2{}^{79}Br_2 = a^2 = 0.256\,9_5$

Abundance of $C_2H_2{}^{79}Br{}^{81}Br = 2ab = 0.499\,9_0$

Abundance of $C_2H_2{}^{81}Br_2 = b^2 = 0.243\,1_5$

Relative abundances: $M^+ : M+1 : M+2 = 1 : 1.946 : 0.946\,3$

Figure 22-12 shows the stick diagram.

Mass Spectrometry

22-18. ^{10}B abundance $\equiv a = 0.199$ ^{11}B abundance $\equiv b = 0.801$

Abundance of $^{10}B_2H_6 = a^2 = 0.039\,6_0$

Abundance of $^{10}B^{11}BH_6 = 2ab = 0.318_8$

Abundance of $^{11}B_2H_6 = b^2 = 0.641_6$

Relative abundances: $M^+ : M+1 : M+2 = 1 : 8.05 : 16.20$

22-19. ^{79}Br abundance $\equiv a = 0.506\,9$ ^{81}Br abundance $+ b = 0.493\,1$

Abundance of $CH^{79}Br_3 = a^3 = 0.130\,25$

Abundance of $CH^{79}Br_2{}^{81}Br = 3a^2b = 0.380\,1_0$

Abundance of $CH^{79}Br^{81}Br_2 = 3ab^2 = 0.369\,7_5$

Abundance of $CH^{81}Br_3 = b^3 = 0.119\,9_0$

Relative abundances: $M^+ : M+1 : M+2 : M+3 = 0.342\,7 : 1 : 0.972\,8 : 0.315\,4$

22-20. (a) Nominal mass of $C_{72}H_{146} = 72 \times 12 + 146 \times 1 = 1\,010$ Da

(b) Monoisotopic mass of $^{12}C_{72}{}^1H_{146} = 72 \times 12 + 146 \times 1.007\,825$

$= 1\,011.14$ Da

(c) $(a+b)^n = (0.989\,3 + 0.010\,7)^{72}$

$= (0.989\,3)^{72} + \dfrac{72}{1!}(0.989\,3)^{71}(0.010\,7)^1 + \ldots$

$= 0.460\,9 + 0.358\,9 + \ldots$

Relative intensity $M/(M+1) = 0.460\,9/0.358\,9 = 1 : 0.778\,7$

22-21. (a) phenobarbital, $C_{11}H_{18}N_2O_3$

$R + DB = c - h/2 + n/2 + 1$

$R + DB = 11 - 18/2 + 2/2 + 1 = 4$

The molecule has one ring + three double bonds.

(b) $C_{12}H_{15}BrNPOS$

$R + DB = c - h/2 + n/2 + 1$

$R + DB = 12 - \dfrac{15+1}{2} + \dfrac{1+1}{2} + 1 = 6$

The molecule has two rings + four double bonds. Note that h includes H + Br, and n includes N + P. S, like O, does not contribute to the count.

(c)

[Structure: H₂C=CH–CH₂⁺ fragment]

A fragment in a mass spectrum

$C_3H_5^+$

$R + DB = c - h/2 + n/2 + 1$

$R + DB = 3 - 5/2 + 1 = 1\tfrac{1}{2}$ Huh?

We come out with a fraction instead of an integer because the species is an ion in which one C makes three bonds instead of four.

22-22. (a)

[Structure: chlorobenzene] C_6H_5Cl: $M^{+\bullet} = 112$

The pair of peaks at $m/z = 112$ and 114 strongly suggest that the molecule contains 1 Cl.

rings + double bonds $= c - h/2 + n/2 + 1 = 6 - 6/2 + 1 = 4$
                                        ↑
                              h includes H + Cl

Expected intensity of M+1 is $\underbrace{1.08(6)}_{\text{carbon}} + \underbrace{0.012(5)}_{\text{hydrogen}} = 6.54\%$

Observed intensity of M+1 = 69/999 = 6.9%

Expected intensity of M+2 = $\underbrace{0.005\ 8(6)(5)}_{\text{carbon}} + \underbrace{32.0(1)}_{\text{chlorine}} = 32.2\%$

Observed intensity of M+2 = 329/999 = 32.9%

The M+3 peak is the isotopic partner of the M+2 peak. M+3 contains ^{37}Cl plus either 1 ^{13}C or 1 ^{2}H. Therefore, the expected intensity of M+3 (relative to M+2) is $1.08(6) + 0.012(5) = 6.54\%$ of predicted intensity of M+2 = $(0.065\ 4)(32.2) = 2.11\%$ of $M^{+\bullet}$.

Observed intensity of M+3 is 21/999 = 2.1%.

(b)

[Structure: 1,4-dichlorobenzene] $C_6H_4Cl_2$: $M^{+\bullet} = 146$

The peaks at $m/z = 146$, 148, and 150 look like the isotope pattern from 2 Cl in Figure 22-7.

rings + double bonds $= c - h/2 + n/2 + 1 = 6 - 6/2 + 1 = 4$

Expected intensity of M+1 is $\underbrace{1.08(6)}_{\text{carbon}} + \underbrace{0.012(4)}_{\text{hydrogen}} = 6.53\%$

Observed intensity of M+1 = 56/999 = 5.6%

Expected intensity of M+2 = $\underbrace{0.005\ 8(6)(5)}_{\text{carbon}} + \underbrace{32.0(2)}_{\text{chlorine}} = 64.2\%$

Observed intensity of M+2 = 624/999 = 62.5%

The M+3 peak is the isotopic partner of the M+2 peak. M+3 contains 1 ^{35}Cl + 1 ^{37}Cl plus either 1 ^{13}C or 1 ^{2}H. Therefore, the expected intensity of M+3 (relative to M+2) is 1.08(6) + 0.012(4) = 6.53% of predicted intensity of M+2 = (0.065 3)(64.2) = 4.19% of M$^{+\cdot}$.

Observed intensity of M+3 is 33/999 = 3.3%.

Expected intensity of M+4 from $C_6H_4{}^{37}Cl_2$ is 5.11(2)(1) = 10.22% of M$^{+\cdot}$. The small contribution from $^{12}C_4{}^{13}C_2H_4{}^{35}Cl^{37}Cl$ is based on the predicted intensity of M+2. It is 0.005 8(6)(5) = 0.174% of 64.2% = 0.11%.

Total expected intensity of M+4 is 10.22% + 0.11% = 10.33% of M$^{+\cdot}$
Observed intensity = 99/999 = 9.9%.

Expected intensity of M+5 from $^{12}C_5{}^{13}CH_4{}^{37}Cl_2$ and $^{12}C_6H_3{}^{2}H^{37}Cl_2$ is based on the predicted intensity of M+4. M+5 should have 1.08(6) + 0.012(4) = 6.53% of M+4 = 6.53% of 10.33% = 0.67%.

Observed intensity = 5/999 = 0.5%.

(c) [benzene ring]—NH$_2$ C_6H_7N: M$^{+\cdot}$ = 93

The peak at m/z = 93 was chosen as the molecular ion, because it is the tallest peak in the cluster and it has plausible isotope peaks at M+1 and M+2. The significant peak at M−1 could be from loss of 1 H. The tiny stuff at M−2 and M−3 could be noise or, possibly, loss of more than 1 H.

With an odd mass, the nitrogen rule tells us that there are an odd number of N atoms in the molecule.

rings + double bonds = $c - h/2 + n/2 + 1 = 6 - 7/2 + 1/2 + 1 = 4$

Expected intensity of M+1 is 1.08(6) + 0.012(7) + 0.369(1) = 6.93%
carbon$$hydrogen$$nitrogen

Observed intensity of M+1 = 71/999 = 7.1%

Expected intensity of M+2 = 0.005 8(6)(5) = 0.17%

Observed intensity of M+2 = 2/999 = 0.2%

(d) $(CH_3)_2Hg$ C_2H_6Hg: M$^{+\cdot}$ = 228

There are six strong peaks in an unfamiliar pattern. Given that only elements from Table 22-1 are admissible, we notice that Hg has six significant isotopes. By convention, we take the lightest isotope, ^{198}Hg, for the molecular ion at m/z = 228. This leaves just 30 Da for the rest of the molecule, which could be composed of two methyl groups.

In computing rings + double bonds, we include Hg as a Group 6 atom (like O or S) because it makes 2 bonds.

rings + double bonds = $c - h/2 + n/2 + 1 = 2 - 6/2 + 1 = 0$.

The peak at m/z = 228 is M$^{+\bullet}$ = $(CH_3)_2{}^{198}Hg$.

Small peaks at m/z = 227 and 226 could arise from loss of one or two H atoms. If $(CH_3)_2{}^{198}Hg$ loses H atoms, then all the species at higher mass, such as $(CH_3)_2{}^{199}Hg$, will also lose H atoms. That is, each isotopic molecule is going to contribute some intensity to peaks of lower mass. It makes no sense for us to get too carried away with the analysis of the isotopic pattern, because each peak derives intensity from peaks at lower and higher mass. The peak at m/z = 229 is M+1, composed mainly of $(CH_3)_2{}^{199}Hg$, with some $(^{12}CH_3)(^{13}CH_3)^{198}Hg + {}^{12}C_2H_5{}^2H^{198}Hg$. Just considering Hg, the predicted intensity, based on M$^+$, is $\frac{16.87}{9.97} \times 100 = 169.2\%$ of M$^{+\bullet}$. The observed intensity is 215/130 = 165% of M$^{+\bullet}$. In this calculation, the fraction $\frac{16.87}{9.97}$ is the ratio of the abundances of ^{199}Hg to ^{198}Hg. The peak at m/z = 230 is M+2, composed mainly of $(CH_3)_2{}^{200}Hg$. The predicted ^{200}Hg isotopic intensity, based on M$^+$, is $\frac{23.10}{9.97} \times 100 = 231.7\%$ of M$^{+\bullet}$.

Observed intensity of M+2 = 291/130 = 224% of M$^{+\bullet}$.

Just considering Hg isotopes, we expect the peaks at M, M+1, M+2, M+3, M+4, and M+6 to have the ratios 9.97 : 16.87 : 23.10 : 13.18 : 29.86 : 6.87 = 1 : 1.69 : 2.32 : 1.32 : 2.99 : 0.69.

Observed intensity ratio = 1 : 1.65 : 2.24 : 1.29 : 2.81 : 0.64.

(e) CH_2Br_2: M$^{+\bullet}$ = 172

The three peaks at m/z = 172, 174 and 176, with approximate ratios 1 : 2 : 1 looks like the pattern from 2 Br atoms in Figure 22-7.

rings + double bonds = $c - h/2 + n/2 + 1 = 1 - 4/2 + 1 = 0$
↑
h includes H + Br

Expected intensity of M+1 is 1.08(1) + 0.012(2) = 1.10%
                              carbon   hydrogen

Observed intensity of M+1 = 12/531 = 2.3%. It is possible that this peak at m/z = 173 also has contributions from $CH^{79}Br^{81}Br$. We have no way to compute the intensity at m/z = 173 if some of this peak comes from

CH^{79}Br^{81}Br. Given this ambiguity, we will just compare the theoretical pattern for 2 Br atoms to the observed pattern:

Theoretical intensity of M+2 = 97.3(2) = 194.6%

Observed intensity of M+2 = 999/531 = 188%

Theoretical intensity of M+4 = 47.3(2)(1) = 94.6%

Observed intensity of M+4 = 497/531 = 93.6%

(f) 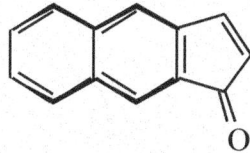 1,10-Phenanthroline, C$_{12}$H$_8$N$_2$: M$^{+\bullet}$ = 180

The strongest peak in the high-mass cluster is at m/z = 180, which could be the molecular ion. It has plausible isotopic peaks at 181 and 182. The significant peak at m/z = 179 could be from loss of 1 H.

The intensity ratio M+1/M$^{+\bullet}$ = 138/999 = 13.8%. We estimate that the number of C atoms is 13.8/1.08 = 12.8.

If the molecule contains 13 C atoms, the formula might be C$_{13}$H$_8$O, which would have 13 – 8/2 + 1 = 10 rings plus double bonds. The expected intensity of M+1 would be 1.08(13) + 0.012 (8) + 0.038(1) = 14.2%. The expected intensity of M+2 would be 0.005 8(13)(12) + 0.205(1) = 1.1%. Observed intensity of M+2 = 9/999 = 0.9%. The formula C$_{13}$H$_8$O fits the data and a conceivable structure is

If the molecule contains 12 C atoms, the formula might be C$_{12}$H$_4$O$_2$, which would have 12 – 4/2 + 1 = 11 rings plus double bonds. A molecule with this many rings + double bonds would be pretty implausible.

If the molecule contains nitrogen, it must contain an even number of N atoms because the molecule has an even mass. A possible formula is C$_{12}$H$_8$N$_2$, which would have 12 – 8/2 + 2/2 + 1 = 10 rings plus double bonds. This turns out to be the correct formula, and the structure is shown at the beginning of this answer. The predicted intensity of M+1 is 1.08(12) + 0.012(8) + 0.369(2) = 13.8%, which is exactly equal to the observed intensity. The expected intensity of M+2 is 0.005 8(12)(11) = 0.8%. Observed intensity = 0.9%.

(g) Ferrocene, $C_{10}H_{10}Fe$: $M^{+\bullet} = 186$

The strongest peak at high mass is at $m/z = 186$, which could be the molecular ion. It has plausible isotopic peaks at 187 and 188. Significant peaks at $m/z = 184$ and 185 could be from loss of H. Calling $M^{+\bullet} = 186$, we find the following ratios of peak intensities:

M–2	M–1	$M^{+\bullet}$	M+1	M+2
8.3	1.6	100	13.2	1.0

From the intensity ratio $M+1/M^{+\bullet} = 13.2\%$, we could estimate that the number of C atoms is 13.8/1.08 = 12.8. From this we could propose formulas like $C_{13}H_{14}O$ or $C_{12}H_{10}O_2$.

Alternatively, noting the significant intensity of M–2, we could propose that the molecule has Fe in it, which, in fact, it does. For the formula $C_{10}H_{10}Fe$, we predict that M–2 will have an intensity of $\frac{5.845}{91.754} \times 100 = 6.37\%$ of $M^{+\bullet}$, which is not terribly far from the observed value of 8.3%. The intensity at M+1 will have a contribution from ^{57}Fe and from ^{13}C and ^{2}H. The ^{57}Fe contribution is 2.119/91.754 = 2.31% of $M^{+\bullet}$. The other contributions are 1.08(10) + 0.012(10) = 10.92%. The total intensity predicted at M+1 is 13.23% and the observed intensity is 13.2%. The predicted intensity at M+2 is $\frac{0.282}{91.754} \times 100$ (from Fe) + 0.005 8(10)(9) (from C) = 0.83%, and the observed intensity is 1.0%.

22-23. The compound is dibromochloromethane:

212	$CH^{81}Br_2^{37}Cl$		94	$CH^{81}Br$
210	$CH^{81}Br_2^{35}Cl + CH^{79}Br^{81}Br^{37}Cl$		93	$C^{81}Br$
208	$CH^{79}Br^{81}Br^{35}Cl + CH^{79}Br_2^{37}Cl$		92	$CH^{79}Br$
206	$CH^{79}Br_2^{35}Cl$		91	$C^{79}Br$
175	$CH^{81}Br_2$		81	^{81}Br
173	$CH^{79}Br^{81}Br$		79	^{79}Br
171	$CH^{79}Br_2$		50	$CH^{37}Cl$
162	$^{81}Br_2$		49	$C^{37}Cl$
160	$^{79}Br^{81}Br$		48	$CH^{35}Cl$
158	$^{79}Br_2$		47	$C^{35}Cl$
131	$CH^{81}Br^{37}Cl$		37	^{37}Cl
129	$CH^{81}Br^{35}Cl + CH^{79}Br^{37}Cl$		35	^{35}Cl
127	$CH^{79}Br^{35}Cl$			

22-24. (a) For the formula $C_9H_4N_2Cl_6$,

rings + double bonds = $c - h/2 + n/2 + 1 = 9 - (4+6)/2 + 2/2 + 1 = 6$,

which agrees with the structure that has 2 rings + 4 double bonds.

(b) Nominal mass = integer mass of the species with the most abundant isotope of each of the constituent atoms. For $C_9H_4N_2Cl_6$,

nominal mass = $(9 \times 12) + (4 \times 1) + (2 \times 14) + (6 \times 35) = 350$.

(c) The sequence m/z 350, 315, 280, 245, and 210 corresponds to successive losses of 35 Da. A logical assignment is $C_9H_4N_2{}^{35}Cl_6^+$, $C_9H_4N_2{}^{35}Cl_5^+$, $C_9H_4N_2{}^{35}Cl_4^+$, $C_9H_4N_2{}^{35}Cl_3^+$, $C_9H_4N_2{}^{35}Cl_2^+$.

22-25. The CO_2 that we exhale is derived from oxidation of the food we eat. The chart shows that the group of plants called C_3 plants has less ^{13}C than the groups called C_4 and CAM plants. If the diet in the United States contains more C_4 and CAM plants and the diet in Europe contains more C_3 plants, then the difference in ^{13}C content of exhaled CO_2 might be explained.

22-26. (a) Mass of proton + electron = 1.007 276 467 + 0.000 548 580
= 1.007 825 047 Da. To the number of significant digits in Table 1, the masses of the proton and electron are equal to the mass of ^1H.

(b) mass of proton + neutron + electron
= 1.007 276 467 + 1.008 664 916 + 0.000 548 580 = 2.016 489 963 Da
mass of ^2H in table = 2.014 10 Da.
The ^2H atom is 0.002 39 Da lighter than the sum of its elementary particles.

(c) Mass difference = (0.002 39 Da) (1.660 5 × 10^{-27} kg/Da)
= 3.97 × 10^{-30} kg
$E = mc^2$ = (3.97 × 10^{-30} kg)(2.997 9 × 10^8 m/s)^2 = 3.57 × 10^{-13} J
mc^2 is the binding energy of a single nucleus. For a mole, the energy is
(3.57 × 10^{-13} J)(6.022 × 10^{23} mol^{-1}) = 2.15 × 10^{11} J/mol = 2.15 × 10^8 kJ/mol.

(d) Binding energy for atom = (13.6 eV)(1.602 18 × 10^{-19} J/eV) = 2.18 × 10^{-18} J
To convert to a mole: (2.18 × 10^{-18} J)(6.022 × 10^{23} mol^{-1}) = 1.31 × 10^6 J/mol = 1.31 × 10^3 kJ/mol. The ratio of the nuclear binding energy to the electron binding energy is (2.15 × 10^8 kJ/mol)/(1.31 × 10^3 kJ/mol)
= 1.64 × 10^5.

(e) $\dfrac{\text{nuclear binding energy}}{\text{bond energy}} \approx (2.15 \times 10^8 \text{ kJ/mol})/(400 \text{ kJ/mol}) = 5 \times 10^5$

22-27. ^{28}Si abundance $\equiv a = 0.922\,30$ ^{29}Si $\equiv b = 0.046\,83$ ^{30}Si $\equiv c = 0.030\,87$

$(a + b + c)^3 = a^3 + 3a^2b + 3a^2c + 3ab^2 + 6abc + 3ac^2 + b^3 + 3b^2c + 3bc^2 + c^3$

Silicon abundance	Polynomial expansion	Relative abundance	Composition	mass	Total intensity at mass
a =	a^3 =	= term value/a^3			
0.92230	0.784543	1.000000	28Si 28Si 28 Si	84	1
b =	3a^2b =				
0.04683	0.119506	0.152326	28Si 28Si 29 Si	85	0.152326
c =	3a^2c =				
0.03087	0.078778	0.100412	28Si 28Si 30 Si	86	0.108146
	3ab^2 =				
	0.006068	0.007734	28Si 29Si 29 Si	86	
	6abc =				
	0.008000	0.010197	28Si 29Si 30 Si	87	0.010328
	3ac^2 =				
	0.002637	0.003361	28Si 30Si 30 Si	88	0.003620
	b^3 =				
	0.000103	0.000131	29Si 29Si 29 Si	87	
	3b^2c =				
	0.000203	0.000259	29Si 29Si 30 Si	88	
	3bc^2=				
	0.000134	0.000171	29Si 30Si 30 Si	89	0.000171
	c^3 =				
	2.94178E-05	0.000037	30Si 30Si 30 Si	90	0.000037
Check: sum of terms in column B =					
	1				

mass:	84	85	86	87	88	89	90
intensity:	1	0.1523	0.1081	0.01033	0.00362	0.000171	0.000037

22-28. At m/z 100, 2 ppm = $(100)(2 \times 10^{-6}) = 0.000\,2$.

At m/z 20 000, 2 ppm = $(20\,000)(2 \times 10^{-6}) = 0.04$.

22-29. An ion of $m/z = 500$ accelerated through a potential difference of V volts attains a velocity $\sqrt{2zeV/m}$. We need mass in kg. 1 Da = (1/12) mass of ^{12}C, so:

$$\dfrac{(1/12) \times (12 \text{ g/mol})}{6.022\,140\,76 \times 10^{23} \text{ mol}^{-1}} = 1.660\,539\,067 \times 10^{-24} \text{ g} = 1.660\,539\,067 \times 10^{-27} \text{ kg}$$

500 Da $\times$ 1.661 $\times$ 10^{-27} kg/Da = 8.30 $\times$ 10^{-25} kg

Mass Spectrometry

$$\text{velocity} = \sqrt{\frac{2zeV}{m}} = \sqrt{\frac{2(1)(1.602 \times 10^{-19}\,\text{C})(5.00 \times 10^3\,\text{V})}{8.30 \times 10^{-25}\,\text{kg}}} = 4.39 \times 10^4\,\text{m/s}$$

To figure out the units, remember that work (joules) = $E \cdot q$ = volts·coulombs. So the product $C \times V = J = m^2 kg/s^2$. Putting these units into the square root gives velocity in m/s.

The time needed to travel 2.00 m is (2.00 m)/(4.39 × 10^4 m/s) = 45.6 μs. If we repeated a cycle each time this heaviest ion reaches the detector, we could collect 1/(45.6 μs) = 2.20 × 10^4 spectra per second.

If we double the mass in the square root to get up to m/z 1 000, the velocity decreases by $1/\sqrt{2}$ and the frequency goes down by $1/\sqrt{2}$ to 1.56 × 10^4 spectra per second.

22-30. We use the equation from the previous problem. The masses of the two ions are

m_{100} = 100 Da × 1.661 × 10^{-27} kg/Da = 1.661 × 10^{-25} kg

$m_{1\,000\,000}$ = 10^6 Da × 1.661 × 10^{-27} kg/Da = 1.661 × 10^{-21} kg

$$\text{velocity} = \sqrt{\frac{2zeV}{m}} = \sqrt{\frac{2(1)(1.602 \times 10^{-19}\,\text{C})(20.0 \times 10^3\,\text{V})}{m}}$$

= 1.964 × 10^5 m/s for 100 Da and 1.964 × 10^3 m/s for 1 000 000 Da

transit time = 2.00 m/velocity = 10.2 μs for 100 Da
= 1.02 ms for 1 000 000 Da

22-31. (a) $\lambda = \dfrac{kT}{(\sqrt{2}\sigma P)} = \dfrac{(1.38 \times 10^{-23}\,\text{J/K})(300\,\text{K})}{(\sqrt{2}(\pi(10^{-9}\,\text{m})^2)(10^{-5}\,\text{Pa}))} = 93\,\text{m}$

(The answer is in meters if you substitute m^2·kg·s^{-2} for J and kg·m^{-1}·s^{-2} for Pa from Table 1-2.)

(b) $\lambda = \dfrac{kT}{(\sqrt{2}\sigma P)} = \dfrac{(1.38 \times 10^{-23}\,\text{J/K})(300\,\text{K})}{(\sqrt{2}(\pi(10^{-9}\,\text{m})^2)(10^{-8}\,\text{Pa}))} = 93\,\text{km}$

22-32. Ions seen in electrospray usually existed in solution prior to electrospray. Atmospheric pressure chemical ionization creates ions in the corona discharge around the high voltage needle.

22-33. In collisional-induced dissociation, ions are accelerated through an electric field and directed into a region with a significant pressure of N$_2$ or Ar. Collisions transfer enough energy to break molecules into fragments. Collision-induced dissociation can be conducted at the entrance to the mass spectrometer or in a collision cell in the middle part of a tandem mass spectrometer.

22-34. A reconstructed total ion chromatogram shows the current from all ions above a selected mass displayed as a function of time. The chromatogram is "reconstructed" by summing the intensities for all observed values of m/z. The total ion chromatogram shows everything coming off the column. An extracted ion chromatogram displays detector current for just one or a few values of m/z as a function of time. The intensity displayed is extracted from the full mass spectrum recorded at each time interval. A selected ion chromatogram also displays detector current for just one or a small number of m/z values. However, for a selected ion chromatogram, the detector is not measuring the signal for all values of m/z in each time interval. The detector is set at just the desired values of m/z and collects that information for the whole time. The extracted ion chromatogram and the selected ion chromatogram are selective for an analyte of interest (plus anything else that gives a signal at the same m/z). The selected ion chromatogram has improved signal-to-noise ratio because all of the time is spent detecting signal at the selected mass.

22-35. In selected reaction monitoring, an ion of one m/z value is selected by the first mass separator. This ion is directed to a collision cell in which it undergoes collisional-induced dissociation to produce fragment ions. One of those fragment ions is then selected by a second mass separator and passed through to the detector. The detector is responding to just one product ion from the selected precursor ion. This technique is called MS/MS because it involves two consecutive mass separation steps. The signal/noise ratio is improved because noise (extraneous signals) is very low. There are few sources of the precursor ion other than the desired analyte, and it is unlikely that other precursor ions of the selected m/z can decompose to give the same product ion being monitored.

22-36. (a) Ibuprofen can readily dissociate to form a carboxylate anion, so I would choose the negative ion mode. It would be harder to form a cation.

The carboxylate anion should exist in neutral solution, since pK_a is probably around 4. In sufficiently acidic solution, the carboxylate will be protonated. I would use a neutral chromatography solvent to ensure a good supply of analyte anions.

(b) The formula of the molecular ion, M^-, is $C_{13}H_{17}O_2^-$. The intensity expected at M+1 is 1.08(13) + 0.012(17) + 0.038(2) = 14.32.
                          carbon   hydrogen   oxygen

22-37. The analysis follows the same steps as Table 22-3. The work is set out below. Peaks A and B give $n_A = 12$ and peaks H and I give $n_H = 19$. The combination of peaks G and H gives $n_G \approx 21$, which makes no sense and will be ignored. Assigning peaks A, B, C... as $n = 12, 13, 14...$ gives the sensible, constant molecular masses in the last column of the table. The mean value, disregarding peak G, is 15 126.

Analysis of electrospray mass spectrum of α-chain of hemoglobin

Peak	Observed m/z $\equiv m_n$	$m_{n+1} - 1.008$	$m_n - m_{n+1}$	Charge = n = $\dfrac{m_{n+1} - 1.008}{m_n - m_{n+1}}$	Molecular mass = $n \times (m_n - 1.008)$
A	1 261.5	1 163.6	96.9	$12.0_1 \approx 12$	15 126
B	1 164.6	—	—	[13]	15 127
C	—	—	—	[14]	—
D	—	—	—	[15]	—
E	—	—	—	[16]	—
F	—	—	—	[17]	—
G	834.3	796.1	37.2	21.4 [18]	14 999
H	797.1	756.2	39.9	$18.9_5 \approx 19$	15 126
I	757.2			[20]	15 124
				mean =	15 100
				mean without peak G =	15 126

22-38. The separation between adjacent peaks is 0.27, 0.28, 0.25, 0.24, 0.24, 0.24, 0.27, 0.23, 0.24, 0.25, 0.26, and 0.24 m/z, giving a mean value of 0.25_1. If species differing by 1 Da are separated by 0.25_1 m/z, the species must carry 4 charges ($z = 4$). The mass of the tallest peak must be 4(1 962.12) = 7 848.48 Da.

22-39. (a)

m/z	z	molecular ion mass (Da)	
38 152.7	2	2 × 38 152.7 = 76 305.4	
25 433.3	3	3 × 25 433.3 = 76 299.9	mean =
19 075.2	4	4 × 19 075.2 = 76 300.8	76 302.0 Da
15 260.4	5	5 × 15 260.4 = 76 302.0	

(b) R = n-$C_{12}H_{25}$ molecular ion mass = 3 × 27 243 = 81 729 Da
R = $CH_2CH_2C_6H_5$ molecular ion mass = 3 × 25 440 = 76 320
R = n-C_6H_{13} molecular ion mass = 3 × 24 876 = 74 628

Molecular mass of thiol ligands		Difference in SR mass from $SC_{12}H_{25}$
–$SC_{12}H_{25}$	201.395	0
–$SCH_2CH_2C_6H_5$	137.225	64.170
–SC_6H_{13}	117.235	84.159

The difference in nanoparticle mass between $Au_x(SC_{12}H_{25})_y$ and $Au_x(SCH_2CH_2C_6H_5)_y$ is 81 729 – 76 320 = 5 409 Da, corresponding to the difference in mass for y thiol ligands.

The difference in mass for one ligand is 64.169 Da.

The number of ligands is therefore y = 5 409 Da/64.17 Da = 84.3.

The difference in nanoparticle mass between $Au_x(SC_{12}H_{25})_y$ and $Au_x(SC_6H_{13})_y$ is 81 729 – 74 628 = 7 101 Da, corresponding to the difference in mass for y thiol ligands.

The difference in mass for one ligand is 84.159 Da.

The number of ligands is therefore y = 7 101 Da/84.159 Da = 84.4.

There appear to be y = 84 thiol ligands in each nanoparticle. The mass of thiol ligand is found in the first column below. Subtracting the mass of thiol from the mass of the nanoparticle gives the mass of Au. Dividing the mass of Au by the atomic mass of Au gives x = number of Au atoms in particle

Mass of thiol = 84 × FM of thiol	Au mass = nanoparticle mass – thiol mass	x = Au atoms = Au mass/atomic mass
84 × 201.395 = 16 917	81 729 – 16 917 = 64 812	329.1
84 × 137.225 = 11 527	76 320 – 11 527 = 64 793	329.0
84 × 117.235 = 9 848	74 628 – 9 848 = 64 780	328.9

Conclusion: the formula of the nanoparticle is $Au_{329}(SR)_{84}$

22-40. Expected intensities for 37:3, whose formula is $[MNH_4]^+ = C_{37}H_{72}ON$

$X + 1 = 0.012n_H + 1.08n_C + 0.369n_N + 0.038n_O$

$= 0.012(72) + 1.08(37) + 0.369(1) + 0.038(1) = 41.2\%$ (observed = 35.8%)

$X + 2 = (0.0058)n_C(n_C - 1) + 0.205n_O$

$= (0.0058)(37)(36) + 0.205(1) = 7.9\%$ (observed = 7.0%)

Expected intensities for 37:3, whose formula is $[MH]^+ = C_{37}H_{69}O$

$X + 1 = 0.012n_H + 1.08n_C + 0.038n_O$

$= 0.012(69) + 1.08(37) + 0.038(1) = 40.8\%$ (observed = 23.0%)

$X + 2 = (0.0058)n_C(n_C - 1) + 0.205n_O$

$= (0.0058)(37)(36) + 0.205(1) = 7.9\%$ (observed = 8.0%)

Expected intensities for 37:2, whose formula is $[MNH_4]^+ = C_{37}H_{74}ON$

$X + 1 = 0.012n_H + 1.08n_C + 0.369n_N + 0.038n_O$

$= 0.012(74) + 1.08(37) + 0.369(1) + 0.038(1) = 41.3\%$ (observed = 40.8%)

$X + 2 = (0.0058)n_C(n_C - 1) + 0.205n_O$

$= (0.0058)(37)(36) + 0.205(1) = 7.9\%$ (observed = 3.7%)

Expected intensities for 37:2, whose formula is $[MH]^+ = C_{37}H_{71}O$

$X + 1 = 0.012n_H + 1.08n_C + 0.369n_N + 0.038n_O$

$= 0.012(71) + 1.08(37) + 0.038(1) = 40.8\%$ (observed = 33.4%)

$X + 2 = (0.0058)n_C(n_C - 1) + 0.205n_O$

$= (0.0058)(37)(36) + 0.205(1) = 7.9\%$ (observed = 8.4%)

22-41. Selected reaction monitoring chooses the molecular ion $^{35}ClO_3^-$ (m/z 83) with mass separator Q1. In collision cell q2, this species could possibly undergo the following decomposition:

$$^{35}ClO_3^- \xrightarrow[\text{collisions}]{\text{high energy}} {}^{35}ClO_2^- + {}^{35}ClO^- + {}^{35}Cl^-$$

$m/z = 83 \qquad\qquad\qquad m/z = 67 \quad\; m/z = 51 \quad\; m/z = 35$

Quadrupole Q3 selects only $m/z = 67$. The measurement is specific for ClO_3^- because there are probably few compounds in water producing ions at $m/z = 83$, and *very few* of them are likely to decompose into $m/z = 67$. None of the species ClO_2^-, BrO_3^-, or IO_3^- can produce $m/z = 83$ to be selected by Q1.

334   Chapter 22

22-42. (a) The amine-reactive group leaves when the mass tag forms an amide bond to an amino group of a peptide. The balance group must have combinations of carbon and oxygen isotopes so that the ion $C_8H_{16}NO^+$ formed when the amide bond cleaves in the mass spectrometer entrance has nominal m/z 146. Mass separator Q1 isolates m/z 146 ion. The reporter mass tag left after collision-induced dissociation in q2 must have a unique value of m/z to be measured by a high-resolution mass spectrometer. The tag identifies the source of the labeled peptide. The relative amounts of each tag are equal to the relative amounts of peptide from each different source.

(b) Cleavage of the amide bond to the peptide leaves behind m/z 146 that is selected by Q1. The rationale for tandem mass tags is that all different tags are contained in the m/z 146 fraction isolated by Q1. In the q2 stage, these ions will break into reporter groups with different values of m/z that are measured by Q3.

(c) Here is one of several answers leading to nominal m/z 115, 117, and 118:

Structure 1: $C_7{}^{13}CH_{16}{}^{15}N^{18}O$, m/z 146.128
↓ q2 collision cell / Q3 mass separation
$C_7H_{16}{}^{15}N$, m/z 115.125

Structure 2: $C_5{}^{13}C_3H_{16}{}^{15}NO$, m/z 146.130
↓ q2 collision cell / Q3 mass separation
$C_5{}^{13}C_2H_{16}{}^{15}N$, m/z 117.131

Structure 3: $C_7{}^{13}CH_{14}D_2{}^{15}NO$, m/z 146.136
↓ q2 collision cell / Q3 mass separation
$C_6{}^{13}CH_{14}D_2{}^{15}N$, m/z 118.141

22-43. (a) Consider the term $A_xC_xm_x$, which applies to the unknown:

$$A_xC_xm_x = \left(\frac{\mu\text{mol isotope A}}{\mu\text{mol isotope A} + \mu\text{mol isotope B}}\right)\left(\frac{\mu\text{mol V}}{\text{g unknown}}\right)(\text{g unknown})$$

$$= \left(\frac{\mu\text{mol isotope A}}{\mu\text{mol isotope A} + \mu\text{mol isotope B}}\right)(\mu\text{mol V})$$

$$= \left(\frac{\mu\text{mol isotope A}}{\mu\text{mol isotope A} + \mu\text{mol isotope B}} \right) (\mu\text{mol isotope A} + \mu\text{mol isotope B})$$

= μmol isotope A in the unknown.

Similarly, $B_X C_X m_X$ = μmol isotope B in the unknown, $A_S C_S m_{S_X}$ = μmol isotope A in the spike, and $B_S C_S m_S$ = μmol isotope B in the unknown. When we mix the unknown and the spike, the isotope ratio is

$$R = \frac{\mu\text{mol A}}{\mu\text{mol B}} = \frac{\mu\text{mol A in unknown} + \mu\text{mol A in spike}}{\mu\text{mol B in unknown} + \mu\text{mol B in spike}}$$

$$= \frac{A_X C_X m_X + A_S C_S m_S}{B_X C_X m_X + B_S C_S m_S}.$$

(b) Cross-multiplying Equation A gives

$$R(B_X C_X m_X + B_S C_S m_S) = A_X C_X m_X + A_S C_S m_S$$

$$R B_X C_X m_X + R B_S C_S m_S = A_X C_X m_X + A_S C_S m_S$$

$$R B_X C_X m_X - A_X C_X m_X = A_S C_S m_S - R B_S C_S m_S$$

$$C_X = \frac{A_S C_S m_S - R B_S C_S m_S}{R B_X m_X - A_X m_X} = \left(\frac{C_S m_S}{m_X} \right) \left(\frac{A_S - R B_S}{R B_X - A_X} \right)$$

(c) A = ^{51}V and B = ^{50}V

Atom fractions in unknown: A_X = 0.997 5 and B_X = 0.002 5

Atom fractions in spike: A_S = 0.639 1 and B_S = 0.360 9

$$C_X = \left(\frac{C_S m_S}{m_X} \right) \left(\frac{A_S - R B_S}{R B_X - A_X} \right)$$

$$\left(\frac{(2.243\ 5\ \mu\text{molV/g})(0.419\ 46\text{g})}{0.401\ 67\ \text{g}} \right) \left(\frac{0.639\ 1 - (10.545)(0.360\ 9)}{(10.545)(0.002\ 5) - 0.997\ 5} \right)$$

= 7.639 4 μmol V/g

(d) $C_X = \left(\dfrac{(2.243\ 5\ \mu\text{molV/g})(0.419\ 46\ \text{g})}{0.401\ 67\ \text{g}} \right) \left(\dfrac{0.639\ 1 - (10.545)(0.360\ 9)}{(10.545)(0.002\ 5 - 0.997\ 5)} \right)$

$$= \left(\frac{(2.243\ 5\ \mu\text{mol V/g})(0.419\ 46\ \text{g})}{0.401\ 67\ \text{g}} \right) \left(\frac{0.639\ 1 - 3.805_7}{0.026_{36} - 0.997\ 5} \right)$$

$$= (2.342_9) \left(\frac{-3.166_6}{-0.971_1} \right) = 7.63_9\ \mu\text{mol V/g}$$

CHAPTER 23
INTRODUCTION TO ANALYTICAL SEPARATIONS

23-1. Three extractions with 100 mL are more effective than one with 300 mL.

23-2. (a) No, distribution constant is unaffected. Shaking increases rate at which partitioning reaches equilibrium, but does not affect position of equilibrium.

(b) Extraction requires mass transfer (i.e., physical movement) of solute between two phases. Shaking separatory funnel forms small droplets of one phase in the other (dispersion). Small droplets shorten the distance that solute must move to transfer from one phase to the other, thus enhancing mass transfer.

23-3. (a) Adjust the pH to 3 so the acid is in its neutral form (CH_3CO_2H), rather than its anionic form ($CH_3CO_2^-$).

(b) Heptane would be a greener alternative.

23-4. (a) The EDTA complex is anionic (AlY^-), whereas the 8-hydroxyquinoline complex is neutral (AlL_3).

(b) EDTA complex is anionic (AlY^-), so we need a hydrophobic cation such as $(C_8H_{17})_3NH^+$ to try to bring hydrophobic AlY^- into the organic solvent.

23-5. The complexation reaction $mHL + M^{m+} \rightleftharpoons ML_m + mH^+$ is driven to the right at high pH by consumption of H^+. This consumption increases the fraction of metal in the form ML_m, which is extracted into organic solvent.

23-6. ML_n is the form extracted into organic solvent. Formation of ML_n is favored by increasing formation constant (β_n). ML_n is also favored by increasing K_a, which increases the fraction of ligand in the form L^-. Increasing K_L decreases the fraction of ligand in the aqueous phase, thereby decreasing the formation of ML_n. Increasing $[H^+]$ decreases the concentration of L^- available for complexation.

23-7. When $pH > pK_{BH^+}$, predominant form is B, which goes into organic phase.
When $pH > pK_a$ for HA, predominant form is A^-, which goes into water.

23-8. (a) $S_{H_2O} \rightleftharpoons S_{heptane}$ $K_D = [S]_{heptane}/[S]_{H_2O} = 4.0$
$[S]_{heptane} = K_D[S]_{H_2O} = (4.0)(0.020\ M) = 0.080\ M$

(b) $\dfrac{\text{mol S in heptane}}{\text{mol S in } H_2O} = \dfrac{(0.080\ M)(10.0\ mL)}{(0.020\ M)(80.0\ mL)} = 0.50$

Introduction to Analytical Separations

23-9. Fraction remaining $= \left(\dfrac{V_1}{V_1 + K_D V_2}\right)^n = \left(\dfrac{80.0}{80.0 + (4.0)(10.0)}\right)^6 = 0.088$

23-10. (a) $D = \dfrac{\text{total conc of all forms in organic}}{\text{total conc of all forms in aqueous}}$

For a weak base, the neutral form B would be extracted into toluene, but the ionic form BH^+ would remain in water. Both B and BH^+ would be present in the aqueous phase.

$$D = \dfrac{[B]_{\text{toluene}}}{[B]_{H_2O} + [BH^+]_{H_2O}}$$

(b) D (distribution ratio) is the quotient of the <u>total</u> concentrations in the phases. Thus, it is a function of secondary equilibria such as acid/base reactions. K_D (distribution constant) is the quotient of the concentrations of the neutral species (B) in the phases. K_D is an intrinsic measure of the hydrophobicity of a compound.

(c) B has $K_b = 1.0 \times 10^{-5}$. Therefore, BH^+ has a $K_a = 1.0 \times 10^{-9}$ or $pK_a = 9.0$

$$D = \dfrac{K_D \cdot K_a}{K_a + [H^+]} = \dfrac{(50.0)(1.0 \times 10^{-9})}{(1.0 \times 10^{-9}) + (1.0 \times 10^{-8})} = 4.5$$

(d) D will be greater at pH 10 because a greater fraction of B is neutral.

23-11. From Equation 23-12, $D \approx \dfrac{[ML_n]_{\text{org}}}{[M^{n+}]_{\text{aq}}} = K_{\text{extraction}} \dfrac{[HL]^n_{\text{org}}}{[H^+]^n_{\text{aq}}}$

Comparing this result to Equation 23-13 gives $K_{\text{extraction}} = \dfrac{K_M \beta_n K_a^n}{K_L^n}$

Constant	Effect on $K_{\text{extraction}}$	Reason
K_M	increase	ML_n is more soluble in organic phase.
β_n	increase	Ligand binds metal more tightly and ML_n is the organic-soluble form.
K_a	increase	Ligand dissociates to L^- more easily, increasing ML_n formation.
K_L	decrease	HL is more soluble in organic phase, where it is not available to react with $M^{n+}(aq)$.

23-12. (a) $D = K_D[H^+]/([H^+] + K_a) = 3 \cdot 10^{-4.00}/(10^{-4.00} + 1.52 \times 10^{-5}) = 2.60$ at pH 4.00. Fraction remaining in water $= q = V_1/(V_1 + DV_2)$
$= 100/[100 + 2.60(25)] = 0.606$. Therefore, the molarity in water is $0.606 \, (0.10 \text{ M}) = 0.060 \, 6$ M.

The total moles of solute in the system is $(0.100 \text{ L})(0.10 \text{ M}) = 0.010$ mol. The fraction of solute in benzene is 0.394, so the molarity in benzene is $0.394(0.010 \text{ mol})/0.025 \text{ L} = 0.16$ M.

(b) At pH 10.0: $D = 1.97 \times 10^{-5}$, $q = 0.999 \, 995 \, 1$, molarity in water $= 0.10$ M, and molarity in benzene $= 2 \times 10^{-6}$ M.

(c) Toluene or xylenes would be greener alternatives with similar properties. Heptane is greener, but its properties are more different from benzene. Experiments would be necessary to determine if heptane or other greener alternatives would be effective extraction solvents.

23-13. $D = C/[H^+]^n$, where $C = K_M \beta_n K_a^n [HL]_{org}^n / K_L^n$

$D_1 = 0.01 = C/[H^+]_1^2$ and $D_2 = 100 = C/[H^+]_2^2$

$D_2/D_1 = 10^4 = [H^+]_1^2/[H^+]_2^2 \Rightarrow [H^+]_1/[H^+]_2 = 10^2 \Rightarrow \Delta pH = 2$ pH units

23-14. (a) Since there is so much more dithizone than copper, it is safe to say that $[HL]_{org} = 0.1$ mM.

$$D = \frac{K_M \beta_n K_a^n}{K_L^n} \frac{[HL]_{org}^n}{[H^+]_{aq}^n} = \frac{(7 \times 10^4)(5 \times 10^{22})(3 \times 10^{-5})^2}{(1.1 \times 10^4)^2} \frac{(1 \times 10^{-4})^2}{[H^+]^2}$$

$= 2.6 \times 10^4$ at pH 1 and 2.6×10^{10} at pH 4

(b) $q = V_1/(V_1 + DV_2) = 100/[100 + 2.6 \times 10^4 (10)] = 3.8 \times 10^{-4}$

23-15. (a) Substituting D for K_D in Equation 23-2 gives

$$q = \text{fraction remaining} = \frac{V_{aq}}{V_{aq} + DV_{org}}$$

$$\% \text{ extracted} = 100(1-q) = 100\left(1 - \frac{V_{aq}}{V_{aq} + DV_{org}}\right) = 100\left(\frac{DV_{org}}{V_{aq} + DV_{org}}\right)$$

(b) Spreadsheet for pH dependence of dithizone extraction

	A	B	C	D	E
1	K_M =	pH	[H^+]	D = Dist. Ratio	% extracted
2	70000	1.0	1.00E-01	2.60E-02	0.05
3	Beta =	2.0	1.00E-02	2.60E+00	4.95
4	5E+18	2.2	6.31E-03	6.54E+00	11.57
5	K_a =	2.4	3.98E-03	1.64E+01	24.73
6	0.00003	2.6	2.51E-03	4.13E+01	45.21
7	K_L =	2.8	1.58E-03	1.04E+02	67.46
8	11000	3.0	1.00E-03	2.60E+02	83.89
9	[HL]$_{org}$ =	3.2	6.31E-04	6.54E+02	92.90
10	0.00001	3.4	3.98E-04	1.64E+03	97.05
11	V_{org} =	3.6	2.51E-04	4.13E+03	98.80
12	2	3.8	1.58E-04	1.04E+04	99.52
13	V_{aq} =	4.0	1.00E-04	2.60E+04	99.81
14	100	5.0	1.00E-05	2.60E+06	100.00
15	C2 = 10^-B2				
16	D2 = ($A$2*$A$4*$A$6^2*$A$10^2)/($A$8^2*C2^2)				
17	E2 = 100*(D2*$A$12)/($A$14+D2*$A$12)				

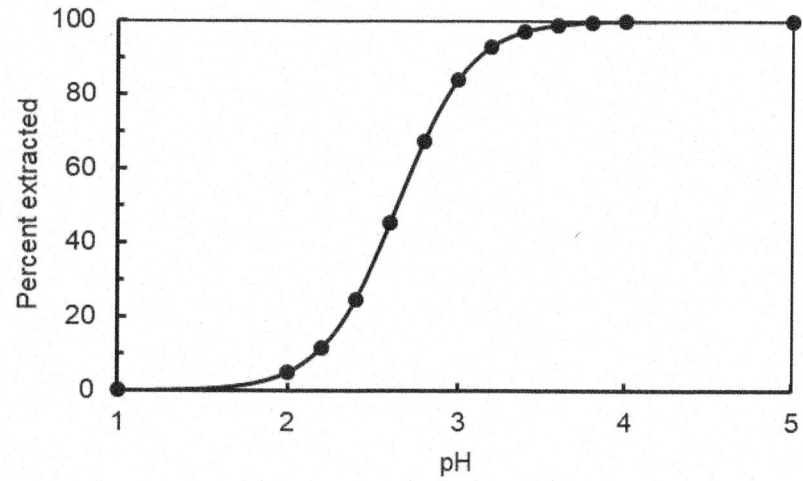

23-16.

	A	B	C	D	E	F
1	Liquid-liquid extraction efficiency					
2						
3	$V_2 =$	50	mL (volume of extraction solvent)			
4	$V_1 =$	50	mL (volume to be extracted)			
5	$K_D =$	2	(Distribution constant = $[S]_2/[S]_1$)			
6	Divide V_2 into n equal portions for n extractions					
7	Theoretical maximum fraction extracted = $1-q_{limit}$ = $1-\exp[-(V_2/V_1)K_D]$					
8		$1-q_{limit} =$	0.865	C8 = 1-EXP(-($B$3/$B$4*$B$5))		
9						
10		individual	q =	1 - q =	% of limiting	
11		extraction	fraction	fraction	fraction	
12	n	volume	remaining	extracted	extracted	
13	1	50	0.333	0.667	77.1	
14	2	25.0	0.250	0.750	86.7	
15	3	16.7	0.216	0.784	90.7	
16	4	12.5	0.198	0.802	92.8	
17	5	10.0	0.186	0.814	94.1	
18	6	8.3	0.178	0.822	95.1	
19	7	7.1	0.172	0.828	95.7	
20	8	6.3	0.168	0.832	96.2	
21	9	5.6	0.164	0.836	96.6	
22	10	5.0	0.162	0.838	97.0	
23	B13 = $B$3/A13 ⇐ V_2/n					
24	C13 = ($B$4/($B$4+B13/1*$B$5))^A13 ⇐ $q = [V_1/(V_1 + (V_2/n)K_D)]^n$					
25	E13 = (D13/$C$8)*100 ⇐ $[(1 - q)/(1 - q_{limit})] \times 100$					

The theoretical limit for fraction extracted is in cell C8. 95% of the theoretical fraction extracted is (0.95)(0.865) = 0.822. This fraction is exceeded with $n = 6$ equal extractions.

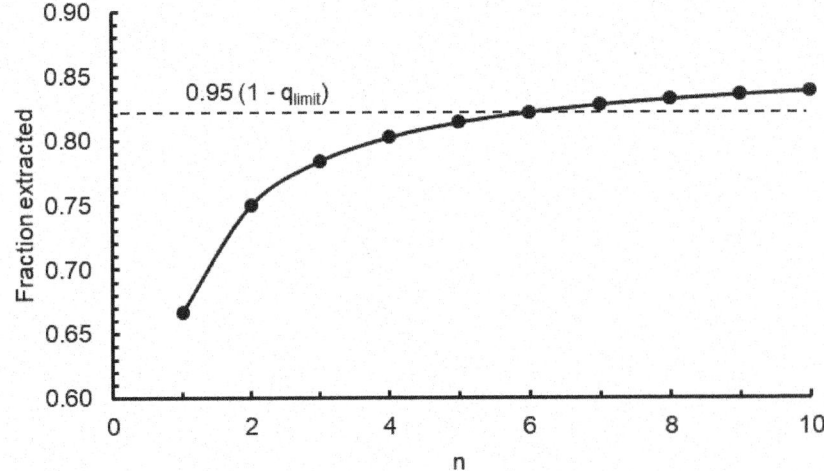

Introduction to Analytical Separations

23-17. (a) Water is denser and so would be bottom layer.

(b) If we mix 10 mL of pure octanol with 10 mL or pure water, the volumes of the two phases would not be 10 mL after some of each dissolves in the other. If each phase is initially saturated with the other, volumes do not change.

(c) The initial aqueous (600 μL) solution was diluted by adding 1 400 μL of water. Corrected absorbance is $A_i = (0.979)[(1\ 400\ \mu L + 600\ \mu L)/(600\ \mu L)] = 3.26_3$. The final solution was not diluted for measurement, so the final corrected absorbance is $(1.056)(1) = 1.056$.

$$K_{ow} = \frac{A_i V_w - A_f V_w}{A_f V_{oct}} = \frac{(3.26_3 - 1.056)10\ \text{mL}}{1.056 \cdot 10\ \text{mL}} = 2.09$$

(d) q = fraction left in aqueous phase = $\dfrac{V_w}{V_w + K_{ow} V_{oct}}$

Increasing V_{oct} decreases q and lowers final absorbance. Lowering V_{oct} increases q which raises final absorbance. Less octanol should be used.

23-18. 1-C, 2-D, 3-A, 4-E, 5-B

23-19. The larger the distribution constant, the greater the fraction of solute in the stationary phase, and the smaller the fraction that is moving through the column.

23-20. (a) $k = \dfrac{\text{time solute spends in stationary phase}}{\text{time solute spends in mobile phase}} = \dfrac{t_R - t_M}{t_M} = \dfrac{t_S}{t_M}$

(b) Fraction of time in mobile phase = $\dfrac{t_M}{t_M + t_S} = \dfrac{t_M}{t_M + kt_M} = \dfrac{1}{1+k}$

23-21. (a) Volume per cm of length = $\pi r^2 \times$ length

$$= \pi \left(\frac{0.461\ \text{cm}}{2}\right)^2 (1\ \text{cm}) = 0.167\ \text{mL}$$

mobile phase volume = $(0.390)(0.167\ \text{mL}) = 0.065\ 1$ mL per cm of column

linear velocity = $u_x = \dfrac{1.13\ \text{mL/min}}{0.065\ 1\ \text{mL/cm}} = 17.4$ cm/min

(b) $t_M = (10.3\ \text{cm}) / (17.4\ \text{cm/min}) = 0.592$ min

(c) $k = \dfrac{t_R - t_M}{t_M} \Rightarrow t_R = kt_M + t_M = 10(0.592) + 0.592 = 6.51$ min

23-22. (a) Linear velocity = (30.1 m)/(2.16 min) = 13.9 m/min

Inner diameter of open tube = 530 μm – 2(3.1 μm) = 523.8 μm
⇒ radius = 261.9 μm.

Volume = πr^2 × length = $\pi(261.9 \times 10^{-4} \text{ cm})^2(30.1 \times 10^2 \text{ cm})$ = 6.49 mL

Volume flow rate = F = (6.49 mL)/(2.16 min) = 3.00 mL/min

(b) $k = \dfrac{t_R - t_M}{t_M} = \dfrac{17.32 - 2.16}{2.16} = 7.02$

$k = t_S/t_M$ (where t_S = time in stationary phase)

Fraction of time in stationary phase = $\dfrac{t_S}{t_S + t_M} = \dfrac{kt_M}{kt_M + t_M} =$

$\dfrac{k}{k+1} = \dfrac{7.02}{7.02+1} = 0.875$

(c) Volume of coating ≈ $2\pi r$ × thickness × length
= $2\pi[(261.9 + 1.55) \times 10^{-4} \text{ cm}](3.1 \times 10^{-4} \text{ cm})(30.1 \times 10^2 \text{ cm})$ = 0.15$_4$ mL

$k = K_D \dfrac{V_S}{V_M} \Rightarrow 7.02 = K_D \dfrac{0.15_4 \text{ mL}}{6.49 \text{ mL}} \Rightarrow K_D = \dfrac{[S]_S}{[S]_M} = 29_5 = 3.0 \times 10^2$

23-23. (a) $\dfrac{\text{Large load}}{\text{Small load}} = \left(\dfrac{\text{Large column radius}}{\text{Small column radius}}\right)^2$

$\dfrac{100 \text{ mg}}{4.0 \text{ mg}} = \left(\dfrac{\text{Large column diameter}}{0.85 \text{ cm diameter}}\right)^2 \Rightarrow$ large column diameter = 4.2$_5$ cm

Use a 40-cm-long column with a diameter near 4.2 cm.

(b) The linear velocity should be the same. Since the cross-sectional area of the column is increased by a factor of 25, the volume flow rate should be increased by a factor of 25 ⇒ F = 5.5 mL/min.

(c) Volume of small column = πr^2 × length = $\pi(0.85/2 \text{ cm})^2(40 \text{ cm})$ = 22.7 mL

Mobile phase volume = 35% of column volume = 7.94 mL

Linear velocity = $\dfrac{40 \text{ cm}}{(7.94 \text{ mL})/(0.22 \text{ mL/min})}$

= 1.11 cm/min for both columns

23-24. $k = K_D \dfrac{V_S}{V_M} = 3\left(\dfrac{1}{5}\right) = \dfrac{3}{5}$ For K_D = 30, $k = 30\left(\dfrac{1}{5}\right) = 6.$

Introduction to Analytical Separations

23-25. (a) $t_M = 0.50$ min $\Rightarrow u_x = \dfrac{L}{u_x} = \dfrac{100 \text{ mm}}{0.50 \text{ min}} = 200$ mm/min $= 3.3$ mm/s

(b) $t_{R(X)} = 1.9$ min; $t'_{R(X)} = t_{R(X)} - t_M = 1.9$ min $- 0.5$ min $= 1.4$ min

$k_X = \dfrac{t_{R(X)} - t_M}{t_M} = \dfrac{1.9 - 0.5}{0.5} = 2.8$, or $k_X = \dfrac{t'_{R(X)}}{t_M} = \dfrac{1.4}{0.5} = 2.8$

$t_{R(S)} = 2.6$ min; $t'_{R(S)} = t_{R(S)} - t_M = 2.6$ min $- 0.5$ min $= 2.1$ min

$k_S = \dfrac{t_{R(S)} - t_M}{t_M} = \dfrac{2.6 - 0.5}{0.5} = 4.2$, or $k_S = \dfrac{t'_{S(X)}}{t_M} = \dfrac{2.1}{0.5} = 4.2$

(c) $\alpha = \dfrac{t'_{R(S)}}{t'_{R(X)}} = \dfrac{2.6 - 0.5}{1.9 - 0.5} = \dfrac{2.1 \text{ min}}{1.4 \text{ min}} = 1.5$, or $\alpha = \dfrac{k_S}{k_X} = \dfrac{4.2}{2.8} = 1.5$

23-26. $K_D = k \dfrac{V_M}{V_S}$

$k = \dfrac{t_R - t_M}{t_M} = \dfrac{433 - 63}{63} = 5.8_7$

$\dfrac{V_M}{V_S} = \dfrac{\pi r^2 \times \text{length}}{2\pi r \times \text{thickness} \times \text{length}} = \dfrac{(103)^2}{2(103.25) \times 0.5} = 102.8$

(In the numerator, r refers to the radius of the open tube $= \frac{1}{2}(207 - 1.0)$ μm $= 103$ μm. In the denominator, r is the radius at the center of the stationary phase, which is $103 + \frac{1}{2}(0.5) = 103.25$ μm.)

Therefore, distribution constant is $K_D = k \dfrac{V_M}{V_S} = 5.8_7(102.8) = 60_3 = 6.0 \times 10^2$.

Fraction of time in stationary phase $= \dfrac{t_S}{t_S + t_M} = \dfrac{kt_M}{kt_M + t_M} = \dfrac{k}{k+1}$

$= \dfrac{5.8_7}{5.8_7 + 1} = 0.85_4 = 0.85$

23-27. (a) Column 2.   (b)   Column 3.   (c)   Column 3.

(d) Column 4 since it provides the highest resolution.

23-28. Resolution = $\dfrac{\sqrt{N}}{4} \dfrac{(\alpha-1)}{\alpha} \dfrac{k_2}{(1+k_2)}$. Plots relating the individual factors N, α, and k_2 versus resolution are shown below.

	A	B	C	D	E	F	G	H
1	N	$\sqrt{N}$		α	$(\alpha-1)/\alpha$		k_2	$k_2/(1+k_2)$
2	0	0.0		1.0	0.00		0	0.000
3	1000	31.6		1.1	0.09		0.5	0.333
4	2000	44.7		1.2	0.17		1	0.500
5	4000	63.2		1.3	0.23		3	0.750
6	6000	77.5		1.4	0.29		6	0.857
7	8000	89.4		1.5	0.33		10	0.909
8	10000	100.0		1.6	0.38		15	0.938
9		B2 = SQRT(A2)			E2=(D2-1)/D2			H2=(G2/(1+G2))

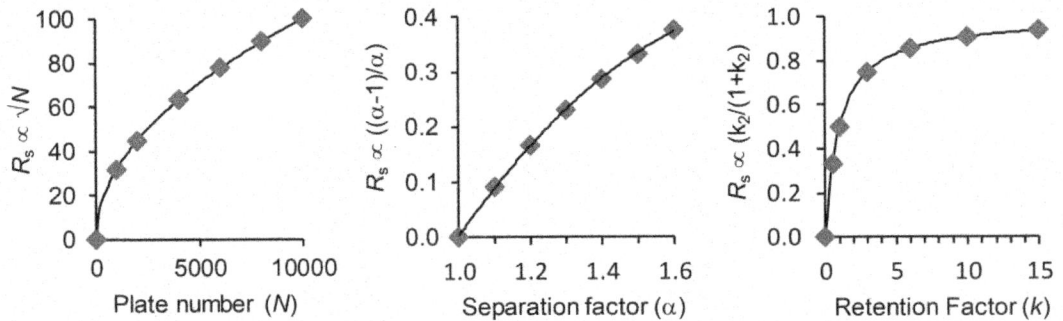

Key observations:

- Resolution increases with plate number, but gradually due to $\sqrt{N}$.
- Resolution increases near proportionally to the amount that the separation factor is greater than 1.
- Resolution increases dramatically with a little retention (e.g., up to $k_2=2$), and then plateaus at large ($k_2 > 10$) retention factors. Large retention factors mean long separation times. Therefore chromatography method development tries to get all compounds in the mixture to have $0.5 \leq k \leq 10$.

23-29. (a) Based on the Stokes-Einstein equation, diffusion coefficients are inversely dependent on the radius of the diffusing molecule. CH_3OH (FM = 32) is smaller than sucrose (FM = 342), and so would be expected to diffuse faster.

(b) Diffusion coefficients in gases are about 10^4 times those in liquids. The diffusion coefficient of H_2O in water is 2.3×10^{-9} m^2/s. The diffusion coefficient of H_2O in air is approximately 10^{-5} m^2/s.

Introduction to Analytical Separations 345

23-30. (a)

Retention factor (k_1)	0	1	2	5	10
Resolution	0	1.14	1.52	1.89	2.07
Retention time t_{R2} (min)	~1.7	~3.7	~5.6	~11.3	~20.9

(b)

Separation factor (α)	1.00	1.10	1.20
Resolution	0.00	1.14	2.08
Retention time (min)	~3.5	~3.7	~3.8$_5$

(c)

Plate number (N)	2 000	4 000	6 000	8 000	10 000	12 000	17 000
Resolution	0.51	0.72	0.88	1.02	1.14	1.24	1.48

Retention time is constant at 3.7 min.

(d)

Plate number (N)	2 000	6 000	17 000
Peak maximum (mAU)	~41	~64	~106

Baseline absorbance ~ 2 mAU

Peak height (mAU)	~39	~62	~104

Higher plate number gives sharper peaks with greater maximum signal. Changes in plate number will affect quantitation using peak height.

23-31. (a) Column 1 (sharper peaks)

(b) Column 2 (large plate height (H) means lower plate number (N) means broader peaks)

(c) Column 1 (less overlap between peaks because they are sharper)

(d) Neither (separation factor, $\alpha = (t'_{RB}/t'_{RA})$ is equal for the two columns

(e) Compound B (longer retention time)

(f) Compound B (longer retention time means greater affinity for stationary phase)

(g) $k_A = \dfrac{8.0-1.3}{1.3} = 5.2$

(h) $k_B = \dfrac{10.0-1.3}{1.3} = 6.7$

(i) $\alpha = \dfrac{10.0-1.3}{8.0-1.3} = 1.3$

23-32. (a) Increasing flow rate decreases the time for longitudinal diffusion to occur, thereby narrowing the bands, decreasing H, increasing N, and increasing resolution.

(b) Decreasing flow rate will narrow the bands and increase resolution.

(c) Multiple flow paths broadening depends on the geometry of the column and the quality of the packing, but does not depend on flow rate. So H, N, and resolution will remain unchanged.

23-33. Plate height described by the van Deemter equation depends on linear velocity. The linear velocity at which solution goes past the stationary phase determines how completely the equilibrium between the two phases is established. The time for equilibration determines the size of the mass transfer term (Cu_x). The extent of longitudinal diffusion depends on the time spent in the mobile phase, which is inversely proportional to linear velocity (B/u_x).

For a constant column diameter, volume flow rate is proportional to linear velocity, and so the terms flow rate and linear velocity are often used interchangeably. For columns of different diameter, different flow rates are needed to get a given linear velocity, and thus a given plate height. So most correctly, plate height depends on linear velocity, not volume flow rate.

23-34. Smaller plate height gives less band spreading: 0.1 mm

23-35. Diffusion coefficients of gases are 10^4 times greater than those of liquids. Therefore, longitudinal diffusion occurs much faster in gas chromatography than in liquid chromatography.

23-36. The optimum linear velocity is that at which longitudinal diffusion (B/u_x) and mass transfer broadening (Cu_x) are equal. The smaller the particle size, the more rapid is mass transfer between mobile and stationary phases, and so the smaller is Cu_x. Hence the optimum linear velocity will be at a higher linear velocity so that B/u_x is decreased to match the smaller Cu_x.

23-37. Minimum plate height is at 1.3 mm/s.

23-38. 1-C_M, 2-A, 3-EC, 4-B, 5-C_S

23-39. Silanization caps hydroxyl groups to which strong hydrogen bonding can occur.

23-40. Isotherms and band shapes are given in Figure 23-21. When a column is overloaded, the excessive amount of solute injected alters the nature of the stationary phase.

In gas chromatography overloading, fronting is observed as the solute becomes more soluble in the stationary phase as solute concentration increases. Consider injection of high concentrations of an alcohol such as hexanol onto a nonpolar stationary phase like polydimethylsiloxane. Partitioning of hexanol into polydimethylsiloxane makes the stationary phase more polar, and thus more attractive for partitioning of more hexanol. This change in polarity of the solute zone in the stationary phase results in greater retention of higher concentrations of hexanol, and gives a non-Gaussian fronting shape.

In liquid chromatography overloading, tailing is observed as the excess injected solute saturates a portion of the retention sites on the column. As more excess solute is injected, more retention sites are saturated and the peak maximum becomes less retained, and gives a non-Gaussian tailing shape.

23-41. (a) Injection of a larger volume of sample will require a plug width (Δt). This increases the variance caused by injection ($=(\Delta t)^2/12$), which will increase the peak width and thus lower the resolution.

(b) With 5.0 mg, the column may be overloaded. That is, the quantity of solute may be too great for the capacity of the stationary phase. Depending on the nature of the interaction, this could lead to the upper nonlinear isotherm in Figure 23-21 which causes fronting or to the lower nonlinear isotherm which causes tailing. Either type of overload broadens bands and decreases resolution.

23-42. $\sigma = \sqrt{2Dt}$ says that the standard deviation of the band is proportional to $\sqrt{t}$. Here is what we know of the rate of diffusion:

time	standard deviation
t_1	$\sigma_1 = 1$
$t_2 = t_1 + 20$	$\sigma_2 = 2$
$t_3 = t_1 + 40$	$\sigma_3 = ?$

From the bandwidths at times t_1 and t_2, we can write

$$\frac{\sigma_2}{\sigma_1} = \sqrt{\frac{t_2}{t_1}} \Rightarrow \frac{2}{1} = \sqrt{\frac{t_1 + 20}{t_1}} \Rightarrow t_1 = \frac{20}{3} \text{ min}$$

For time t_3: $\dfrac{\sigma_3}{\sigma_1} = \sqrt{\dfrac{t_3}{t_1}} \Rightarrow \dfrac{\sigma_3}{1} = \sqrt{\dfrac{\frac{20}{3}+40}{\frac{20}{3}}} \Rightarrow \sigma_3 = 2.65$ mm

23-43. (a) $N = \dfrac{5.55\, t_R^2}{w_{1/2}^2} = \dfrac{5.55\,(9.0\text{ min})^2}{(2.0\text{ min})^2} = 1.1_2 \times 10^2$ plates

(b) $(10\text{ cm})/(1.1_2 \times 10^2 \text{ plates}) = 0.89$ mm

23-44. (a) Measurement with a ruler gives $B/A = 2.1$

(b) $N = \dfrac{41.7(t_R/w_{0.1})^2}{(B/A)+1.25} = \dfrac{41.7(900\text{ s}/44\text{ s})^2}{2.1+1.25} = 5.2 \times 10^3$ plates

(c) To use the equation $N = (t_R/\sigma)^2$, we need to find the standard deviation of the peak. The width at 1/10 height is $22 + 22 = 44$ s, which we are told is equal to 4.297σ. Therefore, $\sigma = (44\text{ s})/4.297 = 10._{24}$ s. $N = (t_R/\sigma)^2 = (900\text{ s}/10._{24}\text{ s})^2 = 7.7_2 \times 10^3$ plates.

The equation for an asymmetric peak from (b) gives
$N = \dfrac{41.7(t_R/w_{0.1})^2}{(B/A)+1.25} = \dfrac{41.7(900\text{ s}/44\text{ s})^2}{(22\text{ s}/22\text{ s})+1.25} = 7.7_5 \times 10^3$ plates

23-45. (a) Resolution $= \dfrac{\Delta t_R}{w_{av}} = \dfrac{5\text{ min}}{6\text{ min}} = 0.83$. This is most like the diagram for a resolution of 0.75.

(b) At resolution $=1.5$ the valley between the peaks does not quite return to the baseline signal. The baseline separation described has resolution > 1.5.

23-46. Since $w = 4V_R/\sqrt{N}$, w is proportional to V_R (if N is constant).
$w_2/w_1 = V_2/V_1 = 127/49 \Rightarrow w_2 = (127/49)(4.0) = 10.4$ mL

23-47. $w_{1/2} = (39.6\text{ s}/60\text{ s/min}) \times 0.66$ mL/min $= 0.436$ mL

$\sigma_{obs}^2 = \left(\dfrac{w_{1/2}}{2.35}\right)^2 = \left(\dfrac{0.436\text{ mL}}{2.35}\right)^2 = 0.034\,4\text{ mL}^2$

$\sigma_{injection}^2 = \dfrac{\Delta V_{injection}^2}{12} = \dfrac{(0.40\text{ mL})^2}{12} = 0.013\,3\text{ mL}^2$

$\sigma_{detector}^2 = (\Delta V)_{detector}^2/12 = (0.25\text{ mL})^2/12 = 0.005\,2\text{ mL}^2$

From Table 23-1, sucrose has $D = 0.52 \times 10^{-9}$ m^2/s $= 3.12 \times 10^{-4}$ cm^2/min

$$\sigma^2_{tubing} = \frac{\pi d_t^4 l_t F}{384 D} = \frac{\pi (0.050 \text{ cm})^4 (20 \text{ cm})(0.66 \text{ cm}^3/\text{min})}{384 (3.12 \times 10^{-4} \text{cm}^2/\text{min})}, = 0.002\,2 \text{ cm}^6$$

$$= 0.002\,2 \text{ mL}^2$$

$$\sigma^2_{obs} = \sigma^2_{column} + \sigma^2_{injection} + \sigma^2_{detector} + \sigma^2_{tubing}$$

$$0.034\,4 \text{ mL}^2 = \sigma^2_{column} + 0.013\,3 \text{ mL}^2 + 0.005\,2 \text{ mL}^2 + 0.002\,2 \text{ mL}^2$$

$$\Rightarrow \sigma_{column} = 0.117 \text{ mL}$$

$$w_{1/2} = 2.35\, \sigma_{column} = 0.275 \text{ mL}$$

$$w_{1/2} = (0.275 \text{ mL})(60 \text{ s/min})/0.66 \text{ mL/min} = 25 \text{ s}$$

23-48. $\alpha = \dfrac{t'_{R2}}{t'_{R1}} = \dfrac{k_2}{k_1} = \dfrac{K_{D2}}{K_{D1}} = \dfrac{18}{15} = 1.2_0$

$$k_2 = K_{D2} \frac{V_S}{V_M} = 18 \left(\frac{1}{3.0}\right) = 6.0$$

$$\text{Resolution} = \frac{\sqrt{N}}{4} \frac{(\alpha-1)}{\alpha} \frac{k_2}{(1+k_2)}$$

$$1.5 = \frac{\sqrt{N}}{4} \frac{(1.2_0-1)}{1.2_0} \frac{6.0}{(1+6.0)} = \frac{\sqrt{N}}{4} (0.14_3) \Rightarrow 1.8 \times 10^3 \text{ plates}$$

23-49 (a) $N_X = \dfrac{16 t_R^2}{w^2} = \dfrac{16(1.9_0 \text{ min})^2}{(0.3_5 \text{ min})^2} = 470$

$N_S = \dfrac{16 t_R^2}{w^2} = \dfrac{16(2.6_0 \text{ min})^2}{(0.4_5 \text{ min})^2} = 530$

(b) Resolution = 1.5

(c) Resolution = $\dfrac{\Delta t_R}{w_{av}}$

$= \dfrac{(2.6-1.9)}{\left(\dfrac{0.3_5 + 0.4_5}{2}\right)} = 1.8$

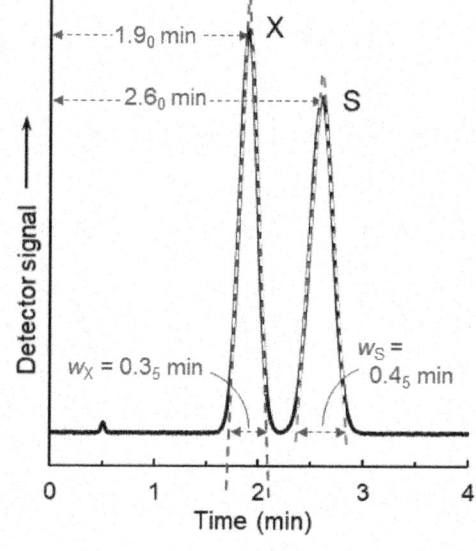

23-50. (a) Resolution = $2.0 = \dfrac{\sqrt{N}}{4} \dfrac{(1.05-1)}{1.05} \dfrac{5.0}{(1+5.0)} \Rightarrow N = 4.1 \times 10^4$ plates

(b) Resolution = $2.0 = \dfrac{\sqrt{N}}{4} \dfrac{(1.10-1)}{1.10} \dfrac{5.0}{(1+5.0)} \Rightarrow N = 1.1 \times 10^4$ plates

(c) Resolution = $2.0 = \dfrac{\sqrt{N}}{4} \dfrac{(1.05-1)}{1.05} \dfrac{10.0}{(1+10.0)} \Rightarrow N = 3.4 \times 10^4$ plates

(d) N can be increased by increasing the column length ($N \propto L$), by decreasing the capillary radius (r) in an open tubular column, or by adjusting the flow rate so that H is minimized. α can be increased by changing solvent in liquid chromatography or by changing the stationary phase in both liquid and gas chromatography. k_2 can be increased by increasing the volume of the stationary phase in an open tubular gas chromatography column or increasing the surface area in a packed liquid chromatography column.

23-51. (a) C_6HF_5: $t'_R = 12.98 - 1.06 = 11.92$ min. $k = 11.92/1.06 = 11.25$

C_6H_6: $t'_R = 13.20 - 1.06 = 12.14$ min. $k = 12.14/1.06 = 11.45$

(b) $\alpha = 11.45/11.25 = 1.018$

(c) $w_{1/2}(C_6HF_5) = 0.124$ min; $w_{1/2}(C_6H_6) = 0.121$ min

If your $w_{1/2}$ are larger, you may have measured half height from the x-axis, rather than from baselines drawn under the peaks.

C_6HF_5: $N_1 = \dfrac{5.55\, t_R^2}{w_{1/2}^2} = \dfrac{5.55(12.98)^2}{0.124^2} = 6.08 \times 10^4$ plates

Plate height = $\dfrac{30.0 \text{ m}}{6.08 \times 10^4 \text{ plates}} = 0.493$ mm

C_6H_6: $N_2 = \dfrac{5.55(13.20)^2}{0.121^2} = 6.60 \times 10^4$ plates

Plate height = $\dfrac{30.0 \text{ m}}{6.60 \times 10^4 \text{ plates}} = 0.455$ mm

(d) $w(C_6HF_5) = 0.220$ min; $w(C_6H_6) = 0.239$ min

C_6HF_5: $N_1 = \dfrac{16\, t_R^2}{w^2} = \dfrac{16(12.98)^2}{0.220^2} = 5.57 \times 10^4$ plates

Introduction to Analytical Separations

$$C_6H_6: \quad N_2 = \frac{16(13.20)^2}{0.239^2} = 4.88 \times 10^4 \text{ plates}$$

(e) $\quad \text{Resolution} = \dfrac{\Delta t_R}{w_{av}} = \dfrac{13.20 - 12.98}{0.229} = 0.96$

(f) $\quad N_{av} = \sqrt{(5.57 \times 10^4)(4.88 \times 10^4)} = 5.21 \times 10^4 \text{ plates}$

$$\text{Resolution} = \frac{\sqrt{N}}{4} \frac{(\alpha - 1)}{\alpha} \frac{k_2}{(1 + k_2)} = \frac{\sqrt{5.21 \times 10^4}}{4} \frac{(1.018 - 1)}{1.018} \frac{11.45}{(1 + 11.45)}$$

$= 0.93$

23-52. Concentration ($c = $ mol/m^3) is computed as a function of distance (x) and time (t) from the center of the band with the equation

$$c = \frac{m}{\sqrt{4\pi D t}} \, e^{-x^2/(4Dt)}$$

where D is the diffusion coefficient (m^2/s) and initial concentration per unit *area* (m) = 10 nmol/(1.96 × 10^{-3} m^2) = 5.09 × 10^{-6} mol/m^2. Diffusion will be symmetric about the origin. Only diffusion in the positive direction is computed below for $t = 60$ s. Other conditions in the graphs are obtained by changing t and the diffusion coefficient D.

	A	B	C
1	Diffusion problem		
2		x (m)	c (mol/m^3)
3	moles =	0.0000	4.637E-03
4	1.00E-08	0.0001	4.518E-03
5	diameter (m) =	0.0002	4.178E-03
6	0.05	0.0003	3.668E-03
7	cross-sectional area (m^2) =	0.0004	3.057E-03
8	0.001963495	0.0005	2.418E-03
9	m (mol/m^2) =	0.0006	1.816E-03
10	5.093E-06	0.0007	1.294E-03
11	D (m^2/s) =	0.0008	8.758E-04
12	1.6E-09	0.0009	5.625E-04
13	t (s) =	0.0010	3.430E-04
14	60	0.0012	1.090E-04
15		0.0014	2.815E-05
16	A10 = A4/A8	0.0016	5.901E-06
17	C3=($A$10/(SQRT(4*PI()*$A$12*$A$14)))	0.0018	1.004E-06
18	*EXP(-(B3^2)/(4*$A$12*$A$14))	0.0020	1.388E-07

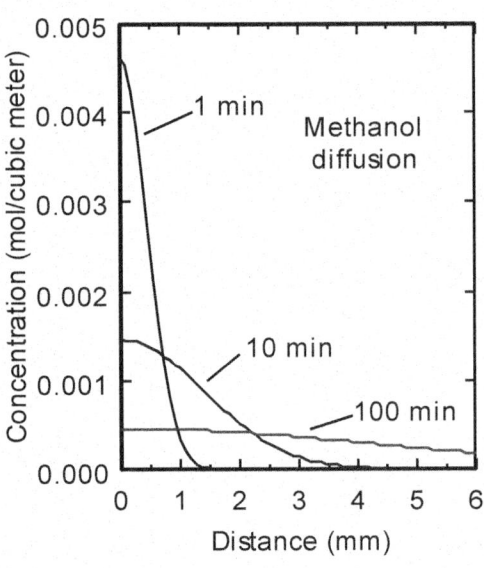
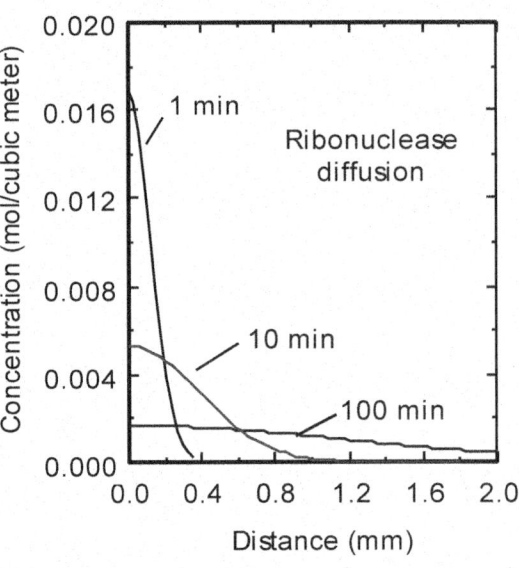

23-53. Plate height = $H_{\text{longitudinal diffusion}} + H_{\text{mass transfer}} = \dfrac{B}{u_x} + (C_S + C_M)u_x$

$$= \dfrac{2D_M}{u_x} + \left(\dfrac{2kd_f^2}{3(k+1)^2 D_S} + \dfrac{1+6k+11k^2 r^2}{24(k+1)^2 D_M}\right)u_x$$

Parameters for 0.25 μm thick stationary phase:

$D_M = 1.0 \times 10^{-5}$ m²/s $\qquad D_S = 1.0 \times 10^{-9}$ m²/s

$d_f = 2.5 \times 10^{-7}$ m $\qquad r = 1.25 \times 10^{-4}$ m $\qquad k = 10$

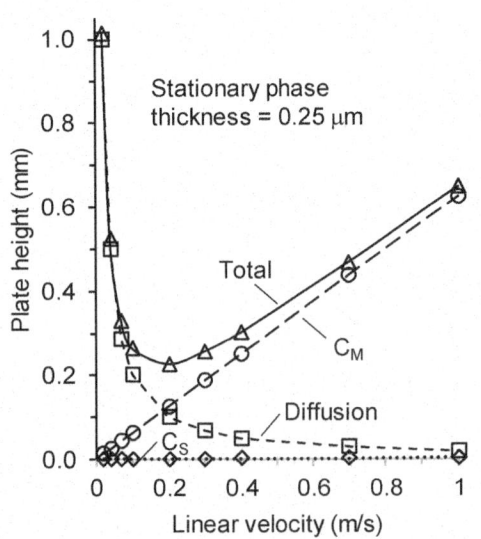
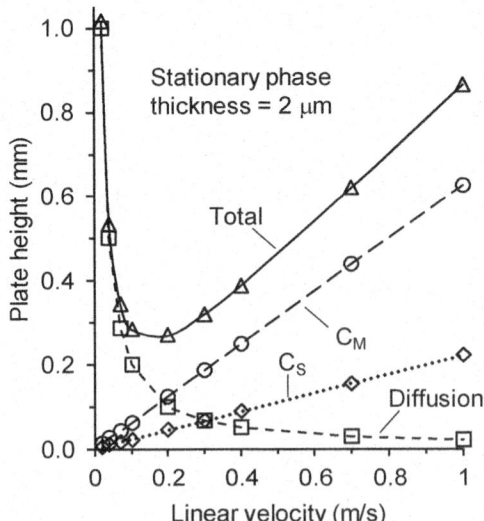

Introduction to Analytical Separations

	A	B	C	D	E	F
1	Plate height calculation for 0.25-μm-thick stationary phase					
2				H(mass transfer)		
3	D_M (m²/s) =	u_x (m/s)	H(diffusion)	C_S term	C_M term	H (total)
4	1.00E-05	0.02	1.0000	0.0001	0.0125	1.0126
5	D_S (m²/s) =	0.04	0.5000	0.0001	0.0250	0.5251
6	1.00E-09	0.07	0.2857	0.0002	0.0437	0.3297
7	k =	0.1	0.2000	0.0003	0.0625	0.2628
8	10	0.2	0.1000	0.0007	0.1249	0.2256
9	d_f (m) =	0.3	0.0667	0.0010	0.1874	0.2551
10	2.50E-07	0.4	0.0500	0.0014	0.2499	0.3012
11	r (m) =	0.7	0.0286	0.0024	0.4373	0.4683
12	1.25E-04	1	0.0200	0.0034	0.6247	0.6481
13	C4 = (2*$A$4/B4)*10^3					
14	D4 = 2*$A$8*$A$10^2*B4/(3*($A$8+1)^2*$A$6)*10^3					
15	E4=(1+6*$A$8+11*$A$8^2)*$A$12^2*B4/(24*($A$8+1)^2*$A$4)*10^3					
16	F4 = C4 + D4 + E4					

For stationary phase thickness = 0.25 μm, plate height contribution from mass transfer in the stationary phase is negligible, as shown in the left graph. If the stationary phase is 2.0 μm thick, plate height from mass transfer in the stationary phase is not negligible, but it is still less than plate height from mass transfer in the mobile phase. C_S and total plate height in the right graph are greater than in the first graph. C_M and longitudinal diffusion terms are unaffected.

23-54. Equation 4-3 shows that the form of a Gaussian curve is $y = Ae^{-(x-x_0)^2/2\sigma^2}$, where A is a constant proportional to the area under the curve, x_0 is the abscissa of the center of the peak, and σ is the standard deviation. We can arbitrarily let $\sigma = 1$, which means that the width at the base ($w = 4\sigma$) is 4. A peak with an area of 1 centered at the origin is $y = 1*e^{-(x)^2/2}$. A curve of area 4 is $y = 4*e^{-(x-x_0)^2/2}$. The resolution is $\Delta x/w$. For a resolution of 0.5, $\Delta x = 0.5*w = 2$. That is, the second peak is centered at $x = 2$ if the resolution is 0.5. Its equation is $y = 4*e^{-(x-2)^2/2}$. Similarly, for a resolution of 1, $\Delta x = 1*w = 4$ and the second peak is centered at $x = 4$. For a resolution of 2, the second peak is centered at $x = 8$. The equations of the curves plotted below are:

Resolution = 0.5: $\quad y = 1*e^{-(x)^2/2} + 4*e^{-(x-2)^2/2}$

Resolution = 1: $\quad y = 1*e^{-(x)^2/2} + 4*e^{-(x-4)^2/2}$

Resolution = 2: $\quad y = 1*e^{-(x)^2/2} + 4*e^{-(x-8)^2/2}$

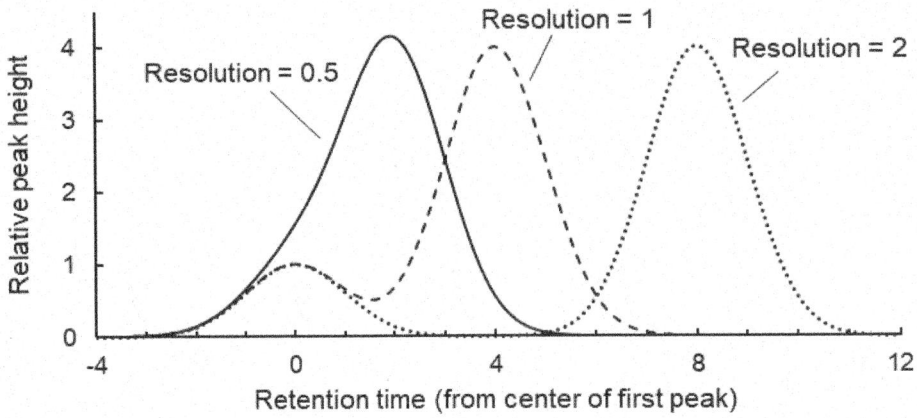

23.55 (a) 0.5 mL/min

(b) Plate number N decreases and plate height H increases. At high flow there is insufficient time for mass transfer between mobile and stationary phase (C-term of van Deemter). Retention times decrease with faster flow.

(c) Plate number N decreased and plate height H increased as flow slowed. At slow flow rates, there is more time for solute to diffuse and broaden—a process known as longitudinal diffusion (B-term of van Deemter). Retention times increase with slower flow rates.

(d) Analyte passing through a packed column finds multiple paths of different length through the column, which causes broadening. Multipath band broadening (A-term of van Deemter) does not depend on flow rate.

23.56. (a)

Retention factor (k)	2.2	4.7	8.3
Plate number (20 µL)	11 411	11 609	11 666
Plate number (100 µL)	7 160	9 761	10 894

Small injection yields plate numbers close to theoretical. Larger injection has lower plate numbers due to extra-column broadening due to injection. Early peaks are affected by extra-column broadening more than late peaks.

Retention factor (k)	2.2	4.7	8.3
Peak height (20 µL)	~26	~10	~5
Corrected peak height (20 µL)	~24	~8	~3
Peak height (100 µL)	~76	~34	~16
Corrected peak height (100 µL)	~74	~32	~14

Peak heights increase with volume injected. Early peak increases only ~3-fold despite 5-fold increase in injection due to broadening caused by large injection volume. Late-eluting peak height increases near 5-fold, as late-eluting peaks are less affected by injection broadening.

(b)
Retention factor (k)	2.2	4.7	8.3
Plate number (20 µL of 100 mg/L)	11 411	11 609	11 666
Plate number (100 µL of 20 mg/L)	7 160	9 761	10 894
Corrected peak height (20 µL of 100 mg/L)	~118	~38	~16
Corrected peak height (100 µL of 20 mg/L)	~74	~32	~14

Both injections load the same mg. Injecting a small volume minimizes extra-column broadening, yielding taller peaks.

(c)
Retention factor (k)	2.2	4.7	8.3
Plate number (100 cm of 127 µm)	11 544	11 651	11 682
Plate number (200 cm of 127 µm)	11 392	11 602	11 664
Plate number (100 cm of 250 µm)	9 720	10 999	11 429

Variance caused by connecting tubing is $\sigma^2_{\text{tubing}} = \dfrac{\pi d_t^4 l_t F}{384D}$.

Variance depends on tubing length, but only to the first power. Increasing length of connecting tubing has only a small effect on plate number. Broadening depends on the 4th power of tubing diameter, so an increase in diameter of connecting tubing has a more dramatic effect on plate number. In both cases, early eluting peaks are more affected than later eluting peaks.

23-57.

	A	B	C	D	E
1	Peak height, area, and weight				
2	Data collected by Caley Craven, University of Alberta				
3	Copy of figure enlarged so 2 mM peak was 10.2 cm in height.				
4			Full width at		Peak
5	Concentration	Peak height	half-height	Peak	weight
6	nitrate (mM)	(cm)	(cm)	area	(g)
7	0.1	0.67	1.02	0.73	0.01060
8	0.2	1.28	1.02	1.39	0.01811
9	0.5	2.89	1.03	3.17	0.04309
10	1	5.88	1.03	6.44	0.08946
11	2	10.18	1.13	12.24	0.16840
12				D7=1.064*B7*C7	

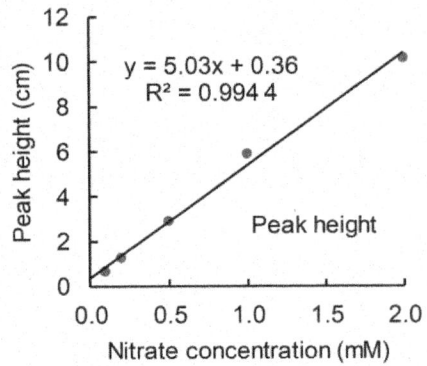
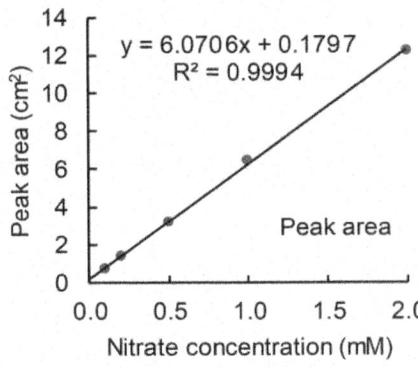

(a) In height-based calibration, there is a pattern to residuals, the intercept is slightly positive, and R^2 is < 0.995. These characteristics suggest that there is subtle curvature to the calibration data.

(b) Area-based calibration has random small residuals (all points on line), smaller intercept, and $R^2 > 0.999$. This calibration is linear.

(c)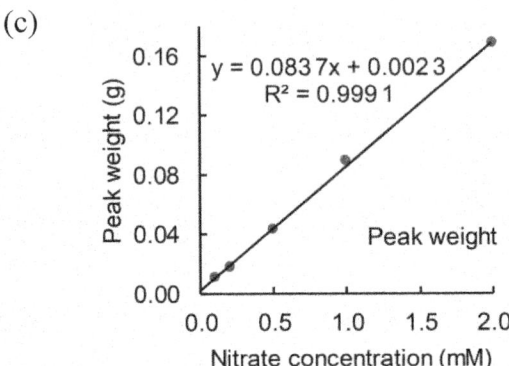

Weight-based calibration has random small residuals (all points on line), small intercept, and $R^2 > 0.999$. This calibration is linear.

(d) Nonlinear detector response or isotherm. Absorbance are high (>2), so deviation from Beer's law is possible. Nonlinear detector response affects peak height more than peak area. Height depends on highest absorbance reading, where any deviation from Beer's law is greatest.

Peak tailing or fronting caused by isotherm nonlinearity would affect the peak height, but not peak area. The 2.0 mM peak has a shorter retention time and wider $w_{1/2}$ than smaller peaks, suggesting that 2.0 mM is overloading column. Retention time, $w_{1/2}$, and N are constant if concentrations are on linear portion of isotherm.

23-58.

	A	B	C	D	E	F
1	Standard addition constant volume least-squares spreadsheet					
2	x	y				
3	Spike (mg/g)	I (x+s) =				
4	$[S]_f$	signal				
5	0.00	15.6				
6	3.12	21.1	Highlight cells B11:C13			
7	7.18	25.5	Type "=LINEST(B5:B10,A5:A10,TRUE,TRUE)"			
8	8.48	30.0	PC: CTRL + SHIFT + ENTER			
9	20.00	48.8	Mac: CTRL + SHIFT + RETURN			
10	38.20	83.4				
11	slope	1.778	14.59	y-intercept		
12	u_m	0.045	0.82	u_b		
13	R^2	0.997	1.42	s_y		
14	x-intercept=-b/m=		-8.21	B14=-C11/B11		
15	n=		6	B15=COUNT(B5:B10)		
16	Mean y =		37.4	B16=AVERAGE(B5:B10)		
17	$\Sigma(x_i-\text{mean }x)^2=$		1004.8	B17=DEVSQ(A5:A10)		
18	$u_x=$		0.62	B18=C13/ABS(B11)*SQRT(1/B15+B16^2/(B11^2*B17))		

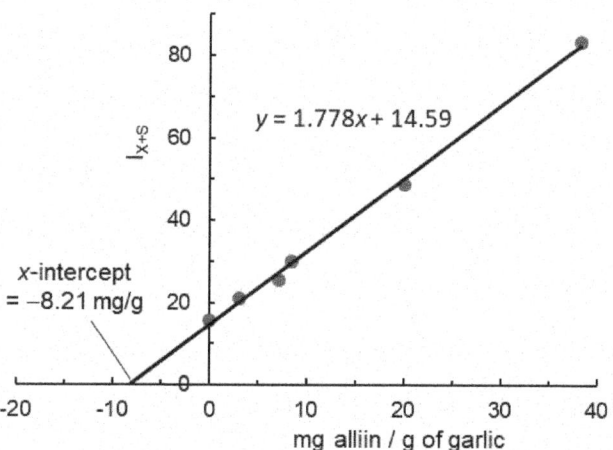

(a) Residuals small and random. Cannot use y-intercept to judge linearity for standard addition graphs. $R^2 > 0.995$. Calibration is linear.

(b) In cells B14 and B18 of the spreadsheet, negative x-intercept of standard addition graph is 8.21 mg alliin/g garlic with a standard uncertainty of 0.62 mg alliin/g garlic. The 95% confidence interval is $tu_x = (2.776)(0.624) = 1.73$ mg alliin/g garlic, where Student's t is taken for 95% confidence and $6 - 2 = 4$ degrees of freedom. A reasonable answer is 8.2 ± 1.7 mg alliin/g.

(c) Two moles of alliin (FM 177.2) produce one mole of allicin (FM 162.3) in the assay. Therefore, the quantity of allicin in garlic is ½(162.3/177.2)(8.21 ± 1.73 mg/g) = 3.76 ± 0.79 mg allicin/g or 3.8 ± 0.8 mg allicin/g garlic.

CHAPTER 24
GAS CHROMATOGRAPHY

24-1. (a) Packed columns offer high sample capacity, while open tubular columns give better separation efficiency (smaller plate height), shorter analysis time, and increased sensitivity to small quantities of analyte.

(b) Wall-coated: liquid stationary phase bonded to the wall of column
Porous-layer: solid stationary phase particles on wall of column

(c) Bonding or cross-linking the stationary phase reduces the tendency for the stationary phase to bleed from the column during use.

24-2. Resolution is proportional to $\sqrt{N}$. $N = L/H$. Open tubular columns eliminate the multiple path term (A) from the van Deemter equation, decreasing plate height H and increasing the plate number, N. Also, the lower resistance to gas flow allows longer columns L to be used with the same elution time.

24-3. (a) Advantages: A narrower column increases the rate of mass transfer between the mobile and stationary phases by decreasing the time needed for solute in the mobile phase to diffuse to the stationary phase. The increased rate of mass transfer lowers plate height H, which increases the plate number N and increases resolution. With lower plate height, a shorter column could be used to obtain the same resolution in a shorter time.
Disadvantages: lower sample capacity; requires high sensitivity detector.

(b) Advantages: larger N and greater resolution.
Disadvantage: longer analysis time.

(c) Advantages: greater retention; greater sample capacity.
Disadvantages: lower N (because the rate of mass transfer in the stationary phase decreases); long retention times for high-boiling point compounds; increased bleeding.

24-4. (a) Poly(dimethylsiloxane) is nonpolar. Based on "like dissolves like," it is most suitable for nonpolar solutes such as nonpolar organic solvents or petroleum products.

(b) Poly(ethylene glycol) is highly polar and so suited for separation of amines and fatty acid methyl esters.

(c) Porous layer open tubular columns are used for separation of compounds such as permanent gases like O_2, N_2, or CO that are too volatile to be adequately retained by a wall coated open tubular column.

24-5. (a) Advantages: Compounds with a wide range of retention characteristics (such as a mixture of low and high boiling compounds of similar polarity) can all have sufficient retention for resolution and yet all be eluted in a reasonable time. Peaks are sharper, which improves sensitivity. Low initial temperature allows use of splitless or on-column injection.

Disadvantages: There may be a sloping baseline if there is stationary phase bleed (decomposition or evaporation of stationary phase at high temperature). Time is needed to cool the column oven between separations.

(b) Higher pressure gives higher flow rate. If pressure is increased during a separation, retention times of late-eluting peaks are reduced. Pressure programming reduces the likelihood of decomposing thermally sensitive compounds.

24-6. (a) Ideal carrier gas should be: inert (not react with samples); low viscosity; compatible with detector.

(b) Diffusion of solute in H_2 and He is more rapid than in N_2. Therefore, mass transfer of solute between mobile phase and stationary phase is faster.

24-7. (a) Split injection is the ordinary mode for open tubular columns. It is best for high analyte concentrations and gas analysis. Splitless injection is useful for trace analysis (dilute solutions). Both split and splitless are tolerant of dirty samples, but dirty samples necessitate more frequent replacement of liners. On-column injection is best for quantitation and for thermally sensitive solutes that might decompose during a high-temperature injection, but requires clean samples.

(b) In solvent trapping, the initial column temperature is low enough to condense solvent at the beginning of the column. Solute is very soluble in the solvent and is trapped in a narrow band at the start of the column. In cold trapping, the initial column temperature is >100°C lower than the boiling points of solutes, which condense in a narrow band at the start of the column. In both cases, elution occurs as the column temperature is raised.

24-8. (a) All analytes

(b) Carbon atoms bearing hydrogen atoms

(c) Molecules with halogens, CN, NO_2, conjugated C=O

(d) P and S and other elements selected by wavelength

(e) P and N (and to a much lesser extent hydrocarbons)

(f) Aromatic and unsaturated compounds

(g) S

(h) Most elements (selected individually by wavelength)

(i) All analytes

(j) All analytes

24-9. The thermal conductivity detector measures changes in the thermal conductivity of the gas stream exiting the column. Any substance other than the carrier gas will change the conductivity of the gas stream. Therefore, the detector responds to all analytes. The flame ionization detector burns eluate in an H_2/O_2 flame to create CH radicals from carbon atoms (except carbonyl and carboxyl carbons), which then go on to be ionized to a small extent in the flame: CH + O → CHO^+ + e^-. Most other kinds of molecules do not create ions in the flame and are not detected.

24-10. A *reconstructed total ion chromatogram* is created by summing all ion intensities (above a selected value of *m/z*) in each mass spectrum at each time interval during a chromatography experiment. The technique responds to essentially everything eluted from the column and has no selectivity at all.
In *selected ion monitoring*, intensities at just one or a few values of *m/z* are plotted versus elution time. Only species with ions at those *m/z* values are detected, so the selectivity is much greater than that of the reconstructed total ion chromatogram. The signal-to-noise ratio is increased because ions are collected at each *m/z* for a longer time than would be allowed if the entire spectrum were being scanned.

Gas Chromatography

Selected reaction monitoring is most selective. One ion from the first mass separator is passed through a collision cell, where it breaks into several product ions that are separated by a second mass separator. The intensities of one or a few of these product ions are plotted as a function of elution time. The selectivity is high because few species from the column produce the first selected ion and even fewer break into the same fragments in the collision cell. This technique is so selective that it can transform a poor chromatographic separation into a highly specific determination of one component with virtually no interference.

24-11. Derivatization uses a chemical reaction to convert analyte into a form that is more convenient to separate or easier to detect. For example, carboxylic acids that might not have enough volatility for gas chromatography can be treated with chlorotrimethylsilane to make the volatile trimethylsilyl ester:

$$\underset{\text{Carboxylic acid}}{\text{RCOH}} + \text{ClSi(CH}_3)_3 \xrightarrow{-\text{HCl}} \underset{\text{Volatile derivative}}{\text{RCOSi(CH}_3)_3}$$

(where the RCOH and RCOSi(CH$_3$)$_3$ groups contain a C=O double bond)

24-12. (a) In solid-phase microextraction, analyte is extracted from a liquid or gas into a thin coating on a silica fiber extended from a syringe. After extraction, the fiber is withdrawn into the syringe. To inject sample into a chromatograph, the metal needle is inserted through the septum and the fiber is extended into the injection port. Analyte slowly evaporates from the fiber in the high-temperature port. Cold trapping is required to condense analyte at the start of the column during slow evaporation from the fiber. If cold trapping were not used, the peaks would be extremely broad because of the slow desorption from the fiber. During solid-phase microextraction, analyte equilibrates between the unknown and the coating on the fiber. Only a fraction of analyte is extracted into the fiber.

(b) In stir-bar sorptive extraction, a thick coating on the outside of a glass-coated stirring bar is used in place of a thin coating on a fiber. After extraction, the bar is placed in a thermal desorption tube where analyte is vaporized and cold trapped for chromatography. The volume of the coating is ~100 times greater in stir-bar sorptive extraction, so the sensitivity is up to 100 times higher for compounds that are not fully extracted by a fiber.

24-13. The goal of purge and trap is to collect *all* of the analyte from the unknown and to inject *all* of the analyte into the chromatography column. Splitless injection is required so analyte is not lost during injection. Any unknown loss of analyte would lead to an error in quantitative analysis.

24-14. In pyrolysis, sample is heated to 600–800°C with no oxygen present. Nonvolatile analytes thermally decompose to volatile species, which are swept by carrier gas into the gas chromatography column where they are cold trapped and then analyzed. The volatile species detected, such as monomers from polymer samples, are characteristic of the analyte and their amount is related to the quantity of nonvolatile analyte.

24-15. The order of decisions is: (1) goal of the analysis, (2) sample preparation method, (3) detector, (4) column, and (5) injection method.

24-16. (a) $t'_R = 8.4 - 3.7 = 4.7$ min; $k = 4.7$ min/3.7 min $= 1.3$

(b) Column radius = 0.16 mm
$\beta = r/2d_f = (0.16 \text{ mm})/(2 \times 1.0 \text{ μm} \times 10^{-3} \text{ mm/μm}) = 80$

(c) $k = K_D/\beta \Rightarrow K_D = k\beta = (1.3)(80) = 104$

(d) The stationary phase and temperature are the same, so the distribution constant would be 104, as in part (c).
$\beta = r/2d_f = (0.16 \text{ mm})/(2 \times 0.5 \text{ μm} \times 10^{-3} \text{ mm/μm}) = 160$
$k = K_D/\beta = 104/160 = 0.65$
$k = t'_R/t_M \Rightarrow t'_R = kt_M = (0.65)(3.7 \text{ min}) = 2.4 \text{ min}$
$t_R = t'_R + t_M = 2.4 \text{ min} + 3.7 \text{ min} = 6.1 \text{ min}$

24-17. (a) $I = 100 \cdot \left[8 + (9-8) \dfrac{\log(12.0) - \log(11.0)}{\log(14.0) - \log(11.0)} \right] = 836$

(b) No change. Based on $k = K_D/\beta$ the retention of all peaks would be similarly affected by the change in phase ratio.

(c) No change. The adjusted retention times of both the reference compounds and the unknown would be similarly affected by the change in length.

Gas Chromatography

24-18. Kováts retention indexes are defined for alkanes as: hexane 600; heptane 700; octane 800. The retention indexes for the other compounds allows prediction of their elution on the columns.

Column (a): hexane < butanol = benzene < 2-pentanone < heptane < octane

Column (b): hexane < heptane < butanol < benzene < 2-pentanone < octane

Column (c): hexane < heptane < octane < benzene < 2-pentanone < butanol

24-19. We reason by analogy with the order of elution of the compounds in the table in the text. For poly(dimethylsiloxane) column (a), the order of elution is

C_6 alkane < C_4 alcohol < C_5 ketone < C_7 alkane

Incrementing the chain length by one CH_2 unit, we predict the elution order

C_7 alkane < C_5 alcohol < C_6 ketone < C_8 alkane

Column (a): 3, 1, 2, 4, 5, 6

For (diphenyl)$_{0.35}$(dimethyl)$_{0.65}$polysiloxane column (b), the order of elution is

C_7 alkane < C_4 alcohol < C_5 ketone < C_8 alkane

Incrementing the chain length by one CH_2 unit, we predict the elution order

C_8 alkane < C_5 alcohol < C_6 ketone < C_9 alkane

Column (b): 3, 4, 1, 2, 5, 6

For poly(ethylene glycol) column (c), the order of elution is

C_9 alkane < C_5 ketone < C_{10} alkane < C_{11} alkane < C_4 alcohol

Incrementing the chain length by one CH_2 unit, we predict the elution order

C_{10} alkane < C_6 ketone < C_{11} alkane < C_{12} alkane < C_5 alcohol

Column (c): 3, 4, 5, 6, 2, 1

24-20. (a) $\Delta H°_{vap} \approx (88 \text{ J mol}^{-1} \text{ K}^{-1}) \cdot T_{bp} = (88 \text{ J mol}^{-1} \text{ K}^{-1})(126° + 273°)$

$= 3.5 \times 10^4 \text{ J mol}^{-1}$

(b) $\ln\left(\dfrac{P_1}{P_2}\right) = -\left(\dfrac{\Delta H_{vap}}{R}\right)\left(\dfrac{1}{T_1} - \dfrac{1}{T_2}\right)$

T_{bp} is the temperature at which the vapor pressure of a solute equals atmospheric pressure (101 325 Pa).

$\ln\left(\dfrac{P_1}{101\ 325\ \text{Pa}}\right) = -\left(\dfrac{3.5 \times 10^4 \text{ J mol}^{-1}}{8.314 \text{ J mol}^{-1} \text{ K}^{-1}}\right)\left(\dfrac{1}{70° + 273°} - \dfrac{1}{126° + 273°}\right) = -1.73$

$\ln P_1 = -1.73 + \ln(101\ 325) = 9.80 \Rightarrow P_1 = e^{9.80} = 1.8 \times 10^4 \text{ Pa} = 0.18 \text{ bar}$

364 Chapter 24

(c) $H°_{vap} \approx (88 \text{ J mol}^{-1} \text{ K}^{-1})(69° + 273°) = 3.0 \times 10^4 \text{ J mol}^{-1}$

$$\ln P_1 = -\left(\frac{3.0 \times 10^4 \text{ J mol}^{-1}}{8.314 \text{ J mol}^{-1}\text{K}^{-1}}\right)\left(\frac{1}{70° + 273°} - \frac{1}{69° + 273°}\right) + \ln(101\,325) = 11.56$$

$P_1 = e^{11.56} = 1.05 \times 10^5 \text{ Pa} = 1.05 \text{ bar (fully vaporized)}$

(d) The lower the vapor pressure, the greater the retention. Fully vaporized compounds (e.g., hexane in part c) are weakly retained on wall coated open tubular columns.

(e) The "gas" in "gas chromatography" refers to the mobile phase (N_2, He or H_2). Solutes are only partially in the vapor phase and spend only a fraction of the time in the mobile phase.

24-21.
$$\left. \begin{array}{l} \log(15.0) = \dfrac{a}{373} + b \\ \log(20.0) = \dfrac{a}{363} + b \end{array} \right\} \Rightarrow a = 1.69_2 \times 10^3 \text{ K} \qquad b = -3.36$$

To solve for a, subtract one equation from the other to eliminate b. Once you have a, substitute it back into either equation and solve for b.

At 353 K: $\log t'_R = \dfrac{1.69_2 \times 10^3 \text{ K}}{353 \text{ K}} - 3.36 \Rightarrow t'_R = 27.1 \text{ min}$

24-22. At 70°C, butanol has a retention time of 9.1 min. Increasing temperature decreases retention. The boiling point of butanol at 1 bar is 117°C. At some temperature above 117°C, butanol will all be in the gas phase in the chromatography column and therefore will have little retention. Its retention time will approach t_M. Since $\log k$ is related to $1/T$, k decreases in an exponential manner from 9.1 min at 70°C toward t_M at some temperature above 117°C.

24-23. (a) $N = 16(t_R/w)^2 = 16(17.0 \text{ min}/0.34 \text{ min})^2 = 4.0 \times 10^4$
$H = L/N = 30 \text{ m} / 4.0 \times 10^4 = 0.000\,75 \text{ m} = 0.75 \text{ mm}$
If your N was lower, you may have incorrectly used the integration tick marks in the figure. To calculate N, the baseline width must be determined by drawing tangents to the sides of the peak.

(b) Resolution = $\Delta t_R/w_{av}$ with width measured by extending tangents to the baseline. Based on measurements with my copy of the figure:
Resolution = 0.68 cm / 0.40 cm = 1.7

Gas Chromatography

24-24. (a) $k = (t_R - t_M)/t_M = (25.9 \text{ min} - 4.2 \text{ min})/4.2 \text{ min} = 5.2$

(b) $N = 5.54(t_R/w_{1/2})^2 = 5.54(25.9 \text{ min}/0.40 \text{ min})^2 = 2.3 \times 10^4$

$N = 16(t_R/w)^2 = 16(25.9 \text{ min}/0.71 \text{ min})^2 = 2.1 \times 10^4$

The difference in N from the two measures may reflect random error in the widths, but might also be due to the slight asymmetry of the docasahexanoic acid peak. Both equations assume a Gaussian peak shape. N based on full-width at half height can over estimate N for asymmetric peaks. If your N is less than 20 000, you may have measured the widths based on the x-axis rather than the extended baseline.

(c) $H = L/N = 60 \text{ m}/(2.2 \times 10^4) = 2.7 \text{ mm}$

24-25. (a) $0.1 \text{ μg/mL} = 0.1 \text{ ng/μL}$

mass analyte injected = $0.1 \text{ ng/μL} \times 1 \text{ μL} = 0.1 \text{ ng} = 100 \text{ pg}$

(b) A split ratio of 100:1 means only 1% of injected sample is loaded onto the column. Therefore, 0.001 ng or 1 pg is loaded using split injection.

(c) Formula mass $C_6H_6 = 78.11 \text{ g/mol}$

Carbon in benzene = $6 \times 12.011 \text{ g/mol} = 72.07 \text{ g C/mol}$

If the peak is 1 s wide, rate of carbon entering the detector is:

Splitless: $100 \text{ pg} \times \dfrac{72.07 \text{ g C/mol}}{78.11 \text{ g/mol}} / 1 \text{ s} = 92 \text{ pg C/s}$

Split: $1 \text{ pg} \times \dfrac{72.07 \text{ g C/mol}}{78.11 \text{ g/mol}} / 1 \text{ s} = 0.92 \text{ pg C/s}$

(d) Detection limit for flame ionization detector is 2 pg C/s. The splitless injection could be detected, but the split injection would be below the detection limit. Split injection is limited to concentrated samples.

24-26. (a) A thin stationary phase permits rapid equilibration of analyte between the mobile and stationary phases, which reduces the C term in the van Deemter equation. A thin stationary phase in a narrow-bore column gives small plate height and high resolution. In a wide-bore column, the large diameter of the column slows down the rate of mass transfer between mobile and stationary phases (because it takes time for analyte to diffuse across the diameter of the column), which defeats the purpose of the thin stationary phase.

(b) Narrow-bore column: plate height = 1/(5 000 m^{-1}) = 2.0 × 10^{-4} m = 200 μm. The area of a length (ℓ) of the inside wall of the column is $\pi d\ell$, where d is the column diameter. The volume of stationary phase in this length is $\pi d\ell t$, where t is the thickness of the stationary phase. For d = 250 μm, ℓ = 200 μm, and t = 0.10 μm, the volume is 1.5$_7$ × 10^4 μm^3. A density of 1.0 g/mL is 1.0 g/cm^3 = 1.0 g/(10^4 μm)^3 = 1.0 g/10^{12} μm^3 = 1 pg/μm^3. The mass of stationary phase in one theoretical plate is (1.5$_7$ × 10^4 μm^3)(1 pg/μm^3) = 1.5$_7$ × 10^4 pg. 1.0 % of this mass is = 0.16 ng.

Wide-bore column: For d = 530 μm, ℓ = 667 μm, and t = 5.0 μm, the volume is 5.5$_5$ × 10^6 μm^3. Mass of stationary phase is (5.5$_5$ × 10^6 μm^3)(1 pg/μm^3) = 5.5$_5$ × 10^6 pg. 1.0% of this mass is = 56 ng.

24-27. (a) At high linear velocity band broadening is dominated by slow mass transfer.

$$C_M = \frac{1+6k+11k^2}{24(k+1)^2} \frac{r^2}{D_M}$$

Mass transfer in the mobile phase depends on inner radius of the column, r, as analyte has to diffuse farther through mobile phase to get to the stationary phase. Diffusion distance, and so C_M, is largest for the 0.53 mm column, and so its plate height increases most for increases in linear velocity.

(b) At low linear velocity, band broadening is dominated by longitudinal diffusion ($B = 2D_M$) which depends on the diffusion coefficient in the mobile phase, but not on inner diameter of the column. So the B-term is constant for a given carrier gas, regardless of the inner diameter.

(c) Optimum velocity is lower for wider columns.

24-28. (a) Oxygen from air could degrade the stationary phase at elevated temperature. Therefore, the temperature must be kept below the point at which oxidation would occur.

(b) The optimum velocity gives the lowest plate height. It is the minimum in each curve. Optimum velocity = 9.3 cm/s for air and 17.6 cm/s for H$_2$. Plate height at optimum velocity = 0.036 cm for air and 0.051 for H$_2$. (Values come from the original publication. You will probably measure somewhat different values from the figure.)

Gas Chromatography 367

(c) Plates = column length/plate height = 3.0 m/0.036 cm = 8 300 for air and 5 900 for H$_2$

(d) Time = column length/optimum velocity = 3.0 m/9.3 cm/s = 32 s for air and 17 s for H$_2$

(e) The two terms describe broadening due to the finite time for solute to diffuse through the stationary phase and the mobile phase. If the stationary phase is sufficiently thin, the time for diffusion through the stationary phase (the C_s term) becomes negligible.

(f) Acceptable flow rates for H$_2$ are higher than for air because solutes diffuse through H$_2$ faster than they diffuse through air. With H$_2$ carrier, solutes can diffuse from the center of the column to the wall more rapidly than they can with air carrier.

24-29. (a) Resolution = $\dfrac{\sqrt{N}}{4} \dfrac{(\alpha-1)}{\alpha} \left(\dfrac{k_2}{1+k_2} \right)$. This equation provides guidance.

Increasing the column length or using a narrower column would increase the number of plates N and thereby increase the resolution.

Increasing the film thickness (decreasing the phase ratio) would increase the retention factor k, and so increase resolution.

Changing polarity of the stationary phase would alter the separation factor α and the retention factor k, and so would change the resolution. But it is less predictable whether that change would increase or decrease in the resolution.

Adjusting the flow rate to the optimum in the van Deemter curve will give the maximum N, and thus better resolution.

The retention factor k is inversely dependent on the column temperature. Decreasing temperature would increase the last term in the resolution equation, and thus increase resolution. Lowering the temperature would have a large effect on resolution if the original k_2 was less than 3.

(b) Increasing column length, using a narrower column, or using a thicker film would all require purchase of a new column. Changing the flow rate or temperature only involve adjusting instrument settings, and so these would be the approaches to explore first.

24-30. (a) $S = [\text{pentanol}] = \dfrac{234 \text{ mg} / 88.15 \text{ g/mol}}{10.0 \text{ mL}} = 0.265_5$ M

$X = [2,3\text{-dimethyl-2-butanol}] = \dfrac{237 \text{ mg} / 102.18 \text{ g/mol}}{10.0 \text{ mL}} = 0.231_9$ M

$\dfrac{A_X}{[X]_f} = F\left(\dfrac{A_S}{[S]_f}\right) \Rightarrow \dfrac{1.00}{[0.231_9 \text{ M}]} = F\left(\dfrac{0.913}{[0.265_5 \text{ M}]}\right) \Rightarrow F = 1.25_4$

(b) I estimate the areas by measuring the height and $w_{1/2}$ in millimeters. Your answer will differ from mine if the figure size in your book is different from that in my manuscript. However, relative peak areas should be the same.

pentanol: height = 40.1 mm; $w_{1/2}$ = 3.7 mm;

area = 1.064 × peak height × $w_{1/2}$ = 15$_8$ mm^2

2,3-dimethyl-2-butanol:

height = 77.0 mm; $w_{1/2}$ = 2.0 mm; area = 16$_4$ mm^2

(c) $\dfrac{164}{2,3\text{-dimethyl-2-butanol}} = 1.25_4 \left(\dfrac{158}{[93.7 \text{ mM}]}\right)$

$\Rightarrow [2,3\text{-dimethyl-2-butanol}] = 77._6$ mM = 78 mM

24-31. $\dfrac{A_X}{[X]_f} = F\left(\dfrac{A_S}{[S]_f}\right) \Rightarrow \dfrac{395}{[63 \text{ nM}]} = F\left(\dfrac{787}{[200 \text{ nM}]}\right) \Rightarrow F = 1.59$

The concentration of internal standard mixed with unknown is

$\dfrac{0.100 \text{ mL}}{10.00 \text{ mL}}(1.6 \times 10^{-5} \text{M}) = 0.16$ μM

$\dfrac{633}{[\text{iodoacetone}]} = 1.59 \left(\dfrac{520}{[0.16 \text{ μM}]}\right) \Rightarrow [\text{iodoacetone}] = 0.12_2$ μM

[iodoacetone] in original unknown = $\dfrac{10.00}{3.00}(0.12_2 \text{ μM}) = 0.41$ μM

24-32. $I = 100 \left[(7 + (10 - 7)\dfrac{\log(20.0) - \log(12.6)}{\log(22.9) - \log(12.6)}\right] = 932$

24-33. (a) Multiple small extractions are more effective than a single large extraction. The fraction remaining in phase 1 after four extractions would be q^4, where q is the fraction remaining after one extraction.

(b) On-column injection is better for thermally unstable molecules. Also, use of on-column injection allows higher sample loading on the column which would aid detection of the odors.

(c) Many of the odorants are polar molecules such as alcohols, esters, aldehydes, etc. Based on like dissolves like, a polar column such as poly(ethylene glycol) would provide best retention and resolution of these polar solutes.

(d) $\beta = r/2d_f = [(0.53 \text{ mm}/2) \times 1\,000 \text{ μm/mm}]/(2 \times 1.0 \text{ μm}) = 132$

(e) A wide-bore column enables greater sample loading onto the column. This higher loading will yield more odorant eluting off the column, and thus increase the chance of smelling it.

24-34. (a) NaCl lowers the solubility of moderately nonpolar compounds, such as ethers, in water. Adding NaCl increases the fraction of the organic compounds that will be transferred to the extraction fiber.

(b) Selected ion monitoring is measuring ion abundance for m/z 73. Only three compounds in the extract have appreciable intensity at m/z 73.

(c) The base peak for both MTBE and TAME is at m/z 73. This mass corresponds to M – 15 (loss of CH_3) for MTBE and M – 29 (loss of C_2H_5) for TAME. Loss of the ethyl group bound to carbon in TAME suggests that the methyl group lost from MTBE is also bound to carbon, not to oxygen. If methyl bound to oxygen were easily lost from MTBE and TAME, we would expect to see the ethyl group bound to oxygen lost from ETBE. There is no significant peak at M – 29 (m/z 73) in ETBE. The following structures are suggested:

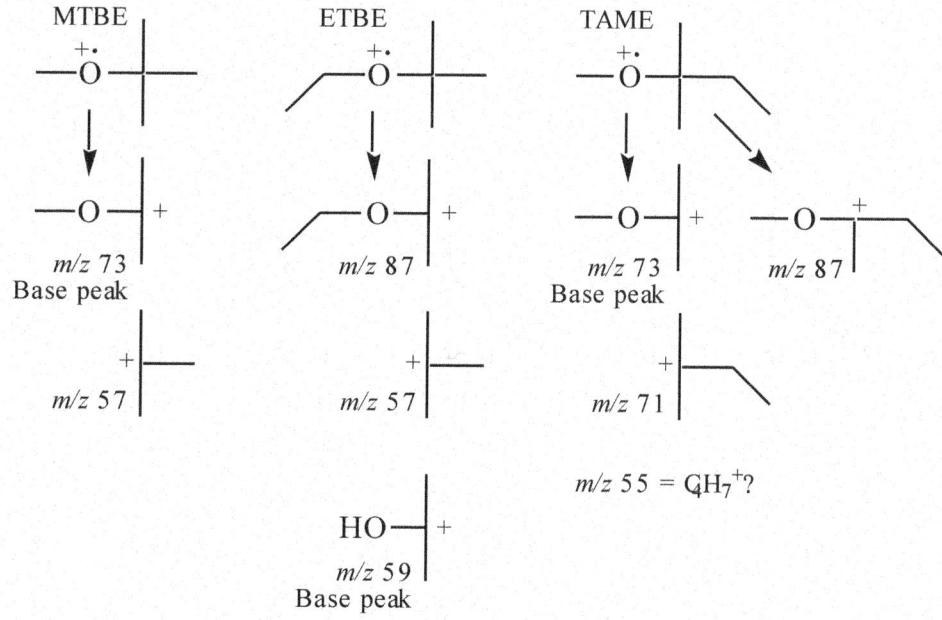

24-35. (a) The vial was heated to increase the vapor pressure of the analyte and the internal standard, so there would be enough in the gas phase (the headspace) to extract a significant quantity with the microextraction fiber.

(b) At 60°C the analyte and internal standard are cold trapped at the beginning of the column. Since desorption from the fiber takes many minutes, we do not want chromatography to begin until desorption is complete.

(c) $C_5H_{10}N^+$, m/z 84 (N-methylpyrrolidinium)

For 5-aminoquinoline, m/z 144 is the molecular ion, $C_9H_8NC_5H_{10}N_2^+$

(d)

	A	B	C	D	E
1	Least-squares spreadsheet				
2		[Nicotine]	Area ratio		
3		12	0.056		
4		12	0.059		
5	Highlight cells B13:C15	51	0.402		
6	Type "=LINEST(C3:C12,	51	0.391		
7	B3:B12,TRUE,TRUE)"	102	0.684		
8	For PC, press	102	0.669		
9	CTRL+SHIFT+ENTER	157	1.011		
10	For Mac, press	157	1.063		
11	CTRL+SHIFT+RETURN	205	1.278		
12		205	1.355		
13	**LINEST output:**	m 0.006401	0.0222	b	
14		u_m 0.000185	0.0234	u_b	
15		R^2 0.9933	0.0409	s_y	
16	n =	10	B16 = COUNT(B3:B12)		
17	Mean y =	0.6968	B17 = AVERAGE(C3:C12)		
18	$\Sigma(x_i - \text{mean } x)^2$ =	48554.4	B18 = DEVSQ(B3:B12)		
19	Measured y =	1.25	<= Input		
20	k = Replicate measures of y =	2	<= Input		
21	Derived x =	191.83	B21 = (B19-C13)/B13		
22	u_x =	5.54			
23	B22 = (C15/ABS(B13))*SQRT((1/B20)+(1/B16)+((B19-B17)^2)/(B13^2*B18))				

Least-squares parameters are computed in the block B13:C15. In cell B19, we insert the mean y value (1.25) for 2 replicate unknowns. The number of replicates is entered in cell B20. The derived value of x is computed in cell B21 and the uncertainty is computed with Equation 4-27 in cell B22.

Answers for the unknowns:

nonsmoker: 78 ± 5 µg/L

nonsmoker with smoking parents: 192 ± 6 µg/L

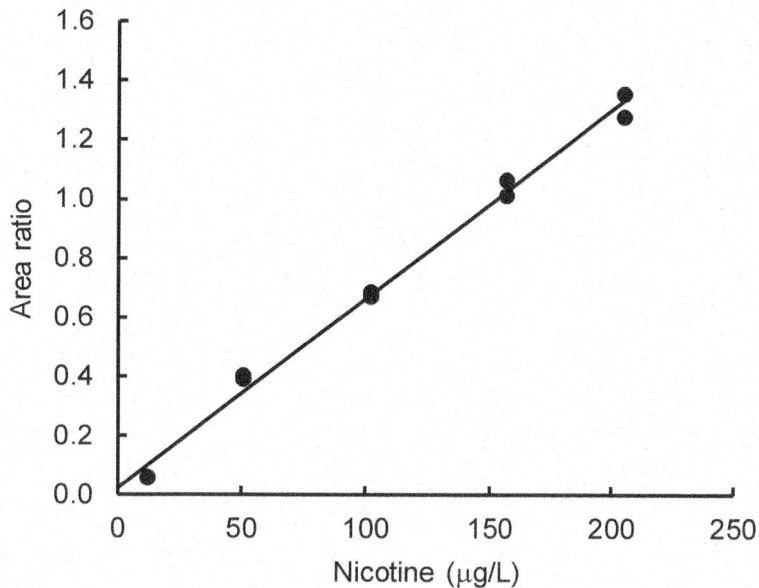

24-36. Nitrite: $[^{14}NO_2^-] = [^{15}NO_2^-](R - R_{blank}) = [80.0\ \mu M](0.062 - 0.040)$

$$= 1.8\ \mu M$$

Nitrate: $[^{14}NO_3^-] = [^{15}NO_3^-](R - R_{blank}) = [800.0\ \mu M](0.538 - 0.058)$

$$= 384\ \mu M$$

24-37. (a) The A term describing multiple flow paths is 0 for an open tubular column. Multiple paths arise in a packed column when liquid takes different paths through the column.

(b) $B = 2D_M$, where D_M is the diffusion coefficient of solute in mobile phase.

(c) $C = C_S + C_M$

$$C_S = \frac{2k}{3(k+1)^2}\frac{d_f^2}{D_S} \qquad C_M = \frac{1 + 6k + 11k^2}{24(k+1)^2}\frac{r^2}{D_M}$$

where k = retention factor
d_f = thickness of stationary phase
r = column radius
D_S = diffusion coefficient of solute in the stationary phase
D_M = diffusion coefficient of solute in the mobile phase

(d) $H = B/u_x + Cu_x$ (u_x = linear velocity)

Plate height is a minimum at the optimum velocity:

$$\frac{dH}{du_x} = -\frac{B}{u_x^2} + C = 0 \Rightarrow u_x \text{(optimum)} = \sqrt{\frac{B}{C}}$$

The minimum plate height is found by plugging this value of u_x (optimum) back into the van Deemter equation:

$$H_{min} = B/u_x + Cu_x = B\sqrt{\frac{C}{B}} + C\sqrt{\frac{B}{C}} = 2\sqrt{BC} = 2\sqrt{B(C_S + C_M)}$$

$$H_{min} = 2\sqrt{(2D_M)\left(\frac{2k}{3(k+1)^2}\frac{d^2}{D_S} + \frac{1+6k+11k^2}{24(k+1)^2}\frac{r^2}{D_M}\right)}$$

$$H_{min} = 2\sqrt{\frac{4k}{3(k+1)^2}\frac{d^2 D_M}{D_S} + \frac{(1+6k+11k^2)2r^2}{24(k+1)^2}}$$

24-38. (a) As $k \to 0$, $H_{min}/r = \sqrt{1/3} = 0.58$

As $k \to \infty$, $H_{min}/r = \sqrt{\frac{1+6k+11k^2}{3(1+k)^2}} \to \sqrt{\frac{11k^2}{3k^2}} = \sqrt{\frac{11}{3}} = 1.9$

(b) As $k \to 0$, $H_{min} = 0.58\, r = 0.058$ mm

As $k \to \infty$, $H_{min} = 1.9\, r = 0.19$ mm

(c) For $k = 5.0$, $H_{min} = r\sqrt{\frac{1+6\cdot 5.0 + 11\cdot 25}{3(36)}} = 1.68\, r = 0.168$ mm

Plate number = $\dfrac{50 \times 10^3 \text{ mm}}{0.168 \text{ mm/plate}} = 3.0 \times 10^5$

(d) $\beta = V_M/V_S$. For a length of column, ℓ, the volume of mobile phase is $\pi r^2 \ell$ and the volume of stationary phase is $2\pi r d_f \ell$. Substituting these volumes into the equation for β gives $\beta = (\pi r^2 \ell)/(2\pi r d_f \ell) = r/2d_f$.

(e) $k = K_D/\beta = 1\,000/\left(\dfrac{100\ \mu m}{2(0.20\ \mu m)}\right) = 4.0$

24-39. The van Deemter equation has the form

$$H = B/u_x + Cu_x = B/u_x + (C_S + C_M)u_x$$

$$B = 2D_M \qquad C_S = \frac{2k}{3(k+1)^2}\frac{d_f^2}{D_S} \qquad C_M = \frac{1+6k+11k^2}{24(k+1)^2}\frac{r^2}{D_M}$$

where k = retention factor = 8.0
d_f = thickness of stationary phase = 3.0×10^{-6} m
r = column radius = 2.65×10^{-4} m
D_S = diffusion coefficient of solute in the stationary phase
D_M = diffusion coefficient of solute in the mobile phase

Experimentally, we find $H = (6.0 \times 10^{-5} \text{ m}^2/\text{s})/u_x + (2.09 \times 10^{-3} \text{ s})u_x$. Therefore, $B = 2D_M = (6.0 \times 10^{-5} \text{ m}^2/\text{s})$, or $D_M = 3.0 \times 10^{-5}$ m^2/s.

From the second term of the experimental van Deemter equation, we know that

$$C_S + C_M = 2.09 \times 10^{-3} \text{ s} = \frac{2k}{3(k+1)^2}\frac{d_f^2}{D_S} + \frac{1+6k+11k^2}{24(k+1)^2}\frac{r^2}{D_M}$$

Inserting the known values of all parameters allows us to solve for D_S:

$$2.09 \times 10^{-3} \text{ s} = \frac{2(8.0)}{3((8.0)+1)^2}\frac{(3.0\times 10^{-6}\text{ m})^2}{D_S} + \frac{1+6(8.0)+11(8.0)^2}{24((8.0)+1)^2}\frac{(2.65\times 10^{-4}\text{ m})^2}{(3.0\times 10^{-5}\text{ m}^2/\text{s})}$$

$$\Rightarrow D_S = 5.0 \times 10^{-10} \text{ m}^2/\text{s}$$

The diffusion coefficient in the mobile phase is $(3.0 \times 10^{-5} \text{ m}^2/\text{s})/(5.0 \times 10^{-10} \text{ m}^2/\text{s}) = 6.0 \times 10^4$ times greater than the diffusion coefficient in the stationary phase. This makes sense, because it is easier for solute to diffuse through He gas than through a viscous liquid phase.

24-40. (a)

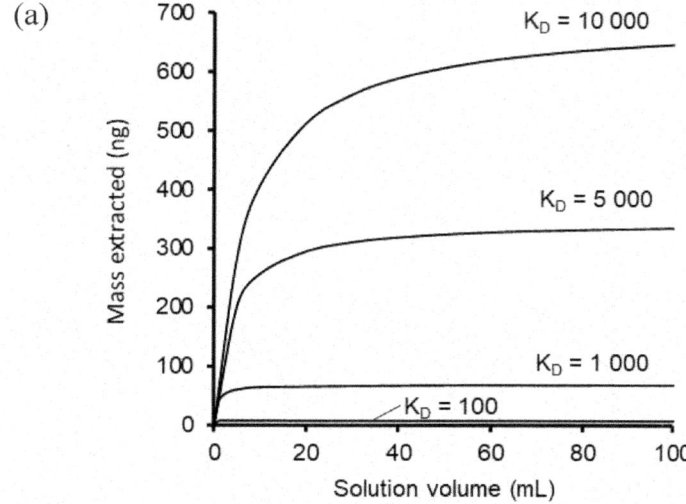

	A	B	C	D
1	Mass of analyte extracted by solid-phase microextraction			
2				
3		V_s (mL)	m (ng)	
4	K_D = distribution	0	0.000	
5	constant =	1	6.455	
6	1.00E+02	2	6.670	
7	V_f = volume of	3	6.745	
8	film (mL) =	4	6.783	
9	6.90E-04	5	6.806	
10	c_o = initial	10	6.853	
11	concentration in	50	6.890	
12	solution (µg/mL) =	100	6.895	
13	0.1	1000	6.900	
14	C4 = 1000*($A$6*$A$9*$A$13*B4)/($A$6*$A$9+B4)			

(b) $m = \dfrac{K_D V_f c_o V_s}{K_D V_f + V_s}$ If $V_s \gg K_D V_f$, $m = K_D V_f c_o$

For $V_f = 6.9 \times 10^{-4}$ mL and $c_o = 0.1$ µg/mL, $m \to (6.9 \times 10^{-5})(K_D)$ µg

For $K_D = 100$, $m \to 6.9$ ng, which agrees with the graph.

For $K_D = 10\,000$, $m \to 690$ ng, which is where the graph is heading, but it will require ~1 L of solution to attain the limiting concentration in the fiber.

(c) The spreadsheet tells us that when $K_D = 100$, 6.85 ng have been extracted into the fiber and when $K_D = 10\,000$, 408 ng have been extracted into the fiber. The total analyte in 10.0 mL is (0.10 µg/mL)(10.0 mL) = 1.0 µg. The fraction extracted for $K_D = 100$ is 6.86 ng/1.0 µg = 0.006 9 (or 0.69%). The fraction extracted for $K_D = 10\,000$ is 0.41 (or 41%).

24-41. (a) Rainwater:

$F_{\text{calculated}} = (0.008/0.005)^2 = 2._6 < F_{\text{table}} = 6.23$ (for 4 degrees of freedom in the numerator and 6 degrees of freedom in the denominator). Since $F_{\text{calculated}} < F_{\text{table}}$, we use the following equations:

$$s_{\text{pooled}} = \sqrt{\dfrac{s_1^2(n_1-1)+s_2^2(n_2-1)}{n_1+n_2-2}} = \sqrt{\dfrac{0.005^2(6)+0.008^2(4)}{7+5-2}} = 0.006_{37}$$

$$t_{\text{calculated}} = \dfrac{|\bar{x}_1 - \bar{x}_2|}{s_{\text{pooled}}}\sqrt{\dfrac{n_1 n_2}{n_1+n_2}} = \dfrac{0.069-0.063}{0.006_{37}}\sqrt{\dfrac{7\cdot 5}{7+5}} = 1.6_1$$

$t_{\text{table}} = 2.228$ for $n_1+n_2-2 = 7+5-2 = 10$ degrees of freedom

Since $t_{\text{calculated}} < t_{\text{table}}$, the difference is not significant.

Drinking water:

$F_{calculated} = (0.008/0.007)^2 = 1.3_1 < F_{table} = 9.61$ (for $n - 1 = 4$ degrees of freedom in numerator and $n - 1 = 4$ degrees of freedom in denominator). Since $F_{calculated} < F_{table}$, we use the following equations:

$$s_{pooled} = \sqrt{\frac{0.007^2(4) + 0.008^2(4)}{5 + 5 - 2}} = 0.007_{52}$$

$$t = \frac{0.087 - 0.078}{0.007_{52}}\sqrt{\frac{5 \cdot 5}{5 + 5}} = 1.8_9 < 2.306. \text{ The difference is \underline{not} significant.}$$

(b) Gas chromatography:

$F_{calculated} = (0.007/0.005)^2 = 1.9_6 < F_{table} = 6.23$ (for $n - 1 = 4$ degrees of freedom in the numerator and $n - 1 = 6$ degrees of freedom in the denominator). Since $F_{calculated} < F_{table}$, we use the following equations:

$$s_{pooled} = \sqrt{\frac{0.005^2(6) + 0.007^2(4)}{7 + 5 - 2}} = 0.005_{88}$$

$$t = \frac{0.078 - 0.069}{0.005_{88}}\sqrt{\frac{7 \cdot 5}{7 + 5}} = 2.6_1 > 2.228. \text{ The difference \underline{is} significant.}$$

Spectrophotometry: (the two standard deviations are equal)

$$s_{pooled} = \sqrt{\frac{0.008^2(4) + 0.008^2(4)}{5 + 5 - 2}} = 0.008_{00}$$

$$t = \frac{0.087 - 0.063}{0.008_{00}}\sqrt{\frac{5 \cdot 5}{5 + 5}} = 4.7 > 2.306. \text{ The difference \underline{is} significant.}$$

24-42. (a) Both. Acetamide was produced by thermal decomposition, but the last sentence shows parts-per-billion levels were found in every food analyzed.

(b) Acetamide observed in control samples of previous studies varied widely; presence of matrix components that could break down to produce acetamide; and literature studies that observed acetamide produced by heating.

(c) Poor sensitivity with mass spectrometric detection; poor retention; broad peaks; and acetamide contaminant in common eluent.

(d) A surrogate analyte behaves identically to the analyte, but is not the analyte. Deuterated acetamide was used for method development because acetamide was found in varying amounts in all foods. Therefore, there was no analyte-free matrix that could be used to prepare matrix-matched standards.

(e) On-column injection does not heat sample as much as split or splitless injection, and so is better for thermally-labile compounds. It was hoped that lower temperatures of on-column injection would avoid thermal decomposition of matrix to form acetamide. Peaks broadened after a few injections, and column plugged. On-column injection needs clean samples.

(f) Page 301 right column details previous studies had successfully used 9-xanthydrol derivatization for acetamide. Last sentence of page 301 states 9-xanthydrol creates products with high molecular weight, which shifts the measured *m/z* to higher values where there is less interference from matrix components. Derivatization section says acidity, temperature, reaction time, and amount of xanthydrol were optimized.

(g) Molecular ions of xanthyl derivative of acetamide, $^2H_3$-acetamide, and propionamide internal standard were monitored. Other compounds might produce these *m/z*, so additional *m/z* are monitored to confirm detection. Ion at *m/z* 196 was a common fragment from analytes and standards due to loss of COR (e.g., $COCH_3$ from xanthyl-acetamide), and *m/z* 181, 168, and 152 were fragment ions from xanthydrol.

24-43. (a) In the "Apparatus and operation" section, the paper states a thermal conductivity detector was used. Oxalate is converted into CO_2. A flame ionization detector does not respond to CO_2. CO_2 does not contain the elements to which the nitrogen-phosphorus, flame photometric or chemiluminescence detectors selectively respond.

(b) "Method calibration, precision and validation" section states the relative standard deviation of repeated measurements was 0.84% and the limit of quantification (LOQ) was 1.95 μmol. Figure 6 is a calibration curve linear to 30 μmol. Therefore the linear range is from 1.95 μmol (LOQ) to 30 μmol.

(c) Five food samples were analyzed by both the gas chromatography method and an enzymatic method.

(d) The introduction of paper lists the following alternate methods: ion chromatography, high performance liquid chromatography, enzymatic methods, capillary electrophoresis, and chemiluminescence, and gives leading references for each. An enzymatic method is also used in the paper to validate the gas chromatography results.

(e) In the "Apparatus and operation" section the paper states a 30-m-long by 0.53-mm-diameter "model GS-Q (J&W Scientific)," but does not detail the nature of that column. CO_2 is a permanent gas and so would not be retained on a wall coated open tubular column. Therefore it must be a porous layer open tubular column. The manufacturer's literature states that the GS-Q is a porous divinylbenzene polymer.

24-44. (a)

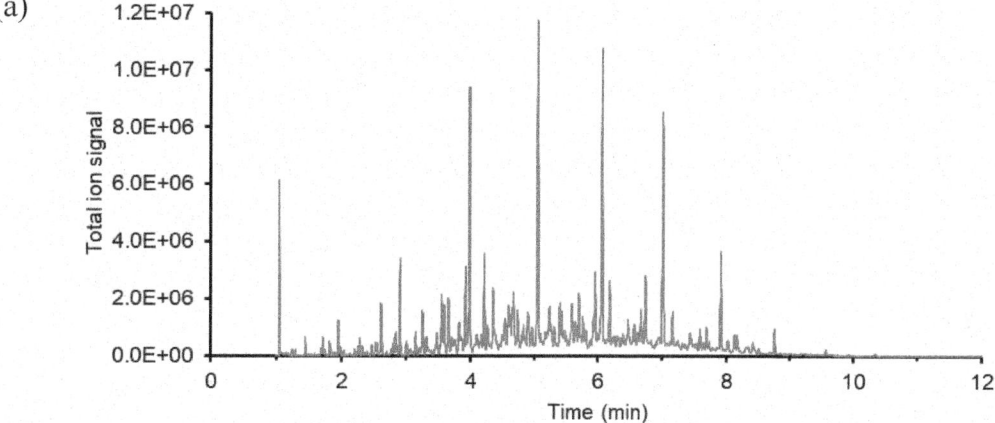

(b)

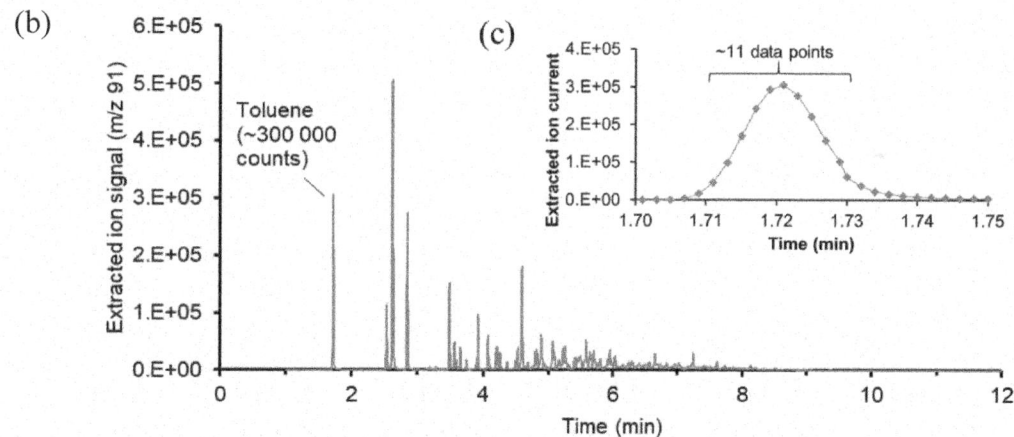

The extracted ion chromatogram is much simpler. The signal has decreased substantially, but removal of overlapping peaks will improve quantitation.

(d)

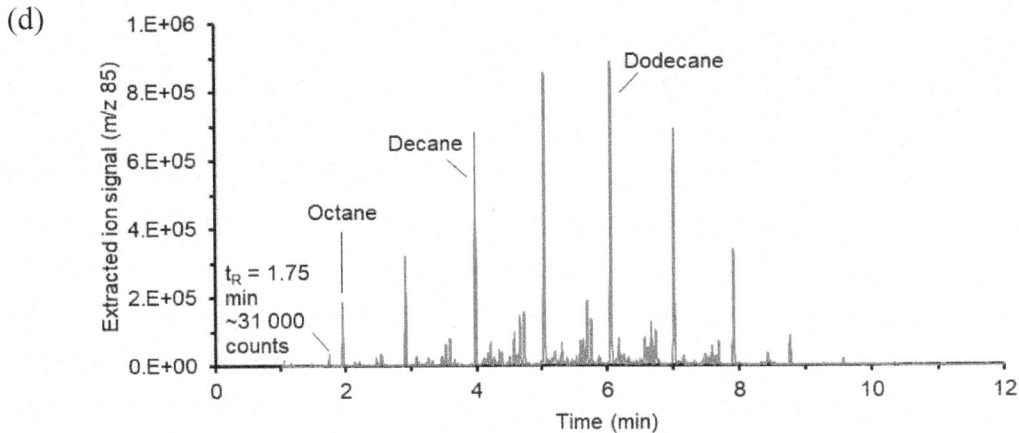

Yes, an alkane elutes at 1.75 min. Its height is ~31 000—about 10% of that of toluene in part (b). The presence of this co-eluting alkane would bias toluene quantitation high by a significant amount.

(e)

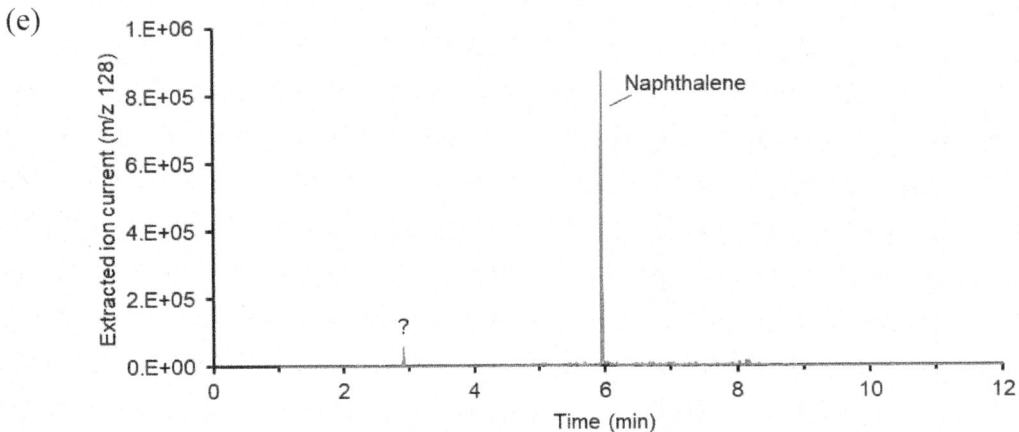

Peak at 2.91 min at m/z 85 characteristic of alkane. No peak at 2.91 min at m/z characteristic of monoaromatics. The peak at 2.91 in the m/z 128 chromatogram is likely from an alkane. Given empirical formula for alkanes is C_nH_{2n+2}, the m/z 128 peak is possibly from nonane. Its retention time fits the alkane pattern in part (d). To check this hypothesis, the retention time of a nonane standard should be determined and the mass spectrum at 2.91 min compared with a mass spectral library.

CHAPTER 25
HIGH-PERFORMANCE LIQUID CHROMATOGRAPHY

25-1. (a) In reversed-phase chromatography, the solutes are nonpolar and more soluble in a nonpolar mobile phase. In normal-phase chromatography, the solutes are polar and more soluble in a polar mobile phase.

(b) A gradient of increasing pressure gives increasing solvent density, which gives increasing eluent strength in supercritical fluid chromatography.

25-2. Solvent is competing with solute for adsorption sites. The strength of the solvent-adsorbent interaction is independent of solute.

25-3. In hydrophilic interaction chromatography, solute equilibrates between the mobile phase and an aqueous layer on the surface of the polar stationary phase. The more water in the eluent, the better the mobile phase competes with the stationary aqueous layer to dissolve a polar solute and elute it from the column.

25-4. (a) Small particles give increased resistance to flow. High pressure is required to obtain a usable flow rate.

(b) Efficiency increases because solute equilibrates between phases more rapidly if the distance that it has to diffuse is smaller. This effect decreases the C term in the van Deemter equation. Also, flow paths between small particles are more uniform, decreasing the multiple path (A) term.

(c) A bonded stationary phase is covalently attached to the support.

25-5. (a) $L \text{ (cm)} \approx \dfrac{N d_p (\mu m)}{3\,000}$

If $N = 1.0 \times 10^4$ and $d_p = 10.0$ μm, $L = 33$ cm

$d_p = 5.0$ μm $\Rightarrow L = 17$ cm; $d_p = 3.0$ μm $\Rightarrow L = 10$ cm

$d_p = 1.5$ μm $\Rightarrow L = 5$ cm

(b) Retention time is proportional to column length. If $t_R = 20.0$ minutes on the 33 cm long column, linear velocity was 33 cm/20.0 min = 1.6_5 cm/min.

$d_p = 5.0$ μm and $L = 17$ cm $\Rightarrow t_R = (17 \text{ cm}/33 \text{ cm})(20 \text{ min}) = 10.3$ min

$d_p = 3.0$ μm and $L = 10$ cm $\Rightarrow t_R = (10 \text{ cm}/33 \text{ cm})(20 \text{ min}) = 6.1$ min

$d_p = 1.5$ μm and $L = 5$ cm $\Rightarrow t_R = (5 \text{ cm}/33 \text{ cm})(20 \text{ min}) = 3.0$ min

380   Chapter 25

(c) $P = \text{(constant)} \times L/d_p^2$ If $P = 4.4$ MPa for $L = 33$ cm and $d_p = 10$ μm, then the constant is 13.33 MPa μm^2 cm^{-1}

$d_p = 5.0$ μm, $L = 17$ cm: $P = (13.33) \times L/d_p^2 = (13.33)(17)/(5.0)^2 = 9.1$ MPa

$d_p = 3.0$ μm, $L = 10$ cm: $P = (13.33)(10)/(3.0)^2 = 14.8$ MPa

$d_p = 1.5$ μm, $L = 5$ cm: $P = (13.33)(5)/(1.5)^2 = 29.6$ MPa

(d) $N = \dfrac{16 t_R^2}{w_b^2} \Rightarrow w_b(\text{minutes}) = \dfrac{4 t_R}{\sqrt{N}}$

$\Rightarrow w_b(\mu L) = w_b(\text{min}) \times 2.0 \text{ mL/min} \times 1\,000 \text{ μL/mL} = w_b(\text{min}) \times 2\,000$

10 μm: $w_b(\text{min}) = \dfrac{4(20.0 \text{ min})}{\sqrt{10^4}} = 0.80_0$ min; $w_b(\mu L) = 1\,600$ μL

5 μm: $w_b(\text{min}) = \dfrac{4(10.3 \text{ min})}{\sqrt{10^4}} = 0.41_2$ min; $w_b(\mu L) = 820$ μL

3 μm: $w_b(\text{min}) = \dfrac{4(6.1 \text{ min})}{\sqrt{10^4}} = 0.24_4$ min; $w_b(\mu L) = 490$ μL

1.5 μm: $w_b(\text{min}) = \dfrac{4(3.0 \text{ min})}{\sqrt{10^4}} = 0.12_0$ min; $w_b(\mu L) = 240$ μL

(e) HPLC instruments can operate in the 7–40 MPa pressure range, but are most reliable below 20 MPa. UHPLC systems can operate at pressures as high as 100 MPa. The 3–10 μm particle columns are within the pressure limits of an HPLC system, but the 1.5 μm particle column should be run on a UHPLC system. Also the variance contribution from extra-column components should be small compared to that of the column.

25-6.   (a) The highly porous nature of HPLC particles greatly increases the surface area of the particles, which in turn increases the mass of sample that may be loaded onto the column without causing overload.

(b) The vast majority of the surface area of an HPLC particle is within the pores of the particle. Pore opening must be large enough to allow the solute to diffuse into the pore. Larger pores are needed to permit macromolecules, such as proteins or polypeptides, to access the surface within the pores. However, the larger the pore, the smaller the surface area (and thus the capacity) of the column. Therefore narrower pores with high surface area are used for smaller molecules.

High-Performance Liquid Chromatography

25-7. Plates $(N) = (15 \text{ cm})/(5.0 \times 10^{-4} \text{ cm/plate}) = 3.0 \times 10^4$

$$N = \frac{5.55\, t_R^2}{w_{1/2}^2} \Rightarrow w_{1/2} = t_R\sqrt{\frac{5.55}{N}} = (10.0 \text{ min})\sqrt{\frac{5.55}{3.0\times 10^4}} = 0.13_6 \text{ min}$$

If plate height = 25 μm, plates = 6 000 and $w_{1/2} = 0.30_4$ min

25-8. (a) $N = \dfrac{5.55\, t_R^2}{w_{1/2}^2} = \dfrac{5.55\,(0.63 \text{ min} \times 60 \text{ s/min})^2}{(2.3 \text{ s})^2} = 1\,500$

$H = L/N = (50 \text{ mm})/(1\,500) = 0.033 \text{ mm} = 33$ μm

Number of particles side-by-side equaling one plate = 33 μm/1.7 μm = 19

(b) Particles equal to one theoretical plate = 4 μm/1.7 μm = 2.4

The column in (a) is being run for maximum speed.

25-9. Silica dissolves above pH 8 and the siloxane bond to the stationary phase hydrolyzes below pH 2. Bulky isobutyl groups hinder the approach of H_3O^+ to the Si–O–Si bond, so the rate of acid-catalyzed hydrolysis is decreased.

25-10. The high concentration of additive binds to the sites on the stationary phase that would otherwise hold on tightly to solutes and cause tailing.

25-11. (a) Your sketch should look like Figure 23-14, in which the asymmetry factor is $B/A = 1.8$, measured at one tenth of the peak height.

(b) Tailing, particularly of ionic compounds, may be due to overload of the column. Try reducing the concentration of sample injected. If all peaks are affected, tailing could be caused by a clogged frit which you might be able to clean by reversing the flow direction. If only amines are tailing, try switching to a Type B silica column or add 30 mM triethylamine to the mobile phase. Tailing of acidic compounds might be eliminated by adding 30 mM ammonium acetate. For unknown mixtures, 30 mM triethylammonium acetate is useful. If tailing persists, 10 mM dimethyloctylamine or dimethyloctylammonium acetate might be effective.

25-12. (a)

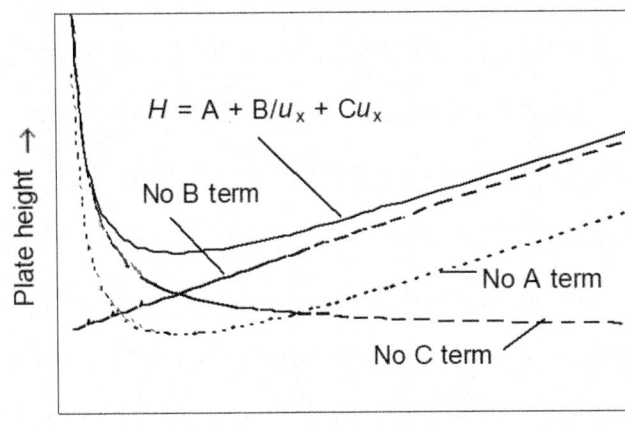

(b) For 1.8-μm particle size, the experimental van Deemter curve looks almost like the curve with no C term in (a) (that is, mass transfer term ≈0). When particle size is small enough, mass transfer between the stationary and mobile phases is very rapid and C term contributes little to peak broadening. The experimental curve for 1.8-μm particles levels off at a smaller plate height than the curves for 5- and 3.5-μm particles. This behavior suggests that the A term (multiple flow paths) is smaller for the smaller particles.

(c) A superficially porous particle has a thin porous shell on a solid inner core. A 2.7-μm superficially porous particle provides resolution similar to that of a 1.8-μm totally porous particle mainly because of decreased multiple path broadening arising from more uniform column packing of superficially porous particles. Resistance to flow depends on the width of the channels between the particles which scales with the overall diameter of the particle. The overall diameter of the superficially porous particle is larger, so its resistance to fluid flow is not as high as that of a small particle.

25-13. (a) $N = \dfrac{5.55\, t_R^2}{w_{1/2}^2} = \dfrac{5.55\,(4.91\text{ s})^2}{(0.27\text{ s})^2} = 1\,835 \approx 1\,800$ for (R)-enantiomer

$N = \dfrac{5.55\,(6.45\text{ s})^2}{(0.41\text{ s})^2} = 1\,373 \approx 1\,400$ for (S)-enantiomer

(b) $w_{1/2\text{av}} = \tfrac{1}{2}(0.27\text{ s} + 0.41\text{ s}) = 0.34\text{ s}$

Resolution $= \dfrac{0.589\, \Delta t_R}{w_{1/2\text{av}}} = \dfrac{0.589\,(6.45\text{ s} - 4.91\text{ s})}{0.34\text{ s}} = 2.7$

(c) Retention factor for *(S)*-enantiomer: $k = \dfrac{(6.45 \text{ s} - 3.85 \text{ s})}{3.85 \text{ s}} = 0.68$

Separation factor: $\alpha = \dfrac{t'_{R2}}{t'_{R1}} = \dfrac{6.45 \text{ s} - 3.85 \text{ s}}{4.91 \text{ s} - 3.85 \text{ s}} = 2.45$

Average $N = \tfrac{1}{2}(1\,835 + 1\,373) = 1\,604$

Resolution $= \dfrac{\sqrt{N}}{4} \dfrac{(\alpha-1)}{\alpha}\left(\dfrac{k_2}{1+k_2}\right) = \dfrac{\sqrt{1\,604}}{4} \dfrac{(2.45-1)}{2.45}\left(\dfrac{0.68}{1+0.68}\right) = 2.4$

(d) $u_x = L/t_M = 3.00 \text{ cm}/3.85 \text{ s} = 0.78 \text{ cm/s}$. A velocity of 7.8 mm/s is at the fast end of the van Deemter velocities. Use of superficially porous particles minimizes mass transfer band broadening, keeping plate height low. Low plate height means short columns generate enough plates for the separation. A short column and high linear velocity yield a fast separation.

25-14. (a) $P \propto \dfrac{1}{d_p^2}$ For two conditions (1 and 2),

$\dfrac{P_2}{P_1} \propto \dfrac{d_1^2}{d_2^2} = \left(\dfrac{3 \text{ μm}}{0.7 \text{ μm}}\right)^2 = 18$. Pressure must be 18 times greater.

(b) $u_x \propto P$, so if linear velocity is increased by a factor of 10, then pressure should increase by a factor of 10.

(c) Mass transfer between the mobile and stationary phase is faster for small particles than for large particles. The optimum velocity for maximum efficiency (highest plate number) increases as the rate of mass transfer increases. In the example cited, the high flow rate is closer to the optimum flow rate than is the low flow rate.

25-15. (a) Normal-phase chromatography

(b) Bonded reversed-phase chromatography

(c) Bonded reversed-phase chromatography with buffered mobile phase

(d) Hydrophilic interaction chromatography (HILIC)

(e) Ion-exchange or ion chromatography

(f) Size-exclusion chromatography

(g) Ion-exchange chromatography with wide pore stationary phase

(h) Size-exclusion chromatography

384   Chapter 25

25-16. 10-μm-diameter spheres: volume = $\frac{4}{3}\pi r^3 = \frac{4}{3}\pi(5 \times 10^{-4} \text{ cm})^3 = 5.24 \times 10^{-10} \text{ cm}^3$

Mass of one sphere = $(5.24 \times 10^{-10} \text{ mL})(2.2 \text{ g/mL}) = 1.15 \times 10^{-9}$ g
Number of particles in 1 g = $1 \text{ g}/(1.15 \times 10^{-9} \text{ g/particle}) = 8.68 \times 10^8$
Surface area of one particle = $4\pi r^2 = 4\pi(5 \times 10^{-6} \text{ m})^2 = 3.14 \times 10^{-10} \text{ m}^2$
Surface area of 8.68×10^8 particles = 0.27 m^2
Since the observed surface area is 300 m^2, the particles must be highly porous.

25-17. (a) Elution order in reversed-phase liquid chromatography is from least to most hydrophobic. Uracil and caffeine have many polar functionalities, and so are essentially unretained. The aromatic ring of phenylethanol makes it more hydrophobic, and so partially retained. Butyl paraben has both a hydrophobic aromatic ring and butyl group, increasing its retention. Anthracene is very hydrophobic, and so strongly retained.

(b) $t_M \approx \dfrac{Ld_c^2}{2F} = \dfrac{(15.0 \text{ cm})(0.46 \text{ cm})^2}{2(1.5 \text{ cm}^3/\text{min})} = 1.0_6$ min

This estimate is close to the ~1 min observed in the opener.

(c) $P = f\dfrac{F\eta L}{\pi r^2 d_p^2} \Rightarrow \dfrac{P_2}{P_1} = \dfrac{\eta_2}{\eta_1}$

$\dfrac{P_{\text{EtOH:H}_2\text{O}}}{P_{\text{ACN:H}_2\text{O}}} = \dfrac{0.002\ 3 \text{ kg/ms}}{0.000\ 80 \text{ kg/ms}} = 2.9; \quad \dfrac{P_{\text{MeOH:H}_2\text{O}}}{P_{\text{ACN:H}_2\text{O}}} = 2.0$

Pressure is almost 3-times higher with ethanol, and 2-times with methanol.

25-18. (a) Since the nonpolar compounds should become more soluble in the mobile phase, the retention time will be shorter in 90% methanol.

(b) At pH 3, the predominant forms are neutral RCO$_2$H and cationic RNH$_3^+$. The amine will be eluted first, since RNH$_3^+$ is insoluble in the nonpolar stationary phase.

(c) Polar compounds become less soluble in the mobile phase as the amount of water is decreased, so retention time will be greater in 90% acetonitrile.

(d) 2-Propanol has a higher eluent strength (0.60) than methyl *t*-butyl ether (0.48). Changing to 60% 2-propanol will increase the eluent strength, so retention times will be shorter in 60% 2-propanol.

High-Performance Liquid Chromatography 385

25-19. (a) Unretained component travels at the solvent velocity, u_x.

$$u_x = \frac{\text{column length}}{\text{transit time}} = \frac{4\,400 \text{ mm}}{(41.7 \text{ min})(60 \text{ s/min})} = 1.76 \text{ mm/s}$$

(b) $k = \dfrac{t_R - t_M}{t_M} = \dfrac{188.1 \text{ min} - 41.7 \text{ min}}{41.7 \text{ min}} = 3.51$

(c) $N = \dfrac{5.55\, t_R^2}{w_{1/2}^2} = \dfrac{5.55\,(188.1 \text{ min})^2}{(1.01 \text{ min})^2} = 192\,000$

$$H = \frac{(440 \text{ cm})(10^4\, \mu\text{m/cm})}{192\,000} = 22.9\ \mu\text{m}$$

(d) $\text{Resolution} = \dfrac{0.589\, \Delta t_R}{w_{1/2\text{av}}} = \dfrac{0.589\,(1.01 \text{ min})}{1.01 \text{ min}} = 0.589$

(e) $\alpha = \dfrac{t'_{R2}}{t'_{R1}} = \dfrac{194.3 \text{ min} - 41.7 \text{ min}}{193.3 \text{ min} - 41.7 \text{ min}} = 1.006_6$

(f) Increasing column length does not change α or k

$$\text{resolution} = \frac{\sqrt{N}}{4}\frac{(\alpha-1)}{\alpha}\left(\frac{k_2}{1+k_2}\right) \Rightarrow \frac{R_2}{R_1} = \frac{\sqrt{N_2}}{\sqrt{N_1}}$$

$$\frac{1.000}{0.589} = \frac{\sqrt{N_2}}{\sqrt{192\,000}} \Rightarrow N_2 = 5.5_3 \times 10^5$$

A column length of 440 cm gave $N = 1.92 \times 10^5$ plates. To obtain $5.5_3 \times 10^5$ plates, the column must be longer by a factor of

$$\frac{5.5_3 \times 10^5 \text{ plates}}{1.92 \times 10^5 \text{ plates}} = 2.8_8.$$

Required length = $(2.8_8)(4.40 \text{ m}) = 12.7$ m

(g) Adjust the flow rate to the optimum velocity in the van Deemter plot to obtain the lowest H and thereby increase N.

Decrease the mobile phase strength to increase the retention factor k. Change the solvent to change the relative retention.

(h) $\text{Resolution} \dfrac{\sqrt{N}}{4}\dfrac{(\alpha-1)}{\alpha}\left(\dfrac{k_2}{1+k_2}\right) = \dfrac{\sqrt{192\,000}\,(1.008_8 - 1)}{4\quad 1.008_8}\left(\dfrac{17.0}{1+17.0}\right) = 0.90$

25-20. (a) On (R,R)-stationary phase, (S)-gimatecan is eluted at 6.10 min. On (S,S)-stationary phase, (S)-gimatecan is retained more strongly and is eluted at 6.96 min. (R)-gimatecan *must have the exact opposite behavior*. It will be eluted at 6.96 min from (R,R)-stationary phase and at 6.10 min from (S,S)-stationary phase.

(b) With (S,S)-stationary phase, we observe a small peak at 6.10 min for (R)-gimatecan. This peak is well separated from the front of the big (S)-gimatecan peak centered at 6.96 min, so the two areas can be integrated and compared with each other. With (R,R)-stationary phase, we see the (S)-gimatecan peak at 6.10 min with no evidence of the minor (R)-gimatecan peak at 6.96 min. The minor peak is lost beneath the tail of (S)-gimatecan. If one enantiomer is in low concentration compared to the other, it is desirable to have the trace enantiomer eluted first. Chromatography on each enantiomer of the stationary phase enables us to unambiguously locate where each enantiomer of gimatecan is eluted, even though we do not have a standard sample of (R)-gimatecan.

(c) For the (S,S)-stationary phase, we have the following information:
(S)-gimatecan: $t_R = 6.96$ min $k = 1.50$
(R)-gimatecan: $t_R = 6.10$ min $k = 1.22$
Relative retention: $\alpha = \dfrac{k_2}{k_1} = \dfrac{1.50}{1.22} = 1.23$

(d) Resolution $= \dfrac{\sqrt{N}}{4}\dfrac{(\alpha-1)}{\alpha}\left(\dfrac{k_2}{1+k_2}\right) = \dfrac{\sqrt{6\,800}}{4}\dfrac{(1.23-1)}{1.23}\left(\dfrac{1.50}{1+1.50}\right) = 2.3$

which is more than adequate for "baseline" separation. Tailing of the peaks creates a little overlap, but it should not be very serious for an equal mixture of the enantiomers.

25-21. Peak areas will be proportional to molar absorptivity, since the number of moles of A and B are equal.

$$\dfrac{\text{Area of A}}{\text{Area of B}} = \dfrac{2.26\times 10^4}{1.68\times 10^4} = \dfrac{1.064\times h_A w_{1/2}}{1.064\times h_B w_{1/2}} = \dfrac{(128)(10.1)}{h_B(7.6)}$$

$\Rightarrow h_B = 126$ mm

25-22. (a) $V_M \approx Ld_c^2/2 = (5.0 \text{ cm})(0.46 \text{ cm})^2/2 = 0.53 \text{ cm}^3 = 0.53 \text{ mL}$

$t_M = V_M/F = (0.53 \text{ mL})/(1.4 \text{ mL/min}) = 0.38$ min for column A
$= (0.53 \text{ mL})/(2.0 \text{ mL/min}) = 0.26_5$ min for column B

(b) Morphine 3-β-D-glucuronide is more polar than morphine because of the added hydroxyl groups and the carboxylic acid. The more polar compound is less retained by the nonpolar reversed-phase column.

(c) $k = \dfrac{t_R - t_M}{t_M} = \dfrac{1.5 - 0.65}{0.65} = 1.3$ for morphine 3-β-D-glucuronide

$k = \dfrac{t_R - t_M}{t_M} = \dfrac{2.8 - 0.65}{0.65} = 3.3$ for morphine

(d) Bare silica is a polar, hydrophilic surface. Morphine should not be retained as strongly as the more polar morphine 3-β-D-glucuronide. The gradient goes to increasing H_2O for increasing polarity (that is, increasing solvent strength) to remove the more strongly adsorbed, more polar compound.

(e) $V_M = F t_M = (2.0 \text{ mL/min})(0.50 \text{ min}) = 1.0 \text{ mL}$

$k^* = \dfrac{t_G F}{\Delta \Phi V_M S} = \dfrac{(5.0 \text{ min})(2.0 \text{ mL/min})}{(0.4)(1.0 \text{ mL})(4)} = 6.2_5 = 6.2$

25-23. (a) Electrical power = current × voltage. Current is the rate of flow of charge through a circuit. It is analogous to the rate of flow of liquid through a column. Voltage is the potential difference driving charge through the wire. It is analogous to the pressure difference driving liquid through a column.

(b) $1 \text{ mL} = 1 \text{ cm}^3 = (10^{-2} \text{ m})^3 = 10^{-6} \text{ m}^3$
$1 \text{ mL/min} = 10^{-6} \text{ m}^3/60 \text{ s} = 1.67 \times 10^{-8} \text{ m}^3/\text{s}$
$3\,500 \text{ bar} = 3\,500 \times 10^5 \text{ Pa} = 3.5 \times 10^8 \text{ Pa}$
power = volume flow rate × pressure drop
$= (1.67 \times 10^{-8} \text{ m}^3/\text{s})(3.5 \times 10^8 \text{ Pa}) = 5.8 \text{ W}$

388   Chapter 25

25-24. (a) Below 210 nm near universal, above 210 nm selective for molecules with an absorbing chromophore

(b) almost any molecule, but not very sensitive

(c) near universal for non-volatile compounds

(d) near universal for non-volatile compounds

(e) compounds that can be oxidized or reduced

(f) molecules that fluoresce

(g) molecules that contain N

(h) ionic or ionizable compounds

25-25. (a) Many chromophores absorb ultraviolet radiation below 210 nm, so the detector is nearly universal below 210 nm. Only some molecules absorb at longer wavelengths. At 254 nm, the ultraviolet detector is selective for those molecules that absorb.

(b) The separation uses gradient elution with increasing methanol in CO_2. CO_2 is transparent above 190 nm. The ultraviolet cutoff for methanol is 205 nm. At 210 nm methanol absorbs light, and absorbance increases with increasing methanol. At 254 nm in the lower trace, methanol does not absorb, and so the baseline absorbance is unaffected by the mobile phase composition.

(c) The dip in the baseline occurs at 1.23 minutes.

$$t_M \approx \frac{L d_c^2}{2F} \approx \frac{(25 \text{ cm})(0.46 \text{ cm})^2}{2 \times 2 \text{ mL/min}} \approx 1.32 \text{ min}$$

25-26. (a)

CocaineH⁺
$C_{17}H_{22}NO_4$
m/z 304

(structure shown with HN⁺–CH₃, CO₂CH₃ group, and O–C(=O)–C₆H₅ ester group on bicyclic ring)

(b) The $C_6H_5CO_2$ group has mass 121 Da. Subtracting 121 from 304 gives 183 Da. m/z 182 probably represents cocaine after loss of $C_6H_5CO_2H$. The structure might be the one below or some rearranged form of it.

High-Performance Liquid Chromatography 389

$$\underset{\substack{C_{10}H_{16}NO_2 \\ m/z\ 182}}{\text{[structure: N-methyl tropane with } CO_2CH_3 \text{ and } +H\text{]}}$$

(c) The ion at *m/z* 304 was selected by mass filter Q1. Its isotopic partner containing ^{13}C at *m/z* 305 was blocked by Q1. Because the species at *m/z* 304 is isotopically pure, there is no ^{13}C-containing partner for the collision-induced dissociation product at *m/z* 182.

(d) For selected reaction monitoring, the mass filter Q1 selects just *m/z* 304, which eliminates components of plasma that do not give a signal at *m/z* 304. Then this ion is passed to the collision cell, in which it breaks into a major fragment at *m/z* 182 which passes through Q3. Few other components in the plasma that give a signal at *m/z* 304 also break into a fragment at *m/z* 182. The 2-step selection process essentially eliminates everything else in the sample and produces just one clean peak in the chromatogram.

(e) The phenyl group must be labeled with deuterium because the labeled product gives the same fragment at *m/z* 182 as unlabeled cocaine.

$$\underset{\substack{C_{17}D_5H_{17}NO_4 \\ m/z\ 309}}{\text{[labeled cocaine structure with }D_5\text{-phenyl]}} \longrightarrow \underset{\substack{C_{10}H_{16}NO_2 \\ m/z\ 182}}{\text{[fragment structure]}}$$

(f) First, we need to construct a calibration curve to get the response factor for cocaine compared to $^2H_5$-cocaine. We expect this response factor to be near 1.00. We would prepare a series of solutions with known concentration ratios [cocaine]/[$^2H_5$-cocaine] and measure the area of each chromatographic peak in the chromatography/atmospheric chemical ionization/selected reaction monitoring experiment. A graph would be constructed, in which (peak area of cocaine)/(peak area of $^2H_5$-cocaine) is plotted versus [cocaine]/[$^2H_5$-cocaine]. The slope of this line is the response factor.

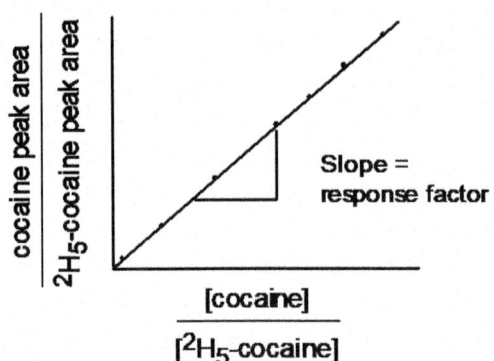

For quantitative analysis, a known amount of the internal standard $^2H_5$-cocaine is injected into the plasma. From the calibration curve, the relative peak areas tell us the relative concentrations of cocaine and the internal standard. From the known quantity of internal standard injected into the plasma, we can calculate the quantity of cocaine.

25-27. (a) Atmospheric pressure chemical ionization gives a prominent peak at m/z 234, which must be MH$^+$. The peak at m/z 84 is probably the fragment $C_5H_{10}N^+$, which might have the structure shown below.

In selected reaction monitoring, m/z 234 is selected by mass filter Q1 and m/z 84 is selected by mass filter Q3 in a triple quadrupole spectrometer.

(b) Deuterated internal standard has the formula $C_{14}H_{16}{}^2H_3O_2N$, with a nominal mass of 236. The protonated molecule is m/z 237. Cleavage of the C-C bond gives the same $C_5H_{10}N^+$ fragment as unlabeled Ritalin. The transition to monitor is m/z 237 → 84.

25-28. (a) To find k, measure the retention time for the peak of interest (t_R) and the elution time for an unretained solute (t_M). Then use the formula $k = (t_R - t_M)/t_M$. The resolution between neighboring peaks is the difference in their retention time divided by their average width at the baseline.

(b) (i) Unretained solutes such as uracil or sodium nitrate could be run and observed with an ultraviolet detector. (ii) t_M is usually the time when the first baseline disturbance is observed. (iii) Alternatively, the formula $t_M \approx L d_c^2 /(2F)$ can be used, where L is the length of the column (cm), d_c is the column diameter (cm), and F is the flow rate (mL/min).

(c) (i) Unretained solutes such as toluene could be run and observed with an ultraviolet detector. (ii) and (iii) from part (b) will also work in HILIC.

(d) $t_M \approx L d_c^2/(2F) = (15)(0.46)^2/(2 \cdot 1.5) = 1.0_6$ min

t_M does not depend on particle size. The estimate is 1.0_6 min for both 5.0- and 3.5-μm particles.

25-29. *Extra-column volume* is the volume of the system (not including the chromatography column) from the point of injection to the point of detection. *Dwell volume* is the volume of the system from the point of mixing solvents to the beginning of the column. Excessive extra-column volume causes peak broadening, particularly of early eluting peaks. In gradient elution, dwell volume determines the time from the initiation of a gradient until the gradient reaches the column. The greater the dwell volume, the more the delay between initiating a gradient and the actual increase of solvent strength on the column.

25-30. A rugged procedure should not be seriously affected by gradual deterioration of the column, *small* variations in solvent composition, pH, and temperature, or use of a different batch of the same stationary phase. A procedure should be rugged so that inevitable, small variations in conditions do not substantially affect the outcome of the separation.

25-31. $0.5 \leq k \leq 20$; resolution ≥ 2; operating pressure ≤ 20 MPa; $0.9 \leq$ asymmetry factor ≤ 1.5

25-32. Run a wide gradient (such as 5%B to 100%B) in gradient time t_G selected to produce $k^* \approx 5$ in Equation 25-8. Measure the difference in retention time (Δt) between the first and last peaks. Use a gradient if $\Delta t/t_G > 0.40$ and use isocratic elution if $\Delta t/t_G < 0.25$. For $0.25 < \Delta t/t_G < 0.40$, either may be appropriate.

25-33. The first steps are to (1) determine the goal of the analysis, (2) select a method of sample preparation, and (3) choose a detector that allows you to observe the desired analytes in the mixture. The next step could be a wide gradient elution to determine whether or not an isocratic or gradient separation is more appropriate. If isocratic separation is chosen, vary %B until criterion for good retention ($0.5 \leq k \leq 20$) is met. If adequate resolution is not attained, try minor adjustments in %B, a different organic solvent of equivalent mobile phase strength, or a different column. Finally, select column length or particle size, either to increase plate number if resolution < 2 or to shorten separation time if resolution >> 2.

25-34. (a) 53% tetrahydrofuran in water

(b) Mix 530 mL tetrahydrofuran and 470 mL H_2O. The total volume of the mixture will not equal 1 L. Do not add any additional tetrahydrofuran or water, as this would alter the composition of the mixture.

(c) Ultraviolet detection might be affected by the high wavelength cutoff (212 nm) of tetrahydrofuran. Tetrahydrofuran is also incompatible with polyether ether ketone plastic components of many HPLC systems.

25-35. Chromatography is conducted with four conditions: (A) high %B, low T, (B) high %B, high T, (C) low %B, high T, and (D) low %B, low T. Based on the appearance of the chromatograms, combinations between the points A, B, C, and D can be explored for further improvement in the separation.

25-36. Peak 5 has a retention time (t_R) of 11.0 min for 50% B. The retention factor is $k = (t_R - t_M)/t_M = (11.0 - 2.7)/2.7 = 3.1$. When B is reduced to 40%, the rule of three predicts $k = 3(3.1) = 9.3$. Rearranging the definition of retention factor, we find $t_R = t_M k + t_M = t_M (k + 1)$. We predict for 40% B $t_R = t_M (k + 1) = (2.7)(9.3 + 1) = 27.8$ min. Observed retention time at 40% B is 20.2 min.

25-37.

	A	B	C	D	E	F	G	H	I	J
1					t_M = 2.7 min					
2	Φ	Retention time t_R (min)			Retention factor $k = (t_R - t_M)/t_M$				log k	
3		peak 6	peak 7	peak 8	peak 6	peak 7	peak 8	peak 6	peak 7	peak 8
4	0.90	4.4	4.4	4.9	0.630	0.630	0.815	-0.201	-0.201	-0.089
5	0.80	4.5	4.5	5.1	0.667	0.667	0.889	-0.176	-0.176	-0.051
6	0.70	5.6	5.6	7.3	1.074	1.074	1.704	0.031	0.031	0.231
7	0.60	8.2	8.2	12.2	2.037	2.037	3.519	0.309	0.309	0.546
8	0.50	13.1	13.6	24.5	3.852	4.037	8.074	0.586	0.606	0.907
9	0.40	24.8	27.5	65.1	8.185	9.185	23.111	0.913	0.963	1.364
10	0.35	37.6	44.2	125.2	12.926	15.370	45.370	1.111	1.187	1.657

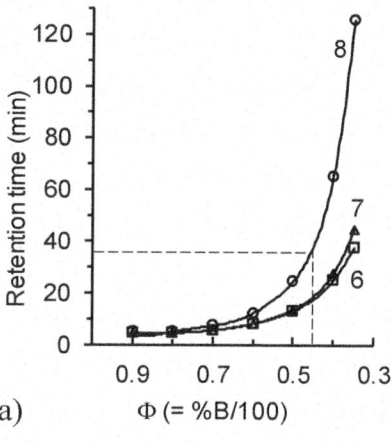
(a)

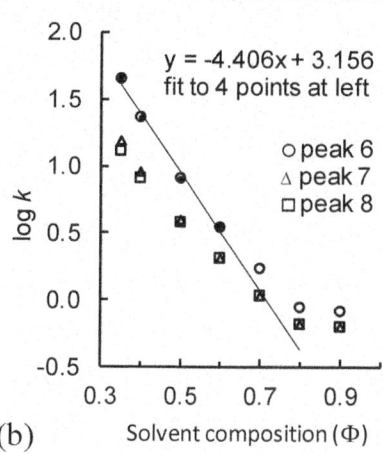
(b)

(a) At 45% B, we could estimate that Peak 8 will be eluted halfway between the times for 40% B and 50% B, which is about 45 min. The fit to the curve above suggests that 36 min is a more realistic estimate.

(b) Columns E–G are the calculated retention factors k, and columns H–J are the log k. The first obvious point is that log k versus Φ does not follow a straight line over a wide range of solvent composition. The straight line going through the four points for Peak 8 from Φ = 0.35 to 0.6 is
log k = –4.406Φ + 3.156. At Φ = 0.45, we compute log k = 1.173 and
k = 14.9. We compute $t_R = t_M (k+1)$ = 42.9 min. If we had only taken the first three points (Φ = 0.35 to 0.5), we would find log k = –4.936Φ + 3.366. At Φ = 0.45, we compute log k = 1.114, k = 13.96, and t_R = 40.4 min.

(c) The gradient goes from 40–70% acetonitrile over 30 minutes. Therefore the mobile phase changes 10% every 10 minutes.

Time (min)	%B	Retention factor peak 6	peak 8
0	40	8.185	23.111
10	50	3.852	8.074
20	60	2.037	3.519
30	70	1.074	1.704

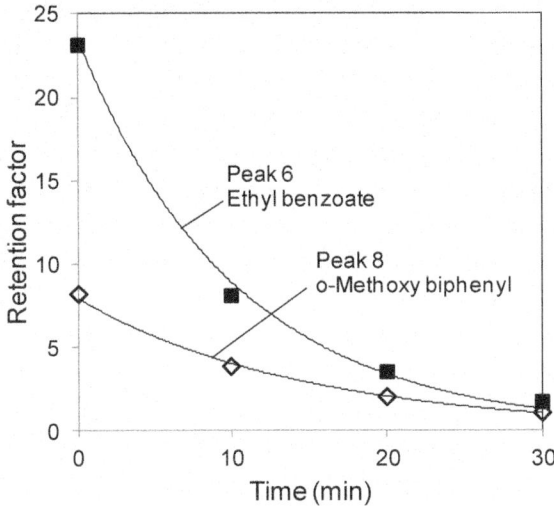

The retention factor of both compounds is very high at the start of the gradient. As the gradient proceeds, retention factors decrease exponentially. At the end of the gradient the compounds are almost unretained.

(d) Compounds spend much of the gradient retained near the column inlet. But by the time a compound elutes, it is almost unretained. At the time of elution, the peak width is similar to that of a weakly retained compound.

25-38. In the nomograph in Figure 25-31, a vertical line at 48% methanol intersects the acetonitrile line at 38%.

25-39. (a) Lower solvent strength usually increases the difference in retention between different compounds. Use a lower percentage of acetonitrile.

(b) In normal-phase chromatography, solvent strength increases as the solvent becomes more polar, which means increasing methyl *t*-butyl ether. We need a higher concentration of hexane to lower the solvent strength, increase the retention times, and probably improve resolution.

25-40. (a) Cumulative impact scales with the mass of solvent used. Fifty separations use 50 times the mass of mobile phase of one separation.

Cumulative impact per run = 18.0

Cumulative impact of 50 runs = 50 × 18.0 = 900

(b) 60% methanol is equivalent mobile phase strength to 50% acetonitrile

Mass methanol per run = 10 min × 1.0 mL/min × 0.60 × 0.786 g/mL = 4.72 g

Cumulative impact per run = (1.91 + 0.43 + 0.32)(4.72 g) = 12.6

Cumulative impact of 50 runs = 50 × 12.6 = 630

(c) Changing diameter from 4.6 to 2.1 mm, reduces volume by a factor of $(2.1/4.6)^2 = 0.208$. F should be reduced by the same factor.

F(small column) = 0.208 × 1.0 mL/min = 0.208 mL/min

Mobile phase is same as (a), so retention factor is same as in (a). With k and u_x constant, retention times are the same as (a), so separation takes 10 min.

Mass per run = 10 min × 0.208 mL/min × 0.50 × 0.792 g/mL = 0.824 g

Cumulative impact per run = (2.72 + 1.06 + 0.77)(0.824 g) = 3.75

Cumulative impact of 50 runs = 50 × 3.75 = 187

(d) Cumulative impact increases with number of separations performed. Reducing the volume of solvent per separation (by using a narrower column) reduces the cumulative impact more than changing to a greener solvent, but solvent choice does have a significant impact.

(e) Particle size and linear velocity are the same in parts (a) and (c), so pressure is the same. Therefore, the higher-pressure limit in UHPLC is not the reason for the change. Narrow columns require smaller extra-column volumes to ensure that broadening outside the column is minimal. UHPLC systems have smaller-extra-column volumes.

25-41. (a) A is a weak acid with pK_a = 7.7 and k_{HA} = 6.5. B is a weak acid with pK_a = 3.7 and k_{HA} = 5.9. C is a weak base with pK_a = 4.7 and k_B = 3.2.

(b) Chromatography of C would be least rugged at pH ≈ 3 to 6.5 because retention changes with pH in the vicinity of pK_a.

(c) Mixtures of an acid and its conjugate base have buffering capacity within pK_a ± 1. To study pH 2–11, buffers of different pK_a are required.

25-42. (a) pH 2.0 provides retention within 0.5 ≤ k ≤ 20. Minor changes in pH would not significantly affect retention.

(b) pH ≈ 4. Lower pH would result in insufficient retention of 4-methylaniline. Higher pH would not provide rugged methods, as retention of benzoic acid and 3-methyl benzoic acid would vary significantly if the pH changed slightly. However, the retention factors for all three compounds would be influenced by small changes in pH.

(c) pH 7.5. pH greater than 8 would provide stable retention of 4-methylaniline and codeine, but little retention of 4-nitrophenol. Also, silica bonded phase columns are typically limited to the pH 2–8 range. pH below 7 would yield inadequate retention of codeine.

25-43. Acetophenone is neutral at all pH values. Its retention is nearly unaffected by pH. For salicylic acid, we expect the neutral molecule, HA, to have some affinity for the C_8 nonpolar stationary phase and the ion, A^-, to have little affinity for C_8. Salicylic acid is predominantly HA below pH 2.97 and A^- above pH 2.97. At pH 3, there is nearly a 1:1 mixture of HA and A^-, which is moderately retained on the nonpolar column. At pH 5 and 7, more than 99% of the molecules are A^-, so retention is weak (small retention factor).

Ionic forms of nicotine ought to have low affinity for the nonpolar stationary phase and the neutral molecule would have some affinity. Abbreviating nicotine as B, the form B is dominant above pH = pK_2 = 7.85. BH^+ is dominant between pH 3.15 and 7.85. BH_2^{2+} is dominant below pH 3.15. B does not become appreciable until pH ≈ 7, so the retention factor is low below pH 7 and increases at pH 7.

25-44. (a) $\Delta t/t_G = (50-22)/60 = 0.47$. Because $\Delta t/t_G > 0.40$, gradient elution is suggested.

(b) At $t = 22$ min, the solvent composition entering the column can be calculated by linear interpolation: $5 + \frac{22}{60}(100-5) = 39.8\%$. At 50 minutes, the composition is $5 + \frac{50}{60}(100-5) = 84.2\%$. A reasonable gradient for the second experiment is from 40 to 85% acetonitrile in 60 min.

25-45. (a) (1) Change the solvent strength by varying the fraction of each solvent. (2) Change the temperature. (3) Change the pH (in small steps). (4) Use a different solvent. (5) Use a different kind of stationary phase.

(b) Use a slower flow rate, a different temperature, a longer column, or a smaller particle size.

25-46. (a) Start with conditions to give $k^* = 5$ and assume that $S = 4$ for molecules in the mixture. $V_M \approx L d_c^2/2 = (15 \text{ cm})(0.46 \text{ cm})^2/2 = 1.5_9$ mL. Particle size does not come into the calculation.

$$t_G = \frac{k^* \Delta \Phi V_M S}{F} = \frac{(5)(0.9)(1.59 \text{ mL})(4)}{(1.0 \text{ mL/min})} = 29 \text{ min}$$

(b) $k^* = \dfrac{t_G F}{\Delta \Phi V_M S} = \dfrac{(11.5 \text{ min})(1.0 \text{ mL/min})}{(0.14)(1.59 \text{ mL})(4)} = 12.9$

The large column has the same length as the small column, but the diameter is increased from 0.46 to 1.0 cm. The volume increases by a factor of $(1.0/0.46)^2 = 4.7$. Therefore, we increase the flow rate and the sample loading by a factor of 4.7. Flow rate = 4.7 mL/min and sample load = 4.7 mg. The gradient time is unchanged at 11.5 min. For the large column, $V_M \approx L d_c^2/2 = (15 \text{ cm})(1.0 \text{ cm})^2/2 = 7.5$ mL and

$$k^* = \frac{t_G F}{\Delta \Phi V_M S} = \frac{(11.5 \text{ min})(4.7 \text{ mL/min})}{(0.14)(7.5 \text{ mL})(4)} = 12.9$$

25-47. Spreadsheet and figure are the same as in the textbook.

25-48. (a) $t_M = 1.74$ min. Retention decreases with increasing % methanol, so methanol is the strong mobile phase component. No peak elutes before 1.74 min. It takes 1.74 min for an unretained compound to elute from the column.

(b) Retention time of last peak is 2.05 min. The retention factor is $k = (t_R - t_M)/t_M = (2.05 - 1.74)/1.74 = 0.18$.

(c) Retention times and retention factors are shown in spreadsheet. Retention factor of last peak increases 2.5-fold for each 10% decrease in methanol.

	A	B	C	D	E	F
1	% Methanol	Φ	t_R, last (min)	k_{last}	$k_n/k_{n+10\%}$	log k_{last}
2	70	0.70	2.05	0.18		−0.749
3	60	0.60	2.50	0.44	2.45	−0.360
4	50	0.50	3.64	1.09	2.50	0.038
5	40	0.40	6.52	2.75	2.52	0.439
6	B2 = A2/100			E3 = D3/D2		
7	D2 = (C2−1.74)/1.74			F2 = LOG10(D2)		

(d) The plot of log k versus % methanol yields a straight line. Interpolating at 45% yields $y = 0.24$, which corresponds to $k = 10^{0.24} = 1.74$. $t_R = (1 + k)t_M = 4.77$ min, which is what is observed with *HPLC Teaching Assistant*.

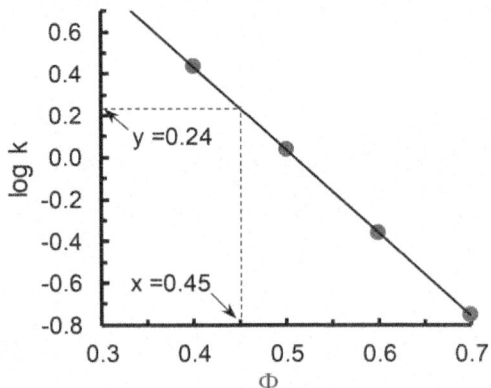

(e) At 30% methanol, only 3 peaks are observed due to coelution of some analytes. The log k versus % methanol plots in the upper right of the Excel screen shows lines for 3 components converge at 30% methanol.

(f) A mobile phase of 52% methanol yields resolution of 1.5 in 3.5 min. Using 45% methanol provides resolution of 1.75, and so would be more rugged. Separation time does increase to ~5 min.

High-Performance Liquid Chromatography

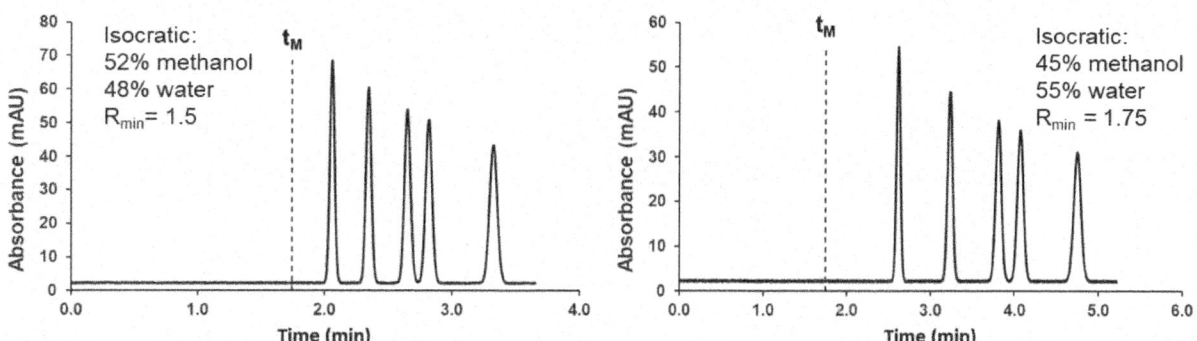

25-49. (a) $t_M \approx \dfrac{Ld_c^2}{2F} = \dfrac{(10.0 \text{ cm})(0.46 \text{ cm})^2}{2(2.5 \text{ mL/min})} = 0.42$ min

There is a negative/positive peak at 0.42 min. Such up/down peaks are caused by differences in refractive index of the injection solvent and eluent.

(b) $t_D = V_D/F = 1.00$ mL / (2.5 mL/min) = 0.40 min

(c) Time gradient gets to detector = $t_M + t_D$ = 0.42 min + 0.40 min = 0.82 min

Baseline starts to rise at ~0.82 min. Shifts in baseline occur if solvent B absorbs differently to solvent A.

25-50. (a) $\Delta t = t_{R5} - t_{R1}$ = 22.69 min – 9.26 min = 13.43 min

$\Delta t/t_G$ = 13.43 min/40 min = 0.336; $0.25 < \Delta t/t_G < 0.40$, so either isocratic or gradient could be used.

(b) t_{R1} = 9.26 min. % Methanol = 10% + (90 – 10)/40 min × 9.26 min = 28.52% ≈ 29%. For t_{R5} = 22.69 min, % methanol is 55.34%. With a gradient from 29–55% over 40 min, t_{R1} = 4.4 min, t_{R5} = 24.7 min, and minimum resolution is 1.3.

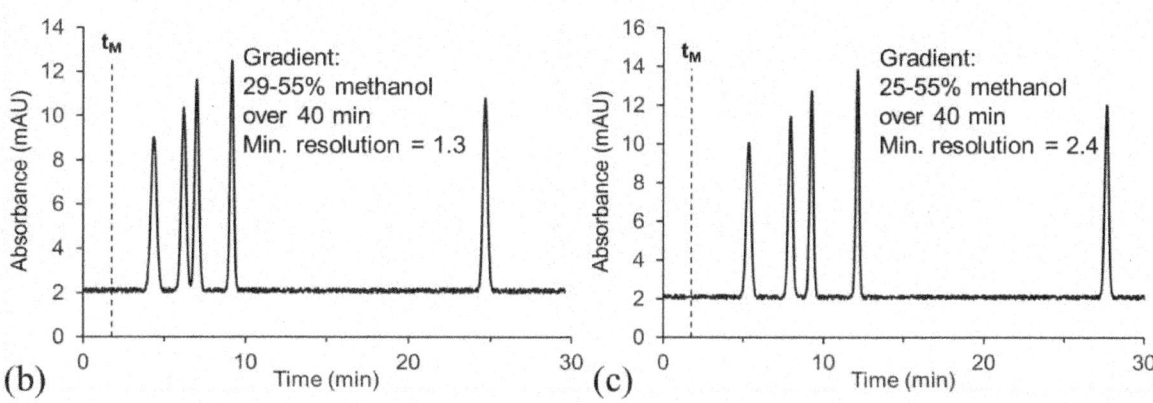

(c) The minimum resolution in gradient (b) are peaks 2 and 3 which are early in the chromatogram. Early peaks are close to t_M. Lowering initial % methanol increases retention of early peaks, and improves resolution. Altering final % methanol has little effect on resolution of early peaks. A gradient of 25–55% methanol over 40 minutes yields retention times from 5.4 to 27.7 min with minimum resolution of 2.4.

(d) A 25–55 methanol gradient over 10 minutes provides resolution ≥1.9.

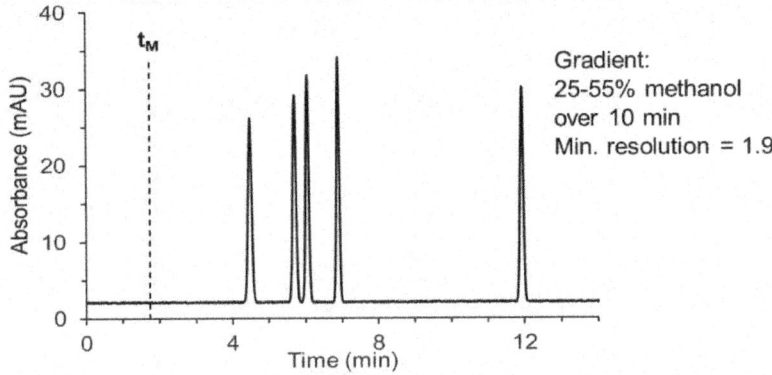

The minimum resolution is close to our target value of 2. The method would be satisfactory.

25-51. (a) Linearity was studied from 0.01–0.50 mg mL^{-1}. Limit of detection was 0.004 mg mL^{-1}.

(b) From the second column of the Introduction on the first page, gas chromatography with flame ionization, nitrogen phosphorus detection, and mass spectrometric detection and HPLC with electrochemical, UV or mass spectrometric detection have been used.

(c) From "HPLC-DAD conditions" on the second page, the column was an Agilent Zorbax Eclipse XDB-RP-C18 of dimensions 150 mm × 4.6 mm inner diameter with 5 μm particles. A guard column was also used.

(d) The first paragraph of the Results and Discussion states peaks were identified by comparing retention time with standards as illustrated in Figure 3, and by comparing the absorbance spectra collected using the photodiode array detector with the spectra of standards (Figure 2).

(e) In the Results and Discussion, the method validation is detailed in a section of that name. International Conference on Harmonization (ICH) guidelines were followed and the method was validated based on the specificity, linearity, limit of detection (LOD), limit of quantification (LOQ), precision, accuracy, and robustness.

(f) Robustness is defined on page 5 of the article, within the Results and Discussion section, in the section titled "Robustness." Robustness is a measure of an analytical method's ability to remain unaffected by small but deliberate variations in the method parameters.

25-52. (a) Abstract states the method can be used within 0.4 to 25 mg mL^{-1}.

(b) The Abstract mentions colorimetric, electrophoretic, and immunological assays. The second column of the Introduction on the first page goes into more detail, also mentioning a fluorometric method.

(c) From the Materials and Methods section on the second page, under the subtitle Equipment, the column is a C_4 (butyl) bonded phase of dimensions 250 mm × 4.6 mm inner diameter with 5 μm particles with a pore diameter of 300 Å. A guard column was also used.

(d) Table 1 shows that the gradient went from 20% of mobile phase B to 60% of B over 10 minutes. There was another 10 minutes devoted to washing the column with 100% mobile phase B and re-equilibrating the column with 20% B.

(e) From the Materials and Methods section on the second page, under the subtitle Method Validation, peak identity, linearity, range, precision, accuracy, robustness, specificity, and sensitivity and were all assessed.

(f) In the Discussion section at the top left of the sixth page it is noted that human serum albumin is a relatively large protein (66–67 kDa). The pores must be big enough to allow entry of such a large molecule. Most of the surface area of the stationary phase is inside the pores.

CHAPTER 26
CHROMATOGRAPHIC METHODS AND CAPILLARY ELECTROPHORESIS

26-1. Increased cross-linking gives decreased swelling, increased exchange capacity and selectivity, but longer equilibration time.

26-2. One way is to rinse a column containing a weighed amount of resin extensively with NaOH to regenerate the columns so that all ion-exchange sites are loaded with OH^-. After thoroughly washing with water to remove excess NaOH, elute the column with a large quantity of aqueous NaCl to displace OH^-. Then titrate the eluate with standard HCl to measure the moles of displaced OH^-.

26-3. Particles pass through 200 mesh (75 μm) sieve and are retained by 400 mesh (38 μm) sieve. 200/400 mesh particles are smaller than 100/200 mesh particles.

26-4. (a) Ion-exchange retention is largely based on ion charge. Pyruvate is bracketed by F^- and Cl^-, so probably –1; 2-oxovalerate is bracketed by Cl^- and NO_2^-, and so also –1; maleate is between CO_3^{2-} and SO_4^{2-} and so is probably –2.

(b) In addition to the charge of an ion, ion exchange also depends on the polarizability of the ion. Iodide is a very polarizable anion. Its electron cloud is greatly deformed by the positive charges on the anion-exchange resin. The resultant induced dipole strongly binds iodide to the resin.

26-5. (a) As pH is lowered the anionic protein becomes protonated, so the magnitude of the negative charge decreases. The protein becomes less strongly retained by the anion-exchange gel.

(b) As the ionic strength of eluent is increased, the protein will be displaced from the gel by the increasing concentration of anions in the eluent.

26-6. The pK_a values are: NH_4^+ (9.24), $CH_3NH_3^+$ (10.64), $(CH_3)_2NH_2^+$ (10.77), and $(CH_3)_3NH^+$ (9.80). If the four ammonium ions are adsorbed on a cation exchange resin at, say, pH 7, they might be separated by elution with a gradient of increasing pH. The anticipated order of elution is $NH_3 < (CH_3)_3N < CH_3NH_2 < (CH_3)_2NH$. We should not be surprised if the elution order were different, since steric and hydrogen bonding effects could be significant determinants of the selectivity coefficients. It is also possible that elution with a constant pH (of, say, 8) might separate all four species from each other.

26-7. Deionized water has had cations and anions removed. Deionized water has been passed through ion exchangers to convert cations to H^+ and anions to OH^-, making H_2O. Nonionic impurities (such as organic compounds) are not removed by this process, but can be removed by activated carbon.

26-8. (a) $V_{seawater} = 10$ mL/min $\times$ 17 h $\times$ 60 min/h $= 10$ L

The Fe^{3+} from 10 L of seawater was eluted from the column using 10 mL of acid. The concentration increased by a factor of

$$\frac{10 \text{ L} \times 1\,000 \text{ mL/L}}{10 \text{ mL}} = 1\,000$$

(b) $[Fe^{3+}]_{seawater} = \dfrac{57 \text{ nM} \times 10 \text{ mL}}{10 \text{ L} \times 1\,000 \text{ mL/L}} = 0.057$ nM $= 57$ pM

(c) 0.2 ppm Fe $= \dfrac{(0.2 \text{ mg/L})/(1\,000 \text{ mg/g})}{55.845 \text{ g/mol}} = 3.6 \times 10^{-6}$ M

$[Fe^{3+}]_{1.5M \text{ } HNO_3} \leq \dfrac{1.5 \text{ M}}{15.7 \text{ M}} (3.6 \times 10^{-6} \text{ M}) = 3.4 \times 10^{-7}$ M $= 340$ nM

apparent $[Fe^{3+}]_{seawater} \leq$ (concentration factor) $\times [Fe^{3+}]_{1.5M \text{ } HNO_3}$
$\leq 1\,000 \times 340$ nM $= 340\,000$ nM (6 million times greater than $[Fe^{3+}]_{seawater}$)

High-purity acids and reagents are essential when performing trace analysis.

26-9. The sum of anion charge in the spreadsheet is $-0.001\,59$ M, and the sum of cation charge is $0.002\,02$ M. Either some of the ion concentrations are inaccurate, or there are other ions in the pondwater that were not detected. For example, there could be large organic anions derived from living matter (such as humic acid from plants) that are not detected in this experiment.

	A	B	C	D	E	F
1	Ion	Formula mass	Concentration		Ion	Equivalents
2		(g/mol)	(µg/mL)	(mol/L)	charge	(mol/L)
3	Fluoride	19.00	0.26	1.37E-05	-1	-1.37E-05
4	Chloride	35.45	43.6	1.23E-03	-1	-1.23E-03
5	Nitrate	62.00	5.5	8.87E-05	-1	-8.87E-05
6	Sulfate	96.06	12.6	1.31E-04	-2	-2.62E-04
7					Sum of anion charge =	-1.59E-03
8	Sodium	22.99	2.8	1.22E-04	1	1.22E-04
9	Ammonium	18.04	0.2	1.11E-05	1	1.11E-05
10	Potassium	39.10	3.5	8.95E-05	1	8.95E-05
11	Magnesium	24.31	7.3	3.00E-04	2	6.01E-04
12	Calcium	40.08	24.0	5.99E-04	2	1.20E-03
13					Sum of cation charge =	2.02E-03

26-10. (a) For ion exclusion, the stationary phase has a fixed charge of the same sign as the ions separated. For the carboxylic acid separation, the column has a fixed negative charge. This would be a cation-exchange column if used for ion exchange chromatography.

(b) At pH 1.9 (0.1% trifluoroacetic acid), oxalic acid is more dissociated than glyoxylic acid, which is more dissociated than lactic acid. The greater the average charge of the compound, the more it is excluded from the negatively charged ion-exchange resin and the more rapidly it is eluted.

(c) Weaker acids would be dissociated even less than lactic acid, and so would penetrate the stationary phase more, and so be more retained.

(d) $\text{Resolution} = \dfrac{0.589 \Delta t_R}{w_{1/2\,av}} = \dfrac{0.589(4.5 \text{ min} - 2.4 \text{ min})}{0.5 \text{ min}} = 2.4_7 \approx 2.5$

Resolution of 2.5 is more than baseline resolution. Peaks are well separated.

(e) Matrix matched external samples could be used, but with various biological fluids being analyzed it would be difficult to exactly match the matrix of the standards with those of the samples. Standard addition would correct for matrix effects. The authors used an internal standard of ^{13}C-labeled oxalic acid. A stable-isotope labeled internal standard corrects for losses in sample handling, and co-elutes with the analyte and so also corrects matrix effects.

26-11. The separator column separates ions by ion exchange, while the suppressor exchanges the counterion with either H^+ or OH^- to neutralize the eluent and reduce its conductivity. After separating cations in the cation-exchange column, the suppressor exchanges Cl^- for OH^- to convert H^+Cl^- eluent into H_2O.

26-12. (a) $K^+(\text{reservoir}) = (0.75)(1.5 \text{ L})\left(2.0\,\dfrac{\text{mol } K_2HPO_4}{L}\right)\left(2.0\,\dfrac{\text{mol } K^+}{\text{mol } K_2HPO_4}\right) = 4.5 \text{ mol}$

$\text{Flow rate} = \left(20 \times 10^{-3}\,\dfrac{\text{mol KOH}}{L}\right)\left(0.001\,0\,\dfrac{L}{\text{min}}\right) = 2.0 \times 10^{-5}\,\dfrac{\text{mol KOH}}{\text{min}}$

$\text{Time available} = \dfrac{4.5 \text{ mol K}}{2.0 \times 10^{-5}\,\dfrac{\text{mol KOH}}{\text{min}}} = 2.25 \times 10^5 \text{ min}$

$\dfrac{2.25 \times 10^5 \text{ min}}{60 \text{ min/h}} = 3.8 \times 10^3 \text{ h}$

(b) A flow of 5.0 mM KOH at 1.0 mL/min provides

$(5.0 \times 10^{-3} \text{ mol KOH/L})(0.001\ 0 \text{ L/min}) = 5.0 \times 10^{-6}$ mol KOH/min

$$\frac{5.0 \times 10^{-6} \text{ mol KOH/min}}{60 \text{ s/min}} = 8.33 \times 10^{-8} \text{ mol KOH/s}$$

One electron provides one OH⁻ at the cathode, so the current must provide 8.33×10^{-8} mol e⁻/s. We multiply by the Faraday constant to convert moles of electrons into coulombs:

$(8.33 \times 10^{-8} \text{ mol e}^-/\text{s})(9.648\ 5 \times 10^4 \text{ C/mol e}^-)$
$= 8.0 \times 10^{-3}$ C/s $= 8.0 \times 10^{-3}$ A $= 8.0$ mA.

To produce 0.10 M KOH at 1.0 mL/min requires 20 times as much current, because the concentration of KOH is 20 times higher than 5.0 mM. The current at the end of the gradient will be (20)(8.0 mA) = 160 mA = 0.16 A.

26-13.

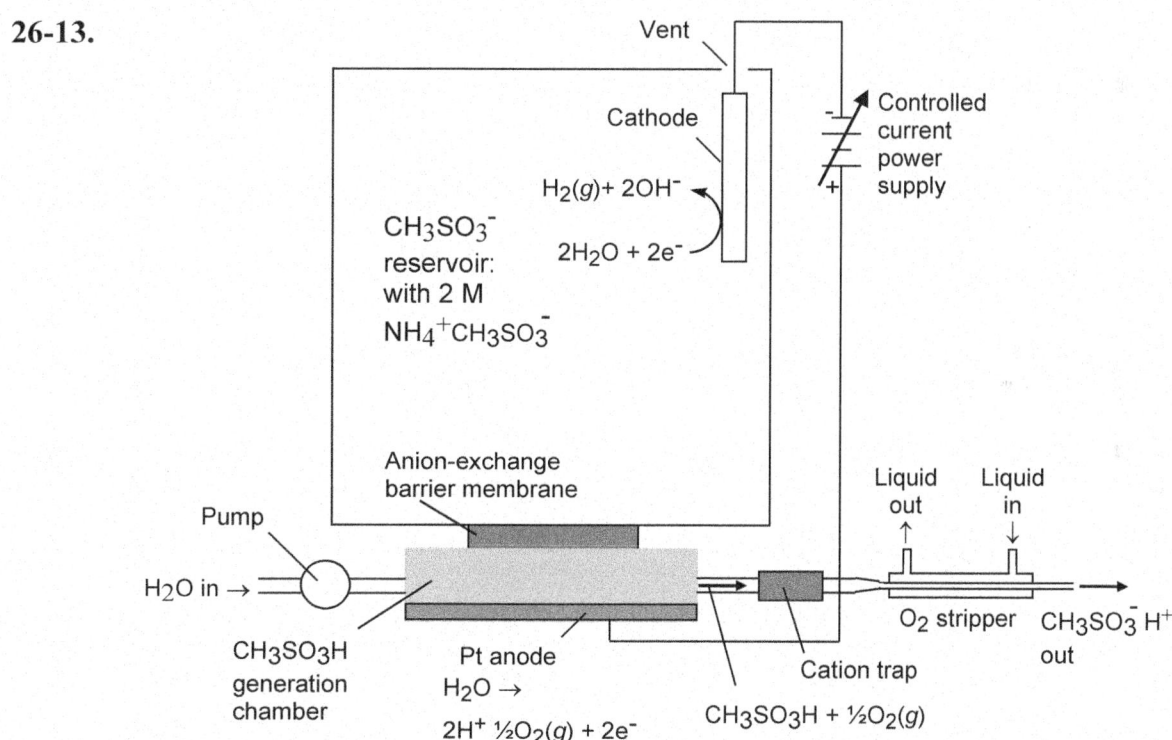

26-14. (a) Suppression converts analyte ions X⁻ into their conjugate acids HX. If HX is a strong acid such as HCl or HBr, it fully dissociates to yield a high conductivity signal. If HX is a very weak acid, such as HCN (pK_a = 9.21) or boric acid (pK_a = 9.237), very little HX deprotonates and so the conductivity signal is low.

(b) In the suppressor, H⁺ replaces Na⁺, to yield the protonated form of the eluent. For carbonate and bicarbonate, the suppression product is H_2CO_3. Carbonic acid is a weak acid (pK_a = 10.329), which dissociates to sufficient ions to increase the background conductivity.

26-15. (a) Conductivity detection: $y = (298 \pm 5)x + (1.1 \pm 1.5)$; $R^2 = 0.998\,5$. Residuals are small, intercept is near zero, and R^2 is almost 0.999. The response is linear.

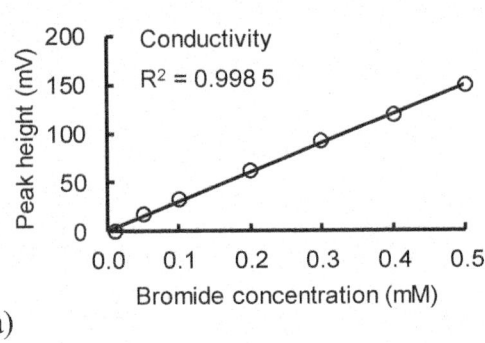
(a)

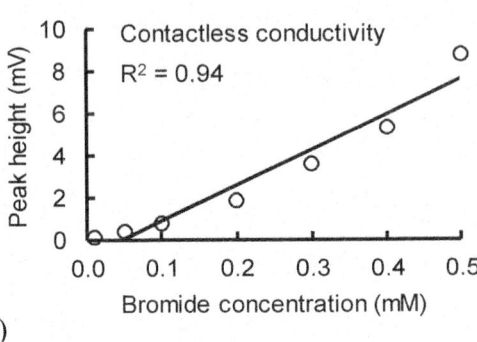
(b)

(b) Contactless conductivity detection: $y = (16.8 \pm 1.8)x + (-0.76 \pm 0.51)$; $R^2 = 0.94$. Residuals are large and show U-shaped pattern. Intercept is far from zero. R^2 is much lower than 0.999. The response is severely nonlinear.

A line through 0.01 to 0.20 mM yields $R^2 = 0.992\,9$, which does not seem too bad. But the intercept (–0.06) is still negative, and a U-shaped residual pattern is still noticeable, although reduced. But wait until you see the answer to (d).

(c) $1.46 = 29.84x^2 + 1.906x + 0.190 \Rightarrow 0 = 29.84x^2 + 1.906x - 1.27_0$

$$x = \frac{-b \pm \sqrt{b^2 - 4ac}}{2a} = \frac{-1.906 \pm \sqrt{(1.906)^2 - 4(29.84)(-1.27_0)}}{2(29.84)}$$

$= -0.241$ or 0.177

A negative concentration makes no sense, so 0.177 mM is the correct answer.

(d) Using the full range calibration in (b), a height of 1.46 mV yields 0.132 mM.

$$\text{Error} = 100 \times \frac{\text{value found} - \text{known value}}{\text{known value}} = 100 \times \frac{0.132 - 0.177}{0.177} = -25.4\%$$

Using the calibration from 0.1 to 0.20 mM yields 0.165 mM. While closer to the known value, the error is still –6.8%.

26-16. This is an example of *indirect detection*. Eluent contains naphthalenetrisulfonate, which absorbs at 280 nm. Charge balance dictates that when one of the analyte anions is emerging from the column, there must be less naphthalenetrisulfonate anion emerging. Since analytes do not absorb as strongly at 280 nm, the absorbance is negative with respect to the steady baseline.

26-17. (a) Sodium octyl sulfate partitions into the nonpolar stationary phase making the particles cation exchangers. The surface charge forms an ion-pair with NE or DHBA. Other ions in the eluent compete with NE or DHBA, and slowly elute them from the column by ion exchange.

(b) Construct a graph of (peak height ratio) vs. (added concentration of NE). The x-intercept gives [NE] = 29 ng/mL.

Added NE	signal
0	0.298
12	0.414
24	0.554
36	0.664
48	0.792

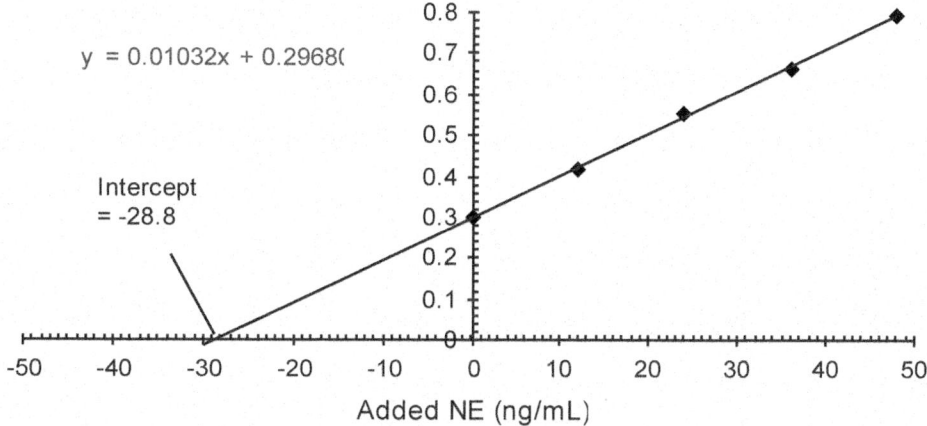

26-18. Mixed-mode chromatography retains analytes by more than one retention mechanism, for example reversed phase and ion exchange. Ion-pair chromatography also retains analytes through reversed phase and ion exchange, but ion-pair chromatography requires an ion-pairing agent in the mobile phase. In mixed-mode chromatography, the stationary phase possesses both reversed phase and ion exchange character, and so no additive needs to be added to the mobile phase.

26-19. (a) Affinity chromatography. The strong and specific interaction between antibodies and antigens yields strong retention on an affinity column which allows the antibody to be cleanly separated from matrix components.

(b) Size exclusion. The protein is too large to fit into the pores of the gel and flows quickly through the column. Small molecules such as salts diffuse into the pores of the gel, and so are eluted later at volume V_M of the column.

(c) Size exclusion. Different polymer chain lengths are eluted at different volumes. Longer chains emerge earlier and shorter chains emerge later. There are so many possible chain lengths that individual chain lengths are not resolved. Only a single broad peak is observed. The width of this unresolved peak reflects the mass distribution of the polymer.

(d) Hydrophobic interaction. The difference in surface hydrophobicity of the two proteins suggests they can be separated by hydrophobic interaction chromatography. The difference in the protein masses is too small to yield a separation by size exclusion chromatography.

26-20. (a) There is a range in which retention volume is logarithmically related to molecular mass. The unknown is compared with standards of known molecular mass.

(b) Molecular mass 10^5 is near the middle range of the 25 nm pore size column.

26-21. (a) $V_t = \pi(0.80 \text{ cm})^2 (20.0 \text{ cm}) = 40.2 \text{ mL}$

(b) $K_{av} = \dfrac{V_R - V_o}{V_M - V_o} = \dfrac{27.4 - 18.1}{35.8 - 18.1} = 0.53$

26-22. Ferritin maximum is in tube 22 (=22 × 0.65 mL) = 14.3 mL = V_o
Ferric citrate maximum is in tube 84 (=84 × 0.65 mL) = 54.6 mL = V_M
Does this value of V_M make sense? The total column volume is $V_t = \pi r^2 \times$ length
= $\pi(0.75 \text{ cm})^2(37 \text{ cm})$ = 65.4 mL, so V_M = 54.6 mL is plausible.
Transferrin maximum = tube 32 = 20.8 mL $\Rightarrow K_{av} = \dfrac{20.8 - 14.3}{54.6 - 14.3} = 0.16$

26-23. (a) $V_0 = 4.7$ mL. The vertical line begins at $\approx 10^6 = 1\,000\,000$ Da.

(b) A vertical line at 9.7 mL intersects the 12.5-nm calibration line at (molecular mass) $\approx 10^4 = 10\,000$ Da.

(c) The vertical drop on the 45-nm curve begins at $\approx 10^4 = 10\,000$ Da.

26-24. (a) The total column volume is $\pi r^2 \times$ length $= \pi(0.39)^2 (30) = 14.3$ mL. Totally excluded molecules do not enter the pores and are eluted in the solvent volume (the interstitial volume) outside the particles. Interstitial volume = 40% of 14.3 mL = 5.7 mL.

(b) The smallest molecules that completely penetrate pores will be eluted in a volume that is the sum of the volumes between particles and within pores = 80% of 14.3 mL = 11.5 mL.

(c) These solutes must be adsorbed on the polystyrene resin. Otherwise, they would all be eluted between 5.7 and 11.5 mL.

26-25. A graph of log (molecular mass, MM) vs. V_R should be constructed.

	log(MM)	V_R(mL)
aldolase	5.199	35.6
catalase	5.322	32.3
ferritin	5.643	28.6
thyroglobulin	5.825	25.1
Blue Dextran	6.301	17.7
unknown	?	30.3

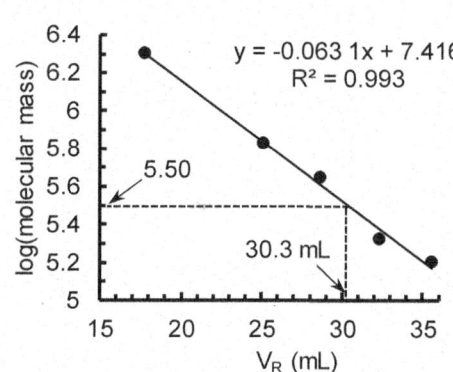

The equation of the graph of K_{av} vs. log (MM) is $y = -0.063\,1x + 7.416$. Inserting $x = 30.3$ gives $y = \log$ (MM) $= 5.50 \Rightarrow$ molecular mass $= 320\,000$ Da

26-26. Hydrophobic regions of the protein are less soluble in water as the salt concentration in the water increases. This decrease in solubility of nonpolar substances in water with increasing salt concentration is known as "salting out." By decreasing the salt concentration, the protein becomes more soluble in the aqueous phase and can be eluted from the column. Eluent strength increases as the salt concentration decreases.

26-27. (a) Electrophoretic mobility is governed by $\mu_{ep} = q/6\pi\eta r$, where r is the radius of the solute. Addition of an alkyl group increases the size of phenol, which increases r, and decreases the electrophoretic mobility. An ethyl group is larger than a methyl group, so the magnitude of the mobility decreases in the order phenol > 4-methylphenol > 4-ethylphenol.

(b) Predicted electrophoretic mobility is $\mu_{ep\ predicted} = (\alpha_{A^-})(\mu_{ep\ A^-})$, where $\alpha_{A^-} = K_a/([H^+] + K_a)$

	A	B	C	D	E
1		pK_a	μ_{A^-}	α_{A^-}	$\mu_{ep,\ predicted}$
2			$m^2/(V \cdot s)$	pH 10.0	pH 10.0
3	Phenol	9.98	-2.99E-08	0.512	-1.53E-08
4	4-Methylphenol	10.27	-2.59E-08	0.349	-9.05E-09
5	4-Ethylphenol	10.22	-2.39E-08	0.376	-8.99E-09
6	2,4,5-Trichlorophenol	6.83	-2.85E-08	0.999	-2.85E-08
7	$\alpha_{A^-} = 10^{-pK_a}/(10^{-pH} + 10^{-pK_a})$			D3 = 10^-B3/(10^-10.0 + 10^-B3)	
8	$\mu_{ep,\ predicted} = (\alpha_{A^-})(\mu_{A^-})$			E3 = D3*C3	

The pH is well above the pK_a of trichlorophenol and so it is fully ionized and its predicted electrophoretic mobility equals its μ_{A^-}. The pH is near the pK_a of phenol, and so its average charge is ~ −0.5, and its predicted electrophoretic mobility is ~50% of its μ_{A^-}. The pK_a of the alkyl phenols are higher, and so they are less ionized at pH 10. So their predicted mobilities are lower.

(c) Order of elution is cations, then neutrals, and then anions, with highest mobility anions last. From part (b), ethylphenol would be eluted first, followed closely by methylphenol, then phenol, and finally the highest mobility trichlorophenol.

26-28. (a) At pH well below the pK_a, a base is fully in its protonated form ($\alpha_{BH^+} = 1$), and so $\mu_{ep\ obs} = \mu_{ep\ BH^+}$. At pH well above the pK_a, the base is in its neutral form, and so $\mu_{ep\ obs} = 0$. The base goes from fully protonated to fully deprotonated in the vicinity of pK_a, so electrophoretic mobility decreases with increasing pH in the vicinity of pK_a.

(b) Ethedrone is smaller, and so is expected to have greater $\mu_{ep\ BH^+}$.

(c) The upper curve has greater $\mu_{ep\ BH^+}$ at low pH, and so must be ethedrone. Lower curve at low pH is methylenedioxypyrovalerone.

(d) At pH = pK_a, half of the base is deprotonated, and so $\mu_{ep\ obs} = 0.5\mu_{ep\ BH^+}$. The $0.5\mu_{ep\ BH^+}$ occurs at a lower pH for the upper curve at low pH. Therefore, pK_a = 8.77 belongs to ethedrone.

(e) Resolution depends on difference in electrophoretic mobility between analytes.

$$\text{Resolution} = \frac{\sqrt{N}}{4}\frac{\Delta\mu_{app}}{\bar{\mu}_{app}}$$

The apparent mobility depends on the electrophoretic and electroosmotic mobilities. The electroosmotic mobility is the same for both compounds, and so $\Delta\mu_{app} = \Delta\mu_{ep\ obs}$. Resolution would be greatest for pH < 7.

26-29. Electroosmosis is the bulk flow of fluid in a capillary caused by migration of the dominant ion in the diffuse part of the double layer toward the anode or cathode.

26-30. At pH 10, the wall of the bare capillary is negatively charged with –Si–O⁻ groups and there is strong electroosmotic flow toward the cathode. At pH 2.5, the wall is nearly neutral with –Si–OH groups and there is almost no electroosmotic flow. The few –Si–O⁻ groups left give slight flow toward the cathode. The aminopropyl capillary also has positive flow at pH 10, but the rate is only about half as great as that of the bare capillary. The negative charge might be reduced because there are fewer –Si–O⁻ groups (because some of them have been converted to –Si–CH$_2$CH$_2$CH$_2$NH$_2$) or because some of the aminopropyl groups are protonated (–Si–CH$_2$CH$_2$CH$_2$NH$_3^+$) at pH 10. At pH 2.5, all the aminopropyl groups are protonated. The net charge on the wall is *positive* and the flow is *reversed*.

26-31. (a) The unknown appears before the electroosmotic flow (methanol), and so must be a cation.

(b) $\mu_{app} = \dfrac{u_{net}}{E} = \dfrac{0.400\ m/86.0\ s}{5.00\times 10^4\ V/m} = 9.30\times 10^{-8}\ \dfrac{m^2}{V\cdot s}$

$\mu_{ep} = \mu_{app} - \mu_{eo}$

$\mu_{eo} = 4.26\times 10^{-8}\ \dfrac{m^2}{V\cdot s}$ was calculated in the Example Mobility Calculations.

$\mu_{ep} = \mu_{app} - \mu_{eo} = (9.30 - 4.26)\times 10^{-8}\ \dfrac{m^2}{V\cdot s} = +5.04\times 10^{-8}\ \dfrac{m^2}{V\cdot s}$

26-32. Arginine is the only amino acid listed with a positively charged side chain. All of the derivatized amino acids have a negative charge because the fluorescent group and the terminal carboxyl group are both negative. Arginine is least negative, so its electrophoretic mobility toward the anode is slowest and its net migration toward the cathode (from electroosmosis) is fastest.

26-33. Under ideal conditions, longitudinal diffusion is the principle source of zone broadening. Even under ideal conditions, the finite length of the injected sample and, possibly, the finite length of the detector contribute to zone broadening. In real electrophoresis, adsorption on the capillary wall and irregular flow paths due to imperfections in the capillary could contribute to zone broadening. For an experimental study of zone broadening, see D. Xiao, T. V. Le, and M. J. Wirth, "Surface Modification of the Channels of Poly(dimethylsiloxane) Microfluidic Chips with Polyacrylamide for Fast Electrophoretic Separations of Proteins," *Anal. Chem.* **2004**, *76*, 2055.

26-34. (a) At pH 2.8, electroosmotic flow will be very small. Anionic analyte will migrate from negative to positive polarity with little effect from the small electroosmotic flow. Reverse polarity places the detector at the positive end of the capillary.

(b) The conductivity of the buffer needs to be higher than the conductivity of the sample so that the sample will stack. At lower buffer concentration, analyte bands will be broader and resolution of heparin from its impurities would be diminished.

(c) High buffer concentration gives high conductivity, high current, and high heat generation. The narrow column reduces the current and the heat generation and makes it easier to cool the entire volume inside the capillary.

(d) Li^+ has lower mobility than Na^+, so the conductivity of lithium phosphate solution will be lower than the conductivity of sodium phosphate solution at the same pH. The lower the conductivity, the higher the electric field required to generate the same current.
High field strength reduces the migration time to shorten the analysis. Also, according to Equation 26-13, plate number increases in proportion to applied voltage.

26-35. Electroosmotic flow can be reduced by (a) lowering the pH, so the charge on the capillary wall is reduced; (b) adding ions such as $^+H_3NCH_2CH_2CH_2CH_2NH_3^+$ that adhere to the capillary wall and effectively neutralize its charge; and (c) covalently attaching silanes with neutral, hydrophilic substituents to the Si—O$^-$ groups on the walls. A cationic surfactant can form a bilayer, which effectively reverses the charge on the wall.

26-36. In the absence of micelles, neutral molecules are all swept through the capillary at the electroosmotic velocity. Negatively charged micelles swim upstream with some electrophoretic velocity, so they take longer than neutral molecules to reach the detector. A neutral molecule spends some time free in solution and some time dissolved in the micelles. Therefore, the net velocity of the neutral molecule is reduced from the electroosmotic velocity. Because different neutral molecules have different partition coefficients between the solution and the micelles, each type of neutral molecule has its own net migration speed. We say that micellar electrokinetic chromatography is a form of chromatography because the micelles behave as a "stationary" phase in the capillary because their concentration is uniform throughout the capillary. Analyte partitions between the mobile phase and the micelles as the analyte travels through the capillary. Micelles are called a pseudostationary phase because they are not stationary and their concentration is constant in the capillary because they are part of the run buffer.

26-37. (a) Volume of sample = cross-sectional area × length
$$= \pi r^2(\text{length}) = \pi(25 \times 10^{-6} \text{ m})^2(0.0060 \text{ m}) = 1.18 \times 10^{-11} \text{ m}^3$$

$$\Delta P = \frac{128 \eta L_t (\text{Volume})}{t \pi d^4} = \frac{128(0.0010 \text{ kg/(m·s)})(0.600 \text{ m})(1.18 \times 10^{-11} \text{ m}^3)}{(4.0 \text{ s})\pi(50 \times 10^{-6} \text{ m})^4}$$

$$= 1.15 \times 10^4 \text{ Pa } (= 1.15 \times 10^4 \text{ kg/(m·s}^2))$$

(b) $\Delta P = h\rho g \Rightarrow h = \dfrac{\Delta P}{\rho g} = \dfrac{1.15 \times 10^4 \text{ kg/(m·s}^2)}{(1\,000 \text{ kg/m}^3)(9.8 \text{ m/s}^2)} = 1.17 \text{ m}$

Since the column is only 0.6 m long, we cannot raise the inlet to 1.17 m. Instead, we could use pressure at the inlet (1.15×10^4 Pa = 0.114 atm) or an equivalent vacuum at the outlet.

26-38. (a) Volume = πr^2(length) = $\pi(12.5 \times 10^{-6}\text{ m})^2(0.006\ 0\text{ m})$ = $2.95 \times 10^{-12}\text{ m}^3$ = 2.95 nL. Moles = $(10.0 \times 10^{-6}\text{ M})(2.95 \times 10^{-9}\text{ L})$ = 29.5 fmol.

(b) Moles injected = $\mu_{app}\left(E\dfrac{\kappa_b}{\kappa_s}\right)t\pi r^2 C = \mu_{app}\left(\dfrac{V}{L_t}\dfrac{\kappa_b}{\kappa_s}\right)t\pi r^2 C$

In order for the units to work out, we need to express the concentration, C, in mol/m^3: $(10.0 \times 10^{-6}\text{ mol/L})(1\ 000\text{ L/m}^3) = 1.00 \times 10^{-2}$ mol/m^3

$$V = \dfrac{(\text{moles})L_t(\kappa_s/\kappa_b)}{\mu_{app}t\pi r^2 C}$$

$$= \dfrac{(29.5\times 10^{-15}\text{ mol})(0.600\text{ m})(1/10)}{(3.0\times 10^{-8}\text{ m}^2/(\text{V}\cdot\text{s}))(4.0\text{ s})\pi(12.5\times 10^{-6}\text{ m})^2(1.00\times 10^{-2}\text{ mol/m}^3)}$$

$$= 3.00 \times 10^3\text{ V}$$

26-39. Electrophoretic peak: $N = \dfrac{16 t_R^2}{w^2} = \dfrac{16(6.08\text{ min})^2}{(0.080\text{ min})^2} = 9.2 \times 10^4$ plates

Chromatographic peak: $N \approx \dfrac{41.7(t_R/w_{0.1})^2}{(B/A + 1.25)}$

$= \dfrac{41.7(6.03\text{ min}/0.37\text{ min})^2}{(1.45 + 1.25)} = 4.1 \times 10^3$ plates

(According to my measurements, both plate numbers are about 1/3 lower than the values labeled in the figure from the original source.)

26-40. (a) Fumarate is a longer molecule than maleate, so we guess that fumarate has a greater friction coefficient than maleate. Electrophoretic mobility is (charge)/(friction coefficient). Both ions have the same charge, so we predict that maleate will have the greater electrophoretic mobility.

(b) Since maleate moves upstream faster than fumarate, fumarate is eluted first.

(c) Since the anions move faster than the electroosmotic flow, the faster anion (maleate) is eluted first.

26-41. (a) pH 2: $u_{neutral} = \mu_{eo}E = \left(1.3 \times 10^{-8} \dfrac{m^2}{V \cdot s}\right)\left(\dfrac{27 \times 10^3 \ V}{0.62 \ m}\right) = 5.6_6 \times 10^{-4}$ m/s

Migration time = (0.52 m)/(5.6$_6$ × 10^{-4} m/s) = 9.2 × 10^2 s

pH 12: $u_{neutral} = \mu_{eo}E = \left(8.1 \times 10^{-8} \dfrac{m^2}{V \cdot s}\right)\left(\dfrac{27 \times 10^3 \ V}{0.62 \ m}\right) = 3.5_3 \times 10^{-3}$ m/s

Migration time = (0.52 m)/(3.5$_3$ × 10^{-3} m/s) = 1.4$_7$ × 10^2 s

(b) pH 2: $\mu_{app} = \mu_{ep} + \mu_{eo} = (-1.6 + 1.3) \times 10^{-8} \dfrac{m^2}{V \cdot s} = -0.3 \times 10^{-8} \dfrac{m^2}{V \cdot s}$

The anion will not migrate toward the detector at pH 2.

pH 12: $\mu_{app} = \mu_{ep} + \mu_{eo} = (-1.6 + 8.1) \times 10^{-8} \dfrac{m^2}{V \cdot s} = 6.5 \times 10^{-8} \dfrac{m^2}{V \cdot s}$

$u_{anion} = \mu_{app}E = \left(6.5 \times 10^{-8} \dfrac{m^2}{V \cdot s}\right)\left(\dfrac{27 \times 10^3 \ V}{0.62 \ m}\right) = 2.8_3 \times 10^{-3}$ m/s

Migration time = (0.52 m)/(2.8$_3$ × 10^{-3} m/s) = 1.8$_4$ × 10^2 s

26-42. (a) The net speed of an ion moving through the capillary by electroosmosis plus electrophoresis is proportional to electric field ($u_{net} = \mu_{app}E$), which, in turn, is proportional to voltage. Increasing voltage by 120 kV/28 kV = 4.3 should increase the speed by 4.3 and decrease the migration time to 1/(4.3) = 0.23 of its original value. Peak 1 has a migration time of 54.36 min at 120 kV and 211.3 min at 28 kV. The ratio is 54.36 min /211.3 min = 0.26.

(b) Plate number is proportional to voltage ($N = \dfrac{\mu_{app} V}{2D} \dfrac{L_d}{L_t}$). Increasing voltage by a factor of 4.3 should increase the plate number by 4.3.

(c) Bandwidth is proportional to the $1/\sqrt{N}$ ($N = L_d^2/\sigma^2 \Rightarrow \sigma = L_d/\sqrt{N}$). Increasing voltage by 4.3 should increase N by 4.3 and decrease bandwidth by factor of $\sqrt{4.3}$ = 2.1. Bandwidth at 120 kV should be 48% as great as bandwidth at 28 kV.

(d) Increasing voltage makes the ions move faster, which gives the peaks less time to undergo longitudinal diffusion broadening. Therefore, the bandwidth is reduced and resolution is increased.

26-43. At low voltage (low electric field), the plate number increases in proportion to voltage. Above ~25 000 V/m, the capillary is probably overheating, which produces band broadening and decreases the plate numbers.

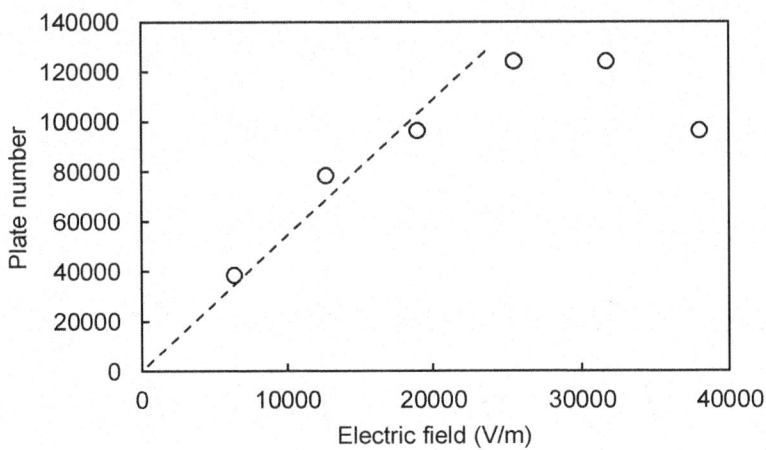

26-44. $N = \dfrac{5.55\, t_R^2}{w_{1/2}^2} = \dfrac{5.55\,(39.9\text{ min})^2}{(0.81\text{ min})^2} = 1.3_5 \times 10^4$ plates

Plate height = 0.400 m/$(1.3_5 \times 10^4$ plates$) = 30$ μm

26-45. $t = \dfrac{L_d}{u_{net}} = \dfrac{L_d}{\mu_{app} E}$

(t = migration time, L_d = length to detector, u = speed, E = field)

$\Rightarrow \mu_{app} = \dfrac{L_d}{tE} = \dfrac{L_d/E}{17.12}$ for Cl$^-$ and $\mu_{app} = \dfrac{L_d/E}{17.78}$ for I$^-$

Therefore, we can write that the difference in mobilities is

$\Delta\mu_{app}(\text{I} - \text{Cl}) = \dfrac{L_d/E}{17.12} - \dfrac{L_d/E}{17.78}$ (L_d/E is an unknown constant)

But we know that $\Delta\mu_{app}(\text{I} - \text{Cl}) = [\mu_{eo} + \mu_{ep}(\text{I}^-)] - [\mu_{eo} + \mu_{ep}(\text{Cl}^-)] =$
$\mu_{ep}(\text{I}^-) - \mu_{ep}(\text{Cl}^-) = 0.05 \times 10^{-8}$ m^2/(s·V) in Table 15-1.

For the difference between Cl$^-$ and Br$^-$ we can say

$\Delta\mu_{app}(\text{Br} - \text{Cl}) = \dfrac{L_d/E}{17.12} - \dfrac{L_d/E}{x}$

and we know that $\Delta\mu_{app}(\text{Br} - \text{Cl}) = 0.22 \times 10^{-8}$ m^2/(s·V) in Table 15-1.

Therefore, we can set up a proportion:

$$\frac{\Delta\mu_{app}(Br-Cl)}{\Delta\mu_{app}(I-Cl)} = \frac{0.22}{0.05} = \frac{\frac{L_d/E}{17.12} - \frac{L_d/E}{x}}{\frac{L_d/E}{17.12} - \frac{L_d/E}{17.78}} \Rightarrow x = 20.5 \text{ min}$$

The observed migration time is 19.6 min. Considering the small number of significant digits in the Δμ values, this is a reasonable discrepancy.

26-46.

	A	B	C	D	E	F
1	Molecular mass by SDS/capillary gel electrophoresis					
2					Relative	
3		Molecular		Migration	migration	
4	Protein	mass (MM)	log(MM)	time (min)	time (t_{rel})	$1/t_{rel}$
5	Marker dye	low		13.17		
6	a-Lactalbumin	14200	4.152	16.46	1.250	0.8001
7	Carbonic anhydrase	29000	4.462	18.66	1.417	0.7058
8	Ovalbumin	45000	4.653	20.16	1.531	0.6533
9	Bovine serum albumin	66000	4.820	22.36	1.698	0.5890
10	Phosphorylase B	97000	4.987	23.56	1.789	0.5590
11	b-Galactosidase	116000	5.064	24.97	1.896	0.5274
12	Myosin	205000	5.312	28.25	2.145	0.4662
13	Ferritin light chain	?		17.07	1.296	0.7715
14	Ferritin heavy chain	?		17.97	1.364	0.7329

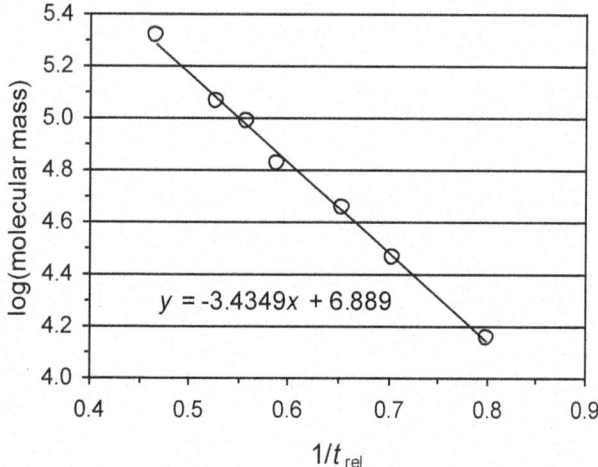

$\log(MM) = (-3.434\,9)/t_{rel} + 6.889 = 4.239$ for $t_{rel} = 1.296$ (ferritin light chain)
$= 4.372$ for $t_{rel} = 1.364$ (ferritin heavy chain)

Molecular mass $= 10^{\log(MM)} = 17\,300$ (ferritin light chain)
$= 23\,500$ (ferritin heavy chain)

Molecular masses observed from amino acid sequences are 19 766 and 21 099 Da

26-47. Resolution $= \dfrac{\sqrt{N}}{4}\dfrac{\Delta\mu_{app}}{\bar{\mu}_{app}} \Rightarrow N = \left(4\,(\text{Resolution})\dfrac{\bar{\mu}_{app}}{\Delta\mu_{app}}\right)^2$

SO_4^{2-}: $\mu_{ep} = -8.27 \times 10^{-8}$ m²/(s·V) in Table 15-1

$\mu_{app} = \mu_{eo} + \mu_{ep} = 16.1 \times 10^{-8} - 8.27 \times 10^{-8} = 7.8_3 \times 10^{-8}$ m²/(s·V)

Br^-: $\mu_{ep} = -8.13 \times 10^{-8}$ m²/(s·V) in Table 15-1

$\mu_{app} = \mu_{eo} + \mu_{ep} = 16.1 \times 10^{-8} - 8.13 \times 10^{-8} = 7.9_7 \times 10^{-8}$ m²/(s·V)

$\bar{\mu}_{app} = \tfrac{1}{2}(7.8_3 \times 10^{-8} + 7.9_7 \times 10^{-8}) = 7.9_0 \times 10^{-8}$ m²/(s·V)

$\Delta\mu_{app} = (8.27 - 8.13) \times 10^{-8} = 0.14 \times 10^{-8}$ m²/(s·V)

$N = \left(4\,(\text{Resolution})\dfrac{\bar{\mu}_{app}}{\Delta\mu_{app}}\right)^2 = \left(4(2.0)\dfrac{7.9_0}{0.14}\right)^2 = 2.0 \times 10^5$ plates

26-48. In the absence of micelles, the expected order of elution is cations before neutrals before anions: thiamine < (niacinamide + riboflavin) < niacin. Since thiamine is eluted last, it must be most soluble in the micelles.

26-49. Carbon atoms labeled with black circles in cyclobarbital and thiopental are chiral, with four different substituents. These compounds are not superimposable on their mirror images. The carbon atom indicated by the diamond in phenobarbital is not chiral because two of its substituents are identical. Cyclodextrin has a chiral pocket, in which these compounds can bind. The equilibrium constant for association of each of the enantiomers of cyclobarbital and thiopental with cyclodextrin will not be the same. Each enantiomer spends a different fraction of time associated with cyclodextrin as it migrates through the capillary. Therefore, cyclobarbital and thiopental will each separate into two peaks. Phenobarbital will only give one peak because it does not have enantiomers.

26-50. (a) Plate height rises sharply at low velocity because bands broaden by diffusion when they spend more time in the capillary. This is the effect of the B term in the van Deemter equation, and it always operates in capillary electrophoresis. Plate height rises gradually at high velocity because solutes require a finite time to equilibrate with the micelles on the column. This is the effect of the C term in the van Deemter equation, and it is absent in capillary electrophoresis but present to a small extent in micellar electrokinetic capillary chromatography.

(b) There should be no irregular flow paths because the micelles are nanosized structures in solution. The large A term most likely arises from extra-column effects, such as the finite size of the injection plug and the finite width of the detector zone.

26-51. (a) *Repeatability* is the spread in results when one person uses one procedure to analyze one sample with the same equipment multiple times. *Reproducibility* is the spread in results when different people in different labs using different equipment follow the same procedure to analyze the same kind of sample. Reproducibility will be significantly poorer than repeatability.

(b) Migration time: mean = 3.916 min, standard deviation = 0.023 min,
relative standard deviation = (0.023 min/3.916 min) × 100 = 0.59%
Peak area: mean = 9 033, standard deviation = 145,
relative standard deviation = (145/9 033) × 100 = 1.6%

(c) Corrected peak area for injection 1 is 8 947/3.946 min = $2\,267._4$ min^{-1}

For injection 2–6: $2\,327._0$, $2\,265._1$, $2\,352._5$, $2\,313._4$, $2\,315._4$ min^{-1}

Mean = $2\,306._8$ min^{-1}, standard deviation = $34._4$ min^{-1}
relative standard deviation = ($34._4$ min^{-1}/ $2\,306._8$ min^{-1}) × 100 = 1.5%

Relative standard deviation for corrected peak area is lower because dividing peak area by the migration time corrects for variation in the speed that the analyte passes through the detector.

26-52. For the acid H_2A, the average charge is $\alpha_{HA^-} + 2\alpha_{A^{2-}}$, where α is the fraction in each form. From our study of acids and bases, we know that

$$\alpha_{HA^-} = \frac{K_1[H^+]}{[H^+]^2 + K_1[H^+] + K_1K_2} \text{ and } \alpha_{A^{2-}} = \frac{K_1K_2}{[H^+]^2 + K_1[H^+] + K_1K_2}$$

where K_1 and K_2 are acid dissociation constants of H_2A. The following spreadsheet finds the average charge of malonic acid (H_2M) and phthalic acid (H_2P) and finds that the maximum difference between them occurs at pH 5.55.

Charge Difference Between Malonic and Phthalic Acids

	A	B	C	D	E	F	G	H	I	J
1	Malonic:			Alpha	Alpha	Alpha	Alpha	Average charges		Charge
2	K1 =	pH	[H+]	HM-	M2-	HP-	P2-	Malonate	Phthalate	Difference
3	1.42E-03	5.52	3.0E-06	0.600	0.399	0.436	0.563	-1.398	-1.562	-0.16392
4	K2 =	5.53	3.0E-06	0.594	0.405	0.430	0.569	-1.403	-1.567	-0.16405
5	2.01E-06	5.54	2.9E-06	0.589	0.410	0.425	0.574	-1.409	-1.573	-0.16413
6	Phthalic:	5.55	2.8E-06	0.583	0.416	0.419	0.580	-1.415	-1.579	-0.16418
7	K1 =	5.56	2.8E-06	0.577	0.421	0.413	0.585	-1.420	-1.584	-0.16417
8	1.12E-03	5.57	2.7E-06	0.572	0.427	0.408	0.591	-1.426	-1.590	-0.16413
9	K2 =	5.58	2.6E-06	0.566	0.433	0.402	0.597	-1.432	-1.596	-0.16404
10	3.90E-06	5.59	2.6E-06	0.561	0.438	0.397	0.602	-1.437	-1.601	-0.16391
11										
12	D3 = $A$3*C3/(C3^2+$A$3*C3+$A$3*$A$5)							C3 = 10^-B3		
13	E3 = $A$3*$A$5/(C3^2+$A$3*C3+$A$3*$A$5)							H3 = -D3-2*E3		
14	F3 = $A$8*C3/(C3^2+$A$8*C3+$A$8*$A$10)							I3 = -F3-2*G3		
15	G3 = $A$8*$A$10/(C3^2+$A$8*C3+$A$8*$A$10)							J3 = I3-H3		

26-53. (a) In ion mobility spectrometry, gaseous ions are generated by irradiating analyte plus reagent gas (such as acetone in air) with high energy electrons (β emission) from radioactive ^{63}Ni. Periodically, ions are admitted into a drift tube by a short voltage pulse applied to an electronic gate (a grid). In the drift tube, ions experience a constant electric field that causes either cations or anions to migrate from the gate to a detector at the other end of the tube. The time to reach the detector is the drift time. Ions drift at a constant speed governed by the driving force of the electric field and the retarding force of friction (drag) by the atmosphere of gas (usually dry air) in the drift tube. Also, gas in the drift tube flows from the detector to the source, further decreasing the migration speed of an ion.

The electric field in ion mobility spectrometry causes ions to migrate from the source to the detector, just as the electric field in electrophoresis causes ions to migrate. Drift time in ion mobility spectrometry is the same quantity as migration time in electrophoresis. The mobility of an ion in liquid or in gas is governed by the charge-to-size ratio. The greater the charge and the smaller the size, the greater the mobility. In liquid or gas, the retarding force is caused by collisions with solvent or gas molecules.

(b)

	A	B	C	D	E	F
1	Ion Mobility Spectrometry					
2						
3	k =		Volts	t_d (s)	$w_{1/2}$ (s)	N
4	1.38065E-23	J/K	100	5.0000	2.68E-01	1.94E+03
5	e =		1000	0.5000	8.47E-03	1.94E+04
6	1.60218E-19	C	2000	0.2500	2.99E-03	3.87E+04
7	T =		3000	0.1667	1.63E-03	5.80E+04
8	300	K	4000	0.1250	1.06E-03	7.73E+04
9	μ =		5000	0.1000	7.59E-04	9.64E+04
10	0.00008	m^2/(sV)	6000	0.0833	5.78E-04	1.15E+05
11	z =		7000	0.0714	4.60E-04	1.34E+05
12	1		8000	0.0625	3.77E-04	1.52E+05
13	L =		9000	0.0556	3.18E-04	1.70E+05
14	0.2	m	10000	0.0500	2.72E-04	1.87E+05
15	t_g =		12000	0.0417	2.10E-04	2.19E+05
16	5.00E-05	s	14000	0.0357	1.69E-04	2.47E+05
17	16kT(ln2)/ez =		16000	0.0313	1.41E-04	2.71E+05
18	2.86707E-01		18000	0.0278	1.22E-04	2.90E+05
19			20000	0.0250	1.07E-04	3.03E+05
20	D4 = $A$14^2/($A$10*C4)					
21	E4 = SQRT($A$16^2+($A$18/C4)*D4^2)					
22	F4 = 5.55*(D4/E4)^2					

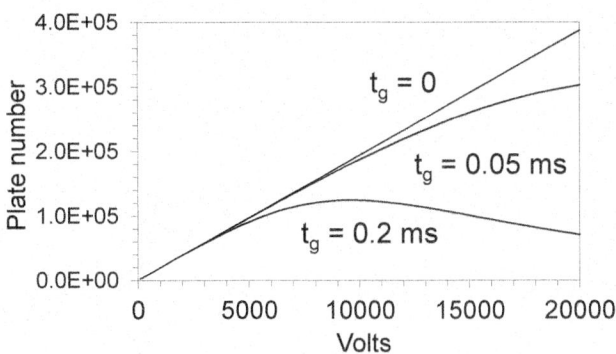

Increasing V increases N by decreasing the drift time, and therefore decreasing the time for diffusion to broaden the peak. Increasing the time that the ion gate is open increases the initial width of the peak, and therefore decreases the plate number. The peak can never be narrower than the pulse that is admitted by the gate. At high voltage, the effect of t_g on plate number overwhelms the effect of t_d. The disadvantage of using a short gate opening time is that fewer ions are admitted to the drift cell and the signal will be weaker.

(c) Decreasing T increases N because diffusional broadening decreases with decreasing temperature.

(d) $N = 5.55 \,(t_d/w_{1/2})^2 = 5.55 \,(0.024\,925\text{ s}/0.000\,154\text{ s})^2 = 1.45 \times 10^5$ plates

Theoretical $w_{1/2}^2 = t_g^2 + \left(\dfrac{16kT \ln 2}{Vez}\right) t_d^2$

$= (5.0 \times 10^{-5}\text{ s})^2 + \left(\dfrac{16(1.38 \times 10^{-23}\text{ J/K})(300\text{ K}) \ln 2}{(12\,500\text{ }V)(1.602 \times 10^{-19}\text{ C})(1)}\right)(0.0249\,25\text{ s})^2$

$= 1.674 \times 10^{-8}\text{ s}^2$

Theoretical $N = 5.55 \,(t_d^2/w_{1/2}^2) = 5.55 \,(0.024\,925\text{ s})^2/(1.674 \times 10^{-8}\text{ s}^2)$

$= 2.06 \times 10^5$ plates

(e) Resolution $= \dfrac{\sqrt{N}}{4} \dfrac{\Delta\mu_{app}}{\bar{\mu}_{app}}$ and $t = \dfrac{L_d}{u_{net}} = \dfrac{L_d}{\mu_{app}E} \Rightarrow \mu_{app} = \dfrac{L_d}{tE}$

$\mu_{app,leucine} = \dfrac{L_d}{tE} = \dfrac{10.0\text{ cm}}{0.0225\text{ s} \times 200\text{ V/cm}} = 2.22\text{ cm}^2/(\text{s} \cdot \text{V})$

$\mu_{app,isoleucine} = \dfrac{10.0\text{ cm}}{0.022\,0\text{ s} \times 200\text{ V/cm}} = 2.27\text{ cm}^2/(\text{s} \cdot \text{V})$

$\bar{\mu}_{app} = \tfrac{1}{2}(2.22 + 2.27) = 2.24\text{ cm}^2/(\text{s} \cdot \text{V})$

$\Delta\mu_{app} = (2.27 - 2.22) = 0.05\text{ cm}^2/(\text{s} \cdot \text{V})$

Resolution $= \dfrac{\sqrt{N}}{4} \dfrac{\Delta\mu_{app}}{\bar{\mu}_{app}} = \dfrac{\sqrt{80\,000}}{4} \dfrac{0.05}{2.24} = 1.6$

CHAPTER 27
GRAVIMETRIC AND COMBUSTION ANALYSIS

27-1. (a) In a<u>d</u>sorption, a substance becomes bound to the surface of another substance. In a<u>b</u>sorption, a substance is taken up inside another substance.

(b) An inclusion is an impurity that occupies lattice sites in a crystal. An occlusion is an impurity trapped inside a pocket in a growing crystal.

27-2. An ideal gravimetric precipitate should be insoluble, easily filterable, pure, and possess a known, constant composition.

27-3. High relative supersaturation often leads to formation of colloidal product with a large amount of impurities.

27-4. Relative supersaturation can be decreased by increasing temperature (for most solutions), mixing well during addition of precipitant, and using dilute reagents. Homogeneous precipitation is also an excellent way to control relative supersaturation.

27-5. Washing with electrolyte preserves the electric double layer and prevents peptization.

27-6. HNO_3 evaporates during drying. $NaNO_3$ is nonvolatile and will lead to a high mass for the precipitate.

27-7. During the first precipitation, the concentration of unwanted species in the solution is high, giving a relatively high concentration of impurities in the precipitate. In the reprecipitation, the level of solution impurities is reduced, giving a purer precipitate.

27-8. In thermogravimetric analysis, the mass of a sample is measured as the sample is heated. The mass lost during decomposition provides some information about the composition of the sample.

424   Chapter 27

27-9. A quartz crystal microbalance consists of a specially cut, thin, disk-shape slice of quartz with gold electrodes on each of the two faces. Application of an oscillating electric field causes the crystal to oscillate at a characteristic frequency. Binding of small masses to the gold electrodes lowers the oscillation frequency. From the change in frequency, we can deduce how much mass was bound.

27-10. $\dfrac{0.214\ 6\ \text{g AgBr}}{187.772\ \text{g AgBr/mol}} = 1.142\ 9 \times 10^{-3}\ \text{mol AgBr}$

$[\text{NaBr}] = \dfrac{1.142\ 9 \times 10^{-3}\ \text{mol}}{50.00 \times 10^{-3}\ \text{L}} = 0.022\ 86\ \text{M}$

27-11. $\dfrac{0.104\ \text{g CeO}_2}{172.114\ \text{g CeO}_2/\text{mol}} = 6.043 \times 10^{-4}\ \text{mol CeO}_2 = 6.043 \times 10^{-4}\ \text{mol Ce}$

$= 0.084\ 66\ \text{g Ce}$

weight % Ce $= \dfrac{0.084\ 66\ \text{g}}{4.37\ \text{g}} \times 100 = 1.94\ \text{wt%}$

27-12. (a) $3\text{BaCl}_2(aq) + 2\text{Na}_3\text{PO}_4(aq) \rightarrow \text{Ba}_3(\text{PO}_4)_2(s) + 6\text{NaCl}(aq)$ (A)
    FM 208.23   FM 163.94   FM 601.92

mol $\text{Ba}^{2+} = (0.502\ \text{g})/(208.23\ \text{g/mol}) = 2.41_{08}\ \text{mmol}$

mol $\text{PO}_4^{3-} = (1.02\ \text{g})/(163.94\ \text{g/mol}) = 6.22\ \text{mmol}$

Reaction A requires 2 mol PO_4^{3-} for 3 mol $\text{Ba}^{2+} = (2/3)(2.41_{08}\ \text{mmol}) =$ 1.61 mmol PO_4^{3-}. There are 6.22 mmol PO_4^{3-}, so BaCl_2 is limiting reagent.

Theoretical yield =

$(0.502\ \text{g BaCl}_2)\left(\dfrac{1\ \text{mol Ba}_3(\text{PO}_4)_2}{3\ \text{mol BaCl}_2}\right)\left(\dfrac{601.92\ \text{g Ba}_3(\text{PO}_4)_2/\text{mol}}{208.23\ \text{g BaCl}_2/\text{mol}}\right) = 0.484\ \text{g}$

Observed yield $= \left(\dfrac{0.733\ \text{g product}}{0.484\ \text{g theoretical}}\right) = 151\%$ (yield exceeds theoretical!)

(b) $\text{BaCl}_2(aq) + \text{Na}_3\text{PO}_4(aq) \rightarrow \text{NaBaPO}_4 \cdot 9\text{H}_2\text{O}(s)$ (B)
    FM 208.23   FM 163.94   FM 417.42

Theoretical yield =

$(0.502\ \text{g BaCl}_2)\left(\dfrac{1\ \text{mol Ba}_3(\text{PO}_4)_2}{1\ \text{mol BaCl}_2}\right)\left(\dfrac{417.42\ \text{g Ba}_3(\text{PO}_4)_2/\text{mol}}{208.23\ \text{g BaCl}_2/\text{mol}}\right) = 1.006\ \text{g}$

Observed yield $= \left(\dfrac{0.733\ \text{g product}}{1.006\ \text{g theoretical}}\right) = 73\%$ (sensible yield!)

(c) $3BaCl_2(aq) + 2Na_3PO_4(aq) \rightarrow Ba_3(PO_4)_2(s) + 6NaCl(aq)$ (A)
    FM 208.23   FM 163.94   FM 601.92

Theoretical yield =

$(0.502 \text{ g BaCl}_2)\left(\dfrac{1 \text{ mol Ba}_3(PO_4)_2}{3 \text{ mol BaCl}_2}\right)\left(\dfrac{601.92 \text{ g Ba}_3(PO_4)_2/\text{mol}}{208.23 \text{ g BaCl}_2/\text{mol}}\right) = 0.484 \text{ g}$

Observed yield = $\left(\dfrac{0.394 \text{ g product}}{0.484 \text{ g theoretical}}\right) = 81\%$ (sensible for $Ba_3(PO_4)_2$)

27-13. One mole of product (206.242 g) comes from one mole of piperazine (86.138 g). Grams of piperazine in sample =
(0.712 9 g of piperazine / g of sample) × (0.050 02 g of sample) = 0.035 66.
Mass of product = $\left(\dfrac{206.242}{86.132}\right)(0.035\,66) = 0.085\,39$ g.

27-14. 2.500 g bis(dimethylglyoximate)nickel(II) = $8.653\,2 \times 10^{-3}$ mol Ni = 0.507 85 g Ni = 50.79% Ni.

27-15. Formula masses: $CaC_{14}H_{10}O_6 \cdot H_2O$ (332.32), $CaCO_3$ (100.09), CaO (56.08).
At 550°C, $CaC_{14}H_{10}O_6 \cdot H_2O$ is converted to $CaCO_3$ (calcium carbonate). 332.32 g of starting material will produce 100.09 g of CaO.

Mass at 550°C = (100.09/332.32)(0.635 6 g) = 0.191 4 g. At 1 000°C, the product is CaO (calcium oxide) and the mass is (56.08/332.32)(0.635 6 g) = 0.107 3 g.

27-16. (2.378 mg CO_2)/(44.009 g/mol) = $5.403\,4 \times 10^{-5}$ mol CO_2 = $5.403\,4 \times 10^{-5}$ mol C = $6.489\,9 \times 10^{-4}$ g C
ppm C = $10^6 (6.489\,9 \times 10^{-4} / 6.234) = 104.1$ ppm

27-17. 2.07% of 0.998 4 g = 0.020 67 g of Ni = 3.521×10^{-4} mol of Ni.
This requires $(2)(3.521 \times 10^{-4})$ mol of DMG = 0.081 77 g.
A 50.0% excess is (1.5)(0.081 77 g) = 0.122 7 g. The mass of solution containing 0.122 7 g is (0.122 7 g DMG)/(0.021 5 g DMG/g solution) = 5.705 g of solution. The volume of solution is 5.705 g/(0.790 g/mL) = 7.22 mL.

27-18. Moles of Fe in product (Fe_2O_3) = moles of Fe in sample.
Because 1 mole of (Fe_2O_3) contains 2 moles of Fe, we can write the equation
$\dfrac{2(0.264 \text{ g})}{159.69 \text{ g/mol}} = 3.306 \times 10^{-3}$ mol of Fe.

This many moles of Fe equals 0.919 2 g of $FeSO_4 \cdot 7 H_2O$. Because we analyzed just 2.998 g out of 22.131 g of tablets, the FeSO$_4$ · 7 H$_2$O in the 22.131 g sample is $\frac{22.131 \text{ g}}{2.998 \text{ g}}$ (0.919 2 g) = 6.786 g. This is the FeSO$_4$ · 7 H$_2$O content of 20 tablets. The content in one tablet is (6.786 g)/20 = 0.339 g.

27-19. (a) Mass of product (CaCO$_3$) = 18.546 7 g – 18.231 1 g = 0.315 6 g

$$\text{mol CaCO}_3 = \left(\frac{0.315 \text{ 6 g}}{100.086 \text{ g/mol}}\right) = 3.153 \times 10^{-3} \text{ mol}$$

The product contains 3.153 mmol Ca = (3.153 × 10^{-3} mol)(40.078 g/mol) = 0.1264 g Ca.

$$\text{wt\% Ca} = \frac{0.126 \text{ 4 g Ca}}{0.632 \text{ 4 g mineral}} \times 100 = 19.98 \%$$

(b) The solutions are heated before mixing to increase the solubility of the product that will precipitate. If the solution is less supersaturated during the precipitation, crystals form more slowly and grow to be larger and purer than if they precipitate rapidly. The larger crystals are easier to filter.

(c) (NH$_4$)$_2$C$_2$O$_4$ provides oxalate ion to prevent CaC$_2$O$_4$ from redissolving. Also, the ammonium and oxalate ions provide an ionic atmosphere that prevents the precipitate from peptizing (breaking into colloidal particles).

(d) AgNO$_3$ solution is added to the filtrate to test for Cl$^-$ in the filtrate. If Cl$^-$ is present, AgCl(*s*) will precipitate when Ag$^+$ is added. The source of Cl$^-$ is the HCl used to dissolve the mineral. All the original solution needs to be washed away, so no extra material is present that would increase the mass of final product, which should be pure CaCO$_3$(*s*).

27-20. (a) 70 kg $\left(\frac{6.3 \text{ g P}}{\text{kg}}\right)$ = 441 g P in 8.00 × 10^3 L. This corresponds to

$$\frac{441 \text{ g P}}{8.00 \times 10^3 \text{ L}} = 0.055 \text{ 1 g/L or 5.5}_1 \text{ mg/100 mL.}$$

(b) Fraction of P in one formula mass is $\frac{2(30.974)}{3 \text{ 596.67}}$ = 1.722%.

P in 0.338 7 g of P$_2$O$_5$ · 24 MoO$_3$ = (0.017 22)(0.338 7) = 5.834 mg

This is near the amount expected from a dissolved man.

27-21. Let *x* = mass of NH$_4$Cl and *y* = mass of K$_2$CO$_3$.

For the first part, 1/4 of the sample (25 mL) gave 0.617 g of precipitate containing both products:

$$\frac{1}{4}\left(\underbrace{\left(\frac{x}{53.489}\right)(337.27)}_{\text{g }\phi_4\text{BNH}_4}^{\text{mol NH}_4\text{Cl}} + \underbrace{\left(\frac{2y}{138.20}\right)(358.33)}_{\text{g }\phi_4\text{BK }(\phi=\text{phenyl}=C_6H_5)}^{\text{mol K}_2\text{CO}_3 \times 2}\right) = 0.617 \text{ g}$$

We multiplied moles of K_2CO_3 by 2 because one mole of K_2CO_3 gives 2 moles of ϕ_4BK. In the second part, half of the sample (50 mL) gave 0.554 g of ϕ_4BK:

$$\frac{1}{2}\underbrace{\left(\frac{2y}{138.20}\right)(358.33)}_{\text{g }\phi_4\text{BK}}^{\text{mol K}_2\text{CO}_3 \times 2} = 0.554 \text{ g} \Rightarrow y = 0.213_7 \text{ g K}_2\text{CO}_3$$

wt% K_2CO_3 = 100 × (0.213₇ g K_2CO_3)/(1.475 g sample) = 14.5 wt% K_2CO_3

Putting $y = 0.213_7$ g into the first equation gives $x = 0.215_7$ g = 14.6 wt% NH_4Cl

27-22. $\underbrace{Fe_2O_3 + Al_2O_3}_{2.019 \text{ g}} \xrightarrow[H_2]{\text{Heat}} \underbrace{Fe + Al_2O_3}_{1.774 \text{ g}}$

The mass of oxygen lost is 2.019 – 1.774 = 0.245 g, which equals 0.015 31 moles of oxygen atoms. For every 3 moles of oxygen there is 1 mole of Fe_2O_3, so moles of $Fe_2O_3 = \frac{1}{3}(0.015\ 31) = 0.005\ 105$ mol of Fe_2O_3. This much Fe_2O_3 equals 0.815 g, which is 40.4 wt% of the original sample.

27-23. Let x = g of $FeSO_4 \cdot (NH_4)_2 SO_4 \cdot 6H_2O$ and y = g of $FeCl_2 \cdot 6H_2O$.

We can say that $x + y = 0.548\ 5$ g. The moles of Fe in the final product (Fe_2O_3) must equal the moles of Fe in the sample.

Moles of Fe in Fe_2O_3 = 2 (moles of Fe_2O_3) = $2\left(\dfrac{0.167\ 8}{159.69}\right)$ = 0.002 101 6 mol.

Mol Fe in $FeSO_4 \cdot (NH_4)_2 SO_4 \cdot 6H_2O = x/392.12$ and
mol Fe in $FeCl_2 \cdot 6H_2O = y/234.84$.

$$0.002\ 101\ 6 = \frac{x}{392.12} + \frac{y}{234.84} \quad (1)$$

Substituting $x = 0.548\ 5 - y$ into Eq. (1) gives $y = 0.411\ 48$ g of $FeCl_2 \cdot 6H_2O$.

Mass of Cl = $2\left(\dfrac{35.45}{234.84}\right)(0.411\ 48) = 0.124\ 23$ g = 22.65 wt%

27-24. (a) Let x = mass of $AgNO_3$ and $(0.4321 - x)$ = mass of $Hg_2(NO_3)_2$ in unknown. Each mol of $AgNO_3$ gives 1/3 mol $Ag_3[Co(CN)_6]$ and each mol of $Hg_2(NO_3)_2$ gives 1/3 mol $(Hg_2)_3[Co(CN)_6]_2$. Mass of both products must equal 0.4515 g:

$$\underbrace{\overbrace{\frac{1}{3}\left(\frac{x}{169.872}\right)}^{\text{mol } Ag_3Co(CN)_6}(538.646)}_{\text{mass of } Ag_3Co(CN)_6} + \underbrace{\overbrace{\frac{1}{3}\left(\frac{0.4321-x}{525.19}\right)}^{\text{mol}(Hg_2)_3[Co(CN)_6]_2}(1633.63)}_{\text{mass of } (Hg_2)_3[Co(CN)_6]_2} = 0.4515$$

$\Rightarrow x = 0.1729$ g = 40.01 wt% (full $x = 0.1728644$)

(b) 0.30% error in 0.4515 g = ± 0.00135 g. This changes the equation of (a) to:

$$\frac{1}{3}\left(\frac{x}{169.872}\right)(538.646) + \frac{1}{3}\left(\frac{0.4321-x}{525.19}\right)(1633.63) = 0.4515\,(\pm 0.00135)$$

$1.056964x + 0.448023 - 1.036850x = 0.4515\,(\pm 0.00135)$

$0.020114x = 0.4515\,(\pm 0.00135) - 0.448023$

$0.020114x = 0.003477\,(\pm 0.00135)$

$$x = \frac{0.003477(\pm 0.00135)}{0.020114} = \frac{0.003480(\pm 38.8\%)}{0.020109} = 0.17\text{ g} \pm 39\%$$

27-25. (a) Balanced equation for overall (31.8%) mass loss:

$$\underset{\text{FM } 298.30 + x(18.015)}{Y_2(OH)_5Cl \cdot xH_2O} \xrightarrow{31.8\% \text{ mass loss}} \underset{\text{FM } 225.81}{Y_2O_3} + \underset{\text{FM }(2+x)(18.015)}{xH_2O + 2H_2O} + \underset{\text{FM } 36.46}{HCl}$$

$$\underbrace{(2+x)(18.015) + 36.46}_{\text{mass lost}} = \underbrace{(0.318)\left[298.30 + x(18.015)\right]}_{31.8\% \text{ of original mass}} \Rightarrow x = 1.82$$

(b) Logical molecular units that could be lost are H_2O and HCl. At ~8.1% mass loss, the product is $Y_2(OH)_5Cl$. Loss of 2 more H_2O would give a total mass loss of

$$\frac{1.82\,H_2O + 2H_2O}{Y_2(OH)_5Cl \cdot 1.82\,H_2O} = \frac{68.82}{331.09} = 20.8\%$$

Loss of HCl from $Y_2(OH)_5Cl$ would give a total mass loss of

$$\frac{1.82\,H_2O + HCl}{Y_2(OH)_5Cl \cdot 1.82H_2O} = \frac{69.25}{331.09} = 20.9\%$$

The composition at the ~19.2% plateau could be either $Y_2O_2(OH)Cl$ (from loss of $2H_2O$) or $Y_2O(OH)_4$ (from loss of HCl).

27-26. (a) $\alpha = \dfrac{\text{mass of KPO}_3}{\text{mass of K(D}_x\text{H}_{1-x})_2\text{PO}_4} = \dfrac{118.069}{136.084 + 2.012\,55x}$

Cross-multiply: $(136.084)\alpha + (2.012\,55)\alpha x = 118.069$

$(2.012\,55)\alpha x = 118.069 - (136.084)\alpha$

Divide by $(2.012\,55)\alpha$:

$x = \dfrac{118.069}{(2.012\,55)\alpha} - \dfrac{(136.084)\alpha}{(2.012\,55)\alpha}$ $\qquad x = \dfrac{58.666\,4}{\alpha} - 67.617\,7$

For fully deuterated material, $x = 1$ and $\alpha = \dfrac{58.666\,4}{x + 67.617\,7} = 0.854\,975$

(b) $x = \dfrac{58.666\,4}{0.856\,7_7} - 67.617\,7 = 0.856\,2_2$

(c) For the function $x = f(\alpha)$, we can write

$$e_x = \sqrt{\left(\dfrac{\delta F}{\delta \alpha}\right)^2 e_\alpha^2}$$

For $x = \dfrac{58.666\,4}{\alpha} - 67.617\,7$, $\dfrac{\delta F}{\delta \alpha} = -\dfrac{58.666\,4}{\alpha^2}$, giving

$$e_x = \sqrt{\left(-\dfrac{58.666\,4}{\alpha^2}\right)^2 e_\alpha^2} = \dfrac{58.666\,4\,e_\alpha}{\alpha^2}$$

(d) For $e_\alpha = 0.000\,1$, $e_x = \dfrac{(58.666\,4)(0.000\,1)}{(0.856\,7_7)^2} = 0.008$

D:H stoichiometry $= x \pm e_x = 0.856 \pm 0.008$

If e_x were 0.001, then $e_x = \dfrac{(58.666\,4)(0.001)}{(0.856\,7_7)^2} = 0.08$ and

D:H stoichiometry $= 0.86 \pm 0.08$

27-27. (a) Formula mass of $YBa_2Cu_3O_{7-x}$ = $666.19 - (16.00)x$

mmol of $YBa_2Cu_3O_{7-x}$ in experiment = $\dfrac{34.397 \text{ mg}}{[666.19 - (16.00)x] \text{ mg/mmol}}$

mmol of oxygen atoms lost in experiment = $\dfrac{(34.397 - 31.661) \text{ mg}}{16.00 \text{ mg/mmol}}$

= 0.171 00 mmol

From the stoichiometry of the reaction, we can write

$$\dfrac{\text{mmol oxygen atoms lost}}{\text{mmol } YBa_2Cu_3O_{7-x}} = \dfrac{3.5 - x}{1}$$

$$\dfrac{0.171\,00}{34.397/[666.19 - (16.00)x]} = 3.5 - x \Rightarrow x = 0.2042$$

(without regard to significant figures)

(b) Now let the uncertainty in each mass be 0.002 mg and let all atomic and molecular masses have negligible uncertainty.

The mmol of oxygen atoms lost are:

$$\dfrac{[34.397(\pm 0.002) - 31.661(\pm 0.002)]\,\text{mg}}{16.00 \text{ mg/mmol}} = \dfrac{2.736(\pm 0.002\,8)}{16.00}$$

$$= 0.171\,00\,(\pm 0.102\%)$$

The relative error in the mass of starting material is $\dfrac{0.002}{34.397} = 0.005\,8\%$

The master equation becomes

$$\dfrac{0.171\,00\,(\pm 0.102\%)}{34.397\,(\pm 0.005\,8\%)/[666.19 - (16.00)x]} = 3.5 - x$$

$0.171\,00\,(\pm 0.102\%)[666.19 - (16.00)x] = (3.5 - x)[34.397\,(\pm 0.005\,8\%)]$

$113.918\,(\pm 0.116) - [2.736\,(\pm 0.002\,79)]\,x$
   $= 120.389\,5\,(\pm 0.006\,98) - [34.397\,(\pm 0.002)]\,x$

$[31.66\,(\pm 0.003\,46)]\,x = 6.471\,5\,(\pm 0.116)$

$= 0.204\,4\,(\pm 1.79\%) = 0.204 \pm 0.004$

27-28. In *combustion*, a substance is heated in the presence of excess O_2 to convert carbon to CO_2 and hydrogen to H_2O. In *pyrolysis*, the substance is decomposed by heating in the absence of added O_2. All oxygen in the sample is converted to CO by passage through a suitable catalyst.

27-29. WO_3 catalyzes the complete combustion of C to CO_2 in the presence of excess O_2. Cu converts SO_3 to SO_2 and removes excess O_2.

27-30. The tin capsule melts and is oxidized to SnO_2 to liberate heat and crack the sample. Tin uses the available oxygen immediately, ensures that sample oxidation occurs in the gas phase, and acts as an oxidation catalyst.

27-31. By dropping the sample in before very much O_2 is present, pyrolysis of the sample to give gaseous products occurs prior to oxidation. This minimizes the formation of nitrogen oxides.

27-32.
$$C_6H_5CO_2H + \tfrac{15}{2} O_2 \rightarrow 7\,CO_2 + 3\,H_2O$$
$$\text{FM } 122.123 \qquad\qquad\qquad 44.009 \quad 18.015$$

One mole of $C_6H_5CO_2H$ gives 7 moles of CO_2 and 3 moles of H_2O. 4.635 mg of benzoic acid = 0.037 95 mmol, which gives 0.265 7 mmol CO_2 (= 11.69 mg CO_2) and 0.113 9 mmol H_2O (= 2.051 mg H_2O).

27-33. $C_8H_7NO_2SBrCl + 9\tfrac{1}{4} O_2 \rightarrow 8CO_2 + \tfrac{5}{2} H_2O + \tfrac{1}{2} N_2 + SO_2 + HBr + HCl$

27-34. 100 g of compound contains 46.21 g C, 9.02 g H, 13.74 g N, and 31.03 g O. The atomic ratios are C : H : N : O =

$$\frac{46.21\text{ g}}{12.011\text{ g/mol}} : \frac{9.02\text{ g}}{1.008\text{ g/mol}} : \frac{13.74\text{ g}}{14.007\text{ g/mol}} : \frac{31.03\text{ g}}{15.999\text{ g/mol}}$$

$$= 3.847 : 8.94_8 : 0.980\,9 : 1.939$$

Dividing by the smallest factor (0.980 9) gives the ratios C : H : N : O = 3.922 : 9.12 : 1 : 1.977. The empirical formula is probably $C_4H_9NO_2$.

27-35.
$$C_6H_{12} + C_2H_4O \rightarrow CO_2 + H_2O$$
$$\text{FM } 84.162 \quad\;\; 44.053 \qquad 44.009$$

Let x = mg of C_6H_{12} and y = mg of C_2H_4O

$$x + y = 7.290.$$

We also know that moles of CO_2 = 6 (moles of C_6H_{12}) + 2 (moles of C_2H_4O), by conservation of carbon atoms.

$$6\left(\frac{x}{84.162}\right) + 2\left(\frac{y}{44.053}\right) = \frac{21.999}{44.009}$$

Making the substitution $x = 7.290 - y$ allows us to solve for y.
$$y = 0.766_3 \text{ mg} = 10.5 \text{ wt\%}.$$

27-36. The atomic ratio H:C is

$$\frac{\left(\dfrac{6.76 \pm 0.12 \text{ g}}{1.008 \text{ g/mol}}\right)}{\left(\dfrac{71.17 \pm 0.41 \text{ g}}{12.011 \text{ g/mol}}\right)} = \frac{6.706 \pm 0.119}{5.925 \pm 0.0341} = \frac{6.706 \pm 1.78\%}{5.925 \pm 0.576\%} = 1.132 \pm 0.021$$

If we define the stoichiometry coefficient for C to be 8, then the stoichiometry coefficient for H is $8(1.132 \pm 0.021) = 9.06 \pm 0.17$.

The atomic ratio N:C is

$$\frac{\left(\dfrac{10.34 \pm 0.08 \text{ g}}{14.007 \text{ g/mol}}\right)}{\left(\dfrac{71.17 \pm 0.41 \text{ g}}{12.011 \text{ g/mol}}\right)} = \frac{0.7382 \pm 0.0057}{5.925 \pm 0.0341} = \frac{0.7382 \pm 0.774\%}{5.925 \pm 0.576\%}$$

$$= 0.1246 \pm 0.0012$$

If we define the stoichiometry coefficient for C to be 8, then the stoichiometry coefficient for N is $8(0.1246 \pm 0.0012) = 0.9968 \pm 0.0096$.

The empirical formula is reasonably expressed as $C_8H_{9.06 \pm 0.17}N_{0.997 \pm 0.010}$.

27-37. The reaction between H_2SO_4 and NaOH can be written

$$H_2SO_4 + 2\text{NaOH} \rightarrow 2H_2O + Na_2SO_4$$

One mole of H_2SO_4 requires two moles of NaOH. In 3.01 mL of 0.01576 M NaOH there are $(0.00301 \text{ L})(0.01576 \text{ mol/L}) = 4.74_4 \times 10^{-5}$ mol of NaOH. The moles of H_2SO_4 must have been $(\frac{1}{2})(4.74_4 \times 10^{-5}) = 2.37_2 \times 10^{-5}$ mol. Because one mole of H_2SO_4 contains one mole of S, there must have been $2.37_2 \times 10^{-5}$ mol of S (= 0.760_5 mg). The percentage of S in the sample is

$$\frac{0.760_5 \text{ mg S}}{6.123 \text{ mg sample}} \times 100 = 12.4 \text{ wt\%}.$$

27-38. (a) Experiment 1: $\bar{x} = 10.16_0$ µmol Cl⁻ $s = 2.70_7$ µmol Cl⁻

95% confidence interval $= \bar{x} \pm \dfrac{ts}{\sqrt{n}}$

$= 10.16_0 \pm \dfrac{(2.262)(2.70_7)}{\sqrt{10}} = 10.16_0 \pm 1.93_6$ µmol Cl⁻

Experiment 2: $\bar{x} = 10.77_0$ µmol Cl⁻ $s = 3.20_5$ µmol Cl⁻

95% confidence interval $= \bar{x} \pm \dfrac{ts}{\sqrt{n}}$

$$= 10.77_0 \pm \frac{(2.262)(3.20_5)}{\sqrt{10}} = 10.77_0 \pm 2.29_3 \text{ μmol Cl}^-$$

(b) $s_{\text{pooled}} = \sqrt{\dfrac{s_1^2(n_1-1)+s_2^2(n_2-1)}{n_1+n_2-2}} = \sqrt{\dfrac{2.70_7^2(10-1)+3.20_5^2(10-1)}{10+10-2}}$

$$= 2.96_6$$

$$t_{\text{calculated}} = \frac{|\bar{x}_1 - \bar{x}_2|}{s_{\text{pooled}}}\sqrt{\frac{n_1 n_2}{n_1+n_2}} = \frac{|10.16_0 - 10.77_0|}{2.96_6}\sqrt{\frac{(10)(10)}{10+10}}$$

$$= 0.46_0 < t_{\text{tabulated}} \text{ for 18 degrees of freedom for 95\%}$$

confidence level (or even for 50% confidence level)

Therefore, the <u>difference is not significant</u>. The result means that addition of excess Cl⁻ prior to precipitation does not lead to additional coprecipitation of Cl⁻ under the conditions of these experiments. (In general, we might expect extra Cl⁻ to lead to extra coprecipitation.)

(c) 10.0 mg of SO_4^{2-} = 0.104_1 mmol = 24.2_{95} mg $BaSO_4$

(d) In Experiment 1, the precipitate includes an additional 10.16_0 μmol Cl⁻ = 5.08 μmol $BaCl_2$ = 1.05_8 mg $BaCl_2$. Total mass of precipitate = 24.2_{95} mg $BaSO_4$ + 1.05_8 mg $BaCl_2$ = 25.35 mg. The increase in mass is $(1.05_8)/(24.2_{95})$ = 4.35%. This represents a large error in the analysis.

CHAPTER 28
SAMPLE PREPARATION

28-1. There is no point analyzing a sample if you do not know that it was selected in a sensible way and stored so that its composition did not change after it was taken.

28-2. "Analytical quality" refers to the accuracy and precision of the method applied to the sample that was analyzed. "High quality" means that the analysis is accurate and precise. "Data quality" means that the sample that was analyzed is representative and appropriate for the question being asked and that the analytical quality is adequate for the intended purpose. If an accurate and precise analysis is performed on an unrepresentative or contaminated or decomposed sample, the results are meaningless.

28-3. (a) $s_o^2 = s_a^2 + s_s^2 = 3^2 + 4^2 \Rightarrow s_o = 5\%$.

(b) $s_s^2 = s_o^2 - s_a^2 = 4^2 - 3^2 \Rightarrow s_s = 2.6\%$.

28-4. $mR^2 = K_s$. $m(6^2) = 36 \text{ g} \Rightarrow m = 1.0 \text{ g}$

28-5. Pass the powder through a 120 mesh sieve and then through a 170 mesh sieve. Sample retained by 170 mesh sieve has a size between 90 and 125 µm. It would be called 120/170 mesh.

28-6. 11.0×10^2 g will contain 10^6 total particles, since 11.0 g contains 10^4 particles. $n_{KCl} = np = (10^6)(0.01) = 10^4$.

Relative standard deviation = $\sqrt{npq}/n_{KCl} = \sqrt{(10^6)(0.01)(0.99)}/10^4 = 0.99\%$.

28-7. (a) $\sqrt{(10^3)(0.5)(0.5)} = 15.8$.

(b) We are looking for the value of z, whose area is 0.45 (since the area from $-z$ to $+z$ is 0.90). The value lies between $z = 1.6$ and 1.7, whose areas are 0.445 2 and 0.455 4, respectively. Linear interpolation:

$$\frac{z-1.6}{1.7-1.6} = \frac{0.45-0.445\,2}{0.455\,4-0.445\,2} \Rightarrow z = 1.647.$$

(c) Since $z = (x - \bar{x})/s$, $x = \bar{x} \pm zs = 500 \pm (1.647)(15.8) = 500 \pm 26$. The range 474–526 will be observed 90% of the time.

Sample Preparation   435

28-8. Use Equation 28-7, with $s_s = 0.05$ and $e = 0.04$. The initial value of t for 95% confidence in Table 4-4 is 1.960. $n = t^2 s_s^2 / e^2 = 6.0$ For $n = 6$, there are 5 degrees of freedom, so $t = 2.571$, which gives $n = 10.3$. For 9 degrees of freedom, $t = 2.262$, which gives $n = 8.0$. Continuing, we find $t = 2.365 \Rightarrow n = 8.74$. This gives $t = 2.306 \Rightarrow n = 8.30$. Use <u>8 samples</u>. For 90% confidence, the initial t is 1.645 in Table 4-4 and the same series of calculations gives $n = $ <u>6 samples</u>.

28-9. (a) $mR^2 = K_S$. For $R = 2$ and $K_S = 20$ g, we find $m = 5.0$ g.

(b) Use Equation 28-7 with $s_s = 0.02$ and $e = 0.015$. The initial value of t for 90% confidence in Table 4-4 is 1.645. $n = t^2 s_s^2 / e^2 = 4.8$.

For $n = 5$, there are 4 degrees of freedom, so $t = 2.132$, which gives $n = 8.1$. For 7 degrees of freedom, $t = 1.895$, which gives $n = 6.4$.
Continuing, we find $t = 2.015 \Rightarrow n = 7.2$. This gives $t = 1.943 \Rightarrow n = 6.7$. Use <u>7 samples</u>.

28-10.

	A	B	C	D
1	Evaluation of the relation $mR^2 = K_s$			
2				
3	m (pg)	R %	R^2	$K_s = mR^2$
4	57	0.057	0.00325	0.185
5	68	0.069	0.00476	0.324
6	110	0.049	0.00240	0.264
7	110	0.045	0.00203	0.223
8	506	0.035	0.00123	0.620
9	515	0.027	0.00073	0.375
10	916	0.018	0.00032	0.297
11	955	0.022	0.00048	0.462
12			average	0.344
13			std dev	0.141

Average value of $K_S =$ 0.34 ± 0.14 pg

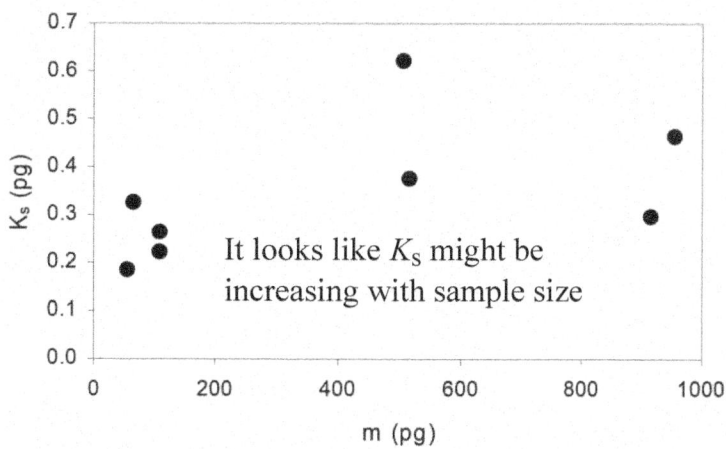

It looks like K_S might be increasing with sample size

The confidence interval is $\pm ts/\sqrt{n}$, where t is Student's t, s is the standard deviation, and n is the number of replicate measurements. If n is the same for all points, then t is the same for all points and the confidence interval is proportional to s. The equation in the text is expressed in terms of s. If the confidence interval is proportional to s, then the same equation should hold for the confidence interval.

28-11. (a) Volume = $(4/3)\pi r^3$, where $r = 0.075$ mm = 7.5×10^{-3} cm.
Volume = 1.767×10^{-6} mL.
Na_2CO_3 mass = $(1.767 \times 10^{-6}$ mL$)(2.532$ g/mL$) = 4.474 \times 10^{-6}$ g.
K_2CO_3 mass = $(1.767 \times 10^{-6}$ mL$)(2.428$ g/mL$) = 4.291 \times 10^{-6}$ g.
Number of particles of Na_2CO_3 = $(4.00$ g$)/(4.474 \times 10^{-6}$ g/particle$)$
  = 8.941×10^5.
Number of particles of K_2CO_3 = $(96.00$ g$)/(4.291 \times 10^{-6}$ g/particle$)$
  = 2.237×10^7.
The fraction of each type (which we will need for part c) is
$p_{Na_2CO_3}$ = $(8.941 \times 10^5)/(8.941 \times 10^5 + 2.237 \times 10^7) = 0.0384$
$q_{K_2CO_3}$ = $(2.237 \times 10^7)/(8.941 \times 10^5 + 2.237 \times 10^7) = 0.962$.

(b) Total number of particles in 0.100 g is $n = 2.326 \times 10^4$.

(c) Expected number of Na_2CO_3 particles in 0.100 g is 1/1000 of number in 100
  grams = 8.94×10^2.
Expected number of K_2CO_3 particles in 0.100 g is 1/1 000 of number in 100
  grams = 2.24×10^4.
Sampling standard deviation = $\sqrt{npq}$ = $\sqrt{(2.326 \times 10^4)(0.0384)(0.962)}$
  = 29.3.
Relative sampling standard deviation for Na_2CO_3 = $\dfrac{29.3}{8.94 \times 10^2}$ = 3.28%.
Relative sampling standard deviation for K_2CO_3 = $\dfrac{29.3}{2.24 \times 10^4}$ = 0.131%.

28-12. (a) Preconcentration factor = $\dfrac{V_{initial}}{V_{final}} = \dfrac{195 \text{ mL}}{5.0 \text{ mL}} = 39$

(b) % recovery = $\dfrac{C_{spiked\ sample} - C_{unspiked\ sample}}{C_{added}} \times 100\%$

$$= \frac{1.9 \text{ µg/L} - 0 \text{ µg/L}}{2.3 \text{ µg/L}} \times 100 = 83\%$$

(c) 95% confidence interval $= \bar{x} \pm \frac{ts}{\sqrt{n}} = 1.9 \text{ µg/L} \pm \frac{(4.303)(0.2 \text{ µg/L})}{\sqrt{3}}$

$$= 1.9 \pm 0.5_0 \text{ µg/L} \Rightarrow 1.4 \text{ to } 2.4 \text{ µg/L}$$

The 95% confidence interval includes the expected value of 2.3 µg/L, so the difference is <u>not</u> significant at the 95% confidence level.

For $n = 6$ measurements, $t = 2.571$ and $\bar{x} = 1.9 \pm 0.2_1$ µg/L, which <u>is significantly less</u> than 2.3 µg/L at the 95% confidence level.

28-13. Metals with reduction potentials below zero [for the reaction $M^{n+} + ne^- \rightarrow M(s)$] are expected to dissolve in acid. These are Zn, Fe, Co, and Al.

28-14. HNO_3 was used first to oxidize any material that could be easily oxidized. This helps prevent the possibility that an explosion will occur when $HClO_4$ is added.

28-15. Barbital has a higher affinity for the octadecyl phase than for water, so it is retained by the column. The drug dissolves readily in acetone/chloroform, which elutes it from the column.

28-16. Cocaine is an amine base. It will be a cation at low pH and neutral in ammonia. The cation at pH 2 is retained by the cation-exchange resin. The neutral molecule is easily eluted by methanol. Benzoylecgonine has an amine and carboxylate functionality. At pH 2, the amine will be protonated and the carboxylic acid should be neutral, so the molecule will be retained by the cation exchange column. At elevated pH, the amine will be neutral and the carboxylate will be negative. The anion is not retained by the cation-exchanger and is eluted with methanol.

28-17. The product gas stream is passed through an anion-exchange column, on which SO_2 is absorbed by the following reactions:

$$SO_2 + H_2O \rightarrow H_2SO_3$$

$$2\text{Resin}^+OH^- + H_2SO_3 \rightarrow (\text{Resin}^+)_2SO_3^{2-} + H_2O$$

The sulfite is eluted with Na_2CO_3/H_2O_2, which oxidizes it to sulfate that can be measured by ion chromatography.

28-18. Large particle size allows sample to drain through the solid-phase extraction column without applying high pressure. In chromatography, small particle size increases the efficiency of separation, but high pressure is necessary to force solvent through the column.

28-19. (a) Solid-phase extraction retains acrylamide while passing many other components in the aqueous extract of the french fries. We want to remove as many other components as possible to simplify the chromatographic analysis. The strong acid of the ion-exchange resin protonates acrylamide and retains it by ionic attraction:

$$R-SO_3H + CH_2=CHCONH_2 \rightarrow R-SO_3^- + CH_2=CHCONH_3^+$$

(b) There are many ultraviolet-absorbing components in addition to acrylamide in the acrylamide fraction obtained from the ion-exchange column. Ultraviolet absorbance is not specific for acrylamide.

(c) For acrylamide, m/z 72 is selected by the mass filter Q1 of the mass spectrometer. This ion dissociates by collisions in q2. The product m/z 55 is selected in Q3 for passage to the detector.

$$\underset{m/z\ 72}{CH_2=CHCONH_3^+} \rightarrow \underset{m/z\ 55}{CH_2=CHC\equiv O^+}$$

$$\underset{m/z\ 75}{CD_2=CDCONH_3^+} \rightarrow \underset{m/z\ 58}{CD_2=CDC\equiv O^+}$$

(d) Even though many compounds are applied to the chromatography column, acrylamide is the only one with m/z 72 that gives a significant reaction product at m/z 55.

(e) Acrylamide gives one peak by selected reaction monitoring of the transition pair m/z 72→55. The internal standard gives just one peak for $^2H_3$-acrylamide monitored by the transition pair m/z 75→58 with the same retention time as acrylamide. The transition pair m/z 72→55 does not respond to the internal standard, and the transition pair m/z 75→58 does not respond to unlabeled acrylamide. We know the concentration of internal standard added to the aqueous extract of the french fries, and we measure the integrated area of the m/z 75→58 peak for the internal standard. We also measure the integrated area of the m/z 72→55 peak for acrylamide. The concentration of acrylamide in the aqueous extract is found by the proportion

$$\frac{[\text{acrylamide}]}{[\text{internal standard}]} = \frac{[\text{area of } m/z\ 72 \to 55 \text{ peak}]}{[\text{area of } m/z\ 75 \to 58 \text{ peak}]}$$

(f) With ultraviolet absorption, the internal standard appears at the same elution time as acrylamide. The molar absorptivity of deuterated internal standard is probably very similar to that of acrylamide, so equal concentrations of internal standard and acrylamide contribute approximately the same integrated area in the chromatogram. With selected reaction monitoring by mass spectrometry, the detector sees either acrylamide or the internal standard, with no interference from the other, even though they are eluted at the same time.

(g) The internal standard is mixed with the aqueous extract from the french fries prior to solid-phase extraction. We expect little isotope effect on the binding of acrylamide to the solid-phase sorbent or the HPLC stationary phase. Therefore, the fraction of acrylamide and the fraction of internal standard that bind to and are recovered from the solid-phase extraction column are equal. Even though neither one is bound or eluted quantitatively, equal fractions of each are bound and eluted. The ratio of acrylamide and internal standard should remain constant throughout the entire procedure.

28-20. (a) *Extraction:* Mix homogenized or powdered, hydrated sample with acetonitrile, NaCl, $MgSO_4$, and buffer. Salts create two phases and drive organic materials out of the aqueous phase. Buffer protects pH-sensitive substances. After centrifugation, collect the organic phase. *Sample cleanup:* Add 1 mL of organic phase to anhydrous $MgSO_4$ (to absorb residual H_2O), "primary secondary amine" sorbent (to absorb anions such as fatty acids), and optional sorbents such as C_{18}-silica and graphitized carbon black (to absorb nonpolar and aromatic substances). After shaking and centrifugation, analyze supernatant liquid by chromatography.

(b) Internal standard is intended to suffer the same losses as analyte in the different steps of the procedure.

(c) Total ion chromatogram shows total signal from all ions eluted at any given short time interval.

(d) The extracted ion chromatogram shows m/z 312 signal taken from the full mass spectrum recorded at any short time interval during elution. In selected ion monitoring, only the signal at *m/z* 312 would be measured and other masses would not be measured. Since the detector spends only a small fraction of the time measuring *m/z* 312 in an extracted ion chromatogram, but full time measuring *m/z* 312 in a selected ion chromatogram, the signal:noise ratio is greater in the selected ion chromatogram.

(e) Selected reaction monitoring.

28-21. (a) Liquid-liquid extraction uses a large volume (≳100 mL) of organic solvent to extract an aqueous phase by continuous distillation. In dispersive liquid-liquid microextraction, a cloudy emulsion is created by a small volume (~10–100 µL) of immiscible organic phase plus ~0.5–1 mL of dispersant solvent in an aqueous sample. After centrifugation, the phases separate and the organic phase is collected.

(b) Disperser solvent is miscible with both the aqueous and organic phases. It lowers the interfacial energy between the two phases, permitting a high-surface-area emulsion with a high rate of mass transfer to be formed. In the absence of disperser solvent, the two phases would just separate.

28-22. In solid-supported liquid-liquid extraction, the aqueous phase is suspended in a microporous medium through which organic solvent is passed to extract analytes. In solid-phase extraction, aqueous sample is passed through a small column containing a chromatographic stationary phase that retains analytes. Impurities and analytes are eluted by a series of washes with small volumes of organic liquid with increasing solvent strength.

28-23. (a) Highest concentration of Ni ≈ 80 ng/mL. A 10 mL sample contains 800 ng Ni = 1.36×10^{-8} mol Ni. To this is added 50 µg Ga = 7.17×10^{-7} mol Ga. Atomic ratio Ga/Ni = $(7.17 \times 10^{-7})/(1.36 \times 10^{-8})$ = 53.

(b) Apparently all the Ni is in solution because filtration does not decrease its total concentration. Since filtration removes most of the Fe, it must be present as a suspension of solid particles.

28-24. One-fourth of the sample (25 mL out of 100 mL) required $(0.011\,44\text{ M})(0.032\,49\text{ L})$ = 3.717×10^{-4} mol EDTA $\Rightarrow (3.717 \times 10^{-4})(4)$ = 1.487×10^{-3} mol Ba^{2+} in sample = 0.204 2 g Ba = 64.90 wt%.

28-25. (a) From the acid dissociation constants of Cr(III), we see that the dominant forms at pH 8 are $Cr(OH)_2^+$ and $Cr(OH)_3(aq)$. The dominant form of Cr(VI) is CrO_4^{2-}.

(b) The anion exchanger retains the anion, CrO_4^{2-}, but permits the $Cr(OH)_2^+$ cation and neutral $Cr(OH)_3(aq)$ to pass through, thereby separating Cr(VI) from Cr(III).

(c) A "weakly basic" anion exchanger contains a protonated amine ($-^+NHR_2$) that might lose its positive change in basic solution. A "strongly basic" anion exchanger ($-^+NR_3$) is a stable cation in basic solution.

(d) CrO_4^{2-} is eluted from the anion exchanger when the concentration of sulfate in the buffer is increased from 0.05 M in step 3 to 0.5 M in step 4.

28-26. One possible cost-saving scheme is to monitor wells 8, 11, 12, and 13 individually, but to pool samples from the other sites. For example, a composite sample could be made with equal volumes from wells 1, 2, 3, and 4. Other composites could be constructed from (5, 6, 7), (9, 10), (14, 15, 16, 17), and (18, 19, 20, 21). If no warning level of analyte is found in a composite sample, we would assume that each well in that composite is free of the analyte. If analyte is found in a composite sample, then each contributor to the composite would be separately analyzed. The disadvantage of pooling samples from n wells is that the sensitivity of the analysis for analyte in any one well is reduced by $1/n$.